John Felter
9/01

OPTICAL FIBER
SENSOR TECHNOLOGY
Volume 2

Optoelectronics, Imaging and Sensing Series

Series editors
Dr. A. T. Augousti, Kingston University, UK
Professor K. T. V. Grattan, City University, UK
Professor G. Parry, Imperial College, London University, UK

Optoelectronics, Imaging and Sensing brings together the best elements of two successful series, the **Optical and Quantum Electronics Series** and the **Sensor Physics and Technology Series**. The new series will focus on exciting new developments and applications in the rapidly changing areas of optoelectronic sensing and imaging technology. The volumes cover both systems and devices, and are aimed at scientists involved in research and development as well as practising engineers. Advanced undergraduate and graduate textbooks are also included, giving tutorial introductions that are essential for those wishing to work in the challenging and multidisciplinary areas of optoelectronics, imaging and sensing. This up-to-date series will include electronic products as well as conventional books, meeting the needs of its users effectively in the most modern formats.

Titles Available:

Electromagnetic Waves
C. G. Someda

Optical Fiber Sensor Technology Volume 2
Devices and Technology
Edited by K. T. V. Grattan and
B. T. Meggitt

Optical and Quantum Electronics Series

1 **Optical Fiber Sensor Technology**
Edited by K. T. V. Grattan
and B. T. Meggitt

2 **Vision Assistant Software**
A practical introduction to image processing and pattern classifiers
C. R. Allen and N. C. Yung

3 **Silica-based Buried Channel Waveguides and devices**
F. Ladouceur
and J. D. Love

4 **Essentials of Optoelectronics**
With applications
A. Rogers

Sensor Physics and Technology Series

1 **Biosensors**
T. M. Cahn

2 **Fiber Optic Fluorescence Thermometry**
K. T. V. Grattan and
Z. Y. Zhang

3 **Silicon Sensors and Circuits**
F. Wolffenbuttel

4 **Ultrasonic Measurements and Technologies**
S. Kocis and Z. Figura

5 **Data Acquisition for Sensor Systems**
H. R. Taylor

Optical Fiber Sensor Technology

Volume 2

Devices and Technology

Edited by

K. T. V. Grattan

Professor of Measurement and Instrumentation
Head, Department of Electrical,
Electronic and Information Engineering
City University, London, UK

and

B. T. Meggitt

Visiting Professor, Department of Electrical,
Electronic and Information Engineering
City University, London, UK

CHAPMAN & HALL
London · Weinheim · New York · Tokyo · Melbourne · Madras

Published by Chapman & Hall, an imprint of Thomson Science, 2–6 Boundary Row, London SE1 8HN, UK

Thomson Science, 2–6 Boundary Row, London SE1 8HN, UK
Thomson Science, Suite 750, 400 Market Street, Philadelphia, PA 19106, USA
Thomson Science, Pappelallee 3, 69469 Weinheim, Germany
Thomson Science, 115 Fifth Avenue, New York, NY 10003, USA

First edition 1998

© 1998 Chapman & Hall

Thomson Science is a division of International Thomson Publishing I(T)P®

Typeset in Times New Roman by AFS Image Setters, Glasgow
Printed in Great Britain by TJ International Ltd, Padstow, Cornwall

ISBN 0 412 78290 1

Apart from any fair dealing for the purposes of research or private study, or criticism or review, as permitted under the UK Copyright Designs and Patents Act, 1988, this publication may not be reproduced, stored, or transmitted, in any form or by any means, without the prior permission in writing of the publishers, or in the case of reprographic reproduction only in accordance with the terms of the licences issued by the Copyright Licensing Agency in the UK, or in accordance with the terms of licences issued by the appropriate Reproduction Rights Organization outside the UK. Enquiries concerning reproduction outside the terms stated here should be sent to the publishers at the London address printed on this page.
    The publisher makes no representation, express or implied, with regard to the accuracy of the information contained in this book and cannot accept any legal responsibility or liability for any errors or omissions that may be made.

A catalogue record for this book is available from the British Library

Contents

List of Contributors		vii
Preface		ix
1	**Classification of optical fiber sensors** K. T. V. Grattan and Y. N. Ning	1
2	**Optical fiber lasers** N. Langford	37
3	**Fiber lasers in optical sensors** B. Y. Kim	99
4	**Multiplexing optical fiber sensors** J. D. C. Jones and R. McBride	117
5	**Progress in optical fiber interferometry** D. A. Jackson	167
6	**Optical fiber speckle interferometry** R. P. Tatam	207
7	**Interferometric vibration measurement using optical fiber** K. Weir and B. T. Meggitt	237
8	**Fiber optic laser anemometry** S. W. James and R. P. Tatam	261
9	**Fiber optic gyroscopes** J. Blake	303
10	**Fiber gratings: principles, fabrication and properties** V. A. Handerek	329
11	**Fiber Bragg grating sensors: principles and applications** Y.-J. Rao	355
12	**Fiber Bragg grating sensors: signal processing aspects** Y. N. Ning and B. T. Meggitt	381
Index		419

Contributors

J. Blake
Department of Electrical Engineering, Texas A&M University,
College Station, Texas 77843-3128, USA

K.T.V. Grattan
Department of Electrical, Electronic and Information Engineering, City University,
Northampton Square, London EC1V 0HB, UK

V.A. Handerek
Department of Electronic and Electrical Engineering, King's College, Strand,
London WC2R 2LS, UK

D.A. Jackson
Department of Physics, University of Kent, Canterbury, Kent CT2 7NZ, UK

S.W. James
Optical Sensors Group, Centre for Photonics and Optical Engineering, School of
Mechanical Engineering, Cranfield University, Cranfield, Beds MK43 0AL, UK

J.D.C. Jones
Department of Physics, Heriot-Watt University, Riccarton, Edinburgh
EH14 4AS, UK

B.Y.Kim
Department of Physics, Korea Advanced Institute of Science and Technology,
373-1 Kusong-dong, Yusong-gu, Taejon, Korea

N. Langford
Department of Physics and Applied Physics, University of Strathclyde,
John Anderson Building, 107 Rottenrow East, Glasgow G4 0NG, UK

R. McBride
Department of Mechanical and Chemical Engineering, Heriot-Watt University,
Riccarton, Edinburgh
EH14 4AS, UK

B.T. Meggitt
Department of Electrical, Electronic and Information Engineering, City University,
London, and EM Technology, PO Box 4191, London SE21 8YY, UK

Y.N. Ning
Department of Electrical, Electronic and Information Engineering, City University,
Northampton Square, London EC1V 0HB, UK

Y.-J. Rao
Applied Optics Group, Physics Department, University of Kent, Canterbury, Kent CT2 7NZ, UK

R.P. Tatam
Optical Sensors Group, Centre for Photonics and Optical Engineering, School of Mechanical Engineering, Cranfield University, Cranfield, Beds MK43 0AL, UK

K. Weir
Department of Physics, Blackett Laboratory, Imperial College of Science, Technology and Medicine, London SW7 2AZ, UK

Preface

Progress in optical fiber sensors

The field of optical fiber sensor technology is one that continues to expand and develop at a rate that could barely have been predicted a few years ago. The wealth of publications appearing in the technical literature and the burgeoning number of papers presented at the now well-established series of national and international conferences, which are attended by a wide selection of technically qualified optoelectronics professionals, gives a clear indication of both the range and scale of the devices and applications now seen in the subject. Such a rapid expansion makes it very difficult for the scientist and engineer, under pressure to be both informed and effective for an employer, to attend all these meetings, selectively read the appropriate literature and be able quickly to gain the knowledge in those specific areas which will give the best advantage for the work in hand. To that end, this volume has been planned and carefully designed to provide an essential overview, and detailed specific information, on those novel and exciting aspects of optical fiber sensor technology that have recently emerged, with particular focus on the devices and the exciting applications of this part of optoelectronic technology in the vast international measurement and instrumentation area.

Optical Fiber Sensor Technology Volume 2: Devices and Technology is a companion volume to the recently published text *Optical Fiber Sensor Technology*, building upon the emphasis in that book on the principles and underpinning technology of optical fiber sensors and to the forthcoming *Optical Fiber Sensor Technology Volume 3: Applications and Systems* which highlights new sensor applications and their associated technology. The new text herein reflects the developments that have occurred during the time since the publication of the first volume, in particular in the maturity of optical fiber sensors and the expansion of the range of successful devices and their associated technology that has happened in a relatively short period. It draws upon a key group of international authors working in the UK, USA and Asia who themselves are at the 'cutting-edge' of the technological developments, are renowned international experts and familiar conference speakers in this field. In this way the text is authoritative and definitive of the most relevant aspects of what is happening in the subject. It draws upon their knowledge of sensor systems and advances in both physical, chemical, biological and biomedical sensing and their experience of both the technological developments themselves and the new areas for their application and use, in what is annually a multi-billion dollar industry. That makes this series of closely linked texts essential for the informed professional by drawing together the distilled experience and knowledge of the highly respected contributors, and will, with the emphasis on both clarity of presentation and readability to the graduate engineer or scientist, whatever discipline was originally studied, be a reference work valuable into the future.

The two new texts (of which this is the first) are structured to complement the comprehensive work described in the first volume and expand upon it in all the topical areas of the subject. The text divides into two main sections: firstly work on sensor classification, fiber laser sources and their applications in sensors, and sensor multiplexing and secondly, interferometry, anemometry and the optical gyroscope and in-fiber gratings and their applications. This allows the third volume to focus upon chapters on specialized sensor techniques for physical measurands, including modeling techniques applied to fiber optic sensors and an important section on chemical, biochemical and biomedical fiber optic sensing.

The first begins with a chapter by Kenneth Grattan and Ya Nong Ning who have developed, for the first time, a comprehensive classification system for all optical fiber sensors, in which context the rest of the material in the book can be clearly seen. The subject of doped optical fiber lasers and their applications has been one of the success stories of recent years offering new and compact high power sources and Nigel Langford presents a comprehensive review of the technology of new lasers based on the major developments in doped fibers which have occurred in the last few years and which will, because of their compact and convenient nature, have a wide impact upon a range of optoelectronic systems. The sensor applications which are opening up for optical fiber lasers are reviewed by Byoung Yoon Kim, who has pioneered the many novel applications of such lasers in optical sensing, and who reports extensively on his own and other research. Other uses of doped fiber luminescence in temperature sensing are covered in a separate chapter in the third volume by Kenneth Grattan and Edward Zhang, together with a detailed review of the use of a wide range of such luminescent methods used to determine this most important industrial parameter, reflecting the complementary nature of the approach. The first section ends with a review by Julian Jones and Roy McBride of the variety of multiplexing techniques available for fiber optic sensors, and their applications.

The second section deals with interferometry, anemometry and the rapidly expanding topic of in-fiber gratings. David Jackson's pioneering work in interferometry has influenced the field for nearly two decades and he has written a chapter highlighting developments and progress in what remains an essential technology in optical fiber sensing. Ralph Tatam and Steve James bring their expertise in two key areas, reflected in separate chapters where fiber optics are making familiar techniques more accessible and easier to use: optical fiber speckle interferometry and laser Doppler velocimetry (LDV).

This is complemented by a review by Kenny Weir and Beverley Meggitt of methods of measurement of vibration using interferometric fiber optic sensors. A review of one of the most successful interferometric-based fiber sensors, the fiber optic gyroscope, follows from Jim Blake, which now having passed its 20th anniversary, is going from strength to strength in a wide range of applications in aerospace, automobiles and even the wider industrial environment. Vınce Handerek brings his expertise and experience in the properties and fabrication of one of the most exciting of recent developments, in-fiber Bragg gratings, in a chapter reviewing some basic principles and fabrication issues, complemented by contributions from Yun-Jiang Rao and Beverley Meggitt and Ya Nong Ning concentrating on the principles and applications, followed by signal processing aspects of fiber Bragg grating sensors, including multiplexing configurations.

Overall the volume is comprehensively cross-referenced by the contributors to both the milestone papers in the specific subject areas and topical presentations at the most major, very recent conferences and from the best of the new technical literature accessible in most technical libraries. Over one thousand such papers are cited by the various authors, and these should provide a wealth of further reference material and very relevant reading to those interested in following up on some of the exciting developments in optical fiber sensor technology, devices and applications.

1
Classification of optical fiber sensors

K.T.V. Grattan and Y. N. Ning

1.1 INTRODUCTION TO CLASSIFICATION METHODS OF OPTICAL FIBER SENSORS

Historically a number of different approaches have been used in the classification and categorization of fiber optic sensors. The reason for the development of an appropriate and effective classification scheme at all lies in the increasing complexity of the wide variety of fiber optic sensor systems which exist today. Several schemes for classification of fiber optic sensors have been developed, from different points of view, ranging from the essentially straightforward methods used in a simple survey, such as those based on the physical quantity to be transduced, through to the use of more precise subdivisions which focus on sensor type, detection systems and radiation properties. Indeed, in the development of the most appropriate scheme for the ordering of material in any survey or review of fiber sensor systems, a particular classification of fiber optic sensor schemes is used in order to emphasize the most important aspects of the subject and thus to reflect that in the essential nature of the text itself.

Early reviews of optical fiber sensor (OFS) technology, such as that by Gaillorenzi [1] undertaken in 1982, could aim to cover most of the important fiber optic sensor systems which had been discussed at the time. With the very rapid progress that has been made in the field, an approach which aims simply to catalog a range of fiber optic sensors now is inappropriate and in practice almost impossible. Other authors have looked at the classification of fiber optic sensors by the modulation scheme used, such as those reviews authored by Medlock [2] or by Spooncer [3], and thus techniques such as those involving intensity, wavelength, polarization, phase or rate may be considered as prime classification features. Such a scheme has a disadvantage of focusing on the technique used rather than upon the measurand itself, and may be less appropriate for most applications where the aim is to find the most appropriate technology, using fiber optics, for the measurement of a specific parameter, such as temperature or pressure. This latter approach, where sensors are categorized according to measurand parameter such as pressure, temperature, flow, vibration, electric or magnetic field, displacement, velocity, chemical parameters and gas partial pres-

Optical Fiber Sensor Technology, Vol. 2. Edited by K. T. V. Grattan and B. T. Meggitt.
Published in 1998 by Chapman & Hall, London. ISBN 0 412 782 901

sure, has also been employed, for example by Grattan [4] in considering the measurement of temperature. This method does, however, have the disadvantage when used more widely that the same or similar techniques are often applied to the measurement of several different parameters and in particular a measurand such as displacement can be transduced to give information on a wide range of other different parameters, especially temperature and pressure. Indeed categorization of sensors has been carried out on the basis of features such as novelty, reflecting recent progress, for example by Grattan [5] or even by nonscientific considerations such as geographical region of sensor production or authorship. These criteria are clearly inappropriate to a rigorous classification where a useful and practical means of system categorization is required which can then be used for subsequent sensor selection. It should be remembered that in the context of this work, the terms 'sensor' and 'transducer' are employed interchangeably, as they are regularly in the literature, although the word transducer more correctly implies a change of energy state in the sensor head itself, whereas the sensor often refers to the whole device.

The most familiar subdivision of fiber optic sensors is into **intrinsic** or **extrinsic** devices, and this has been used widely for example by Udd [6] and Spooncer [3]. Jackson [7], in his work, adds to this subdivision by extracting so-called **external** sensors from the extrinsic category, defining these as sensor devices which are fiberized versions of open air-path optical systems such as laser Doppler anemometers or noncontact vibration measurement systems. They have the major advantage over conventional instrumentation that the flexibility of the optical link allows the sensors to be used in situations where access is difficult. Such techniques have been widely used, for example in fiber optic pyrometry for remote temperature measurement such as for ovens and furnaces. **Extrinsic** (including external) sensors are normally defined as those in which the light wave is guided by the fiber but the interaction between the light and the quantity in the measurement takes place outside the fiber itself. This takes advantage of the flexibility of the fiber, its low attenuation and increased mechanical stability when compared to a bulk optical design. This class of device has been successfully used for a number of sensor types ranging from the relatively simple example of the displacement-based Fotonic® sensor to the more sophisticated using, for example, electro-optic techniques. With OFSs, it is often their essential simplicity and relative cost effectiveness which make them particularly attractive for use for a specific measurement.

Figure 1.1 shows a schematic comparison between extrinsic and intrinsic sensor methods. This latter class of **intrinsic** or all fiber optic sensors are those where the sensor action actually takes place within the fiber itself. Often this category is divided into those called **direct** sensors (Annovazzi-Lodi and Merlo [8]), such as the Faraday sensors or fiber gyroscopes where the measurand acts directly on the fiber, or by contrast the **indirect** sensors. In such devices the measurand is locally transduced into another quantity acting on the fiber, with the consequence of the sensing being a two-step process, resulting in an indirect method. For example, there is a class of current sensor which relies upon the stress exerted in

CLASSIFICATION METHODS OF OPTICAL FIBER SENSORS

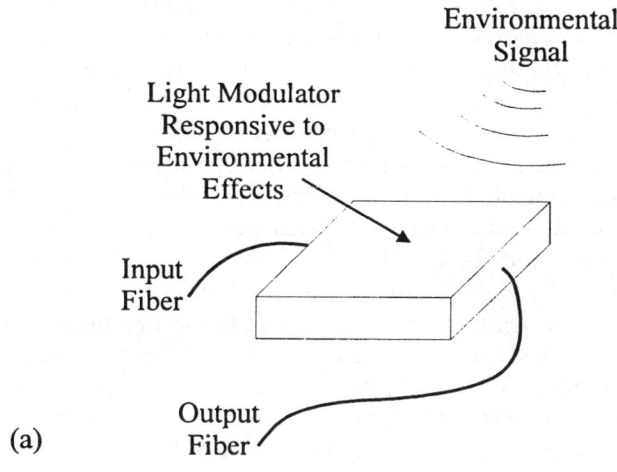

(a)

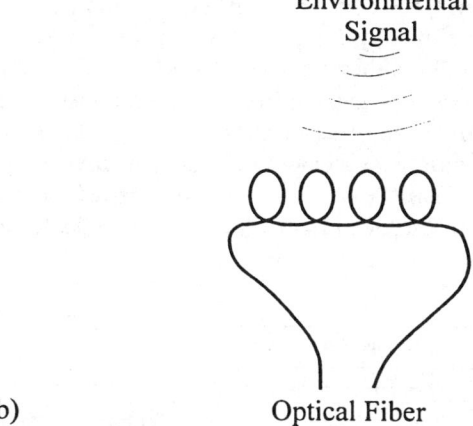

(b)

Fig. 1.1 Schematic comparison between (a) extrinsic and (b) intrinsic sensor methods (after Udd, 1991).

the fiber by a magnetostrictive coating that is sensitive to the measurand, illustrating such indirect sensor action.

A further major subsection of the intrinsic device is the **interferometric** sensor type. A number of variations of interferometric sensor schemes have been developed and applied to the measurement of a wide range of parameters and although they represent a subcategory of the basic intrinsic sensor device, they warrant particular consideration in their own right. A key factor in the use of intrinsic sensors is the optical guidance nature of the fiber itself, whether it be single or multi-mode and in particular in the relation to interferometric sensors, the temporal degree of coherence of the light in the sensor itself is important, as to whether it be high coherence, low coherence or even incoherent light, which will make a difference to the operation of the device. This distinction arises from the different optical sources used in the sensors themselves.

It is useful to consider the use of several different schematic representations to enable these distinctions to be seen more clearly. Following the pattern of Udd [6], a tree representing subdivisions of both extrinsic and intrinsic fiber optic sensors is shown in Figures 1.2 and 1.3, with Figure 1.4 reflecting the degree of diversity of the subdivision of interferometric fiber optic sensors. The devices considered are intrinsically totally passive sensors, i.e. those which do not require electric power at the sensor head, although a separate group of hybrid sensors exists including bulk, micro-optic or integrated optic elements where an additional power source is used, for example when local electrical powering is provided, often using transduction from optical radiation at the sensor head itself.

In order to be aware of and examine the diversity of the use of fiber optic sensors more fully, the wide range of measurements which can be addressed by fiber optic sensors can be seen, as is tabulated in Figure 1.5 from the work of Jackson [7] where the use of different types of fibers to measure a number of parameters is revealed. This is complemented by Figure 1.6 from the work of Spooncer [3] showing an illustration of the subcategories of one specific group, i.e. multi-mode OFSs in terms of intensity, wavelength or time modulation, as examples.

Further scope for classification of sensors exists using a basis of whether the sensor is making a **single point measurement** i.e. a specific measurement at a particular point in space, or offers the possibility of **distributed measurement**, such as can be achieved with the use of optical time domain reflectometry (OTDR)

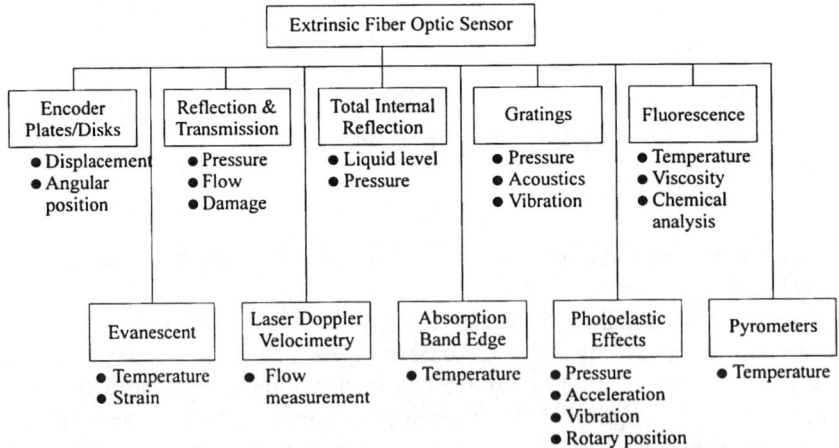

Fig. 1.2 Extrinsic fiber optic sensor applications (after Udd, 1991).

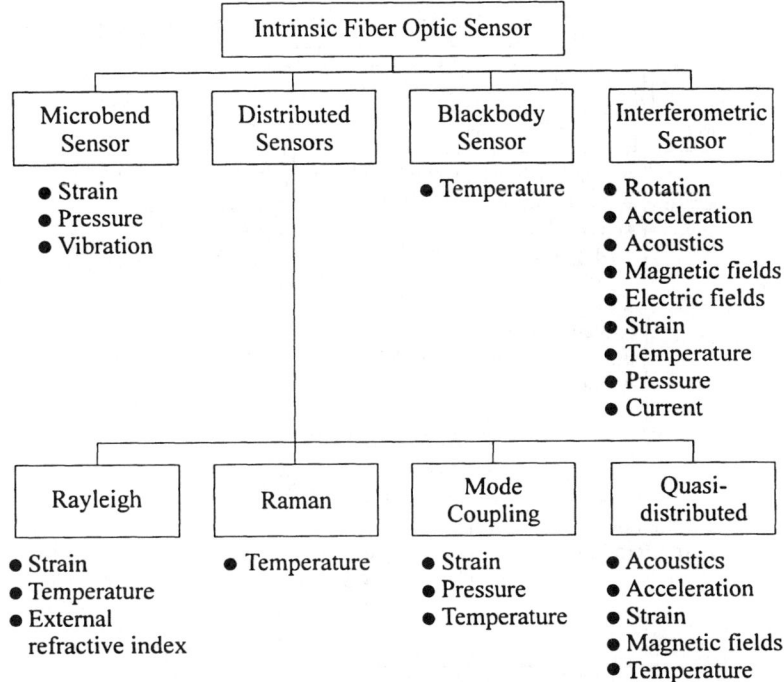

Fig. 1.3 Intrinsic fiber optic sensor applications (after Udd, 1991).

techniques. A number of fiber optic sensor techniques can be used for distributed measurement, in which case the transduction effect can occur (and in principle be measured) at any point along a length of fiber and both the parameter itself and its spatial position can be determined. This involves accurate time-of-flight measurements of the optical pulses used along fibers, usually resulting in spatial resolutions of a meter or so with nanosecond pulse technology. The most successful embodiment of this scheme is in the York Technology distributed temperature sensing device, discussed in more detail elsewhere. A variant of the last category is the **quasi-distributed** sensor, where a series of point sensors are linked together to provide a number of specific and predetermined measurement points along a single fiber loop.

Thus, as can be seen from the foregoing, the practice of classifying fiber optic sensors is a nontrivial exercise, and to be effective to be used, for example with a computerized data base accessible to automated searching, a more fundamental and essential classification approach is required. To both the fiber optic sensor designer and the fiber optic sensor user, there is a need to know (and usually minimize) the degree of complexity of the system and to understand the way in which the components are linked together to produce the optimal overall sensing scheme. An efficient classification scheme can aid this process, especially when aided by modern computer technology.

CLASSIFICATION OF OPTICAL FIBER SENSORS

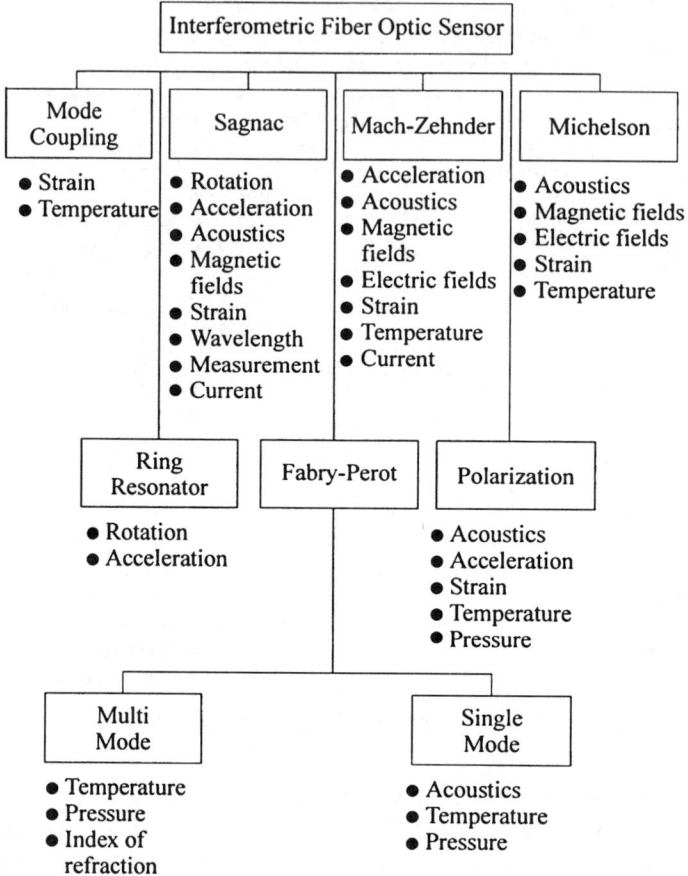

Fig. 1.4 Interferometric fiber optic sensor applications (after Udd, 1991).

1.2 CLASSIFICATION IN CARTESIAN GEOMETRY

In order to develop a fuller overview of the variety of OFS devices in use and under study today, it is worthwhile to develop a new classification of these techniques. By examining the structure of a typical OFS, it can be seen as logical to classify the techniques according to three basic and separate criteria:

1. Source of illumination used;
2. Type of fiber employed;
3. Optical modulation/sensor techniques used.

Such a classification springs almost instinctively from the nature of the term 'optical fiber sensor': the **optical** source, coupled to a transmitting **fiber** to allow the light to undergo transduction or modulation in the **sensor** process.

CLASSIFICATION IN CARTESIAN GEOMETRY

Multimode or external monomode fibers				Monomode fiber
Conventional		Optically powered	Multimode fiber intensity	interferometric and polarimetric
Particle size `}` Turbidity pH	Intensity	Displacement Pressure Temperature	Pressure Displacement Strain	Temperature Displacement Strain
Pressure Displacement Position Magnetic field Temperature Gas Chemical Vibration Level Optical radar	Intensity or Phase	Magnetic field	Flow Switch Force Temperature Vibration Distributed temperature	Magnetic field Acceleration Force Rotation Flow Pressure Vibration Acoustic waves Magnetic field Electric field Chemical
Laser velocimetry Vibrometry Holography	Phase			

Fig. 1.5 Summary of the range of measurements which can be addressed via fiber optic sensors (after Jackson, 1995).

A Cartesian coordinate representation, shown in Figure 1.7, is used to group these three 'variables', where a spatial framework for the classification process can be employed and different types of sensor can then be fitted into such a way

Intensity		Wavelength	Time
Digital	Analog		
On–off Multiple position Encoders Moiré fringes	Mask or shutter Fiber displacement Variable attenuators Reflective	Dispersing devices (grating prisms) Filters (color, interference) Fluorescence	Mechanical resonators Electrical resonators Fluidic resonators Decay-time Echo-sensors Doppler effect
	FTIR Refractive index Defocusing Deformable components Evanescent coupling Microbending Polarization Scattering	Thermal radiation (bolometers) Birefringence Band-edge	

Fig. 1.6 Classification of subcategory of multi-mode optical fiber sensors (after Spooncer, 1992).

8 CLASSIFICATION OF OPTICAL FIBER SENSORS

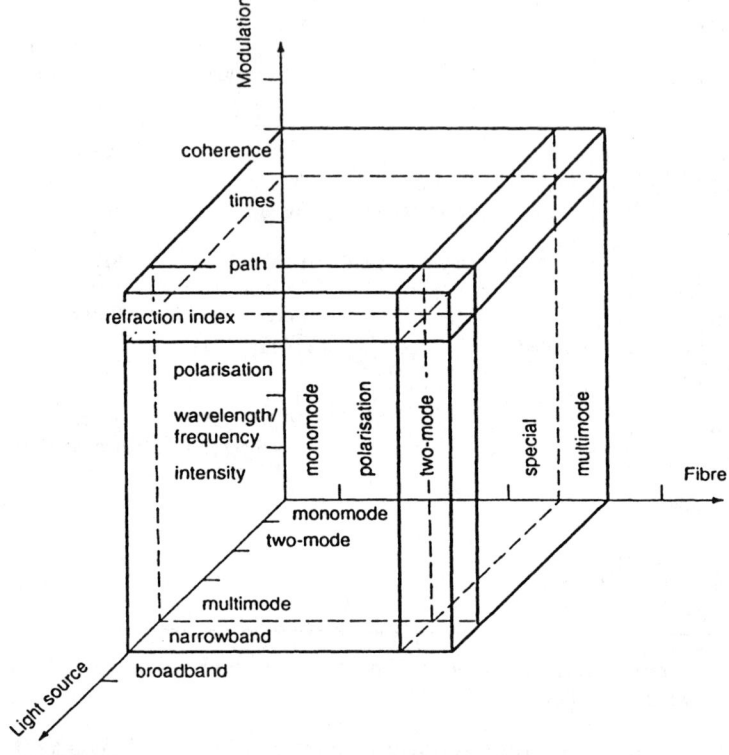

Fig. 1.7 3-D Cartesian coordinate representation scheme.

as best to illustrate their comparative features. Such a classification for generalized sensors, whether electrical, mechanical or otherwise, was originally discussed by Stein [9] and expanded by Middelhoek and his coworkers [e.g. 10], but here it is applied in a rather different way to one specific class of sensor rather than to the general family of all sensors. The approach is flexible in that new subcategories of the three basic criteria can readily be added, and new sensor systems can be placed in this representation.

1.2.1 Source of illumination used

The x-axis of the geometrical space is used to represent light sources which may then be characterized by their spectral distribution, as shown in Figure 1.7. There are at least five common and convenient classifications of light sources that have been reported for use in various types of OFS:

1. Single-mode laser source (e.g. single-mode He–Ne laser, single-mode laser diode);

2. Two-mode or 'few-mode' light source (e.g. a laser with various cavity configurations or a dual-wavelength light source such as a light-emitting diode (LED) and some laser diode or doped fiber sources);
3. N-mode light source ($N > 2$) (e.g. multi-mode laser, such as an Ar ion laser/ laser diode);
4. Narrow-band continuous spectrum light source (e.g. LED or superluminescent LED);
5. Broad-band continuous spectrum light source (e.g. tungsten lamp, mercury lamp, continuous output xenon lamp, Er-doped fiber sources).

1.2.2 Type of fiber employed

On the y-axis of the geometrical space, the various fibers that may be employed are classified by their mode-propagation characteristics. The number of modes transmitted and the polarization-related characteristics can be used on the basis of a division of fiber types into at least five groups, as shown in Figure 1.7:

1. Single (or mono-) mode fiber;
2. Polarization-maintaining fiber;
3. Two-mode fiber;
4. Multi-mode fiber;
5. Specialized fibers, e.g. doped fibers, liquid-core fibers, fibers containing photo-induced optical elements, e.g. Bragg gratings.

1.2.3 Optical modulation/sensor techniques

As any modulation of light may be detected by a change of the light intensity, the so-called optical modulation can only be achieved by encoding the variation of the measurand in one of the parameters of the light beam itself (or its associated electric field), this being sensed by a detector sensitive to that intensity variation. On the z-axis of the space, as shown in Figure 1.7, is a characteristic of OFS systems based on the parameters of the interaction electric field, E, given by $E = E_0 \exp(j(\omega t + 2\pi n L/\lambda))$, where $|E_0|$ is the amplitude, the state of polarization (SOP) of the light can be determined by the azimuth of the electric field vibration, ω is the optical frequency, n the refractive index of the medium, L the optical path, λ the wavelength and σ the wave number ($1/\lambda$). The modulation of the field can be categorized into at least seven groups:

1. Intensity modulation (I–$E.E$);
2. Polarization modulation (I–E_0);
3. Wavelength/frequency modulation (I–λ/ω);
4. Path modulation (I–L);
5. Refractive index modulation (I–n);

6. Time modulation ($I-t$);
7. Coherence length modulation ($I-k$), where k is the fringe visibility.

The modulation types 3–6 are termed 'phase modulations' because their variation will cause a change of the phase of the light propagating in the optical fiber. It should be pointed out that the phase of an optical wave is defined with respect to a particular reference and thus a useful phase modulation for sensing can only be achieved, in practice, for a light source with fixed and known initial phase conditions. The random variation in the initial phase over a period of time (the detector response time) will result in a degradation of, for example, the fringe visibility, k (as discussed later) in an interferometer used to detect the phase condition or its change, and will make the detection of the phase modulation particularly difficult. In other words, a phase-modulation sensor can only effectively be operated with the relative optical path differences inside the coherence length of the optical source.

By using this classification system, as illustrated in Figure 1.7, the many OFS techniques described can more readily be surveyed and examined, and their relationship to one another more readily seen. Such a representation can also more clearly indicate those areas where sensor schemes have already been proposed and/or exploited and its examination may show both areas of inactivity or those that are underdeveloped and point to new techniques which could be further exploited as the technology to achieve them matures. It should be noticed that a particular OFS can employ more than one type of light source, modulation technique or fiber at any one time. Such types of OFS can then be represented by more than one 'cubic unit' in the geometrical space in Figure 1.7. If an OFS 'family' is represented by a number of different 'cubic units', such a family may be of a multiplexed type [11], while if the same cubic units represents a number of types of OFS, this may be a distributed type of OFS [12].

1.3 LIGHT SOURCES: ASSOCIATED OPTICAL CHARACTERISTICS

Light sources can be classified, for example according to their spectral or coherence characteristics, into at least five groups. The coherence characteristic of a light source is a measure of the extent to which a phase relationship is maintained both across the beam (spatial coherence) and along the beam (temporal coherence). In an OFS, the spatial coherence of the light wave is often not particularly important because of the small cross-sectional area of the fiber core, but in contrast, the temporal coherence of the optical radiation is often extensively considered in selecting the appropriate light source for a particular sensor. In a multi-mode fiber, the spatial coherence of the modes is lost, whilst in a true single-mode fiber it is, by definition, preserved. A useful indicator of the temporal coherence is the coherence length of the source or the resulting fringe visibility of the interference fringe in a detection interferometer, the relationship between the coherence length and the spectral width of the light source being written as [13]

LIGHT SOURCES: ASSOCIATED OPTICAL CHARACTERISTICS

$$L_c = 1/\Delta\omega = \lambda^2/\Delta\lambda.2\pi \quad \text{and} \quad \omega = 2\pi/\lambda \quad \text{where} \quad \Delta\omega = 2\pi\lambda/\lambda^2$$

where L_c is the coherence length, λ the wavelength of light and $\Delta\lambda$ (or $/\Delta\omega$) the spectral width of the light source.

1.3.1 Single- (or mono-)mode light sources

Many types of single-mode laser or laser diode belong to this group of sources. Because their bandwidth is very narrow, their coherence lengths are usually very long compared to the optical path inside the OFS itself. For a suitably chosen He–Ne gas laser, the coherence length is of the order of a few hundred meters, for example, and a typical spectral width for a single longitudinal mode in such a laser is about 1 MHz, which corresponds to a coherence length of the order of 300 m [13]. For a typical laser diode such as that used in a compact disc player or CD-ROM drive, a very short coherence length is desired and used, to avoid phase-sensitive effects in the reading of data from the optical disc. The spectral width can be produced to achieve $\sim$ 100 MHz, resulting in L_c having a value up to a few meters [14].

1.3.2 Two-mode light sources

Such light sources can be produced relatively easily for use in an OFS by, for example, 'ramping' a single-mode laser diode at two different values of drive current [15], by combining two or more single-mode lasers which have different wavelengths [16] or by the use of band-pass filters to obtain several wavelengths from an LED [17] or a doped fiber source. The coherence length obtained in each of these cases is, in general, much less than that of a single-mode laser. Obviously, in the latter two cases, the coherence lengths of the wavelengths generated will be independent of each other, and may be equivalent to those obtained from separate sources. However, when observing fringes in a detection interferometer, the fringe visibility changes periodically with the change of optical path difference, resulting from the relatively short coherence length of the source.

1.3.3 N-mode (multi-mode) light sources

This group of light sources covers a wide range of multi-mode solid-state (doped insulator), liquid or gas lasers or the semiconductor laser diode. For instance, the He–Ne gas laser with a cavity length of 0.5 m has a mode separation of 300 MHz and the width of the lasing mode at 633 nm is typically about 1.5 GHz, with the number of longitudinal modes in the laser output being of the order of four or five [13]. By operating a single-mode laser diode with a drive current under the threshold for laser action, it may conveniently be used as an N-mode light source as well [18], as the coherence characteristics deteriorate severely under these circumstances. The coherence characteristics of this type of light source are not easily described because the interference effects of multi-mode light are much more complex than those of light from a mono-mode laser. Thus interference fringes will appear in some spatial regions but not in others. This is due to the superposition of the fringes which are generated by each longitudinal mode of the device [19]. As a result of this superposition, those interference regions will occur periodically along the optical path difference. The concept of a 'coherence region' has been introduced in some papers to explain this and the coherence characteristics of these sources have been discussed in greater detail [19, 20]. The laser diode is available routinely at several wavelengths over the range ~630–1550 nm, and laboratory devices operating in the blue/green and blue regions of the spectrum, developed for the new generation of CD players, are now revolutionizing the potential for compact, low power OFS devices. In summary the devices range from low coherence length, simple structures with a wavelength FWHM of 2–5 nm, available at prices from a few tens of dollars (depending upon the power delivered), through distributed feedback lasers with subnanometer linewidths and at a price that rules them out for most OFS applications, to quantum-well devices with the linewidth performance of a He–Ne laser. The convenience of this semiconductor laser package is mitigated by the current high cost (in the region of thousands of dollars per item).

Other laser devices may be employed for OFS systems, e.g. the Ar ion laser or the He–Cd laser. Both these devices are expensive, physically quite large and require power supplies that use a considerable level of electrical input, but they do provide a continuous (and thus capable of modulation) output in the blue or ultraviolet (UV) part of the spectrum. This is a region where LEDs or laser diodes are very weak and for sensors such as those employing luminescence, where excitation at short wavelengths is preferable, they are a useful source. In addition, early work on distributed sensors employed these sources, due to their high power (several watts being routinely available) and favorable nonlinear scattering characteristics. The coherence length of a typical Ar ion laser is likely to be quite long, comparable to that of the He–Ne laser.

1.3.4 Narrow-band continuous spectrum light source

This is another type of light source widely used for many different kinds of OFS. The ordinary LEDs or SRDs (superradiant diodes) are examples of this group with importance for OFS use, available in the spectral region from the visible, especially the red to the infrared (IR). The width of their output spectrum is about 20–80 nm [21], and as a consequence their coherence length is of the order of 50 μm or less [22]. Due to its relatively large emitting area, the coupling efficiency of light from such an LED into a typical fiber is a major problem for many OFS devices sensitive to the light intensity used. However, in recent years, several methods have been introduced to improve the coupling efficiency, such as using a pigtailed LED or by using an ELED (extended LED) for which pigtailed and nonpigtailed versions are available. In addition, discharge lamps can produce a series of spectral lines of varying widths, distributed across the spectrum according to the energy levels present within the gas itself. The low-pressure mercury lamp, for example, gives a series of lines from UV into the visible spectrum, but there is some difficulty in coupling the divergent output from such a device into an optical fiber. This is a significant problem for all lamp sources, although output optical powers may be such that efficient coupling, whilst preferable, is not essential. A range of low-pressure gas discharge lamps yields specific lines over a wide range of the spectrum. However, there are often problems with the bulk of the lamp and power supply in OFS applications, and the need for lamp cooling. Invariably, in cases where the lamp is inefficient or delivering low optical power at the wavelength of interest, good optical coupling via a carefully chosen lens combination is required. This will usually increase the physical bulk of the source and may limit the application to OFS devices.

1.3.5 Broad-band continuous spectrum light sources

This group of light sources is well suited to a number of OFS applications, particularly those requiring excitation at wavelengths of less than ~ 550 nm, and especially in the UV part of the spectrum ($\lambda < 350$ nm) where solid-state sources are largely unavailable and the use of lasers is uneconomic. Additionally, these sources can offer a wide wavelength spread which is useful for sensors using wavelength-encoding techniques. Modulation of the light can be achieved using simple 'chopper' techniques (limited to about a few kHz) or using an electro-optic approach, but these are usually very inefficient with such sources. As an example, a millimeter-sized tungsten–halogen incandescent lamp is available [23], the optical output being in the spectral range from wavelengths in the near UV and in the visible from about 400 nm upwards, into the IR part of the spectrum, where substantial power can be obtained. Discharge lamps such as the xenon lamp or mercury lamp (at low or high pressure) are widely employed for emission on either discrete spectral lines or as a broad band over the region from the vacuum UV ($\lambda < 200$ nm) to longer wavelengths. The coherence characteristics of this group of light sources are usually very poor, for example the coherence length of a typical white light source is only three or four wavelengths. Details of such sources are discussed in a number of well-known reference texts [14], as their use in conventional analytical instrumentation has been familiar for many years. However, their mechanical instability and physical bulk make them unsuitable for many OFS applications, in addition to which such lamps are often inexpensive but yield a substantial percentage of their power output as heat.

1.4 FIBER-STRUCTURE CHARACTERISTICS OF THE OFS

Current OFS technology has arisen largely on the basis of optical components and especially optical fibers made available as a result of the explosive growth in their use for communications purposes. Whilst the technology for producing ultra-low-loss mono-mode fibers has been developed in recent years to a highly sophisticated level, this same technology is also applicable to the production of fibers for sensor purposes, where, however, other fiber characteristics are needed. Thus the OFS development market is fortunate in being able to call upon relatively cheap and readily available technology to produce fibers with specific characteristics to suit sensing purposes. Whilst fibers produced for communications purposes are preferred to have as low a sensitivity as possible to external effects, even ordinary silica fibers are used in many distributed sensor applications, due to their sensitivity to temperature, for example, through changes in nonlinear scattering processes. As these OFS techniques have developed rapidly, different types of fiber have been manufactured and the number of these types will continue to increase to meet new and special applications. The fibers considered here are characterized by the nature of their structure (such as the core radius, the refractive index distribution along a fiber or across its core, the numerical aperture (NA), polarization performance), and other considerations. The NA is determined in terms of the refractive indices of the core and cladding respectively, and θ_m is the maximum angle of incidence to retain guiding of light, as shown in Figure 1.8.

1.4.1 Mono-mode fiber

A mono-mode optical fiber with a Vebert (V) number < 2.405, defined as

$$V = (2\pi a/\lambda)(n_1^2 - n_2^2)^{1/2}$$

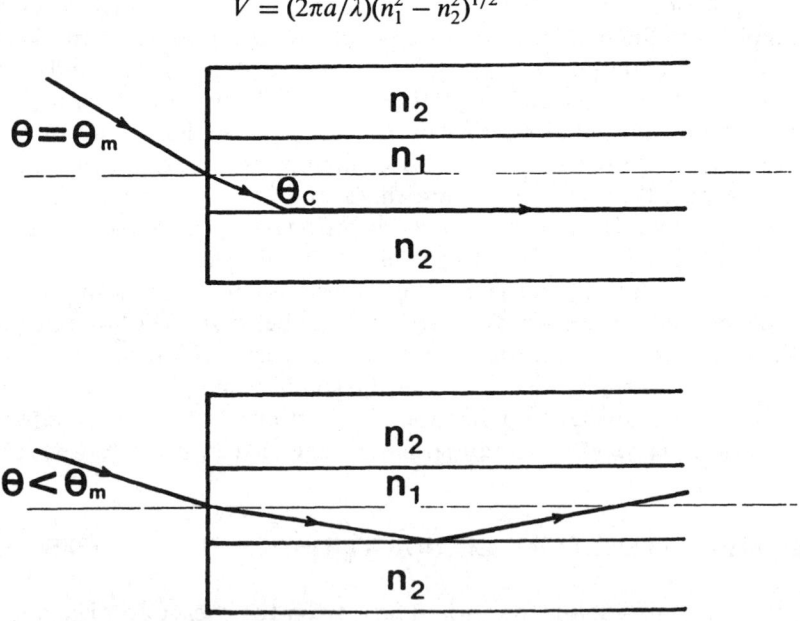

Fig. 1.8 Illustration of propagation of light in a fiber: θ_m = maximum angle of incidence; θ_c = critical angle in the fiber.

where a is the fiber core diameter, λ the wavelength of light guided in the fiber and n_1 and n_2 the refractive indices of fiber core and the outer cladding respectively, is widely used in OFSs based on phase modulation, to encode the measurand. The main characteristic of this is the maintenance of the optical phase relationship along the guided beam in the fiber. A typical mono-mode fiber designed to operate at a wavelength of 633 nm has an NA value of about 0.25, a core radius of a few micrometers, with a cladding radius of between 40 and 50 μm [21]. The use of mono-mode fibers in sensing has been reviewed in some detail by Jones [24]. In an ideal fiber which is perfectly circular, two orthogonal polarization states are generated and the state of polarization of the guided wave (the HE_{11} mode) propagates along the fiber unchanged. Unfortunately, in the practical fiber, such ideal conditions do not exist. As a result of its dominant intrinsic and any additional extrinsic birefringence, the two guided waves at two orthogonal polarization states will develop a phase difference between them after propagating a certain distance along the fiber. The distance over which a phase difference of 2π occurs is called the beat length (L_b), which can be written as

$$L_b = \lambda/(n_x - n_y)$$

where n_x, n_y are the effective refractive indices with x and y being the axes of a noncircularly symmetric fiber core, and λ is the free-space wavelength. Typical single-mode fibers have L_b in the centimeter region, dependent upon their physical characteristics [13]. As birefringence is dependent upon temperature, any variation of the environmental temperature will result in a variation in the optical phase difference between the two propagating polarization states. Hence the state of polarization (SOP) of the output light will depend to some extent upon the ambient temperature. Thus this kind of fiber can only be used where the polarization state can be neglected in the sensing operation.

1.4.2 Polarization-maintaining fiber

Many optical fiber sensors rely upon the use of polarization effects, and the maintenance of the SOP of light propagating in fiber is important. In order to overcome the problem of a variable state of polarization in a mono-mode fiber, considerable effort has been expended to develop polarization-maintaining fiber. This characteristic is achieved during the manufacturing process by inducing stresses in the material itself. There are two categories of polarization-maintaining fiber (PMF) available, linear polarization-maintaining fiber (LPMF) and circular polarization-maintaining fiber (CPMF). In the former category, only one of two orthogonal polarization states (HE_x or HE_y) can be maintained at the output of the fiber, whilst in the CPMF a round fiber is twisted to produce a difference between the propagation constants of the clockwise and counterclockwise circularly polarized HE_{11} modes. Figure 1.9 shows in more detail the classification of PMF and the subject is discussed in greater detail by, for example, Okoshi [25].

16 CLASSIFICATION OF OPTICAL FIBER SENSORS

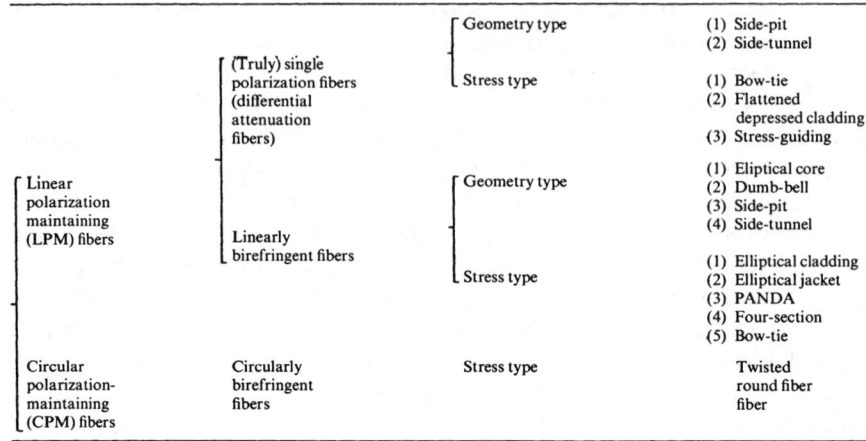

Fig. 1.9 Classification of various polarization-maintaining fibers (after Okoshi, 1987).

1.4.3 Two-mode fiber

Fiber allowing the propagation of only a few modes represents a new type, recently developed, which can provide a number of independent transmission channels in a single-strand fiber. A number of investigations have been carried out to understand better the modal and polarization characteristics of these fibers [26, 27]. Although the multichannel capacity of few-mode fiber has not yet been fully exploited due to mode-control problems, two-mode fiber has been used in several types of OFS [28, 29]. It is usual to derive the two excited modes from a single source to ensure a known coherence relationship between them. It is very difficult to excite and maintain two modes independently. A two-mode fiber with a V number between 2.4 and 3.8 can guide two spatial modes, the LP_{10} and LP_{11} modes [30]. The polarization and intensity distribution for a highly elliptical core, two-mode fiber (operated with LP_{10} and LP_{11}) can be kept unchanged for wavelengths between 488 and 633 nm [31].

1.4.4 Multi-mode fiber

Multi-mode optical fiber, with a V number significantly greater than 2.4, has a much larger core size and sometimes also much larger refractive index differences between core and cladding. The number of modes (M) guided by the fiber is determined from the V number and is given by $4V^2/\pi$. The core diameter of a multi-mode fiber is usually larger than 50 µm which means it has more than 100 times the core cross-sectional area of a mono-mode fiber. This factor is also reflected in the material costs and thus in the price of the fiber itself. This is especially significant for long lengths or specialized fibers. A large index difference between core and cladding yields a large NA, but may be difficult to fabricate. As a result of the larger core area and a large value of NA, multi-mode fiber presents a greater launching efficiency from an optical source. Since a multi-mode fiber can support a large number of modes which will propagate along different optical paths in the fiber, modal noise may be generated if a multi-mode interferometer is subjected to vibration, especially when a high coherence light source is used. In this way, multi-mode fibers are less suited to being coupled to high coherence light sources when phase information is required as the basis of the measurement. However, multi-mode fibers are often used in an OFS in applications where high light intensities are required, particularly in those that employ intensity-modulation techniques. There is flexibility to use a wide variety of optical sources, with suitable coupling optics. In many cases, light signals can be launched into fibers with reasonable efficiency and often incoherent sources show a high degree of optical power over a usable spectral band, so inefficient coupling still results in adequate light levels in the optical fiber for many sensor purposes. Multi-mode OFSs based on coherence modulation have been produced [22, 32].

1.4.5 Specialized fibers

A number of types of special fibers have been made for particular use in the field of OFSs, and are designed with only limited applicability in optical fiber communications systems. However, the use of fiber lasers as optical sources and subsequent laser amplifiers as part of an optical communications system may bring some such special fibers into the communications field. The difference between these and other kinds of fiber lies mainly in the modified shape of the fiber core, the refractive index distribution along or across the fiber (e.g. photoinduced effects) or the material from which it is made. As the varied techniques used in OFSs are developing steadily, the number of special fibers in use continues to increase. However, only a few of the wide variety of specialist fibers are considered below to illustrate the broad nature of the field.

(a) D-shaped fiber

When light is guided along a fiber with D-shaped cross-section, the result is the exposure of the evanescent field near the core of fiber. As most of the cladding on one side of the fiber is removed, the thickness of the material that is left is very small (as shown in Figure 1.10(a)), and light modulation can be achieved in the evanescent field near the core. The techniques for manufacturing D-shaped fiber are described in Millar *et al.* [33] and Dyott *et al.* [34]. One of the main potential uses of single-mode D-shaped fiber is for polarization-holding directional couplers, where the process of coupling the guiding region, either through etching or by fusion, should be greatly simplified with their use [35].

(b) Hollow-section fiber

Hollow-section fibers are a further development of the D-shaped fiber. They may be produced by having a single longitudinal aperture at a fixed distance from the fiber core. Figure 1.10(b) shows such an acrylate-coated metal–glass fiber, with

18 CLASSIFICATION OF OPTICAL FIBER SENSORS

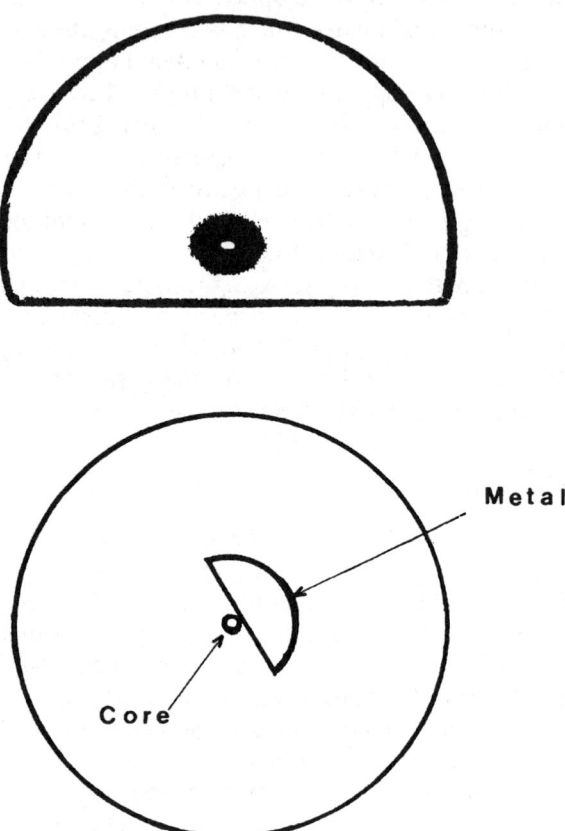

Fig. 1.10 (a) Cross-section of D-shaped fiber; (b) schematic of a composite metal–glass fiber.

a numerical aperture of about 0.16, a cut-off wavelength for transmission of about 1.25 µm and a distance between the core and hollow section of about 3 µm. It can give a high performance in optical polarization, e.g. for a 5 cm length of such polarizing fiber, the extinction ratio is about 40 dB over the spectral region 1300–1600 nm [36], which is useful in sensor applications.

(c) Fiber doped with rare earths

As an alternative to modifying the shape of the optical fiber, new kinds of specialized fiber have been developed by modifying the materials from which the fiber is normally fabricated (pure silica) or by adding additional materials. Fibers doped with rare earths have a number of special properties, due to the introduction of ions such as Nd, Er or Dy into the light-guiding regions of the fiber. Those properties are exploited to construct different 'in-fiber' sensor devices, to improve the measurement region of the OFS, or even to form new kinds of devices such as fiber lasers, fiber amplifiers, a fiber with an increased Verdet constant or a fiber showing an increased Kerr effect and nonlinear optical coefficients [37]. An important application has been the distributed temperature sensor with a rare-earth-doped fiber which has been developed recently. The sensitivity reported for it is better than 1 °C over the temperature region −200 to +100 °C and higher, with a spatial resolution of 3.5 m [38].

(d) Liquid-core fiber

Liquids have been used as core materials for fibers where the special characteristics of the liquid have been required for a specific sensor purpose. Thus the refractive index or scattering properties of the core material may be tailored to a particular application through the use of a specially chosen liquid. An important early example is in the work of Hartog [39], where a fiber filled with an organic liquid with significant scattering properties was used as the basis of an early distributed temperature sensor. However, problems exist in the fabrication of long lengths of such fiber and the possible hazards from spillage of the core material. Additionally, such fibers are expensive to fabricate, and are not widely acceptable in industrial installations.

(e) Photoinduced effects in fibers

The most important manifestations of this phenomenon is the production of Bragg gratings in single-mode optical fibers doped with high levels of germanium. The gratings are produced or 'written' into the fiber, through the side, by imaging two laser beams (usually in the UV part of the spectrum) which overlap to form an interference pattern within the fiber itself. The resultant bright and dark fringes thus produced give rise to the grating which operates as a conventional longitudinal grating via the narrow-band reflection from this region which has been created now to have a periodic variation of the refractive index of the single-mode fiber used. In a sensor configured with this type of fiber, the center or Bragg wavelength of the grating is dependent upon the scale length of the period of the grating and the mean refractive index. In sensor use, for example to measure strain, this will cause a change in the dimensions of the grating, and thus its center wavelength, which can readily be detected, for example when the fiber is incorporated in a so-called 'smart structure' [40].

1.5 SENSOR MODULATION TECHNIQUES

The modulation techniques employed in optical fiber sensors vary widely in their measurand function, performance and position in the sensor system itself. The basic techniques considered here, however, are those in which the light itself is modulated. This process may be categorized into several groups, according to the parameters of the electric field or the optical radiation varied. More than one type of modulation technique may be employed in one sensor; for example, path-length modulation is often associated with refractive-index modulation, and this is still a dominant mechanism in many OFSs.

1.5.1 Intensity modulation

Intensity-modulation techniques, employed widely in the earliest types of OFS, show the virtues of simplicity, reliability and achievability at low cost. The variation of the light intensity at the probe itself indicates the encoded change of an applied measurand, such as temperature, pressure [41], acceleration [42], or 'on/off' state in a simple switch. Various schemes to produce this type of intensity modulation have been produced, for example fiber displacement in the early Fotonic® sensor, shutter modulation, reflective schemes, fiber loss, intensity of coupling via the evanescent field, absorption, light scattering, digital encoding and modulation, amongst others. However, a basic drawback is the lack of an accurate reference signal in many cases, although several internal referencing schemes [43, 44] have been proposed. This intensity-based approach is not widely used in modern types of OFS, except for simple switches, although such devices tend to be comparatively inexpensive.

1.5.2 Polarization modulation

Polarization modulation is another interesting and very valuable technique for measurand encoding and signal recovery in many types of OFS. The basic principle of operation is that the relevant physical quantity to be measured is transduced into a polarization change through an appropriate interactive effect, e.g. the electro-optic (Faraday) or elasto-optic effects. The measured field can be inside the fiber (intrinsic sensor) or outside the fiber (extrinsic sensor), or even both situations may apply in one system. Historically, these were some of the first modulation schemes to be developed and many new sensors based on polarization modulation have been developed in the last few years. These are, in particular, sensors for measuring current and magnetic field, temperature and stress, amongst others, many of which have been reported in detail in the literature [45].

1.5.3 Wavelength modulation/encoding

Wavelength modulation/encoding techniques are very valuable as a means by which a nonintensity-dependent measurement may be made using optical fibers. Wavelength division multiplexing (WDM) has been discussed by a number of authors [46, 47] and relies on the encoding of the measured information in terms of a specific wavelength of light received from the transducer. This frequently means, in a real sensor, the use of a wide-band optical source with the transducer acting as a wavelength encoder, depending on its perturbation by an external parameter. Several displacement/rotation transducers have recently been developed [48, 49], primarily for testing in the aerospace environment, using gratings or filters which are mechanically rotated or moved in a linear fashion to yield the appropriate output.

There are two further kinds of wavelength modulation in common use in interferometric sensors. The first is the so-called 'frequency modulated continuous wave' (FMCW) technique [50–52]. In this case a single-mode laser diode is driven by a linear current ramp, and the wavelength (frequency) of the output is varied as the change of the drive current in an unbalanced interferometer. The variation of wavelength of the diode output causes a phase change, and therefore the intensity received from the interferometer will change and yield information on the optical path difference in the interferometer and thus information on the measurand. A second example is modulation either by the Doppler shift from a moving body or a frequency modulator such as a Bragg cell in the experimental optical arrangement [53]. A considerable amount of effort has been extended in recent years to the development of such systems, using fibers where appropriate [54, 55], and it has become established as an important technique for the measurement of fluid flow and surface velocity [56, 57], often with medical measurement applications, e.g. blood flow [58]. A number of different applications of wavelength modulation using Bragg gratings for strain sensor have been discussed by Dunphy et al. [59].

1.5.4 Optical-path-difference modulation

This modulation is one of the most commonly used modulation schemes in optical fiber sensors, usually of the interferometric type. The change of optical path difference is caused by a physical displacement in an interferometer, operated at the quadrature position of the output or in a 'zero path length difference' mode, and the linear displacement range is less than $\lambda/4$, corresponding to a value of < 200 nm for a typical laser light source. However, when the displacement is greater than $\lambda/2$, the fringes will appear periodically, each fringe representing a $\lambda/2$ change in optical path difference, corresponding to a change of $2N$ rad. By counting the number of the fringes, the displacement change can be calculated from the simple equation

$$\Delta L = N\lambda/2$$

where ΔL is the change of optical path difference, N the number of fringes counted and λ the wavelength of the light source. A very wide range of sensors based on optical path modulation has been developed. The measurands vary from distance [60], to vibration [61], to those associated with a change of displacement, such as pressure [62]. A number of parameters can be configured so as to produce a displacement and thus be amenable to measurement by this sensitive technique.

1.5.5 Refractive index modulation

Modulation of the material refractive index and its use for measurement purposes can be achieved both within the actual optical fiber and extrinsically in sensor systems. In the first case, measurands such as temperature or pressure may be used to vary the index difference between the core and the cladding of a fiber, and thus the amount of light 'leaking' from core to cladding causes a change in the output light intensity observed. In the second case, an external sensor element may be employed, for instance, by using a lithium niobate ($LiNbO_3$) crystal as a temperature sensor. The refractive index of such a sensor element is modulated by the measurand, and therefore the optical path within the material is changed. When an interferometer with a single-mode light source is employed to determine this in an optical fiber sensor, the change of optical path difference can be used to vary the phase difference of the radiation in the interferometer and thus the output intensity observed. The fractional change in optical length (l) in a silica fiber is given by $dn/dT = 10^{-5} K^{-1}$ [4] and the phase shift by $2\pi nl/\lambda$ which is approximately 1 rad K^{-2} cm^{-1} for a wavelength of 633 nm, the wavelength of the He–Ne laser. Other materials less commonly used for optical fibers will show different characteristics. An approach using index modulation outside an optical fiber has been reported by Scheggi et al. [63] in a sensor application. The use of index modulation for Bragg grating generation and sensor use has been discussed earlier.

1.5.6 Time modulation

With the use of coherent light sources, a temporal modulation can be transduced into a path-length modulation through the use of the relationship $\Delta L = c \Delta t$ where ΔL is the optical path length traversed by light during a period of time Δt and c is the speed of light. Therefore time modulation can be considered to be equivalent to path-length modulation under these circumstances. With the use of an incoherent light source, time modulation, also known as rate modulation, is the form of modulation achieved using a low frequency applied to the light intensity or to produce pulses of known temporal duration. Information conveyed in the time domain can be highly dependent on the measurand using the correct sensor encoding. For example, the rate information from a rotating object such as a turbine, in a vortex-shedding flow meter or from a rotating shaft, can be delivered to an optical detector by optical fiber links to detect the optical radiation containing the information signal. The detected signal can then be converted into a digital signal.

A number of OFS techniques based on time modulation have been developed, such as a quartz resonator hybrid OFS for displacement measurement [64] and pressure [65] and temperature sensors [66] based on fluorescent time decay. By the use of correlation techniques, the rotation of a shaft can be monitored directly from the light scattered by the shaft [61]. In such an application the signal received due to the reflection from the shaft is subjected to autocorrelation, and the rotation rate can be then calculated. Agreement to within the few percent accuracy of a mechanical device has been observed [61].

1.5.7 Coherence-length modulation

'Coherence length' is a term used as a measure of the temporal coherence of a light source. The 'coherence region' of an interferometer output is the region where phase modulation can be achieved, e.g. in a Michelson interferometer, this being less than twice the coherence length, L_c. When the optical path difference (OPD) in an interferometer is less than the coherence length of the light source employed, the fringe visibility [24] of the output obtained is in the region between 1 and 0, and phase modulation can be achieved. However, when the OPD is greater than the coherence length, the fringe visibility is zero and the phase modulation becomes undetectable. In this latter case, a coherence technique can be used to 'shift' an unbalanced interferometer into a 'balanced' or a 'near balanced' region. The basic function of this modulation is to compensate the large OPD which is observed in order to shift the coherence region from its original position to where phase modulation can be achieved.

When coherence-length modulation is introduced, it is necessary that a second detector interferometer is employed. The OPDs of the two interferometers are set to be larger than the coherence length of the light source, with the difference between them being less than the coherence length. Under this condition, two interferometers give out an interference intensity distribution which shows the so-called 'fringes of superposition' [67]. The scheme used in this modulation is sometimes called 'path-matched differential-interferometry' or 'white-light interferometry' and has been widely exploited in interferometric fiber optic sensors. The most commonly used low-coherence light sources for such arrangements are the LED, superluminescent diode or multi-mode laser diode, in addition to broad-band sources, and significant work has been done recently to develop OFSs operating by means of coherence-length modulation. As an example, sensors for pressure measurement [22], remote displacement [60, 68], acceleration [69] and flow speed [70] have been reported recently.

1.6 IMPLEMENTATION OF THE CLASSIFICATION SCHEME ON A KNOWLEDGE-BASED SYSTEM (KBS)

1.6.1 The knowledge-based system (KBS)

Using such a classification system, the authors believe that many of the OFS techniques described can be more readily surveyed and examined and their interrelationships clearly seen. Representative examples of some of the different schemes discussed are illustrated in the reference matrix, Table 1.1 [71–107], and discussed in detail in the work of Ning *et al.* [108]. Furthermore, implementation of the scheme in a KBS is relatively easy, and it can be used to point to or

Table 1.1 Examples of different units of OFS space (numbers refer to illustrative references to the technique)

Light source	Intensity modulation	Length modulation	Wave-length/frequency modulation	Polarization modulation	Refraction index modulation	Time modulation	Coherence modulation
Mono-mode fiber							
Mono-mode	86	71	92	79	96	103	88
Two-mode			105	81			58
Multi-mode		89	102				
Narrow-band		85			95		
Broad-band				82			
Polarization-maintaining fiber							
Mono-mode	87	101	84		83		
Two-mode							
Multi-mode		90					
Narrow-band							
Broad-band							
Two-mode fiber							
Mono-mode		78					106
Two-mode							
Multi-mode				100			
Narrow-band						77	
Broad-band							107
Multi-mode fiber							
Mono-mode	91	97					
Two-mode	40						
Multi-mode	76						
Narrow-band	72	99			93		
Broad-band	98		94				
Special fiber							
Mono-mode	104						
Two-mode							
Multi-mode	75						74
Narrow-band	73						
Broad-band							

IMPLEMENTATION OF CLASSIFICATION SCHEME ON A KBS 25

generate new techniques in a graded order according to their feasibility, as discussed by El-Hami *et al.* [109]. The KBS employed by El-Hami [109] was a modified version of a program originally developed for conceptual design of instrument systems [110,111], written in PROLOG and running on a Sun SPARC workstation, but which could easily be transferred to other machines supporting PROLOG, e.g. an Apple Macintosh. The knowledge was been modified according to the type of knowledge and its classification being employed. The knowledge base contained information about the existing and possible optical fibers, sources, modulation techniques and measurands and the relationships between them.

To operate such a KBS, initially a user would specify the features required of an OFS and the KBS would then point to suitable techniques in a graded order according to their feasibility, which could be further studied for ultimate selection by the user. The requirements presented to the KBS are taken from the functional requirements specified [109].

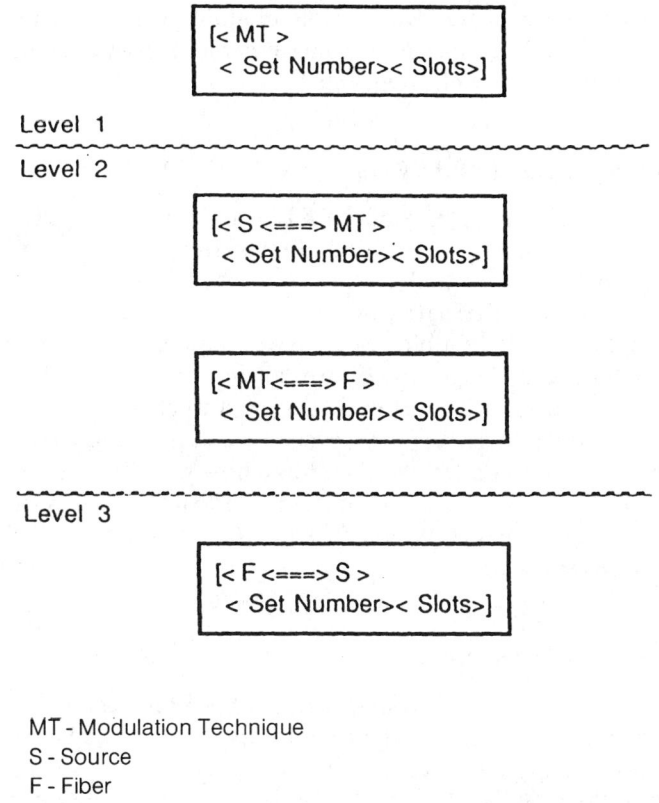

Fig. 1.11 Illustration of the three-level hierarchy for the knowledge base (after El-Hami *et al*, 1993).

1.6.2 Representation of the KBS scheme

It is important for this application to represent the expert's knowledge in a concise and efficient manner. The four most common forms of knowledge representation are semantic networks, predicate calculus, production rules and frames, the last of which is the style seen as more appropriate for this application [112]. Representation of the previously described scheme is based on a compact and efficient knowledge base developed and organized into a hierarchical frame-based structure. The knowledge base is designed to have a three-level hierarchy with frames at each level having a common internal structure, as shown in Figure 1.11, showing the relationship between the various levels of the hierarchies. The first consists of a number of frames, each consisting of a number of sets of slots, associated with a 'modulation technique' (MT). At the second level, frames correspond to the existing relationships between 'optical sources' (S) and MT and MT and F (for fibers). At the third level, frames are associated with the existing relationships between F and S, containing the same kind of information as those on level 1. This scheme allowed an easy expansion to the knowledge base whenever necessary, and each frame consisted of a number of slots containing information related to that frame, where the information is represented at each level. Further details are discussed by El-Hami *et al.* [109].

1.6.3 Operating principle of the KBS

To show how the knowledge base of the KBS can be used, a simple example is employed as illustrative of the operation of the system. This has been chosen so that the operation of the method is relatively clear and the result is capable of cross-evaluation by an alternative means: assessment by an expert and, of course, the results obtained should be capable of duplication by, for example, an expert panel whose knowledge would form all or part of the stored knowledge used in the KBS. Such an illustration is designed to show that in this case the KBS produces a valid and useful result, even in the limited and restricted circumstances of the evaluation undertaken. It shows how the information in the KBS can flow and build up, in order to point both to possible and then suitable techniques, using the functional requirements and how they are evaluated, utilizing a spreadsheet-based software.

1.6.4 Decision evaluation procedure

The decision evaluation procedure employed by El-Hami *et al.* [109] operated as follows. An evaluation matrix, in which alternatives (sources, fibers and modulation techniques) occupy the columns and the attributes (nonfunctional requirements) occupy the rows, is produced. The preference of one alternative over another with reference to each attribute is expressed in terms of numbers (utility points). These range from 1 to 10, where 1 represents the lowest preference and

IMPLEMENTATION OF CLASSIFICATION SCHEME ON A KBS

Table 1.2 Representative designs for OFS temperature transducers (after El-Hami *et al.*, 1993)

Code letter	Mechanisms of OFS	Ref	'S' code	'F' code	'MT' code
A	Optical pyrometer	114	BBCLS	MMF	IM
B	Decay-time approach	113	NMLS	MMF	FM
C	Interferometer	115	TMLS	MMF/SMF	PLM
D	Fabry–Pérot in fiber	116	SMLS	PMF	IM
E	Distributed sensor	117	NMLS	MMF	TM

S codes: BBCLS, broad-band light source; NMLS, N-mode light source; TMLS, two-mode light source; SMLS, single-mode light source.
F codes: MMF, multi-mode fiber;SMF, single-mode fiber; PMF, polarization-maintaining fiber.
MT codes: Im, intensity modulation; FM, frequency modulation; PLM, path-length modulation; tm, time modulation.

10 the highest. Using the evaluation matrix, the user can then be provided with graphical results illustrating a full comparison of the techniques discussed, in a comprehensible form.

1.6.5 Illustration based upon temperature sensor selection

In order to show the results of a procedure based on the selection of an appropriate type of OFS temperature monitor, a simple test of the KBS was reported [109]. For reasons of convenience, several different and representative OFS designs labelled A, B, C, D and E were chosen for evaluation purposes, and these are spelt out in Table 1.2 [113–117]. They represent a series of methods, of different types, either used in commercial instruments or proposed in the literature by a range of authors for temperature measurement with OFS techniques. The data from the 'expert panel' were applied via the KBS and the results shown.

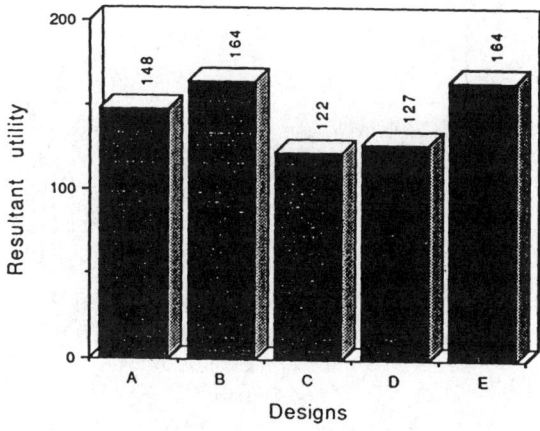

Fig. 1.12 Cross-comparison of selected designs for OFS (after El-Hami *et al*, 1993).

28   CLASSIFICATION OF OPTICAL FIBER SENSORS

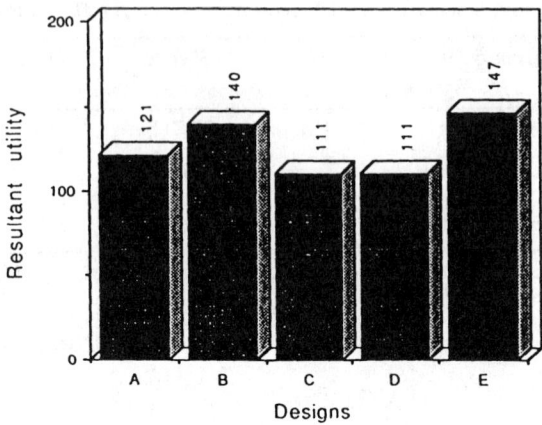

Fig. 1.13 Reevaluated designs (after El-Hami *et al*, 1993).

In Figure 1.12, on the basis of the data in the KBS, the selected designs were cross-compared. It is clearly seen that designs B and E show a higher score than design A and designs A and D are preferable to design C, and so on. Thus the user of the KBS supplied with the data has a basis for selection, in this simple example, of an OFS temperature sensor. Further, if, for example under certain circumstances, the cost of a design is not the overriding factor, a new set of evaluated designs could be produced, as shown in Figure 1.13. This clearly illustrates that, under the new circumstances, design E is preferred and design A has a higher rating than designs C and D. The included degree of preference (weighting factor) can lie anywhere in the range 1–10 (maximum). Figure 1.14 shows a chart corresponding to that of Figure 1.12 but with the further OFS effect of the technique being, for example, extrinsic or intrinsic taken into account.

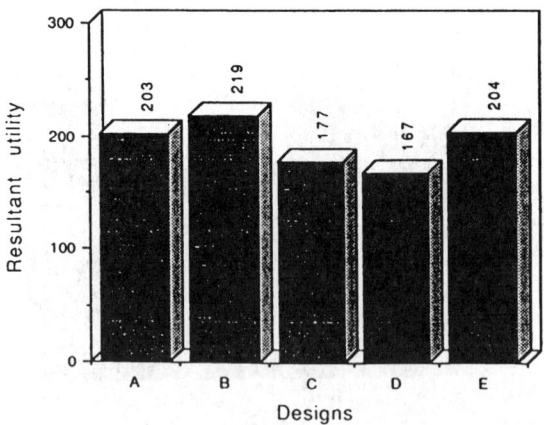

Fig. 1.14 Comparison of result with the 'expert choice' for OFS design (after El-Hami *et al*, 1993).

The former is usually simpler to operate, with less expensive and complex sources: the latter provides spatially resolved data in many cases. A comparison of the result of this limited, yet illustrative, exercise with the 'expert choice' is seen in the sensor design obtained from Figure 1.14. This clearly shows that now design B is the preferable one when all the factors are taken into consideration. This represents a design which is, in fact, the basis of several systems that have either been proposed by manufacturers or developed into commercial exploitation [118–121].

1.7 SUMMARY AND DISCUSSION

In this approach, the construction of an extended logical and systematic classification of OFS devices and appropriate techniques have been described, and a multi-dimensional coordinate scheme has been used to illustrate the classification graphically. Based on the model, a spatial framework for the classification process was employed and fitted with a variety of sensors to illustrate their comparative features. It is believed that such a representation can indicate those areas where OFS schemes have already been proposed or may further be exploited, in a graded order according to their feasibility. An implementation technique for the classification scheme on a KBS has been outlined.

Several points worthy of note emerge, and examples of the different schemes which have been discussed in the literature are seen in the reference matrix, Table 1.1. Most work done in past years, in interferometric-based sensors for example, has concentrated on two major areas. The first is the 'single-mode light source, mono-mode fiber and phase modulation' category, which is represented using all types of interferometers. The second is the 'broad-band, continuous spectrum light source, multi-mode fiber, with intensity modulation' type, which covers many of the incoherent types of OFS employed. A number of examples have been shown to reveal other applications, including the use of low-coherence light sources such as the LED or the multi-mode laser diode. Such work emphasizes the recent tendency toward the exploitation of these types of light sources, which are both convenient and inexpensive, and operate at low voltages.

It is clear that a considerable body of work has been undertaken to develop OFS techniques, many of which are discussed in detail in subsequent chapters. However, there are still many 'blank cubes' which may be filled inside the OFS 'techniques space'; in other words, this OFS techniques space has not yet been fully exploited. It is not expected that this space will be filled immediately, if ever, but this may gradually change as new kinds of light source, optical fiber or sensor approach emerge. However, it is likely that due to the basic physical mechanism and for sound commercial reasons, many cubes will remain 'blank' and perhaps in only a few areas will there be real commercial success with OFS techniques, where their unique advantages are seen. This categorization is designed to draw attention to those areas where devices both do and do not exist and represents a basis for the sensors described in the body of the text and for new areas of the technology which may emerge.

REFERENCES

(References are chosen to reflect early use of the techniques considered, or to cite review or reference works which highlight the methods discussed in the text.)

1. T. G. Gaillorenzi, Optical fiber sensor technology, *IEEE Journal of Quantum Electronics*, **QE 8** (1982) 626–60.
2. R. S. Medlock, Fibre optic intensity modulation sensors, OFS NATO ASI Series, Series E: *Applied Sciences*, No. 132 (1987), pp. 123–124.
3. R. C. Spooncer, Fibre optics in instrumentation, in: *Handbook of Measurement Science* Vol. 3, Ed: P. H. Sydenham & R. Thorn, Wiley, Chichester, UK (1992), pp. 1691–1720.
4. K. T. V. Grattan, Fibre optic sensors – the way forward, *Measurement, J. Int. Meas. Confed.*, **5** (1987) 122.
5. K. T. V. Grattan, New developments in sensor technology – fibre and electro-optics, *Meas. Control*, **22** (1989) 165–175.
6. E. Udd, *Fiber Optic Sensors*, Wiley InterScience, New York (1991).
7. D. A. Jackson, Overview of fiber sensor developments, in: *Optical Fiber Sensor Technology*, Ed: K. T. V. Grattan & B. T. Meggitt, Chapman & Hall, London (1995), pp. 1–9.
8. S. Annovazzi-Lodi and S. Merlo, Single mode fiber optic sensors, in: *Single-Mode Optical Fiber Measurement – Characterization and Sensing*, Ed: G. Cancellieri, Artec House, Boston (1993), pp. 261–329.
9. P. Stein, Sensors as information systems, *Research/Development*, (1970), pp. 30–40.
10. S. Middelhoek and D. C. van Duyn, Classification of solid-state transducers, *Australasian Instrumentation and Measurement Conf.*, Adelaide, Australia, Nov. 14–17 (1989), pp. 1–4.
11. H. Ishio, J. Minowa and J. Nosu, Review and status of wavelength-division multiplexing technology, *IEEE J. Lightwave Technol.*, **LT-2** (1984) 448–463.
12. A. R. Mickelson, O. Klevhus and M. Eriksrud, Back-scatter readout from serial microbending sensors. *IEEE J. Lightwave Technol.*, **LT-2** (1984) 700–709.
13. Proc. Workshop on single fibre sensor technology, University of Kent, Canterbury, p. 42 (unpublished).
14. J. Dakin and B. Culshaw, *Optical Fiber Sensors: Principles and Components*, Artec House, Boston (1989), pp. 151, 198.
15. K. Ho Tate and D. Kamatani, Reflectometry with high spatial resolution and no moving parts by means of source-coherence modulation, *Proc. OFS'89 Paris* (1989), p. 64.
16. B. T. Meggitt, A. W. Palmer and K. T. V. Grattan, Fibre optic sensor using coherence properties processing aspects, *Int. J. Optoelectron.*, **3** (1988) 451.
17. D. J. Webb, J. D. C. Jones and D. A. Jackson, Extended range interferometry using a coherence-tuned synthesized dual-wavelength technique with multimode fibre link. *Electron. Lett.*, **24** (1988) 1173–1174.

18. C. H. F. Velzfl and R. P. Brouwer, Output power and coherence length of stripe geometry double-heterostructure semiconductor laser in incoherent feedback, *IEEE J. Quantum Electron.*, **QE-I5** (1979) 782–786.
19. Y. N. Ning, K. T. V. Grattan, B. T. Meggitt and A. W. Palmer, Characteristics of laser diodes for interferometric use, *Appl. Opt.*, **28** (1989) 3657–3661.
20. K. T. V. Grattan, A. W. Palmer, Y. N. Ning and B. T. Meggitt, Interferometric sensors – extension of range using low coherence light from laser diode, *Proc. EFOC/LAN*, Amsterdam, The Netherlands, June (1989), p. 375.
21. R. Kist, Sources and detectors for fibre optic sensors, OFS NATO ASI Series, Series E: *Applied Sciences*, No. 132 (1987), p. 267.
22. G. Beheim, K. Fritsch and R. N. Poorman, Fibre-linked interferometric pressure sensor, *Rev. Sci. Instrum.*, **58** (1987) 1655–1659.
23. Manufacturer's data, Micro-Gluhlampen-Gesellschaft, Hamburg, Lanster, 1, D-2057.
24. J. D. C. Jones, Monomode fibre optic sensors, in: *Optical Methods in Engineering Metrology* Ed: D. C. Williams, Chapman & Hall, London (1993), pp. 415–464.
25. T. Okoshi, Polarization phenomena in optical fibres, OFS NATO ASI Series, Series E: *Applied Sciences*, No. 132 (1987), pp. 227–242.
26. B. Y. Kim, Few mode fibre devices, *Proc. Int. Conf. Optical Fiber Sensors*, New Orleans, LA, USA (1988), Tech. Digest Series, Vol. 2, Part 1, pp. 146–149.
27. S. Y. Huang, J. N. Blake and B. Y. Kim, Mode characteristics of highly elliptical core two-mode fibre perturbations, *Proc. Int. Conf. Optical Fiber Sensors*, New Orleans, LA, USA (1988), Tech. Digest Series, Vol. 2, Part 1, pp. 14–17.
28. J. N. Blake, B. Y. Kim, H. E. Engan and H. J. Shaw, Analysis of inter-modal coupling in a two-mode fibre with periodic microbends, *Opt. Lett.*, **12** (1987) 281.
29. G. Meltz, J. R. Dunphy, W. W. Morey and E. Snitzer, Cross-talk fibre-optic temperature sensor, *Appl. Opt.*, **22** (1983) 464.
30. W. V. Sorin, B. Y. Kim and H. J. Shaw, Phase-velocity measurements using prism output coupling for single and few-mode optical fibers, *Opt. Lett.*, **11** (1986) 106.
31. B. Y. Kim, J. M. Blake, S. Y. Huang and H. J. Shaw, Use of highly elliptical core fibers for two-mode fiber devices, *Opt. Lett.*, **12** (1987) 729.
32. T. Bosselmenn, Multimode fibre coupled white light interferometer position sensor, OFS NATO ASI Series, Series E: *Applied Sciences*, No. 132 (1988), p. 429.
33. C. A. Millar, B. J. Ainslie, M. C. Brierley and S. P. Craig, Fabrication and characterisation of D-fibres with an accurately controlled core/flat distance, *Electron. Lett.*, **22** (1986) 322–324.
34. R. B. Dyott, J. Bello and V. A. Henderek, Indium-coated D-shaped fiber polarizer, *Opt. Lett.*, **12** (1987) 287–289.
35. R. B. Dyott and P. F. Shrank, Self-locating elliptically cored fiber with an accessible guiding region, *Electron. Lett.*, **18** (1982) 980.
36. L. Li, G. Wylandgowski, D. N. Payne and R. D. Birch, Broadband metal/glass single-mode fibre polarisers, *Electron. Lett.*, **22** (1986) 1020–1022.
37. J. Dakin and B. Culshaw, *Optical Fibre Sensors: Principles and Components*, Artec House, Boston (1989), p. 249.
38. M. C. Farries, M. E. Fermann, R. I. Laming, S. B. Poole, D. N. Payne and A. P. Leach. Distributed temperature sensor using Nd^{3+} doped fibre, *Electron. Lett.*, **22** (1986) 418–419.
39. A. H. Hartog, A distributed temperature sensor based on liquid core fibers, *IEEE J. Lightwave Technol.*, **LT1** (1983) 498–501.

40. R. M. Measures, Fiber optic strain sensing, in: *Fiber Optic Smart Structures* Ed: E. Udd (1995), pp. 171–247.
41. G. Martens, J. Kordts and G. Weidinger, A photo-elastic pressure sensor with loss-compensated fibre link, *Proc. OFS '89*, Springer in Physics, Vol. 44, (1989), p. 458.
42. L. Jonsson and B. Hok, Multimode fibre-optic *Proc. 2nd Int. Conf. Optic Fibre Sensors*, Stuttgart, FRG, 5–7 Sept. (1984), SPIE, Vol. 514, pp. 191–194.
43. R. C. Spooncer, Fibre optics in physics and chemical sensors, *Institute of Measurement and Control Conf.*, Harrogate, UK, Nov. (1985), Institute of Measurement and Control, London.
44. B. Culshaw, Optical system and sensors for measurement and control, *J. Phys. E: Sci. Instrum.*, **16** (1983), 987–988.
45. Y. N. Ning, Z. P. Wang, A. W. Palmer, K. T. V. Grattan and D. A. Jackson, Recent progress in optical current sensing techniques, *Rev. Sci. Instrum.*, **66** (1995) 3097–3111.
46. H. Ishio, J. Minowa and J. Nosu, Review and status of wavelength-division multiplexing technology, *IEEE J. Lightwave Technol.*, **LT-2** (1984) 448–463.
47. G. Winzer, Wavelength multiplexing components – a review of single-mode devices and their applications, *IEEE J. Lightwave Technol.*, **LT-2** (1984) 369–378.
48. M. C. Hutley, Wavelength encoded optical fibre sensors, *Proc. 2nd Int. Conf. OFS*, Stuttgart, FRG (1984), SPIE, Vol. 514, pp. 111–116.
49. B. E. Jones and S. J. Collier, Optical fibre sensors using wavelength modulation and simplified spectral analysis, *J. Phys. E: Sci. Instrum.*, **17** (1984) 1240–1241.
50. H. Ghafoori-Shiraz and T. Okoshi, Fault location in optical fibers using optical frequency domain reflectometry, *IEEE J. Lightwave Technol.*, **LT-4** (1986) 316.
51. S. A. Al-Chalabi, B. Culshaw, D. B. N. Davies, I. P. Giles and D. Uttam, Multiplexed optical fibre interferometers: an analysis based on a radar system, *Proc. IEE* **132A** (1985) 150–156.
52. D. Uttam and B. Culshaw, Precision time domain reflectometry in optical fibre systems using a frequency modulated continuous wave ranging technique, *IEEE J. Lightwave Technol.*, **LT-3** (1985) 971.
53. R. B. Dyott, Fibre-optic Doppler anemometer, *IEEE J. Microwaves Opt. Acoust.*, **20** (1978) 13–18.
54. T. T. Nguyen and L. N. Birch, A fiber-optic laser Doppler anemometer, *Appl. Phys. Lett.*, **45** (1984) 1163.
55. D. A. Jackson, J. D. C. Jones and R. K. Y. Chan, A high-power fibre optic laser Doppler velocimeter, *J. Phys. E: Sci. Instrum.*, **17** (1984) 977–980.
56. K. Kyuma et al., Laser Doppler with a novel optical fibre probe, *Appl. Optics*, **20** (1981) 2424–2427.
57. Y. N. Ning, K. T. V. Grattan, A. W. Palmer and A. E. Baruch, A heterodyne vibration sensor using coherence length modulation technique, *IEE Proc. J.*, **138**, 393–395.
58. M. D. Stern, Laser Doppler velocimetry in blood and multiply-scattering fluids: theory, *Appl. Optics*, **24** (1985) 1968.
59. J. R. Dunphy, G. Meltz and W. W. Morey, Optical fiber Bragg grating sensors: a candidate for smart structure applications, in: *Fiber Optic Smart Structures* Ed: E. Udd (1995), pp. 271–285.
60. G. Ulbers, A sensor for dimensional metrology with an interferometer in integrated optics technology, *Proc. OFS '89*, Springer Proceedings in Phys., Vol. 44 (1989), p. 240.

REFERENCES

61. K. T. V. Grattan, A. W. Palmer, B. T. Meggitt and Y. N. Ning, Use of laser diode source in low coherence operation in fibre optic sensors, *Australasian Instrumentation and Measurement Conf.*, Adelaide, Australia, Nov. 14–17 (1989), pp. 305–308.
62. D. Trouchet, B. Laloux and P. Graindorge, Prototype industrial multi-parameter F.O.S. using white light interferometry, *Proc. OFS '89*, Springer Proc. in Phys., Vol. 44 (1989), pp. 227–233.
63. A. M. Scheggi, M. Brenci, G. Conforti, R. Faliai and G. P. Preti, Optical fibre thermometer for medical use, *1st Int. Conf. Optical Fibre Sensors*, IEE Conf. Publication, Vol. 221 (1983), pp. 13–16.
64. S. M. McGlade and G. R. Jones, Optical sensors for displacement measurement, *Colloq. Digest, Proc. Control*, London, Jan. (1984), IEE, 1984/7.
65. B. Hok and L. Jonsson, Pressure sensor with fluorescence decay as information carrier, *Proc. 2nd Int. Conf. OFS*, Stuttgart, FRG, Sept. 5–7 (1984), SPIE, Vol. 514, pp. 391–394.
66. K. T. V. Grattan, R. K. Selli and A. W. Palmer, Ruby decay-time fluorescence thermometer in a fiber-optic configuration, *Rev. Sci. Instrum.*, **59** (1988) 1328–1335.
67. M. Born and E. Wolf, *Principles of Optics*, Oxford University Press, Oxford (1980), pp. 360–367.
68. G. Behiem, Remote displacement measurement using a passive sensor with a fiber-optic link, *Appl. Opt.*, **24** (1985) 2335–2340.
69. C. J. Zarobila, J. B. Freal, R. L. Lampman and C. M. Davies, Two fibre-optic accelerometers, *Tech., Digest, Proc. OFS Conf.*, New Orleans, LA, USA (1988), Vol. 2, Part 1, p. 296.
70. B. T. Meggitt, Y. N. Ning, K. T. V. Grattan, A. W. Palmer and W. Boyle, Fibre optic anemometer using an optical delay cavity technique, *Eighth Int. Conf. Fibre Optics and Opto-electronics*, London, Apr. (1990), SPIE, Vol. 1314, pp. 321–329.
71. F. Farahi *et al.*, Optical-fibre flammable gas sensor, *J. Phys. E: Sci. Instrum.*, **20** (1987) 435–436.
72. F. P. Milanovich *et al.*, Remote detection of organo-chlorides with a fibre optic based sensor, *Anal. Instrum. (USA)*, **15** (1986) 542–558.
73. I. Aeby, Non-destructive measurement in advanced composite materials and structures using a fibre optic sensing system, *Proc. SPIE*, Vol. 986 (1989), pp. 140–147.
74. A. Safaai-Jazi and R. O. Clau, Synthesis of interference patterns in few-mode optic fibers, *Proc. SPIE*, Vol. 986 (1989), pp. 180–185.
75. P. B. Macedo *et al.*, Development of porous glass fibre optic sensors, *Proc. SPIE*, Vol. 986 (1989), pp. 200–205.
76. J. K. Zienkiewicz, Self-reference fibre optic methane detection system, *Proc. SPIE*, Vol. 992 (1989), pp. 192–197.
77. G. W. Fehrenbach, Fibre optic temperature measurement system based on luminescence decay time, *Tech. Mess.*, **56** (1989) 85–88.
78. F. Gonthier *et al.*, Circular symmetry modal interferometers in optical fibres, *Ann. Telecommun. (France)*, **44**, (1989) 159–166.
79. T. G. Giallorenzi *et al.*, Optical fibre sensor technology, *IEEE J. Quantum Electron.*, **QE-18** (1982) 626.
80. M. Corke *et al.*, All fibre Michelson thermometer, *Electron. Lett.*, **19** (1983) 471.
81. P. Alchavanleilabady, A dual intermometer implemented in parallel on a single birefringent monomode optic fibre, *J. Phys. E: Sci. Instrum.*, **19** (1986) 143.

82. S. A. Al-Chalabi et al., Partially coherent sources in interferometric sensor, *Proc. 1st Int. Conf. Optic Fibre Sensors*, London (1988), pp. 132–135.
83. H. L. W. Chan et al., Polarimetric optical fiber sensor for ultrasonic power measurement, *IEEE 1988 Ultrasonic Symp. Proc.*, Vol. 1 (1988), pp. 599–602.
84. H. Tsuchida et al., Polarimetric optical fibre sensor using a frequency stabilized semiconductor laser, *IEEE J. Lightwave Technol.*, **LT-7** (1989) 799–803.
85. A. V. Belov et al., The measurement of chromatic dispersion in single-mode fibre by interferometric loop, *IEEE J. Lightwave Technol.*, **LT-7** (1989) 863–868.
86. E. Wittendorp-Rechenmann, *Radiat. Prot. Dosim. (UK)*, **23** (1988) 199–202.
87. H. Takahara, F. Togashi and T. Aragaki, Ultrasonic sensor using polarization-maintaining optical fiber, *Can. J. Phys.*, **66** (1988) 844–846.
88. F. Farahi, N. Takahashi, J. D. C. Jones and D. A. Jackson, Multiplexed fibre-optic sensors using ring interferometers, *J. Mod. Opt.*, **36** (1988) 337–348.
89. A. S. Georges, High sensitivity fiber-optic accelerometer, *Opt. Lett.*, **15** (1989) 251–253.
90. H. Takahashi et al., Laser diode interferometer for small displacement and sound pressure measurements, *Conf. Precision Electromagnetic Measurements*, Tsukuba, Japan, June (1988), pp. 135–136.
91. E. Theocharous, Differential absorption distributed thermometer, *1st Int. Conf. Optical Fibre Sensors*, Apr. (1983), IEE, London, pp. 10–12.
92. M. L. Henning et al., Optical fibre hydrophones with down lead insensitivity, *1st Int. Conf. Optical Fibre Sensors*, Apr. (1983), IEE, London, pp. 25–27.
93. K. Spenner et al., Experimental investigations on fibre optic liquid level sensors and refractometer, *1st Int. Conf. Optical Fibre Sensors*, Apr. (1983), IEE, London, pp. 75–78.
94. E. R. Cox and B. E. Jones, Fibre optic colour sensors based on Fabry–Perot interferometry, *1st Int. Conf. Optical Fibre Sensors*, Apr. (1983), IEE, London, pp. 122–126.
95. K. Shibata, A fibre optic electric field sensor using the electrooptic effect of bismuth germinate, *1st Int. Conf. Optical Fibre Sensors*, Apr. (1983), IEE, London, pp. 164–166.
96. S. C. Rashleigh, Polarimetric sensors: exploiting the axial stress in high birefringement fibres, *1st Int. Conf. Optical Fibre Sensors*, Apr. (1983), IEE, London, pp. 210–213.
97. A. J. A. Brunsma and J. A. Vogel, Ultrasonic non-contact inspection system with optical fibre methods, *Appl. Opt.*, **27** (1988) 4690–4695.
98. K. Iwamoto and I. Kamato, Pressure sensors using optical fibres, *Appl. Opt.*, **29** (1989) 375–377.
99. T. Okamoto and I. Y. Yamaguchi, Mutlimode fibre optic Mach–Zehnder interferometer and its use in temperature measurement, *Appl. Opt.*, **27** (1988) 3085–3087.
100. S. Y. Huang et al., Mode characteristics of highly elliptical core two-mode fibres under perturbation, *Tech. Digest, Proc. Int. Conf. Optical Fiber Sensors (OFS-88)*, (IEEE/OSA) 1988, p. 14.
101. E. Sakaoka et al., Polarisation maintaining fibers for fiberoptic gyroscopes, *Tech. Digest, Proc. Int. Conf. Optical Fiber Sensors (OFS-88)*, (IEEE/OSA) 1988, p. 18.
102. D. T. Dong and K. Hotate, Frequency division multiplication of optical fibre sensors using an optical delay line with a frequency shifter, *Tech. Digest, Proc. Int. Conf. Optical Fiber Sensors (OFS-88)*, (IEEE/OSA) 1988, p. 76.

REFERENCES

103. A. D. Kersey and A. Dandridge, Tapped serial interferometric fibre sensor array with time division multiplexing, *Tech. Digest, Proc. Int. Conf. Optical Fiber Sensors (OFS-88)*, (IEEE/OSA) 1988, p. 80.
104. H. Hosokawa *et al.*, Integrated optical micro-displacement sensor using a Y junction and a polarization maintaining fibre, *Tech. Digest, Proc. Int. Conf. Optical Fiber Sensors (OFS-88)*, (IEEE/OSA) 1988, p. 137.
105. H. Koseki *et al.*, Optical heterodyne gyroscope using two reversed fibre coils, *Tech. Digest, Proc. Int. Conf. Optical Fiber Sensors (OFS-88)*, (IEEE/OSA) 1988, p. 168.
106. G. Kotrotsios and D. Pariaux, White light interferometry for distributed sensing on dual mode fibres, *Proc. OFS '89*, Springer Proceedings in Physics, Vol. 44 (1989).
107. S. Y. Huang, H. G. Park and B. Y. Kim, Passive quadrature phase detector for coherence fibre optic system, *Proc. OFS '89*, Springer Proceedings in Physics, Vol. 44 (1989), p. 38.
108. Y. N. Ning, K. T. V. Grattan, W. N. Wang and A. W. Palmer, A systematic classification and identification of optical fibre sensors, *Sensors and Actuators A*, **29** (1991) 21–36.
109. M. El-Hami, L. Finkelstein, K. T. V. Grattan and A. W. Palmer, Expert system application using optical fibre sensor classification data, *Sensors and Actuators A*, **39** (1993) 181–189.
110. M. K. Mirza, F. J. R. Neves and L. Finkelstein, A knowledge-based system for design concept generation of instruments, *Measurement*, **8**, (1990) 7–11.
111. L. Finkelstein, R. Ginger, M. El-Hami and M. K. Mirza, Design concept generation for instrument systems: a knowledge-based system approach, *Measurement*, **11** (1993) 45–53.
112. P. S. Sell, *Expert Systems – a Practical Introduction*, Macmillan, London (1985).
113. Z. Zhang, K. T. V. Grattan and A. W. Palmer, Sensitive fibre optic thermometer using Cr:LiSAF fluorescence for bio-medical sensing applications, *Proc. Int. Conf. Optical Fibre Sensors*, Monterey, CA, USA (1992), pp. 93–96.
114. Accufibre Corp., Vancouver, Canada, publicity data (1986).
115. D. A. Jackson, Monomode optical fibre interferometers, *J. Phys. E: Sci. Instrum.*, **18** (1985) 987.
116. M. Corke, J. D. C. Jones, A. D. Kersey and D. A. Jackson, Dual Fabry–Pérot interferometer implemented in parallel on a single mode optical fibre, *Tech. Digest, 3rd Int. Conf. Opt. Fibre Sensors (OFS '85)*, San Diego, CA, USA (1985), p. 128.
117. J. P. Dakin, D. J. Pratt, G. W. Bibby and J. N. Ross, Distributed optical fibre Raman temperature sensor using a semiconductor light source and detector, *Electron. Lett.*, **21** (1985) 569.
118. K. A. Wickersheim and R. V. Alves, Fluoroptic thermometry: a new RF-immune technology, in *Biomedical Thermometry*, Alan Liss, New York (1982), p. 547.
119. *Techbrief: Fibre Optic Decay Time Temperature Sensor*, Manufacturer's publicity data, GEC, Wembley, UK (1986).
120. M. Hirano, Characteristics and application for some types of fibre-optical temperature sensor, *Tech. Digest, 4th Int. Conf. Opt. Fibre Sensors (OFS 86)*, Informal Workshop, Tsukaba Science City, Japan, Oct. (1986), VII1–VII8.
121. R. R. Scholes and J. G. Small, Fluorescent decay thermometer with biological applications, *Rev. Sci. Instrum.*, **51** (1980) 882.

2
Optical fiber lasers

N. Langford

2.1 Introduction

The idea of using optical fiber geometry for laser and optical amplification purposes was proposed and demonstrated in 1961 by Snitzer [1] who used a 300 μm core diameter fiber doped with neodymium as the laser active gain medium. Since the development of this laser a range of fiber lasers has been constructed using either rare earth ions [2] or optical nonlinearities [3] as the optical gain medium. Although efficient laser action has been observed from nonlinear optical fiber lasers, the greatest research emphasis has been placed on the refinement of rare earth doped systems. This has been stimulated by the observation that significant concentrations of rare earth ions could be introduced into the core of standard telecommunications grade fiber without degrading the guiding properties of the fiber [4]. Strong absorption bands are created by the rare earth ion, whilst the low loss waveguiding characteristics of the fiber are maintained at the emission wavelength of the rare earth ion. Indeed, today, rare earth doped optical fibers are an accepted and important class of laser active gain medium with a wide variety of experimental applications ranging from the remote sensing of magnetic fields [5] through to ultrashort pulse generation for coherent optical communications [6] and high resolution spectroscopy [7]. Although extensive research effort has been directed at the development of efficient erbium-doped fiber lasers for the low loss third telecommunications window of standard silica fiber at 1.55 μm [8], a selection of rare earth ions has been used as gain media giving a wavelength coverage from 0.38 μm [9] to 3.9 μm [10]. These transitions have been observed in several different types of fiber host such as silica-based fibers [11], fluorozirconate or ZBLAN fibers [12], lead germanate fibers [13], lead silicate fibers [14] and tellurite glass fibers [15].

In this chapter I will concentrate on rare earth ion doped fiber lasers. The chapter will be organized as follows. The first section will discuss the advantages of using the fiber geometry and outline why rare earth ions are suitable as laser active gain media. The next section will discuss potential resonator geometries for optical fiber lasers and this will be followed by a section describing the various rare earth ions that have been observed to operate in a fiber host. An outline of the applications of fiber lasers ranging from their use as single frequency sources through to optical sensors will then be given.

2.2 OPTICAL FIBERS AND RARE EARTH IONS

2.2.1 The optical fiber geometry for laser applications

Irrespective of the type of gain medium, for a laser to oscillate it must reach threshold. This corresponds to the situation whereby the saturated optical gain is equal to the losses present in the laser. To reach this threshold condition, energy must be supplied to the gain medium and it will be assumed that this energy is supplied by another laser. The threshold level is determined by the intensity of the pump radiation and by how effective the gain medium is at absorbing the pump radiation [16]. For a low threshold it is necessary to have an intense pump signal which is strongly absorbed by the gain medium. Ideally a significant portion of the intense pump signal should be absorbed over the entire length of the gain medium.

If a bulk gain medium is considered and a Gaussian pump beam of diameter $2w$ and wavelength λ, is focused into it by a lens of focal length f, as shown in Figure 2.1, the minimum diameter spot $2s$ that can be obtained is given by equation 2.1 [17]:

$$2s \approx \frac{2\lambda f}{\pi w} \qquad (2.1)$$

It can be seen that the spot diameter is directly proportional to f. Thus using a short focal length lens will result in a small spot giving a high intensity. There is, however, a downside to using a short focal length lens. The effective absorption length is determined, in principle, by the confocal parameter $2z_0$, of the Gaussian mode in the gain medium [17]. The confocal parameter defines the distance over which the diameter of the mode increases to $\sqrt{2}$ of its minimum value $2s$ and scales as f^2 (see equation 2.2):

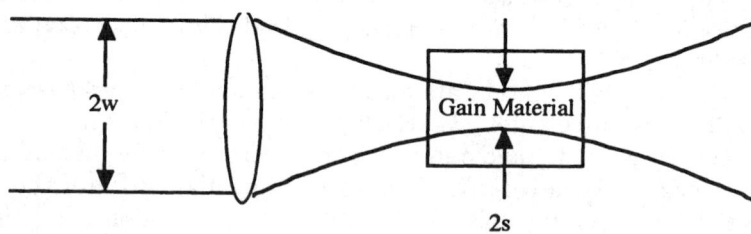

Fig. 2.1 The focussing of a Gaussian mode of diameter $2w$ into a bulk gain medium using a lens of focal length f.

$$2z_0 \approx \frac{\lambda f^2}{\pi w^2} \quad (2.2)$$

So a short focal length lens implies that there will be a short absorption length between the high intensity signal and the bulk material. As mentioned already, the optimum case is when the effective absorption length is equal to the length of the gain medium. So for a given length of gain medium there is an optimal focal length that can be used. This limits the intensity. To increase the intensity the length of the gain medium could be reduced but the dopant concentration would have to be increased to give the same absorption length and this can introduce a different set of problems such as concentration quenching or ion–ion interactions [18].

A typical monomode gain fiber has a core diameter of the order of 5–10 μm and so a high intensity can be established in the core for a modest average power (<10 mW). Once the pump radiation is coupled into the fiber core it is guided along the entire length of the fiber whilst maintaining the intensity. This results in a long interaction length between the intense pump radiation and the gain medium enabling high optical gains to be established for low average pump powers. For example, a small signal gain of 30 dB has been achieved with a pump power as low as 8 mW from an erbium-doped fiber amplifier [19].

Another advantage of the high intensity in the fiber core is that it can bleach the ground state of three level transitions allowing continuous wave operation to be attained on transitions that have previously operated in pulsed mode in bulk materials. An important example of this is the 1.55 μm transition associated with erbium-doped silica fibers [20].

The electronic pathways associated with the absorption and emission processes occurring in the rare earth ion result in the creation of heat, which in bulk materials can be detrimental to the performance of the laser. In the fiber geometry there is a large surface area to volume ratio which allows for the efficient dissipation of heat from the fibre core.

2.2.2 The spectroscopy of rare earths

To date the only ions that have been observed to oscillate in a fiber geometry have belonged to the lanthanide group of rare earths. These ions possess several distinct features which make them attractive for use as laser active gain media. They exhibit sharp absorption and emission bands ($\Delta\lambda \approx 20$ nm) with low transition cross-sections, typically 10^{-21} cm^2 [21], and long excited state lifetimes, typically 10^{-5} s [21]. This means that the rare earths are efficient at storing energy and so can form high gain amplifiers and lasers. The absorption bands of most rare earth ions are closely matched to the emission wavelengths associated with recently developed high power laser diodes, resulting in compact laser systems with high electrical to optical conversion efficiencies.

The sharp absorption and emission bands arise from the interaction of the rare earth ion with its surrounding environment. Neutral atoms of the lanthanide group have the electronic configuration [xenon]$5d^y6s^24f^n$, with y being 0 or 1 and n an integer running from 0 to 14, and it is the partially filled 4f electron shell which is responsible for the optical activity of the rare earth ion. As the atomic number associated with each lanthanide atom increases, the spatial extent of the 4f wavefunction decreases (lanthanide contraction) so that when the atomic number is equal to that of neodymium the spatial peak of the 4f wavefunction falls inside that of the 5s and 5p wavefunctions. Trivalent ionization of the rare earth removes the 5d and 6s electrons, if y is equal to 1, or the 6s electrons and one of the 4f electrons if y is 0. The 5s and 5p electronic shells are unaffected by the ionization process and so they effectively shield the 4f electrons from the surrounding environment. The 4f electron shell has a series of energy levels associated with it which are designated by the spin S, total angular momentum J, and orbital angular momentum L, quantum numbers associated with the ion. These levels are referred to as LSJ levels or LSJ multiplets and are written as $^{2S+1}L_J$. In a neutral environment these multiplets are degenerate and electric dipole transitions between different multiplets are forbidden. When the rare earth ion is introduced into an optical fiber the ion experiences the local environment through the crystal field of the glass host. Although the 4f electron shell is effectively shielded from the host crystal field by the 5s and 5p electron shell there is a sufficiently weak interaction between the 4f electrons and the crystal field to influence the optical properties of the rare earth ion. Firstly, the degeneracy is lifted, resulting in a series of discrete energy levels. These levels are then homogeneously broadened by a coupling of the level to the lattice phonons associated with the host glass matrix. This homogeneous broadening allows for the tunable operation observed from most fiber lasers. Secondly, the crystal field allows radiative electric dipole transitions to occur between different LSJ multiplets. Because of the weak nature of the interaction between the 4f electrons and the crystal field, the oscillator strength associated with the electric dipole transitions is small, typically of the order of 10^{-6}. The emission cross-section for an optical transition is directly proportional to the oscillator strength [16]. This means that transitions associated with rare earth ions have small emission cross-sections, typically of the order of 10^{-21} cm^2 [21]. Thus any energy absorbed by the rare earth ion is effectively stored, resulting in the formation of high optical gains and hence efficient lasers and amplifiers.

As discussed already, the coupling of the LSJ levels to the phonons associated with the host glass matrix results in a homogeneous broadening of the level. This phonon coupling also has important consequences on the spectroscopy of the rare earth ion and depends on the type of glass used as the host matrix. The quantum efficiency of a transition η determines how many pump photons are converted into lasing photons and is defined by the ratio of the sum of all the radiative decay rates A_r to the sum of all the decay rates including both radiative and nonradiative processes W_{nr}

$$\eta = \frac{\sum A_r}{\sum A_r + \sum W_{nr}} \quad (2.3)$$

If the energy gap, ΔE, associated with a particular transition is greater than the phonon energy of the glass then a multiphonon transition occurs and W_{nr} is given by

$$W_{nr} = B \exp(-\alpha \Delta E) \quad (2.4)$$

with B and α constants determined by the glass host. If the nonradiative decay rate W_{nr} is comparable to the radiative decay rate the quantum efficiency of the transition is greatly reduced. In fact depending on the magnitude of W_{nr}, the transition of interest may decay solely by nonradiative processes. For fluorozirconate glasses the nonradiative decay rates at a defined energy gap are much smaller than those observed for silica glasses [22]. As a consequence of this, radiative transitions can occur at much longer wavelengths in a fluorozirconate glass than in a silica glass. For example, the longest wavelength observed from a silica glass fiber laser is 2.1 µm [23] whilst a wavelength of 3.9 µm has been recorded for a fluorozirconate glass [10]. Also because the nonradiative decay rates in fluorozirconate fibers are much lower than those in silica fibers a greater range of oscillating transitions are observed for fluorozirconate type glasses than for silica type glasses and these will be discussed in section 2.4.

2.2.3 Doped fiber manufacture

To exploit the advantages offered by the fiber geometry the rare earth ions have to be incorporated into the fiber core and a variety of schemes have been developed based around vapor deposition techniques. For a full review of these processes the reader is referred to Craig-Ryan and Ainslie [24]. The basic methodology for the incorporation of the rare earth ions into the fiber core is as follows. A standard modified chemical vapor deposition technique is used to create a hollow tube of glass containing the core and cladding layers [25]. The rare earth is then introduced into the core by either filling the hollow tube with an aqueous solution of the rare earth of interest (solution doping) [4], or passing a rare earth vapor along the hollow core (volatile halide doping) [26]. Once the rare earth is incorporated into the core layer the hollow tube is then collapsed to form the preform and then pulled to create the doped fiber. Refractive index profiling measurements of the preform indicate that the rare earth ions remain confined to the core glass when the hollow tube is collapsed to form the preform [27].

2.3 FIBER LASER RESONATORS

The ability to establish a high optical gain in the core of a rare earth doped optical fiber for modest optical pump powers has allowed the development of a variety of different resonator geometries, examples of which are shown in

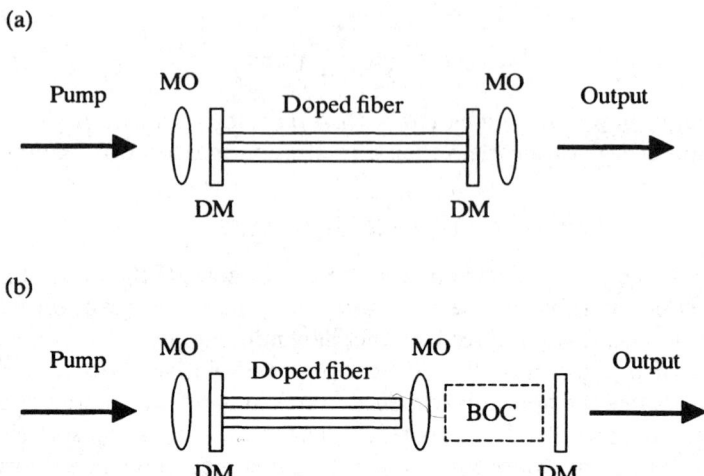

Fig. 2.2 (a) The simplest type of fiber laser cavity; (b) extended fiber laser cavity (DM = dielectric mirror, MO = microscope objective, BOC = bulk optic components).

Figures 2.2 and 2.3. The simplest form of cavity that can be configured, Figure 2.2(a), consists of a length of doped fiber the ends of which are brought into contact with dichroic dielectric mirrors which reflect the oscillating light but transmit the pump signal. This cavity can be extended as illustrated in

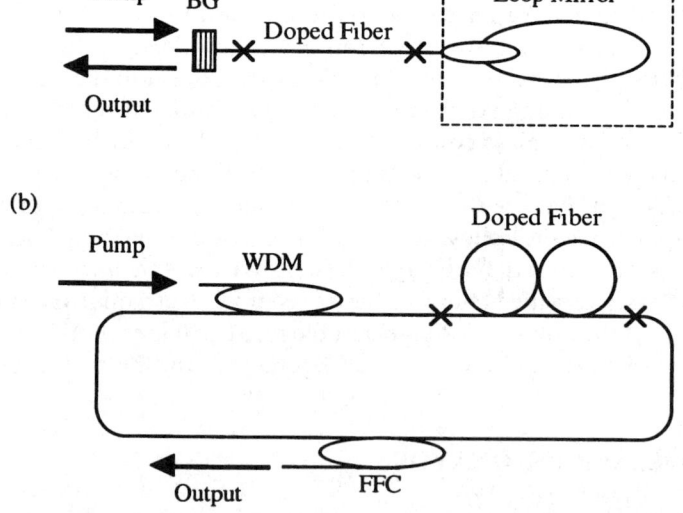

Fig. 2.3 All-fiber laser resonators: (a) linear cavity using a Bragg reflector, BG, and a loop mirror; (b) integrated ring fiber cavity (WDM = wavelength division multiplexer and FFC = fused fiber coupler).

Figure 2.2(b) to include bulk optical elements such as modulators, tuning filters, polarizers and Q-switches. The high optical gain means that these bulk components can be introduced without degrading the overall performance of the laser; however, Fresnel reflections from the free fiber end and the intracavity elements can reach laser threshold and so form subcavities, which can affect the performance of the extended laser. These problems can be avoided in a variety of ways such as index matching the free end, splicing the free end to a high-return loss-angled end and antireflection coating the intracavity elements.

To meet the needs of the telecommunications and sensing industries there has been a wide range of integrated fiber optic components developed which replicate bulk optic components [28]. For example, a wavelength division multiplexer is the fiber optic analog of a dichroic mirror. These devices are inherently compatible with rare earth doped fibers and can be fusion spliced to the doped fiber of interest to form 'all fiber' resonators as illustrated in Figures 2.3(a) and (b). These can take the form of linear cavities as depicted in Figure 2.3(a) whereby one end mirror is formed by a fiber loop mirror [29] and the other mirror is a fiber Bragg reflector [30] which acts as a dichroic mirror allowing the pump light to be coupled into the optical cavity and a fraction of the intracavity circulating intensity to be extracted.

Integrated ring cavities can also be constructed as shown in Figure 2.3(b). In this case the pump radiation is introduced into the cavity by a wavelength division multiplexer and an output from the cavity is obtained using a simple fused fiber coupler. Again integrated devices such as fiber Fabry–Pérot filters and fiber isolators can be spliced into the cavity to create tunable unidirectional lasers.

2.4 RARE EARTH DOPED FIBER LASER TRANSITIONS

2.4.1 Basic transition pathways

The most common form of laser active transition is the four-level transition and this is depicted in Figure 2.4(a). The basic electronic pathway is as follows. The population in the ground state, level 0, absorbs photons from the pump field and undergoes a stimulated transition to level 3. From this level the population undergoes a series of rapid phonon-assisted nonradiative decays to populate level 2. This level acts as a store for the population, the metastable state. A radiative decay then occurs between level 2 and level 1. Once in level 1 the population then undergoes another sequence of nonradiative decays to repopulate the ground state from which the population can recommence the cycle. For a three-level transition (Figure 2.4(b)), the transitions are similar except that the radiative decay occurs directly between the metastable state and the ground state. Such pathways are responsible for the 1.05 μm transition of Nd (4 level) and the 1.55 μm transition in Er (3 level) respectively.

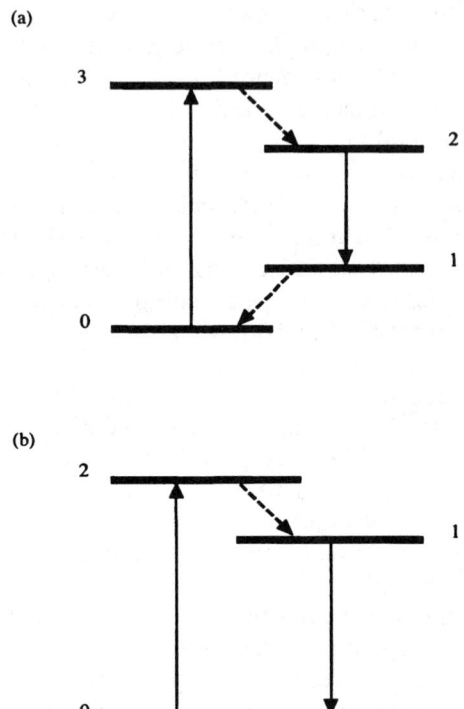

Fig. 2.4 Population pathways for (a) four-level transition and (b) three-level transition. The solid arrows correspond to radiative processes and the dashed arrows correspond to nonradiative processes.

There exists another possible sequence of pathways known as up-conversion transitions which can result in both visible and near infrared wavelengths being obtained from doped fiber lasers. Such up-conversion processes are illustrated in Figures 2.5(a) and (b). The basic pathway is as follows. Population in the ground state, level 0, again absorbs pump photons and undergoes a stimulated transition up to level 2. From this level the population decays via a sequence of nonradiative decays to level 1. The population in level 1 then absorbs more pump photons and undergoes a transition to level 3. From this level the population then decays radiatively back to the ground state level 0. As level 3 is at a higher energy than level 2 the emitted photon is of a shorter wavelength than the absorbed pump photon, hence, this is an up-conversion process. The pathway illustrated in Figure 2.5(b) corresponds to the up-conversion process which produces photons of longer wavelength than the pump.

As most rare earth ions possess many energy levels there is a vast range of different transitions that can occur. This next section will outline some of the transitions which have been observed to oscillate for a variety of rare earth ions.

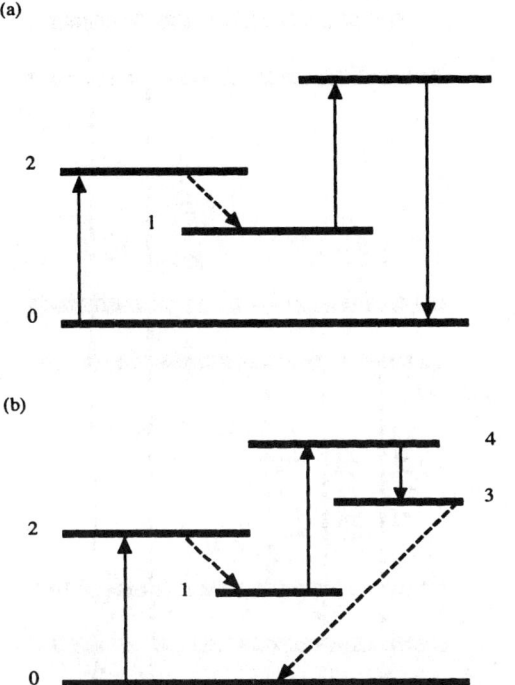

Fig. 2.5 Population pathways for up-conversion processes: (a) corresponds to the case where the emission wavelength is shorter than pump wavelength, (b) the case where emission wavelength is longer than the pump wavelength.

2.4.2 Neodymium doped fiber lasers

The energy levels associated with the triply ionized neodymium ion are shown in Figure 2.6. The most common transition is the $^4F_{3/2}$–$^4I_{11/2}$ transition which is well characterized and has been observed to oscillate in several different fiber hosts [4, 11, 12, 14, 15, 31]. The Nd^{3+} ion exhibits several absorption bands which peak at 520, 590 and 820 nm. The 820 nm absorption band is of particular interest as it can be accessed by high power AlGaAs laser diodes resulting in lasers which exhibit high electrical power to optical power conversion efficiencies.

The first reported case of laser oscillation in a fiber geometry was made by Snitzer who used a 300 μm core diameter Nd^{3+} doped fiber [1]. This work was extended by Stone and Burrus [32] who used a 1 cm long single mode fiber which was optically excited by either a pulsed argon ion or dye laser. By varying the concentration of the neodymium ion the laser could be made to laser oscillate simultaneously at 1.06 and 1.08 μm. The short optical path lengths associated with this laser cavity meant that the advantages obtained from the waveguiding nature of the doped fiber were not fully exploited.

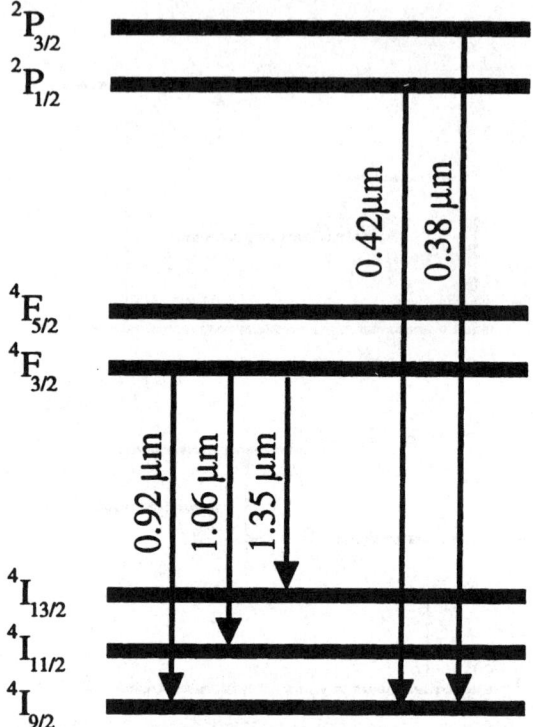

Fig. 2.6 Energy level diagram for neodymium ions. Only the radiative transitions are illustrated.

Continuous-wave operation of a single mode Nd-doped fiber laser was first reported by Mears et al. [20] using an optical resonator similar to that illustrated in Figure 2.2(a). A 2 m long gain fiber was used and the absorbed pump power threshold was measured as 100 μW. An output power of 1 mW for an absorbed power of 5.6 mW was recorded corresponding to a slope efficiency of 20%.

Since this report laser oscillation on the $^4F_{3/2}-^4I_{11/2}$ transition has been observed for a variety of different resonator configurations. Using an extended cavity similar to that illustrated in Figure 2.2(b) except with the end mirror replaced by a diffraction grating, Alcock et al. [33] demonstrated that a Nd^{3+} fiber laser could be tuned from 1.055 to 1.14 μm. All fiber resonators have been constructed whereby the bulk dielectric end mirrors have been replaced by dielectric mirrors coated directly onto the cleaved fiber ends [34]. The dielectric mirrors in turn have been replaced by fiber Bragg reflectors to form a compact laser resonator. Using a pair of matched fiber Bragg reflectors with one at each end of a length of doped fiber Ball and coworkers [35] were able to construct a fiber laser which delivered output powers of 12 mW for 265 mW of incident pump power. Although the Bragg reflectors defined the wavelength by temperature tuning the reflectors the oscillating wavelength could be tuned by 1 nm around 1.085 μm. Miller et al. [36] developed another form of all fiber resonator using fiber Sagnac loop mirrors as the cavity reflectors. The fused fiber couplers used to form the Sagnac loop mirrors were not optimized for the peak of the gain curve of the Nd ion in the silica fiber host. As a consequence of this the slope efficiency of the laser was poor ($<2\%$). Bidirectional ring Nd fiber lasers have also been constructed using either fused fiber [4] or polished couplers [37] to form the ring. With the polished coupler arrangement Chaoyu et al. [37] were able to obtain an average output power of 0.5 mW from one arm of the ring laser for a maximum pump power of 8 mW. Adjusting the polished coupler allowed the laser to be tuned over 60 nm.

Most fiber lasers are constructed from monomode doped fiber which is weakly birefringent and this intrinsic birefringence has been used to define the laser wavelength. For example, the inclusion of a polarizer into an extended fiber laser cavity allows tuning over several nanometers [38]. This work was extended by replacing the polarizer with a length of standard polarization-preserving fiber [39]. For this laser system multiple wavelength operation was observed with a spectral separation between adjacent wavelength peaks of up to 4 nm. Broadband tunability of Nd-doped fiber lasers has also been demonstrated from a laser containing a wavelength filter based on a nematic liquid crystal cell [40]. Tunability over 17 nm for a voltage of 2 V applied to the cell was reported.

For the fiber lasers described above the pump sources have ranged from argon ion lasers through to semiconductor diode lasers. The semiconductor diode laser approach to optical excitation is the most appealing. There are, however, problems associated with semiconductor laser diode pumping. To couple the pump light efficiently into the fiber core the pump mode should be a circular Gaussian TEM_{oo} mode [41]. For most low-power (<500 mW) semiconductor diode lasers the beam produced exhibits a high degree of asymmetry and requires significant reshaping to produce a symmetric mode. High-power laser diodes consist of a series of emitters which produce a highly distorted pump mode which is difficult to reshape and couple effectively into the fiber core. This limits the amount of pump light that can be coupled into the doped fiber and hence restricts the output power that can be obtained from the fiber laser. A simple approach to overcoming this problem is to double clad the fiber as depicted in Figure 2.7. Here the doped core is surrounded by an undoped inner cladding with a refractive index lower than the core refractive index and the inner cladding is then surrounded by an outer cladding made from a glass with a lower refractive index than the inner cladding. The pump light is then coupled into the inner cladding and is confined to the inner cladding by total internal reflection at the interface between the two claddings. As the pump light propagates down the inner cladding it is absorbed by the rare earth ion in the fiber core. This scheme was developed by Snitzer et al. [42] who demonstrated a pump to signal conversion efficiency of 50% for a Nd fiber laser and has subsequently been extended [43–46] to produce output powers as high as 9.2 W from

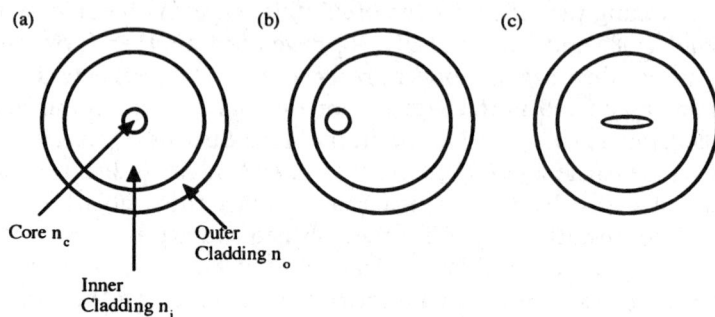

Fig. 2.7 Double clad fiber geometeries: (a) concentric core and cladding; (b) offset core; (c) elliptical core fibers.

a continuous wave Nd fiber laser. This laser had a slope efficiency of more than 25% but the laser mode was twice diffraction-limited [46].

An alternative approach to high-power operation of fiber lasers has also recently been demonstrated [47]. Using a fiber bundle containing ~200 Nd-doped multimode phosphate fibers and flashlamp excitation, output powers of 100 W at repetition rates of 100 Hz have been observed.

Although high-power and efficient laser action has been demonstrated for the $^4F_{3/2}-^4I_{11/2}$ transition there are several other Nd transitions which have oscillated in both silica and fluorozirconate fibers [33, 48–52]. The $^4F_{3/2}-^4I_{9/2}$ transition is a quasi-three-level transition and requires the I ground state to be bleached such that laser threshold can be achieved. Even so laser action has been observed at 938 nm [48] with a modest pump power threshold of 1.9 mW of absorbed pump power. With 10 mW of pump power the laser produced an output power of 3 mW giving an overall conversion efficiency with respect to absorbed pump power of 36%. Alcock et al. [33] showed that this transition could be tuned over 45 nm.

The $^4F_{3/2}-^4I_{13/2}$ transition peaks at 1.34 μm. This wavelength has important implications for telecommunications and optical sensors networks as it corresponds to the zero dispersion window of standard silica fiber. Initial attempts to make this transition oscillate in Nd-doped silica fibers failed and this failure was attributed to the effects of excited state absorption (ESA) whereby population in the $^4F_{3/2}$ level absorbs photons at 1.3 μm and is excited to the $^4G_{7/2}$ level [53]. This problem was partly overcome by Hakimi et al. [49] who determined that the effects of ESA could be minimized by the addition of P_2O_5 to the fiber core. With a fiber that had a 25% P_2O_5 concentration in the core, laser action was observed at a wavelength of 1.36 μm.

Although the problem of ESA occurs in a fluorozirconate glass it has been found that the ratio of the calculated line strength of the emitting transition to the ESA transition at 1.3 μm has a value of 2 in a fluorozirconate fiber in comparison to a value of 0.92 in a silica fiber [50]. Hence the fluorozirconate fiber should be a more suitable host for the observation of efficient laser action on the $^4F_{3/2}-^4I_{13/2}$ transition. Laser action has indeed been observed on the 1.35 μm transition by Miniscalo et al. [50] who used a 2.8 cm long fiber with a fiber core of 50 μm diameter which was optically excited at 514.5 nm by an argon ion laser. The threshold power was 670 mW and a maximum output power of 10 mW was obtained from the resonator for a pump power of 1200 mW giving a slope efficiency of 2.6%. Brierley and Hunt [54] showed the full potential of Nd fluorozirconate fibers by using a 795 nm laser diode to optically excite the gain fiber. A threshold of 60 mW was measured with a slope efficiency of 57%. Komukai et al. [55] have subsequently demonstrated that the $^4F_{3/2}-^4I_{13/2}$ transition in a fluorozirconate fiber can be tuned over 33 nm, covering the wavelength range 1.315–1.348 μm.

The Nd ion also exhibits a series of visible transitions [12] and recently two of these have exhibited laser action. The $^2P_{3/2}-^4I_{11/2}$ transition has produced 0.5 mW of violet light at a wavelength of 412 nm [51] whilst the $^4D_{3/2}-^4I_{11/2}$ transition has generated 74 μW of light at 381 nm [52]. Both these transitions were optically excited at a wavelength of 590 nm and required ∼ 300 mW of pump power to produce the reported output powers.

2.4.3 Erbium-doped fiber lasers

This particular rare earth dopant has attracted significant research interest and the reason for this can be understood by looking at the energy levels associated with the triply ionized ion (Figure 2.8). The $^4I_{13/2}-^4I_{15/2}$ transition in a silica fiber host luminesces at a wavelength of 1.55 μm which is well matched to the minimum loss window of standard telecommunications fiber [56]. The inherent compatibility of erbium-doped silica fiber with this fiber means there is little difficulty in splicing erbium-doped fibers into present-day fiber-based communications or sensor networks. As with neodymium, erbium has also been incorporated into the core of fluorozirconate fibers and a wide range of optical transitions have been observed to oscillate [57–61] using both direct and upconversion pumping schemes.

The $^4I_{13/2}-^4I_{15/2}$ is a three-level transition with the $^4I_{15/2}$ level being the ground state. Thus in order for this transition to reach laser threshold and oscillate in a continuous wave fashion the $^4I_{15/2}$ level must be bleached. The high intensities that can be generated in the fiber core for modest pump powers ensure that this condition is satisfied.

Mears et al. were the first to demonstrate continuous laser action at 1.55 μm [20]. By optically exciting a 0.9 m length of doped fiber using an argon ion laser in a cavity similar to that of Figure 2.2(b) with a Littrow mounted diffraction grating as the end mirror they were able to obtain output powers of 0.4 mW with a tuning range extending from 1.528 to 1.542 μm and from 1.544 to 1.555 μm. O'Sullivan et al. [62] further optimized the 514.5 nm pumping of erbium-doped fiber lasers to the point whereby they were able to generate 96 mW of average output power for 600 mW of incident pump power. These data were recorded with the doped fiber maintained at liquid nitrogen temperatures.

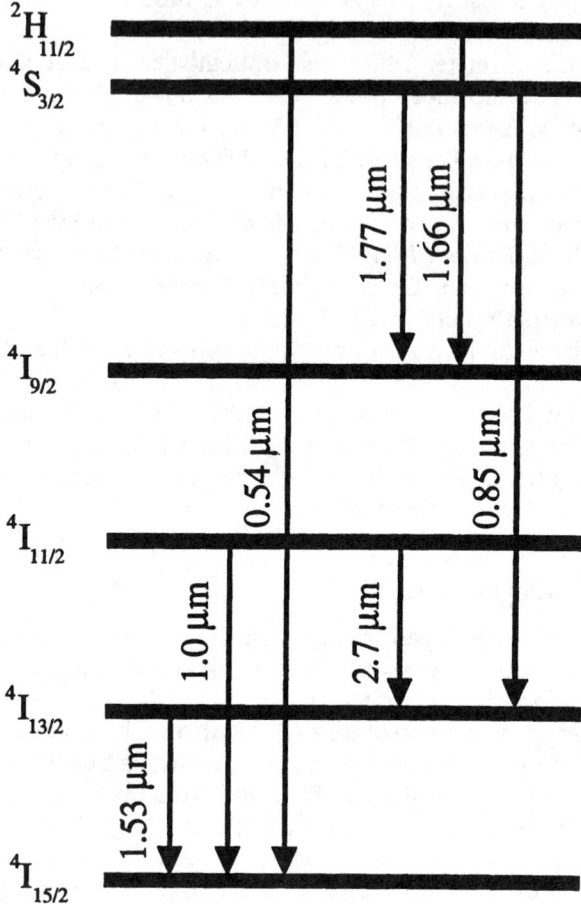

Fig. 2.8 Energy level diagram for erbium ions. Only the radiative transitions are illustrated.

Cooling the fiber helped to depopulate the upper sublevels of the $^4I_{15/2}$ level allowing efficient operation.

As with the neodymium ion, the erbium ion also exhibits several absorption bands which peak at 520, 650, 800, 980 and 1480 nm and optical pumping at all of these wavelengths has been studied. The 800 nm absorption band is attractive as there is a wealth of high-power laser diodes available. Indeed Wyatt *et al.* [63] used a self-injection locked-laser diode array to excite an Er-doped fiber and were able to generate output powers as high as 8 mW with a slope efficiency of 13.5%. There is, however, a serious drawback to using 800 nm pump light. At this wavelength there is a problem with excited state absorption as the population stored in the $^4I_{13/2}$ level can absorb a pump photon and populate the $^2H_{11/2}$ level which removes population from the laser cycle. The 514.5 nm pump wavelength also exhibits a strong excited state absorption and detailed investigations have been made to identify pump bands which are relatively free from excited state absorption [64, 65].

One such pump band is centered around 530 nm and this has allowed optical excitation at 528 and 532 nm [66–68]. Using an argon ion laser operating at 528 nm output powers as high as 70 mW for 360 mW incident pump power with a tuning range extending over 1.52–1.57 nm have been recorded [66]. The quantum efficiency associated with this pump wavelength is 90%, indicating that this pump wavelength is relatively free from excited state absorptions. The 532 nm wavelength can be generated from frequency doubled Nd:YAG lasers. Indeed Gasponstev et al. [68] have used such a laser oscillating at 532 nm to excite a double clad erbium fiber laser and generate output powers as high as 3.9 W.

The pump band at 980 nm has also been shown to be free of excited state absorption and the development of both Ti:sapphire and strained-layer diode lasers that operate at this wavelength has allowed studies to be made of performance characteristics of erbium fiber lasers excited at this wavelength [69, 70]. For example, Wyatt [70], was able to generate output powers as high as 270 mW for 550 mW of pump power from a Ti:sapphire laser. This corresponds to a slope efficiency of 50% and more importantly a quantum efficiency of 96%, again indicating that this pump wavelength is free of excited state absorptions.

Zyskind et al. [71] pointed out that the sublevels of the $^4I_{13/2}$ manifold could be accessed with 1480 nm pump light and Kimura et al. [72] generated output powers as high as 8 mW for 70 mW of incident pump power. Using the cavity illustrated in Figure 2.9, Langford [73] was able to generate 15 mW of light at the peak of the erbium gain curve using 75 mW of incident pump power.

Although the 1.55 µm transition in erbium-doped fibers is a three-level transition the $^4I_{15/2}$ ground state exhibits many sublevels [74] which result in a broadband luminescence. To exploit the full wavelength potential of the 1.55 µm transition several different tuning schemes have been developed. As with neodymium-doped fibers, the simplest approach is to use a cavity similar to that of Figure 2.2(b) and replace the end mirror with a diffraction grating. In this way Wyatt [70] has demonstrated tunability over 60 nm extending from 1520 to 1580 nm whilst Langford [73], using the cavity illustrated in Figure 2.9(a), was able to tune a 1480 nm laser diode pumped erbium fiber laser from 1520 to 1600 nm, as shown in Figure 2.9(b). The bulk diffraction grating can be replaced with a fiber Bragg reflector [75–79] and by applying a compressive stress to a fiber grating Ball and Morey [78] have been able to demonstrate continuous tuning of an erbium-doped fiber laser over 32 nm. By incorporating a fiber grating in an optical circulator Pan and Shi [79] have been able to tune the oscillating wavelength of their erbium fiber laser over 18 nm.

Fiber Bragg filters, by their design, are reflection devices and for certain applications a transmission filter may be more suitable for wavelength selection. A selection of different transmission filters have been developed to meet the needs of the communications and sensing communities and these include

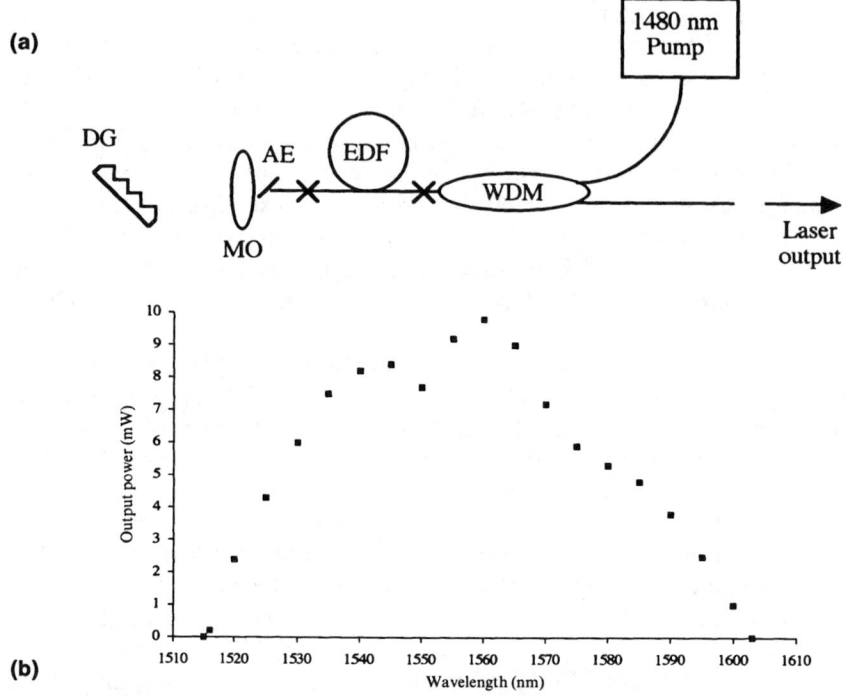

Fig. 2.9 (a) Simple linear cavity erbium doped fiber laser (DG = bulk diffraction grating, MO = microscope objective, EDF = erbium doped fiber and WDM = wavelength division multiplexer); (b) tuning range obtained.

the integration of bulk optic components such as Fabry–Pérot étalons [80] and interference filters [81] as well as polished fiber-waveguide devices [82]. These components have been included in erbium-doped fiber lasers and have allowed broadband tuning of the oscillating wavelength. Vahala et al. [83–87] have used cavities containing two fiber Fabry–Pérot filters and have demonstrated tuning from 1520 to 1560 nm [83] and also multiple wavelength operation [85] of such fiber lasers with a tuning range of 35 nm being observed. Integrated fiber interference filters have also been used to define the oscillating wavelength of erbium fiber lasers with a tuning range extending from 1520 to 1570 nm [86].

Although fiber interference filters and Fabry–Pérot filters are effective wavelength selective elements, in order for them to operate, light must exit the fiber core, propagating through the filter component and then reenter the fiber core. This makes them difficult to manufacture. Recently, however, a continuous fiber filter has been developed which relies on the coupling between a fiber which has been side polished to expose the core and a multimode waveguide. Wavelength selectivity occurs because light can only couple from the fiber core to the multimode waveguide when the fiber mode is in resonance with a mode of the waveguide [84]. Using such a device Gloag et al. [87] have been able to discretely tune an erbium fiber laser over 30 nm extending from 1535 to 1565 nm.

A variable ratio coupler has been used by Scrivener et al. [88] to form a 'continuous fiber' ring cavity and by varying the effective coupling ratio, demonstrated tuning over three distinct wavelength ranges in a 70 nm band.

Electronically tunable fiber lasers have also been demonstrated. Using an acousto-optic modulator as a wavelength-selective element Wysocki et al. [89] have been able to tune a linear cavity erbium fiber laser from 1529 to 1534 nm and from 1542 to 1563 nm. Acousto-optic modulators have also been used to continuously tune a fiber laser made from polarization maintaining fiber [90]. Similar tuning ranges have been observed for a fiber laser containing a liquid crystal étalon filter as the tuning element [91]. By applying a voltage of between 2.1 and 3.3 V to the étalon the laser could be tuned over 45 nm covering the 1523–1568 nm spectral window.

The 1.55 µm transition has attracted significant research interest but the many energy levels of the erbium ion in a fiber host allow for laser action to be observed at several alternative wavelengths. To date these alternative oscillating wavelengths have been observed only in fluorozirconate fibers.

As discussed in section 2.2, fluoride fibers have a lower phonon energy in comparison to silica fibers. This feature allows long wavelength transitions ($\lambda > 2.2$ µm) that would decay nonradiatively in silica fiber hosts to decay radiatively in fluoride hosts. For the erbium ion this means that the $^4S_{3/2}-^4F_{9/2}$ transition which radiates at 3.5 µm can oscillate. Hofle and Tobben [92, 93] have shown that efficient laser action can be obtained from a 12 cm long fluoride fiber with a core radius of 5 µm. When this laser was excited at 650 nm the pump power threshold was 76 mW.

Another mid-infrared wavelength that has been generated from an erbium fluoride fiber system is 2.7 µm. This wavelength results from the $^4I_{11/2}-^4I_{13/2}$ transition. An interesting feature of this transition is that it is self-terminating because the lower level, $^4I_{13/2}$, has a longer lifetime than the upper laser level, $^4I_{11/2}$, 14 ms compared to 9 ms. Thus a population inversion and hence laser action would not be expected to occur. In fact continuous wave operation of this transition has been observed by several research groups [57, 58, 94–99]. The reason for this is attributed to excited state absorption of the pump wavelength which depopulates the $^4I_{13/2}$ level, thereby allowing a population inversion to be established. A variety of pump wavelengths have been used to excite this transition and these include 476.5 nm [94], 802 nm [97] and 980 nm [98]. Schneider [99] has also shown that if both the 1.55 and 2.7 µm transitions are allowed to oscillate simultaneously then output powers as high as 1.2 mW can be obtained from the fiber laser at 2.7 µm. The dual oscillating wavelength approach is successful as the 1.55 µm radiation depopulates the $^4I_{13/2}$ level again allowing a population inversion to be established.

One wavelength range of interest to the remote sensing community is 1.67 µm as this wavelength corresponds to an overtone of methane.

Fluorozirconate fibers doped with erbium have been shown to oscillate in this wavelength region. The $^2H_{11/2}-{}^4I_{9/2}$ transition gives rise to light at 1.66 μm whilst the $^4S_{3/2}-{}^4I_{9/2}$ transition is responsible for a signal occurring at 1.72 μm [59, 100]. Both argon ion and semiconductor diode lasers have been used to excite these transitions; however, as both transitions are self-terminating only pulsed laser action has been observed.

Continuous wave laser action has also been observed at 1.0 μm [60, 100, 101] corresponding to the $^4I_{11/2}-{}^4I_{15/2}$ transition, 0.85 μm [61] arising from the $^4S_{3/2}-{}^4I_{13/2}$ transition and, more recently, visible laser action at 540 nm ($^4S_{3/2}-{}^4I_{13/2}$ transition) has been observed [102–104]. Allain et al. [102] have observed green output powers of 50 mW with slope efficiencies of 15% for a 970 nm pumped fluorozirconate fiber laser. Similarly Massicott et al. [104] have observed green laser operation with thresholds as low as 1 mW of pump power from an 801 nm laser diode.

2.4.4 Thulium-doped fiber lasers

Although the erbium ion has been the subject of extensive research because it exhibits a transition at 1.55 μm there have also been studies made to extend the wavelength coverage of rare earth doped fiber lasers. A fiber laser oscillating around the eye-safe wavelength of 2.0 μm would be of interest as it could be used for atmospheric pollution monitoring and coherent radar applications. One rare earth which exhibits a transition at this wavelength is thulium and this ion has been observed to oscillate in silica, fluorozirconate and lead germanate host fibers [13, 105–117].

The energy levels associated with radiative transitions of the thulium ion are illustrated in Figure 2.10 and, as can be seen, a variety of different wavelengths can be obtained from the thulium ion. The first thulium transition to oscillate was the $^3F_4-{}^3H_6$ transition and it occurred in a silica fiber host [105]. The electronic pathways associated with this transition indicate that this is a three-level transition. Efficient laser action, however, has been observed from this transition for a variety of pump wavelengths [105–109]. Excitation of a silica-based fiber using a pump wavelength of 800 nm from a laser diode has resulted in the generation of output powers as high as 44 mW for pump powers of 167 mW [107]. This corresponds to a slope efficiency of 36% with an internal conversion efficiency of 84%. The internal conversion efficiency is limited by the loss associated with the silica fiber host at the oscillating wavelength and not to an excited state absorption [107]. By simply varying the core constituents of the host silica fiber the thulium ion demonstrates a wide tuning range extending from 1.65 to 2.06 μm. A fiber with an $Al_2O_3-SiO_2$ core could be tuned from 1.77 to 2.06 μm whilst a host fiber with a GeO_2-SiO_2 core could be tuned from 1.65 to 1.85 μm [108]. These lasers were tuned using bulk diffraction gratings as the wavelength-selective element. The bulk grating can be readily replaced by intracore fiber Bragg reflectors and using such a device Boj et al. [111] have successfully demonstrated tuning from 1.9 to 2.1 μm.

RARE EARTH DOPED FIBER LASER TRANSITIONS

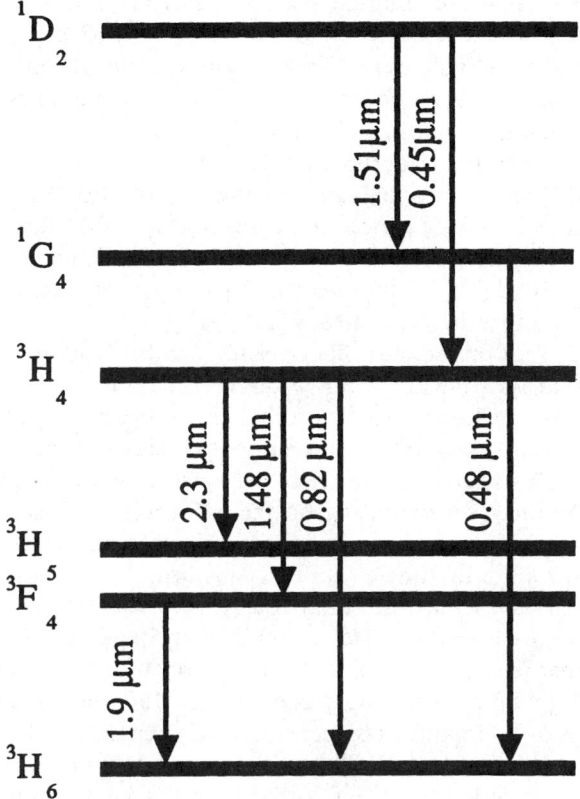

Fig. 2.10 Energy level diagram for thulium ions. Only the radiative transitions are illustrated.

The thulium ion exhibits a broad absorption band at 1.2 μm which extends to 1.05 μm. This allows excitation by high-power Nd:YAG lasers operating at 1.064 μm. Using such a laser single transverse mode output powers of 1.36 W have been observed from a thulium-doped silica fiber laser. Because of the low pump absorption the laser threshold and slope efficiency were measured as 600 mW of absorbed pump light and 36% respectively [112].

Laser oscillations have been observed around 1.9 μm from fluorozirconate fibers. Smart et al. [109] observed continuous wave operation at 1.9 μm. The performance of the laser was significantly improved by forcing the laser to oscillate at 2.3 μm simultaneously. This wavelength corresponds to the $^3H_4$–$^3H_6$ transition which is part of the electronic pathway associated with the 1.9 μm transition. The pump photons at 790 nm populate the $^3H_4$ level and a series of nonradiative decays then populate the upper laser level of the 1.9 μm transition, $^3F_4$ level. By using a radiative process to depopulate the $^3H_4$ level the rate at which the population fills the $^3F_4$ level is greatly increased, resulting in a more efficient laser [109]. This work has been extended by Percival et al. [113]

who observed near quantum limited efficiencies from a laser oscillating at 1.925 μm when the laser oscillated simultaneously at 1.82 and 2.31 μm. The output at 1.82 μm was tunable over 40 nm. Single wavelength operation of the 2.3 μm transition has also been observed from fluorozirconate fibers [114] with tuning ranges extending from 2.25 to 2.5 μm [115].

As the 1.9 μm transition is a three-level transition it is possible, as with erbium, to excite directly into the upper sublevels of the $^3F_4$ manifold and induce efficient laser action. This pumping scheme was first demonstrated by Yamamoto et al. [116] who excited at a wavelength of 1.58 μm and observed a slope efficiency of 70%. This work has subsequently been extended to give a laser which exhibits a threshold of 340 μW [117, 118].

Thulium-doped fluorozirconate fibers also exhibit radiative transitions which occur around the 1.48 μm window thereby making them an alternative source for telecommunications applications. The 1.48 μm transition is essentially a four-level transition and the light emitted arises from the $^3H_4-^3F_4$ transition [119–122]. This transition was first demonstrated by Allain et al. [119] who use a krypton ion laser operating on the 676 nm line. This transition is a self-terminating transition because the upper state lifetime is of the order of 1 ms and the lower state lifetime is about 10 ms long. If, however the 1.9 μm transition is forced to oscillate simultaneously with the 1.48 μm transition— cascaded laser action—then the $^3F_4$ level is effectively depopulated and the transition is no longer self-terminating. This approach was demonstrated by Percival et al. [121] who obtained output powers of 130 and 115 mW for oscillating wavelengths of 1.475 and 1.88 μm respectively. Using a Nd:YAG laser as the pump source Miyajima et al. [122] have been able to exploit a simple up-conversion process to generate 1 W of output power from a thulium laser operating at 1.47 μm. The laser was tunable from 1.445 to 1.51 μm.

Up-conversion processes have also been exploited to give a laser operating at 0.81 μm [123]. Again a Nd:YAG laser was used as the pump and an output power of 1.2 W has been observed at 0.81 μm. The oscillating wavelength was tuned over 13 nm from 803 to 816 nm.

Blue laser action has also been observed from thulium-doped fluorozirconate fibers [124–127]. Light at 0.455 and 0.48 μm has been observed which correspond to the $^1D_2-^3H_4$ and $^1G_4-^3H_6$ transitions respectively. Grubb et al. [125] have obtained output powers as high as 60 mW on the 0.48 μm transition.

2.4.5 Holmium-doped fiber lasers

Along with thulium, which, as discussed above, is a suitable source for light in the eye-safe spectral window of 2.0 μm, holmium has also been used as an active ion to give light in the same wavelength range. Hanna et al. [128] observed laser action on the $^5I_7-^5I_8$ level in a silica-based fiber corresponding to a wavelength of 2.04 μm. Slope efficiencies of 1.8% have been observed for this transition. This transition has also been observed in fluorozirconate fibers [129, 130]. Indeed with reference to Figure 2.11 it can be seen that there are other oscillating

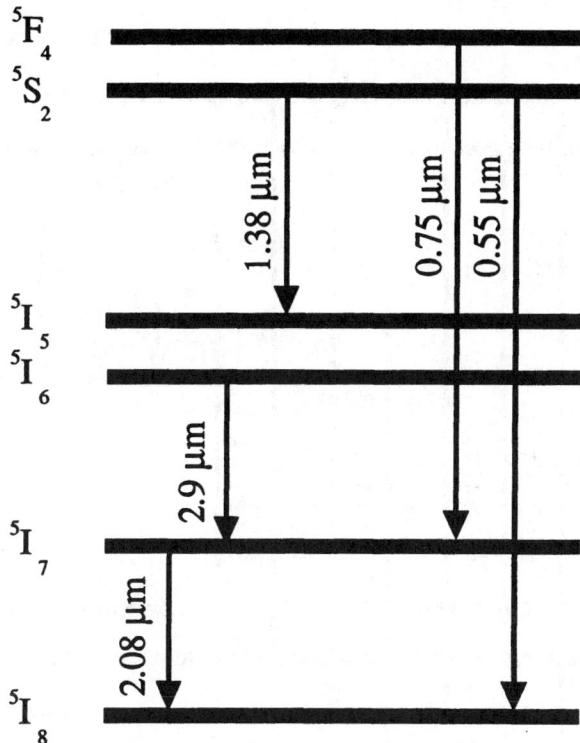

Fig. 2.11 Energy level diagram for holmium ions. Only the radiative transitions are illustrated.

wavelengths which have been observed from holmium-doped fluorozirconate fibers. For example, Wetenkamp has observed laser action at 2.9 μm ($^5I_6 - ^5I_7$ transition) [131] and Schneider ($^5I_3 - ^5S_2$ transition) [10] has seen laser action at 3.9 μm.

Shorter wavelength transitions have also been observed. Brierley et al. [129] have seen laser action at 1.38 μm ($^5S_2 - ^5I_5$ transition) which is of practical importance to the telecommunications industry. Allain et al. [132] built a laser which generated light at 750 nm ($^5F_4 - ^5I_7$ transition). For 370 mW of pump power 2 mW of output power was obtained.

The shortest wavelength observed from holmium-doped fluorozirconate fiber lasers is 0.55 μm ($^5S_2 - ^5I_8$ transition) [132]. This is another example of an up-conversion laser. A slope efficiency of ~20% was observed for this laser with output power of 10 mW. This transition was tunable from 0.54 to 0.553 μm. Funk et al. [133] have investigated the pump wavelength requirements for a green fiber laser and found that the optimum wavelength for pumping the fiber laser was 646.7 nm. This is an interesting wavelength as high-power laser diodes which operate at this wavelength have recently been developed. At this pump wavelength an output power of 12 mW was obtained.

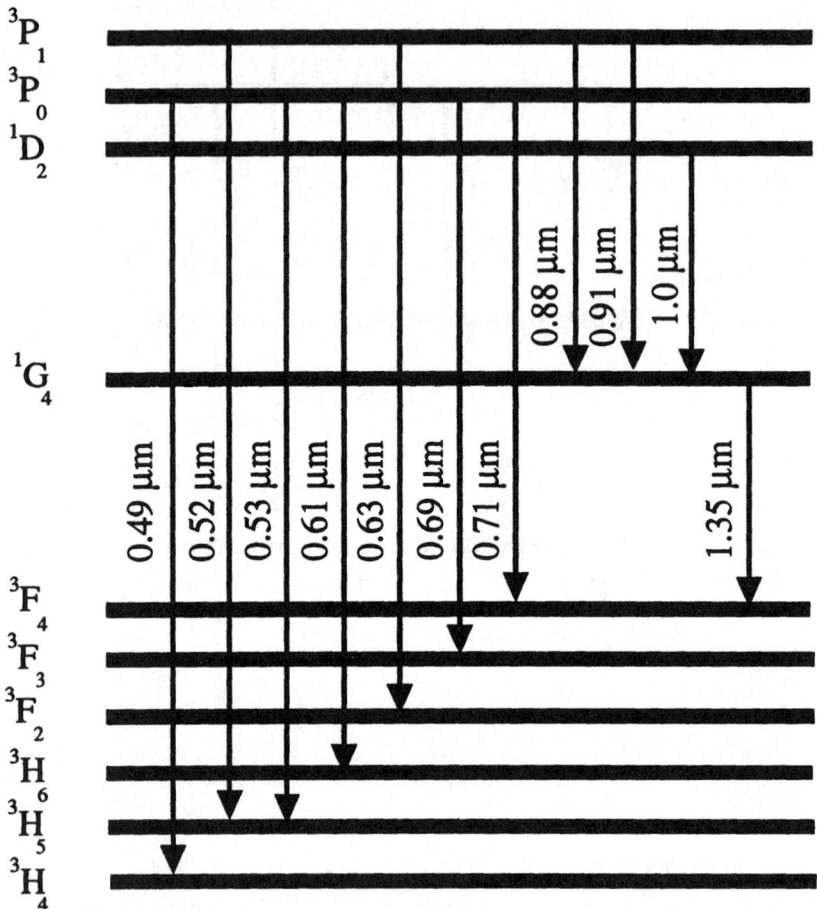

Fig. 2.12 Energy level diagram for praseodymium ions. Only the radiative transitions are illustrated.

2.4.6 Praseodymium-doped fiber lasers

To extend the wavelength coverage of rare earth doped fiber lasers praseodymium has been investigated as a laser active ion. The lasing transitions associated with this ion are depicted in Figure 2.12. Since Allain *et al.* [134] demonstrated that efficient laser action could be observed in the red and near infrared spectral regions, several other research groups have investigated the praseodymium ion [135–137]. Allain and coworkers used an argon ion laser to excite the praseodymium ion in the host fluorozirconate fiber and observed light at 610, 635, 695, 715, 880 and 910 nm. The 610, 635, 695, 715 and 910 nm transitions all start from the $^3P_0$ level and terminate on the $^3H_6$, $^3F_2$, $^3F_3$, $^3F_4$ and $^1G_4$ levels respectively, whilst the 695 and 880 nm transitions start on the level $^3P_1$

and finish on the $^3F_4$ and $^1G_4$ levels. The 635 nm transition generated an output power of 100 mW for a pump power of 400 mW. Smart et al. [136] obtained an output power of 250 mW from the 635 nm transition when pumped with 800 mW of light at 476 nm. The praseodymium ion can also oscillate in the blue–green spectral region at 490 nm ($^3P_0-^3H_4$ transition) and the green spectral window at 520 and 530 nm ($^3P_1-^3H_5$ and $^3P_0-^3H_5$ transitions respectively). Smart et al. [135] were able to exploit up-conversion processes to access these wavelengths. The gain fiber was excited using two distinct pump wavelengths, 0.835 and 1.01 μm. Using these pump wavelengths output powers as high as 185 mW were recorded on the 635 nm transition.

Praseodymium has also attracted interest from the communications industry as it exhibits a transition at 1.38 μm ($^1G_4-^3H_5$ transition). Lasing on the transition has been observed in fluorozirconate-doped fibers [138–140]. Using dielectric mirrors that have been coated directly onto the ends of a fluorozirconate fiber, output powers as high as 50 mW for 300 mW of pump power have been observed [140].

Silica-based praseodymium fiber lasers have also been observed to oscillate in the 1.05 μm ($^1D_2-^3F_4$ transition) spectral region [141]. In fact a tuning range of some 86 nm has been observed.

2.4.7 Ytterbium-doped fiber lasers

The energy levels associated with the triply ionized ytterbium ion are shown in Figure 2.13. Ytterbium is an interesting system as it has only two energy levels. Thus high concentrations of the ion can be introduced into the core of an optical fiber without problems such as ion–ion interactions [142] affecting the overall performance of the system. Although ytterbium only exhibits two energy levels, $^2F_{5/2}$ and $^2F_{7/2}$, these are significantly broadened by the interaction of the ion with the host lattice to give broad levels which allow for wide tunability. The first reported silica fiber laser containing ytterbium was made by Hanna et al. [143]. Using 40 mW of 840 nm pump light an output power of 4 mW tunable over the range 1.015–1.14 μm was observed. By careful selection of the gain fiber length Armitage et al. [144] demonstrated that ytterbium could oscillate at 980 nm and obtained an output power of 40 mW corresponding to a slope effi-

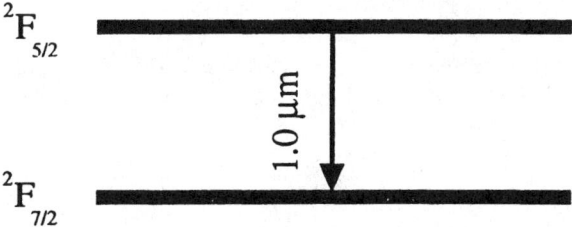

Fig. 2.13 Energy level diagram for ytterbium ions. Only the radiative transitions is illustrated.

ciency of 40%. The broad absorption band associated with the ytterbium ion allows for excitation around 1.05 μm which is a wavelength produced by Nd:YLF and Nd:YAG lasers. Optical excitation at this wavelength has resulted in the generation of output powers as high as 0.5 W at 1.12 μm [145]. The oscillating wavelength has also been defined by fiber Bragg gratings [146, 147]. In fact the output from these lasers has been used to excite a blue upconversion fiber laser [147]. The ytterbium ion has also oscillated in a fluorozirconate host fiber [148, 149] whereby output powers as high as 100 mW were observed at 1.02 μm and the laser was tunable over the wavelength range 0.975–1.055 μm.

As the ytterbium ion will oscillate at 980 nm it is an ideal codopant for erbium-doped fiber systems. The excited ytterbium ion will transfer the stored energy to the erbium ion. This allows for efficient pump absorption, and hence efficient laser operation at the erbium wavelength, without using extensive lengths of erbium-doped fiber. Ytterbium:erbium fiber lasers have been excited at either 980 nm or 1.05 μm and have resulted in highly efficient laser operation at 1.55 μm [150–153]. In fact output powers of 100 mW have been recorded for a Yb:Er codoped system pumped by a laser diode array operating at 960 nm [153].

Ytterbium has also been used as a codopant for both praseodymium-doped [154, 155] and thulium-doped [156] fluorozirconate fibers. Xie and Gosnell observed output powers of 300 mW at 635 nm from a Yb:Pr up-conversion fiber laser as well as improved power performance at the shorter wavelengths associated with the Pr ion [155].

It should be noted that codoping of other rare earth fiber lasers has been investigated. For example, the performance of an erbium-doped fluorozirconate fiber laser operating at 2.7 μm has been improved through the addition of praseodymium to the fiber core [157, 158]. Output powers of 30 mW with a broader tuning range in the 2.7 μm spectral region have been observed. Research has also been directed at the codoping of holmium fiber laser systems whereby the thulium ion is used to sensitize the holmium-doped fiber [159, 160]. Improved output powers of 250 mW were recorded at the holmium laser wavelength of 2.024 μm by pumping the fiber at 830 nm in the thulium absorption band.

2.5 SINGLE FREQUENCY FIBER LASERS

If the laser resonators illustrated in Figures 2.2 and 2.3 have a round trip optical path length p, a series of longitudinal modes are established with a frequency spacing Δv given by

$$\Delta v = \frac{c}{p} \qquad (2.5)$$

where c is the speed of light. A typical fiber laser may have a gain bandwidth of 500 GHz and an optical path length of 2 m giving a mode spacing of

150 MHz, and so many longitudinal modes may oscillate. The way in which the longitudinal modes are controlled influences the performance of the laser. If a coherent phase relationship is maintained between the oscillating modes a mode-locked pulse train is obtained from the laser and this will be discussed in the next section. If only a single mode is allowed to oscillate then a continuous wave output is obtained from the fiber laser with a well-defined frequency.

Narrow linewidth operation of fiber lasers has attracted significant interest not only from the communications industry for wavelength division multiplexed networks but also from the sensing community. From equation 2.5 it can be seen that the oscillating frequency is determined by the optical path length of the laser. If the path length changes then the laser frequency changes and this change in frequency can be monitored. If the laser is mounted on a material which responds to an external stimulus, for example a magnetostrictive element, the oscillating frequency will vary as the sensing material responds to changes in its local environment. To exploit this feature there has been significant work aimed at the development of single frequency fiber lasers. The majority of this effort has been directed at the low loss window of standard silica communications fiber at 1.55 µm using erbium as the laser active ion; however, single frequency operation of both neodymium and ytterbium fiber lasers has also been reported. This section will give an overview of the various approaches that have been used to force single frequency operation from rare earth doped fiber lasers.

Narrow linewidth operation of Nd^{3+}-doped fiber lasers has been studied [161–167]. Jauncey and coworkers [161] butt-coupled a length of Nd-doped silica fiber to a fiber Bragg reflector which forced the laser to operate at a wavelength of 1.084 µm with a linewidth of 16 GHz. This laser operated in a standing wave configuration and suffered from problems associated with spatial hole burning which can result in many modes oscillating simultaneously. One way of overcoming spatial hole burning is to reduce the optical path length of the laser. By doing this Jauncey *et al.* [162] were able to force the laser to operate on a single longitudinal mode with instrument resolution limited linewidths of 1.3 MHz.

Another way of eliminating the effects of spatial hole burning is to introduce a phase modulator into the laser cavity [163–165]. The phase modulator periodically alters the effective optical path length of the cavity which induces a periodic shift of the laser frequency. This frequency shift effectively averages out the spatial hole burning process. Linewidths of 40 MHz have been observed using this approach [164]. By including a second phase modulator at the other end of the cavity which is driven in antiphase to the first modulator the instantaneous linewidth has been reduced to 5 kHz [165].

One of the problems associated with spatial hole burning is that it can induce longitudinal mode hops. Sabert [167] has shown that these mode hops in a linear cavity fiber laser can be inhibited by including several fiber étalons in the cavity and using suitable electronic feedback to lock the laser frequency to a transmission maximum of one of the étalons.

Although single frequency neodymium fiber lasers have applications in high resolution spectroscopy for sensing applications it is better to operate at 1.55 µm as this corresponds to the low loss window of standard silica fiber and so reduces the power requirements for the sensor. As mentioned already, there have been many different approaches to generating narrow linewidth radiation from erbium-doped fiber lasers.

As with neodymium fiber lasers, one of the simplest approaches to generating single frequency radiation is to use a wavelength sensitive mirror, such as a diffraction grating or a Bragg reflector. Using a bulk diffraction grating as the wavelength selective element Gilbert [7] was able to generate light with a linewidth of 1.6 MHz from a linear cavity configuration. The oscillating frequency was locked to an absorption line of acetylene to produce a narrow linewidth fiber laser with a precisely defined wavelength.

The bulk diffraction grating is a reflective device and is normally used in a standing wave cavity geometry, which as discussed already, can lead to spatial hole burning. If a unidirectional travelling wave cavity is configured then spatial hole burning effects can be eliminated. One way of forming a unidirectional travelling wave cavity is to modify the cavity illustrated in Figure 2.3(b). The gain medium and an optical isolator, to ensure unidirectional operation, are placed inside the loop mirror, and the wavelength selective element can then be placed at the other end of the cavity. With such a cavity and a bulk diffraction grating O'Cochlain and Mears [168] were able to generate linewidths of the order of 300 kHz which were tunable over the range 1.525–1.568 µm. Gloag et al. [169] extended this idea by replacing the bulk diffraction grating with a Bragg reflector and were able to generate resolution-limited linewidths of 35 kHz. The laser resonator used in this work and the output from the interferometer used to measure the laser linewidth are depicted in Figure 2.14.

Fiber Bragg gratings have also been used to force single frequency operation from standing wave erbium fiber lasers [170–179]. To avoid problems with spatial hole burning the cavity length of the laser is kept as short as possible such that only a few longitudinal modes fall within the reflectivity profile of the Bragg reflectors. Using this approach Ball et al. [170] have observed linewidths of less than 47 kHz from their fiber laser. Zyskind et al. [171] constructed a similar cavity which was only 1 cm long and generated an output power of 51 µW. The operating characteristics of these short cavity lasers are limited by relaxation oscillations [171, 175] which can be eliminated through the use of suitable electronic feedback [175]. Dual single frequency operation of such fiber lasers can be achieved by writing two sets of Bragg reflectors into the core of the gain fiber, one pair defining one oscillating wavelength, the other pair another, and using this approach Chernikov et al. [173, 174] have been able to produce a fiber laser generating two discrete wavelengths separated by 59 GHz, each oscillating with a measured linewidth of 16 kHz. The frequency stability associated with the two wavelengths was measured to be 3 MHz.

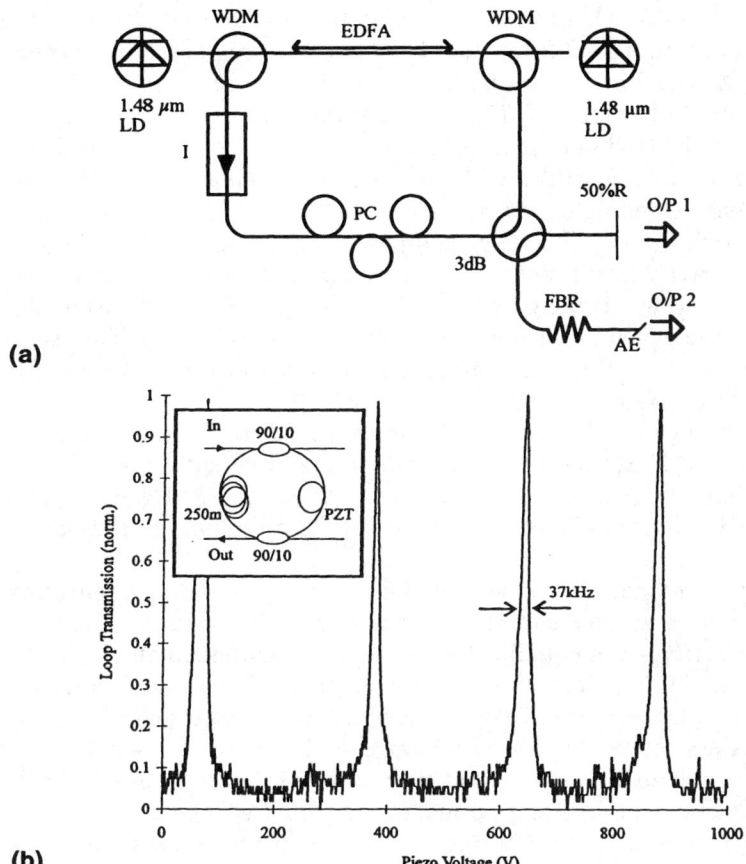

Fig. 2.14 (a) Simple unidirectional fiber loop mirror cavity used for obtaining single frequency operation at 1.55 μm (LD = laser diode, WDM = wavelength division multiplexer, EDFA = erbium doped fiber amplifier, PC = polarization controller, FBR = fiber Bragg reflector, 50%R = 50% reflecting mirror, 3 dB = fused fiber coupler and AE = angled end); (b) interference fringes from the interferometer shown in inset.

Although these short single frequency fiber lasers are effective sources of narrow linewidth radiation the output power that can be generated is low, typically less than 1 mW. This is in part due to the low absorption of the pump light supplied to the laser. Increasing the erbium concentration to improve the pump absorption is not viable for the following reason. As the ion concentration increases, the distance between neighboring ions in the fiber core decreases and at a critical separation the ions start to influence each other (ion–ion interactions) [180]. These ion–ion interactions result in an up-conversion process which depletes the metastable $^4I_{13/2}$ level thereby removing population from the laser cycle.

There are two distinct ways of increasing the laser output power. One is to amplify the output signal from the fiber laser and this has been successfully demonstrated with output powers as high as 60 mW being attained from the fiber laser/amplifier system [176]. An alternative approach is to codope the erbium fiber with ytterbium [181, 182]. This has been tried and narrow linewidth operation (300 kHz linewidth), with output powers as high as 20 mW, has been observed from a standing wave laser cavity [182].

The approach of Chernikov et al. [174] to produce dual wavelength operation from a short cavity length fiber laser was to use two sets of Bragg gratings to define the oscillating wavelengths. A similar effect can be achieved using sampled gratings and Ibsen et al. [183] have generated a narrow linewidth fiber oscillating over several wavelengths. This device may have applications in interrogating multipoint sensor systems.

It is well known that distributed feedback (DFB) semiconductor lasers are efficient sources of narrow linewidth radiation and recently DFB fiber lasers [184–186] have been made. Linewidths as narrow as 15 kHz with amplified output powers of 5.4 mW have been observed for an erbium fiber laser system [184].

The short standing wave cavity containing two fibre Bragg reflectors has been shown to be an efficient way of overcoming the limitations imposed by spatial hole burning. An equally effective way, as mentioned above, is to use a unidirectional ring configuration. The abundance of integrated optical isolators, fused couplers and wavelength selective devices make the construction of fiber ring lasers relatively straightforward. A linewidth of 1.4 kHz has been observed from a unidirectional ring fiber laser by Iwatsuki et al. [187]. The cavity was made from polarization-maintaining fiber which helped to minimize the effects of external perturbations, such as temperature fluctuations, which can induce longitudinal mode hops. The wavelength was defined by an integrated interference filter.

Yue et al. [188] also used an integrated interference to tune a single frequency ring laser. This laser was constructed from standard nonpolarization-maintaining fiber and was susceptible to longitudinal mode hops. These can be eliminated by locking the laser mode to the transmission maximum of a thin étalon. Including a thin étalon in the cavity of a ring fiber laser is difficult. Zhang and Lit [189], however, showed that a simple ring structure made from a pair of fused fiber couplers has the same frequency response as a thin étalon. By including this equivalent étalon in the laser resonator with suitable electronic feedback, longitudinal mode hops could be completely eliminated to give a laser producing linewidths of the order of 4 kHz and output powers as high as 12 mW [86].

Dawson et al. [83] configured a ring resonator which contained two integrated fiber Fabry–Pérot filters. One filter was used to define the wavelength and the other, with suitable electronic feedback, was used to prevent mode hops so producing a stable single frequency laser. This laser was capable of

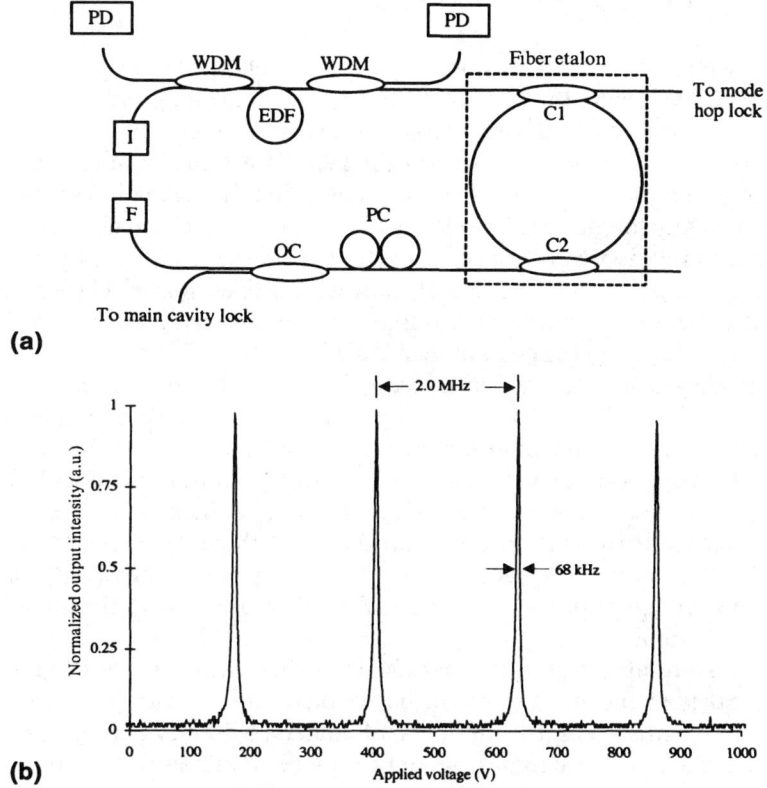

Fig. 2.15 (a) Unidirectional ring cavity showing the effective fiber etalon used to force stable single frequency operation from the laser (PD = pump laser diode, EDF = erbium doped fiber, WDM = wavelength division multiplexer, C1, C2 = 3 dB fused fiber couplers, OC = output coupler, PC = polarization controllers, F = filter and I = isolator); (b) resolution interference fringes observed from the fiber laser.

generating light which was tunable over 40 nm with a measured linewidth of 3 kHz [83, 84].

Both of the approaches taken by Yue et al. [86] and Dawson et al. [83] require the fiber which houses the filter to be broken. This introduces the potential for optical feedback from the broken ends which can lead to the formation of standing waves in the cavity. Gloag et al. [190] have developed a single frequency fiber laser that contains a continuous fiber filter. The cavity configuration used in this laser is illustrated in Figure 2.15 and resolution-limited linewidths of 35 kHz with output powers as high as 5 mW are routinely observed from this laser system. The oscillating frequency of this fiber laser has been stabilized to a confocal etalon and a root mean square frequency jitter of 3 kHz has been recorded [191].

2.6 MODE-LOCKED FIBER LASERS

The above section described various ways of controlling the output spectrum of a fiber laser by forcing the laser to oscillate on a single longitudinal mode. It is possible, however, to make a fiber laser operate simultaneously on many longitudinal modes. When a laser is turned on there is a finite time (the buildup time), before the steady-state oscillating condition is reached. During this buildup time many longitudinal modes can oscillate and each mode may have a different phase relationship. Hence the output from the laser as a function of time will vary in a random fashion. If, however, a fixed phase relationship is established between the longitudinal modes the time-varying output from the laser behaves in a well-described manner. As all the modes have the same phase relationship they coherently superimpose resulting in the formation of a periodic train of pulses. The temporal spacing between successive pulses is determined by the inverse of the longitudinal mode frequency spacing and the pulse duration is inversely proportional to the number of longitudinal modes that are phase locked, and hence the spectral width. Thus a short pulse will have a broad spectrum. This makes ultrashort pulses ideal for distributed wavelength-sensing applications. This approach has been used by Kersey and Morey [192]. In fact the intrinsic fiber compatibility of mode-locked fiber lasers make them the ideal choice for such applications.

Fiber lasers are an interesting class of laser. The shape and flexibility of a doped optical fiber means that it can be spooled onto a drum allowing long optical path lengths (tens to hundreds of meters) without a large physical separation of the cavity mirrors. If an optical pulse circulates in the cavity long interaction lengths between the pulse and gain medium are possible. This can have important consequences for the performance of the mode-locked laser as linear and nonlinear optical processes, which can strongly influence the pulse formation process, can be enhanced.

The main linear process which has to be considered arises from the wavelength dependence of the refractive index of the fiber core. This wavelength dependence results in group velocity dispersion which is quantified by the dispersion parameter D of a fiber [193]. D in turn is related to the material dispersion parameter $\partial^2 n/\partial \lambda^2$ by

$$D = -\frac{\lambda}{c}\frac{\partial^2 n}{\partial \lambda^2} \qquad (2.6)$$

where λ is the wavelength and n the refractive index. The effect of group velocity dispersion on a short pulse is twofold. Firstly, it causes a temporal broadening of the pulse as the different frequency components associated with the pulse propagate along the fiber with different velocities, and secondly, a linear frequency chirp is imposed on the pulse spectrum the sign of which depends on the sign of $\partial^2 n/\partial \lambda^2$. It is well known that there is a wavelength (the zero dispersion wavelength [193]) where the material dispersion parameter changes sign and this can influence the pulse as it propagates around the fiber cavity.

Along with the material dispersion there exists another source of dispersion in the fiber laser cavity which arises from the laser active ion itself [194]. The electronic transitions associated with active ion can introduce atomic dispersion. The magnitude and sign of this atomic dispersion depend on how far the oscillating frequency is from the resonance frequency of the transition and the presence of this atomic dispersion can modify the dispersive properties of the fiber [194].

The nonlinear optical processes occur because of the intensity dependent nature of the refractive index of the fiber core material which manifests itself as the optical Kerr effect. The two main nonlinear processes are self-phase modulation [195] and cross-phase modulation [196]. The effects of self-phase modulation are to broaden and impose a nonlinear frequency chirp on the pulse spectrum. The temporal profile of the pulse is, however, unaffected by self-phase modulation.

It is well known that standard telecommunications grade optical fiber is weakly birefringent. For a pulse propagating in an optical fiber, cross-phase modulation induces a nonlinear coupling of light from one polarization axis of the fiber to the other, nonlinear polarization rotation [197]. If two pulses at different wavelengths are propagating in the fiber then cross-phase modulation induces a nonlinear coupling between the two pulses as the first pulse influences the refractive index experienced by the second, thereby changing its phase.

The interplay between the linear and nonlinear processes has interesting repercussions for the pulse. For wavelengths longer than the dispersion minimum the frequency chirp imposed on the pulse by self-phase modulation is compensated by group velocity dispersion and the pulse transforms into an optical soliton [198]. A soliton is a pulse that will propagate, in a lossless system, without change in its shape or phase. In order for this transformation to occur the pulse circulating in a fiber cavity must be of a certain power level, the soliton power P_s, and must traverse a certain length of fiber, the soliton period Z_0 and these quantities are given by equations 2.7 and 2.8:

$$P_s = 0.776 \frac{\lambda^3 D A}{\pi^2 c n_2 t_p} \tag{2.7}$$

$$Z_0 = 0.322 \frac{\pi^2 c t_p}{\lambda^2 D} \tag{2.8}$$

where t_p is the pulse duration, n_2 the optical Kerr coefficient, A the effective area of the fiber and the λ wavelength. To produce a soliton with a short temporal duration the dispersion parameter D must be close to zero. Although solitons are extremely robust, if a soliton is perturbed in any way, such as experiencing gain or loss or a change in the fiber dispersion, it will respond to the perturbation by shedding radiation in the form of a dispersive wave [199].

If the pulse wavelength is less than the dispersion minimum then the chirps imposed by self-phase modulation and group velocity dispersion add together to produce a broad pulse which is linearly chirped [200]. This chirp can be removed by propagating the pulse through a suitable dispersive delay line, thereby exploiting soliton-type pulse-shaping kinetics. The delay lines can be formed from bulk optical components such as diffraction grating or prisms [201, 202], or through a suitably chirped fiber Bragg grating [203].

To exploit the advantages associated with the soliton pulse-shaping process, pulses must propagate in the fiber laser cavity and so the fiber laser has to be mode-locked. To generate this mode-locked state the longitudinal modes must be made to communicate their phase information and this can be done in one of two ways. The first is to use an active technique which requires the use of an intracavity modulator [204] driven by an external frequency source to force the modes to communicate. The second route relies on the use of a passive mode-locking element [205].

The modulators necessary for active mode-locking fall into two distinct categories which are either amplitude or phase modulators but both operate in the same generic way. For ease of understanding, consider a unidirectional ring laser that oscillates at a frequency v with a modulator operating at a frequency Ω present in the cavity. For a single round trip of the cavity the light passes through the modulator and the action of the modulator shifts light from v to $v \pm \Omega$. For a second transit through the modulator additional frequency components at $\pm 2\Omega$ are generated. If the modulator frequency Ω is equal to the longitudinal mode spacing, Δv, or an integer multiple of the longitudinal mode spacing, light is coupled from one longitudinal mode to the next. As the components at $v \pm \Omega$ were derived from the original mode at v, they have the same phase relationship and so can coherently superimpose.

Bulk and integrated active modulators have been made from acousto-optic and electro-optic materials. There are, however, problems associated with coupling light from the fiber to the modulator and then back to the fiber which can cause high insertion losses so reducing the efficiency of the laser. It has been shown that all optical versions of amplitude and phase modulators can be constructed which rely on the nonlinear process of cross-phase modulation [206, 207]. These devices are shown in Figure 2.16 and operate as follows. The all optical amplitude modulator is formed from a loop mirror which is biased to transmit low-power signals. A pulse train, produced by another mode-locked laser, is coupled into the loop mirror by a wavelength division multiplexer and this pulse train periodically biases the loop mirror into a reflection mode at the repetition rate of the input pulse train. Thus the loop mirror appears as an amplitude modulator.

The all optical phase modulator is formed by propagating a pulse train from a second laser directly through the laser cavity of interest. The phase shift induced by the injected pulses periodically modulates the cavity length of the laser. By matching the repetition rate of the pulse train to the cavity mode-spacing, ultrashort pulses can be generated.

In both types of modulators the pulse train which induces the cross-phase modulation is at a different wavelength to the generated pulse train. Owing to

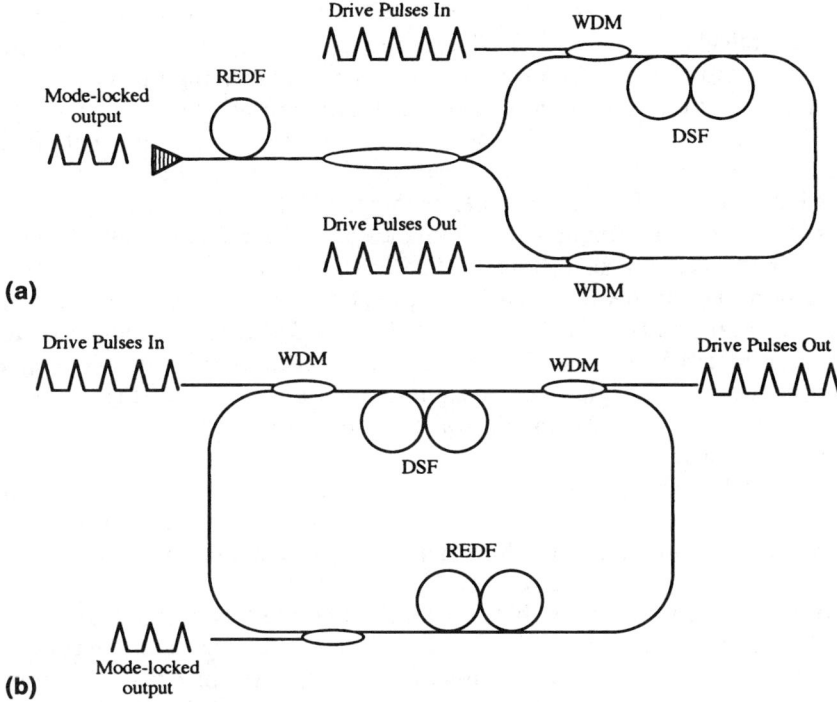

Fig. 2.16 (a) All fiber version of an amplitude modulator; (b) all-fiber version of a phase modulator (REDF = rare earth doped fiber, DSF = dispersion shifted fiber, WDM = wavelength division multiplexer).

the effects of dispersion the two pulse trains travel with different velocities and after a certain length of fiber they will temporally separate—an effect known as walk-off. This walk-off can be minimized by using pulse trains with wavelengths close to the dispersion zero of fiber.

Irrespective of the active mode-locking mechanism used, an external drive source is used to engender the necessary modal communication and it is imperative that the drive and longitudinal mode-spacing frequencies are matched at all times or else the laser performance is severely degraded.

A simple way of overcoming this problem is to passively mode-lock the fiber laser and this is done by including a saturable absorber in the laser cavity [204]. The saturable absorber can be thought of as an intensity-dependent switch which is transparent for high powers and absorbing for low powers. As with the all optical phase and amplitude modulators described above, the optical non-linearities of self-phase and cross-phase modulation can be exploited to form effective saturable absorbers.

A saturable absorber based on self-phase modulation is the nonlinear loop mirror. This device is formed by making a loop mirror with either an asymmetric coupler [208–212] or including gain in one of the arms [213, 214].

Irrespective of the approach taken, different intensities propagate clockwise and counterclockwise around the loop. Self-phase modulation ensures that these signals acquire a different phase shift. Thus, depending on the intensity of the light coupled into the loop mirror, it can transmit or reflect the light. This gives the loop mirror an intensity-dependent reflection coefficient. If the loop mirror is biased such that it transmits low powers and reflects high powers it acts as an effective saturable absorber [215].

Cross-phase modulation can also be exploited to form a saturable absorber. As mentioned already, cross-phase modulation can induce a nonlinear polarization rotation to occur because light is coupled from one polarization axis to the other. If a linear polarizer is placed after a length of fiber this nonlinear polarization process can be exploited to ensure that low-power signals are rejected by the polarizer whilst high-power signals are transmitted by the polarizer. Again this gives a power-dependent transmission profile and so corresponds to a saturable absorber.

Passive mode-locking is an extremely powerful technique for mode-locking and can be combined with active processes to produce a hybrid mode-locked laser which can produce ultrashort pulses with a well-defined modulation frequency.

As discussed already, the high optical gain presented by a fiber laser allows the inclusion of bulk optical components into the cavity. This feature has allowed the mode-locking of a selection of rare earth ion doped fiber lasers with both bulk amplitude and phase modulators [216–230]. Alcock et al. [216] introduced an acousto-optic amplitude modulator into the cavity of a neodymium fiber laser oscillating at $1.08\,\mu m$ and produced pulses as short as 1 ns (detector limited) at a repetition frequency of 20.73 MHz. The behavior of this laser was severely affected by intracavity étalons formed between the fiber end and the intracavity microscope objective. By index matching the free end of their neodymium fiber laser Phillips et al. [217] were able to produce 50 ps duration pulses.

Hofer et al. [220] used an amplitude modulator based on a piezoelectric element to mode-lock a neodymium fiber laser. The oscillating wavelength of this laser is below the dispersion minimum wavelength of the fiber and so soliton pulse shaping would not be expected to assist the pulse-shaping dynamics. To exploit the advantages associated with soliton shaping a dispersive delay line based on two diffraction gratings was incorporated into the cavity and pulses as short as 2.4 ps were produced from the fiber laser. These pulses were accompanied by satellite pulses the origin of which was attributed to incomplete soliton pulse shaping. By using a nonlinear loop mirror as a saturable absorber the satellite pulses could be completely eliminated and pulses as short as 125 fs were produced [220]. Interestingly enough once the pulses were formed the modulator could be turned off and the pulses remained. This indicated that the saturable absorber was dominating the mode-locking process.

Hofer et al. [221, 222] also showed that hybrid mode-locking of the neodymium laser could be achieved using a combination of active mode-locking and nonlinear polarization rotation. With suitable dispersion, compensation pulses as short as 40 fs were generated from this laser. Again when the modulator was turned off stable pulse formation was observed. Measurements made of the timing jitter associated with this laser indicate that it is well mode-locked, with timing jitters as low as 4 fs being recorded [221].

Erbium-doped fiber lasers have also been hybridly mode-locked using a combination of acousto-optic mode-locking and nonlinear polarization rotation. Fermann et al. [223] have been able to generate pulses as short as 180 fs from a linear cavity erbium fiber laser. To ensure that the net cavity dispersion was close to zero, a dispersive delay was included in the cavity and adjusted to compensate for the intrinsic soliton-supporting dispersion associated with the doped fiber in the laser cavity. This work was subsequently extended by using an active fiber which was polarization maintaining and pulses as short as 200 fs were generated with pulse energies of 100 pJ [224]. The use of the polarization-preserving fiber made the laser less susceptible to environmental fluctuations which can degrade the overall performance of the mode-locked laser.

Bulk phase modulators have also been used to mode-lock neodymium, erbium, erbium : ytterbium and praseodymium-doped fiber lasers. For example, Phillips et al. have generated pulses as short as 20 ps from an f.m. mode-locked neodymium fiber laser [225]. With this laser a frequency mismatch was observed between the drive frequency applied to the modulator and the frequency corresponding to the longitudinal mode-spacing. This frequency mismatch was attributed to atomic dispersion induced by the neodymium ion.

Praseodymium fiber lasers operate at 1.35 µm and so are suitable as pulse sources for soliton communications networks. Pataca et al. [226] have used a bulk phase modulator to mode-lock such a fiber laser at a frequency of 420 MHz and produced 30 ps with 1 mW average powers. The output and pulse durations obtained from this laser were consistent with those expected for soliton formation (equations 2.7 and 2.8).

Smith et al. [67] have mode-locked an erbium fiber laser using a bulk phase modulator. The laser had a 10 m long cavity and the oscillating wavelength was defined by a bulk birefringent filter. The modulator operated at a frequency of 420 MHz which is many times greater than the fundamental cavity mode-spacing of the resonator, in this case 10 m. This means that the modulator does not couple light from one mode to the next but instead couples light to the modes which are separated by 42 longitudinal mode frequencies. This corresponds to harmonic mode-locking. Once the modulator drive frequency and cavity length were in resonance 17 ps pulses were generated. When the tuning element was removed and the pump power adjusted 3 ps soliton pulses were observed from the laser as illustrated in Figure 2.17.

Davey et al. [227] used a phase modulator to initiate pulse formation in a hybridly mode-locked fiber laser. The cavity used in this work is depicted in Figure 2.18. It consisted of a 2 m length of doped fiber and a 10 m length of undoped standard telecommunications fiber. The modulator operated at a frequency of 480 MHz and so the laser was harmonically mode-locked. The saturable absorber was based on nonlinear polarization rotation and initially

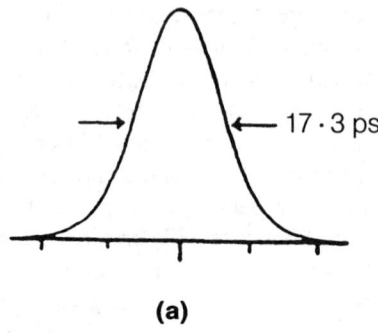

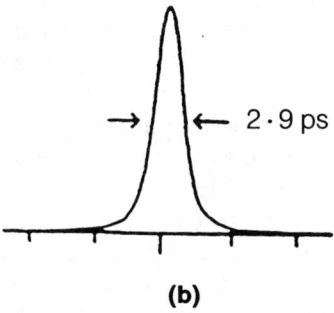

Fig. 2.17 Pulses from an actively mode-locked fiber laser containing a bulk phase modulator; (a) pulses from the laser containing a three-plate birefringent filter; (b) pulses obtained when the pump power is set to the soliton power level (used with permission of K. Smith).

pulses as short as 0.9 ps were routinely observed. When the laser made the transition from continuous wave to mode-locked operation the polarization state of the laser rotated by $\sim 5°$ [228]. The authors found that the pulse duration emitted from the fiber laser was dependent on the length of undoped fiber in the cavity. With a 2 m length of undoped fiber in the laser cavity pulses as short as 0.4 ps were generated [229]. Although this approach successfully generated ultrashort pulses several problems were encountered. Firstly, the pulse trains generated from the laser did not occur at the repetition rate of the modulator (2 ns), but appeared with random timing intervals between successive pulses which repeated after one cavity round trip (100 ns) [230]. As the pump power was reduced the number of pulses circulating in the cavity per round trip reduced. Both this feature and the random timing were attributed to the nature of the soliton [230]. To become a soliton, the pulse must have a certain energy and to support pulses at every temporal position defined by the modulator would require an energy level which the gain medium cannot supply.

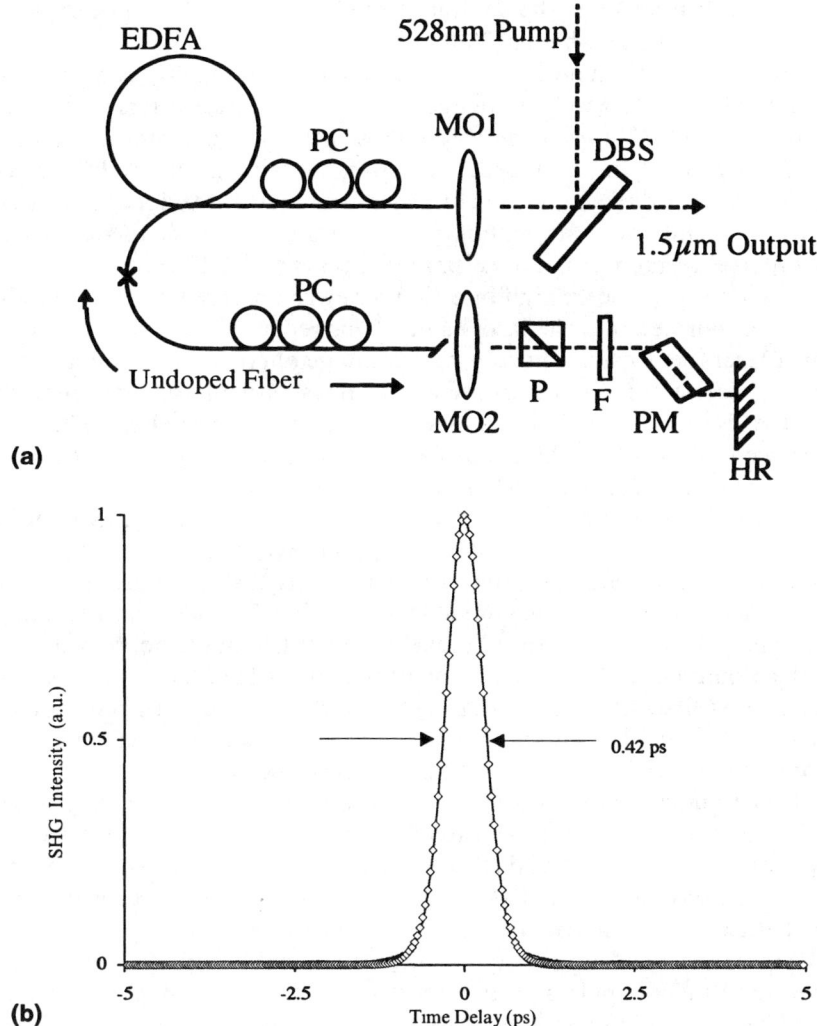

Fig. 2.18 (a) An example of a hybridly mode-locked fiber laser resonator (DBS = dichroic beamsplitter, MO1, MO2 = microscope objectives, EDFA = erbium doped fiber amplifier, PC = polarization controllers, P = polarizer, F = filter, PM = phase modulator and HR = high reflector); (b) the pulses obtained from the resonator (solid line experimental data and diamonds computed fit to such an electric field profile).

Secondly, discrete spectral side-bands were observed in the frequency domain. The origin of the frequency side-bands was attributed to perturbations of the optical soliton. As the soliton propagates around the cavity it is amplified in the gain medium and suffers a loss at the output coupler. This means that the energy of the pulse is constantly varying as it propagates around the cavity.

This energy variation causes the soliton to shed radiation which in turn leads to the formation of the side-bands [200].

Integrated-optic amplitude and phase modulators have also been used to actively mode-lock both neodymium and erbium-doped fiber lasers [231–245]. Geister and Ulrich [232] have generated pulses with durations ranging from 25 to 50 ps from a neodymium-doped fiber laser containing an $LiNbO_3$ phase modulator. This was done by butting up one of the ends of the doped fiber laser to the phase modulator. One major problem associated with this approach arises from the refractive index mismatch between the fiber core and the $LiNbO_3$. As a consequence a high insertion loss was observed when the modulator was incorporated into the laser cavity. This reduced the overall efficiency of the fiber laser thereby minimizing the advantage obtained from miniaturization. An alternative approach is to use a fiber phase modulator that alters the phase of the light whilst it is maintained within the laser cavity [233]. With such a device Phillips et al. [234, 235] were able to generate pulses with durations of 80 ps at a frequency of 417 MHz.

Kafka et al. [236] used an integrated amplitude modulator to mode-lock a ring erbium fiber laser. The laser generated 4 ps pulses and the power associated with these pulses indicated that soliton dynamics assisted the pulse formation process. Schlager et al. [237] also mode-locked a ring fiber laser at 40 MHz to produce 18 ps pulses which were subsequently amplified and compressed using soliton pulse shaping to 250 fs by propagating along a length of standard telecommunications fiber [238]. By carefully controlling the birefringence of the laser cavity Schlager et al. [239, 240] were able to make the laser operate at several different wavelengths each of which was mode-locked.

The telecommunications industry has been one of the main driving forces behind the development of mode-locked fiber lasers as ultrashort pulses could form the basis for future all optical communications systems. Although fiber lasers can be successfully mode-locked at low repetition frequencies (10–500 MHz) there is a need for ultrahigh repetition rate optical pulses (>10 GHz). To this end a variety of schemes has been studied with the aim of producing ultrashort pulses at high repetition rates [241–245]. Takada and Miyazawa have used an integrated amplitude modulator and generated pulses at repetition rates of 30 GHz from an erbium fiber laser [242]. Ultrashort pulses have been successfully generated at high repetition rates. Yoshida et al. [243] were able to generate pulses as short as 1.8 ps at a frequency of 20 GHz directly from a ring erbium fiber laser. These pulses were subsequently compressed to a duration of 170 fs by propagation through a dispersion-decreasing erbium fiber amplifier. Davey et al. [244, 245] used an integrated phase modulator and produced pulses at a repetition rate of 14 GHz. These pulses had durations of 30 ps and were tunable over the entire emission bandwidth of the erbium ion.

To overcome the insertion losses associated with integrated amplitude and phase modulators all optical modulators have been configured [207, 208]. Using an all optical amplitude modulator Nelson et al. [207] were able to generate 8 ps pulses from an erbium-doped fiber laser. All optical phase modulators have been used to mode-lock erbium, erbium:ytterbium and praseodymium-doped fiber lasers [207, 246–251]. For example, 1.2 ps pulses have been observed at 10 GHz from an erbium-doped fiber laser [246] and 2 ps from an ytterbium: erbium system [248].

Irrespective of the active mode-locking technique, an external signal must be used to force the longitudinal modes to communicate. As pointed out earlier, a saturable absorber can be used to passively mode-lock a laser and this approach has been successfully used for a selection of different doped fiber lasers [252–280].

Ober *et al.* [252] have shown that by periodically moving the end mirror of a dispersion compensated linear cavity neodymium fiber laser, passive mode-locking can be initiated. This laser generated output pulses as short as 40 fs with energies as high as 1 nJ. Langford [253] observed 1 ps pulse from a linear cavity erbium fiber laser at a repetition rate of 4 MHz. As with the system described by Ober *et al.* [252] mode-locking could only be initiated by perturbing the cavity end mirror.

Passive mode-locking has also been observed using a technique known as coupled cavity [254] or additive pulse mode-locking [255]. The mode-locked laser is divided into two distinct portions; the linear cavity, which contains the gain medium, and the nonlinear cavity. The nonlinear cavity acts as an intensity-dependent mirror which preferentially reflects high intensity signals and so can be regarded as a saturable absorber. Pulse formation occurs when the optical path length of the nonlinear cavity is an integer multiple of the linear cavity optical path length. Using this approach Sargsan *et al.* [256, 257] have generated 32 ps duration pulses from neodymium-doped fiber lasers. Shi *et al.* [258] have also applied this technique to mode-lock praseodymium-doped fiber lasers and have observed pulses as short as 45 ps.

For both the linear cavity passively mode-locked fiber lasers and the additive pulse mode-locked fiber lasers, mode-locking can only be initiated when the end mirror of the cavity is perturbed. This arises from problems associated with spatial hole burning in gain medium due to the standing wave cavity geometry. As with the single frequency fiber lasers discussed above, the effects of spatial hole burning can be overcome by constructing unidirectional ring lasers.

Self-starting passive mode-locking has been observed from fiber ring lasers using either a nonlinear loop mirror [259–279] or by exploiting nonlinear polarization rotation [280–288] as the saturable absorber, and schematics of these cavities are illustrated in Figure 2.19. These studies have mainly been directed at erbium and erbium:ytterbium doped fiber lasers although praseodymium [289, 290] and thulium [291] fiber lasers have been mode-locked using these techniques. As with the hybridly mode-locked erbium fiber laser described earlier, these passively mode-locked erbium lasers exhibit the same dependence on the overall cavity dispersion [273, 274] and display the same problems associated with the quantized nature of the soliton [267], i.e. randomly spaced pulse trains and strong spectral side-band formation. However, these lasers have been

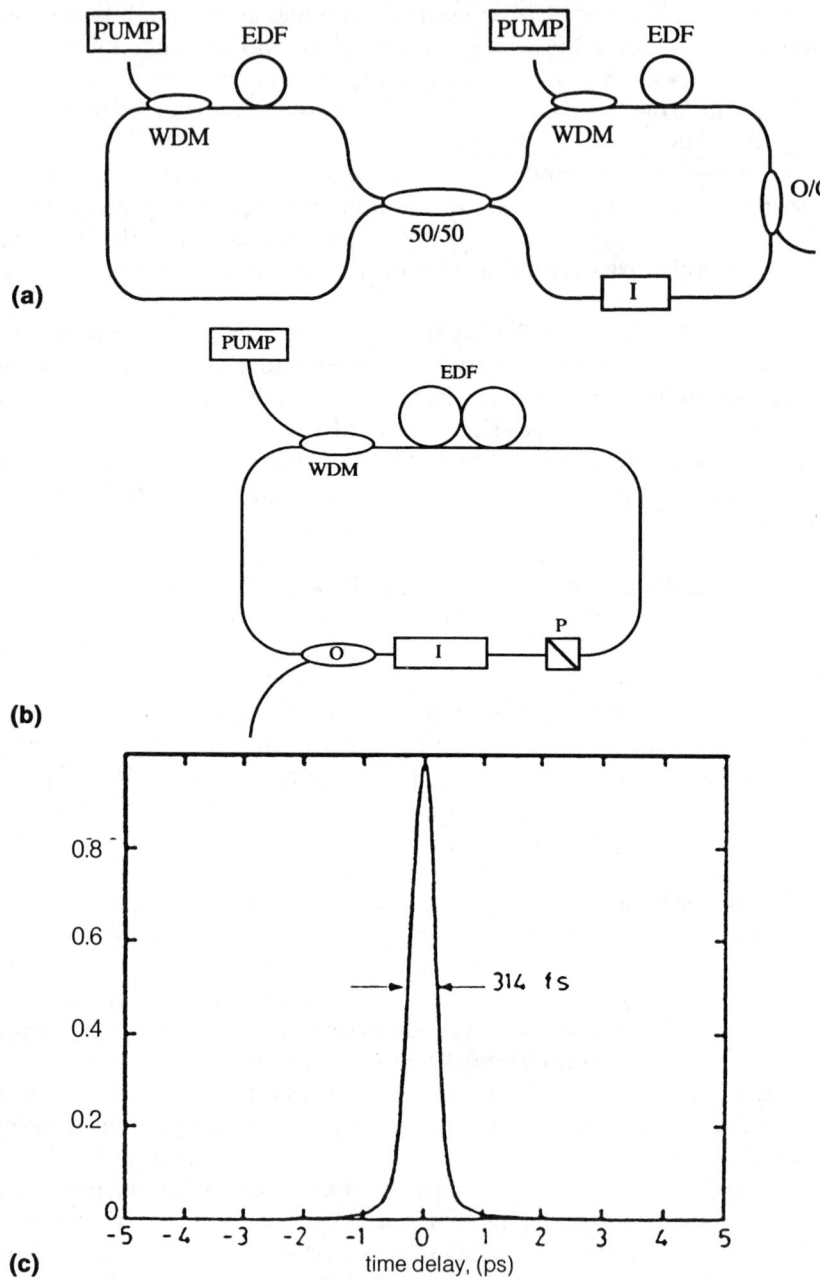

Fig. 2.19 Schematics of passively mode-locked fiber laser cavities: (a) a cavity using a nonlinear loop mirror; (b) a cavity reliant on nonlinear polarization rotation as the saturable absorber (WDM = wavelength division multiplexer, EDF = erbium doped fiber, I = isolator and P = polarizer); (c) a typical pulse obtained from the cavity depicted in (a) (used with the permission of I. N. Duling).

effective at generating ultrashort pulses. For example, Haus *et al.* [287] have shown that by suitable engineering of net cavity dispersion, pulses as short as 50 fs can be obtained from an erbium-doped fiber laser.

As mentioned earlier, chirped fiber Bragg reflectors can be used to control the overall dispersion of an optical resonator. Indeed these devices have been incorporated into erbium, neodymium and praseodymium doped fiber laser systems [251, 292–294]. Pulses as short as 6 ps were observed from the neodymium fiber laser [292]. Using a chirped Bragg grating as the end mirror of an actively mode-locked erbium fiber laser Kean *et al.* [293] observed soliton pulse shaping and spectral side-band formation when the chirped grating was inserted into the cavity such that it had a soliton-supporting dispersion. By turning the chirped grating around, the laser then operated in a nonsoliton mode producing pulses as short as 25 ps but without the observation of spectral side-bands. Pataca *et al.* [251] have used a chirped grating to assist the mode-locking dynamics of a praseodymium fiber laser and have observed the generation of both bright and dark optical solitons.

Most of the passive mode-locking techniques have relied on saturable absorbers made by exploiting the intrinsic intensity dependence of the refractive index of the optical fiber. Semiconductor saturable absorbers have also been used to mode-lock fiber lasers [295–303]. Loh *et al.* [295] used an end mirror made out of a semiconductor saturable absorber to passively mode-lock a neodymium fiber laser. To minimize the effects of group velocity dispersion the gain fiber was only 6 cm long. Pulses as short as 4 ps at a wavelength of 1.053 μm were obtained from the fiber laser. Using a similar cavity configuration Ober *et al.* [296] have been able to produce 260 fs from their Nd fiber laser. Erbium-doped fiber lasers have also been mode-locked using semiconductor materials as saturable absorbers with pulses as short as 320 fs being generated. These pulses had energies of 40 pJ [300].

The pulse energies associated with the mode-locked fiber lasers are typically of the order of picojoules and there has recently been research directed at the amplification of the pulses produced from fiber lasers to generate high energy light sources [304–308]. The favored technique is a process known as chirped pulse regenerative amplification [304] and this has been applied to both neodymium and erbium-doped fiber lasers with pulses as short as 300 fs being generated with energies as high as 10 μJ for a neodymium system [305] and 100 nJ in an 800 fs pulse from an erbium system [307].

2.7 Q-SWITCHED FIBER LASERS

The above section described the mode-locked operation of a variety of fiber lasers and although these lasers were able to generate pulses with high peak powers the energy content of the pulses was low. The pulse energy can be increased by regenerative amplification but this is a complex process. A more effective way of generating energetic pulses from a fiber laser is to Q-switch the laser [309].

The Q of an optical cavity is defined as the ratio of the energy stored in the cavity to that lost per round trip and if the Q of an optical cavity is varied in a precise manner energetic pulses can be generated. The basic method of generating a Q-switched pulse from a laser is as follows. The losses in the cavity are set to such a level that the laser cannot reach threshold, a low Q state, and the gain medium is optically excited and a population inversion is established. As the laser cannot reach threshold the stored population inversion builds up to a level that would not be achieved under normal operating conditions. The losses in the cavity are then switched to a value where the round trip gain exceeds the cavity losses—the high Q state. The light in the cavity, present due to spontaneous emission, is rapidly amplified to form an intense burst of radiation that can saturate the gain medium driving the gain below the threshold level, thereby inhibiting laser action. This rapid amplification and saturation process results in the formation of an optical pulse. As the output pulse is of a much shorter duration than the pumping interval a very energetic pulse is formed.

Rare earth doped fibers are ideal for Q-switched applications. The small gain cross-sections mean that the gain medium is effective at storing the pump energy supplied to it. To this end both silica-based and fluorozirconate-doped fibers have been Q-switched, generating energetic pulses in the 1, 1.55 and 2.7 µm spectral windows.

One of the first reports of the Q-switching of fiber lasers was made by Alcock *et al.* [310] who used an acousto-optic Q-switch to generate 200 ns duration pulses from a neodymium-doped fiber laser. The peak power associated with these pulses was 8 W and the repetition frequency was 100 Hz. Similar performance was observed from an erbium fiber laser operating at 1.55 µm. Pulses with durations ranging from 60 to 100 ns (Figure 2.20) with peak powers close to 2 W were routinely observed [311].

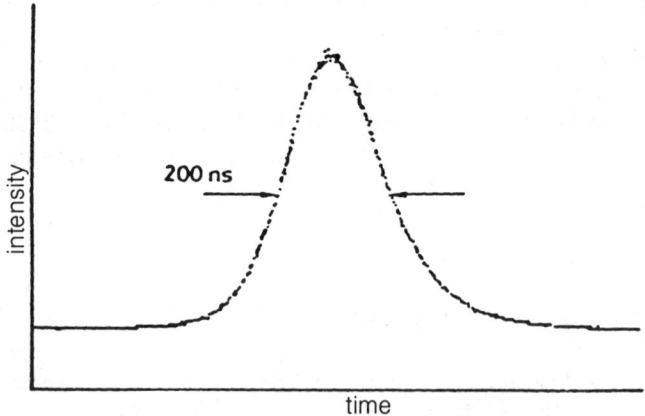

Fig. 2.20 A typical pulse from a Q-switched fiber laser (used with the permission of A. I. Ferguson).

Since these early experiments the performance of the Q-switched fiber lasers has been greatly improved [312–327]. For example, neodymium-doped fiber lasers Q-switched pulses with peak powers in excess of 1 kW have been generated at 1.054 μm with pulse durations of the order of 4 ns [317]. Q-switched erbium-doped fiber lasers [318–323] have been refined to generate pulses with peak powers of 400 W.

Fiber lasers operating at eye-safe wavelengths have also been Q-switched. An erbium laser operating at 2.7 μm has successfully generated pulses as short as 100 ns with peak powers of 2 W [324] whilst a thulium fiber laser has generated pulses of 4 W peak power and 130 ns duration [325–327].

2.8 FIBER LASERS AS SENSORS

Although intense research activity has been directed at the development of rare earth doped fiber lasers for the communications industry there have also been studies made into the use of fiber lasers for optical fiber based sensing systems. One of the most recent advances in optical fiber sensing has been in the use of fiber Bragg gratings as the sensing element. These structures are normally probed using a broadband low power (<100 μW) light source. As the Bragg grating has a narrow reflection profile only a small percentage of the light incident on the sensor is reflected. This can result in problems with extracting the optical signal from background noise. The ability to incorporate Bragg grating into fiber lasers as discussed in section 2.5, allows the construction of high-power (>1 mW) optical sensors thereby alleviating the signal to noise problems, which are extremely sensitive to external perturbations, such as strain and temperature.

There have been several approaches to the development of fiber laser based strain sensors [328–335]. One of the earliest demonstrations was by Melle *et al.* [328] who used a linear cavity. One of the ends of the fiber was reflection coated to form a mirror and the other end of the fiber was spliced to a fiber Bragg reflector. When the Bragg grating is placed under strain, the reflectivity characteristics of the device are altered. This in turn changes the oscillating wavelength of the fiber laser. By detecting the change in wavelength a measure of the applied strain can be obtained. Using this approach strains of 5.4 μstrain have been measured.

As mentioned in section 2.5, narrow linewidth fiber lasers can be configured by using matched pairs of fiber Bragg reflectors as the end mirrors of the laser cavity and these lasers have been used to produce both single point and multipoint sensors [329, 330]. Instead of using the Bragg reflector to sense the change in the environment, the actual laser cavity serves as the sensor. A change in the optical path length induces a change in the oscillating frequency (equation 2.6), so by monitoring the change in wavelength, the environmental perturbation can be monitored. The multipoint sensor consists of a series of fiber lasers each made from Bragg reflectors peaking at different wavelengths. Ball and coworkers [329] compared the performance of two fiber laser based sensors. A 30 mm single longitudinal mode laser and a 250 mm long multiple longitudinal mode laser which oscillated on several longitudinal modes were compared. As expected, the longer cavity sensor gave a higher frequency resolution. When the short laser cavity was used as a strain sensor, a wavelength change of 1.2 nm millistrain^{-1} was recorded. The response of the systems to thermal changes was found to be 0.01 nm °C^{-1}. Although the fiber laser sensors responded linearly to the applied perturbation the resolution of each sensor was ultimately limited by longitudinal mode-hops in the laser.

An alternative approach to strain sensing, demonstrated by Kersey and Morey [332], used a modified version of the cavity illustrated in Figure 2.3. The erbium gain fiber, an integrated optical isolator and fiber Fabry–Pérot étalon filter were included in the loop mirror section of the cavity and the cavity was completed by a series of Bragg reflectors each separated by a length of fiber. Each Bragg reflector had a different operating wavelength and a reflection bandwidth narrower than the Fabry–Pérot étalon. The laser would only oscillate when a filter pass band was in resonance with a Bragg reflector thereby forming a distributed sensor. This allowed one Bragg reflector to be interrogated at a time. A strain was then applied to the Bragg reflector of interest and the shift in oscillating wavelength detected. Using this approach a minimum strain of 25 µstrain was measured. The spectral response of the wavelength-measuring system determined the minimum strain level that could be measured. By replacing the wavelength-monitoring system with an unbalanced Mach–Zehnder interferometer a more sensitive detector was constructed.

Recently Koo *et al.* [5] have shown that magnetic fields can be detected using an active fiber laser sensor. A single frequency fiber laser was attached to a magnetostrictive element. This element exhibits a quadratic dependence on the applied field and so can be used to detect either AC or DC magnetic fields. By applying a linear DC bias to the magnetostrictive element AC fields as low as 0.9 pT $\sqrt{Hz^{-1}}$ at 30 kHz applied field were observed. When used as a DC magnetometer fields as low as 300 pT $\sqrt{Hz^{-1}}$ were observed.

The above sensors rely on changes in the oscillating wavelength to provide information on the applied perturbation to the active sensor. Recently it has been shown that the polarization properties of fiber lasers can be exploited to produce a polarimetric sensor. Most fiber lasers are weakly birefringent. This birefringence arises from the fact that the doped fiber exhibits two polarization axes, the fast and slow axes with different refractive indices. This results in the longitudinal mode-spacing being different for the light propagating on the fast axis in comparison to that of light propagating on the slow axis. The frequency difference between the two sets of longitudinal modes results in a beat frequency. By applying a perturbation to the cavity which alters the birefringence of the laser cavity, the beat frequency will change. By monitoring this change in frequency the applied perturbation can be quantified.

Several variants of this system have been developed [333–335] using different doped fibers as the sensing fiber. For a broadband neodymium fiber laser a frequency change of 4.8 kHz g^{-1} of applied stress was observed [333]. This work has been extended to the erbium wavelength region using two different cavity configurations. The first sensor was constructed from two lengths of polarization-maintaining erbium-doped fiber which were spliced together. At the splice the polarization axes of the fibers were rotated. The effect of this rotation was to reduce the polarization mode beat frequency. With this approach frequency shifts of 124 kHz °C^{-1} cm^{-1} and 137 kHz μm^{-1} were observed for applied thermal and strain perturbations [335].

Aligning the polarization axes of the doped fiber at the splice is difficult and recently Kim et al. [335] have developed a new form of polarimetric sensor which uses nonpolarization-preserving doped fiber which is fusion spliced to a short length of polarization-preserving fiber. This length of fiber acts as the sensor. A frequency shift due to temperature of 867 kHz °C^{-1} cm^{-1} was measured for this laser.

2.9 CONCLUSIONS

In this chapter I hope to have given the reader an indication of the extent to which rare earth doped fiber lasers have become an accepted form of laser active gain medium. The diverse range of applications for which rare earth doped fiber lasers are being employed is a testament to their appeal. It is envisaged that as time progresses the developments made in rare earth fiber laser technology will see them replace semiconductor diode lasers in communications-based systems and optical data storage systems and become the accepted choice for high-power laser applications.

REFERENCES

1. E. Snitzer; 'Optical maser action of Nd^{3+} in a barium crown glass', *Physical Review Letters*, **72**, 36, 1961.
2. See for example N. Langford and A. I. Ferguson; 'Rare earth doped silica fibre lasers', Ch. 3 in *Principles of Modern Optical Systems*, Vol. II, Eds. I. Andonovic and D. Uttamchandani, Artec House, Boston, 1993.
3. See for example G. P. Agrawal; *Non-Linear Optical Fibres*, Academic Press, New York, 1987.
4. S. B. Poole, D. N. Payne and M. E. Fermann; 'Fabrication of low loss optical fibres containing rare earth ions', *Electronics Letters*, **21**, 737, 1985.
5. K. Koo, A. Kersey and F. Bucholtz; Postdeadline paper PD41, 'Fibre Bragg grating laser magnetometer' in the *Proceedings of the Conference on Lasers and Electro-Optics*, Baltimore, USA, 1995.
6. J. R. Taylor; Paper CWM3, 'Soliton fibre lasers' in the *Proceedings of the Conference on Lasers and Electro-Optics*, Baltimore, USA, 1995.
7. S. L. Gilbert; 'Frequency stabilisation of a tunable erbium-doped fibre laser', *Optics Letters*, **16**, 150, 1991.

8. See for example M. J-F. Digonnet and E. Snitzer; 'Nd^{3+} and Er^{3+}-doped silica fibre lasers' in *Rare Earth Doped Fibre Lasers and Amplifiers*, Ed. M. J-F. Digonnet, Dekker, New York, 1993.
9. D. S. Funk, J. W. Carlson and J. G. Eden; Paper CMB2, 'Room temperature upconversion UV and violet lasing in Nd^{3+}:ZBLAN' in the *Proceedings of the Conference on Lasers and Electro-Optics*, Baltimore, USA, 1995.
10. J. Schneider; 'Fluoride fibre laser operating at 3.9 μm', *Electronics Letters*, **31**, 1250, 1995.
11. See for example D. C. Hanna and A. C. Tropper; 'Silica fibre laser oscillators', Ch. 7 in *Optical Fibre Lasers and Amplifiers*, Ed. P. W. France, Blackie, Glasgow, 1991.
12. See for example J. S. Sanghera and I. D. Aggarwal; 'Rare earth doped heavy metal fluoride glass fibres' in *Rare Earth Doped Fibre Lasers and Amplifiers*, Ed. M. J-F. Digonnet, Dekker, New York, 1993.
13. J. R. Lincoln, W. S. Brocklesby, C. J. MacKechnie, J. Wang, R. S. Deol, D. C. Hanna and D. N. Payne; 'New class of fibre laser based on lead-germanate glass', *Electronics Letters*, **28**, 1021, 1992.
14. J. Wang, L. Reekie, W. S. Brocklesby, Y. T. Chow and D. N. Payne; 'Fabrication spectroscopy and laser performance of a Nd^{3+} lead-silicate glass-fibre', *Journal of Non-Crystalline Solids*, **180**, 207, 1995.
15. J. S. Wang, D. P. MacHewirth, F. Wu, E. Snitzer and E. M. Vogel; 'Neodymium-doped tellurite single-mode fibre laser', *Optics Letters*, **19**, 1448, 1994.
16. See for example A. E. Siegman; *Lasers*, p. 123, University Science Books, 1988.
17. See for example P. W. Milonni and J. H. Eberly; *Lasers*, pp. 486–495, Wiley Interscience, New York, 1988.
18. J. L. Wagener, P. F. Wysocki, M. J. F. Digonnet and H. J. Shaw; 'Modelling of ion-pairs in erbium-doped fibre amplifiers', *Optics Letters*, **19**, 347, 1994.
19. M. Shimizu, M. Yamada, M. Horiguchi, T. Takeshita and M. Okayasu; 'Erbium-doped fibre amplifiers with an extremely high gain coefficient of 11.0 dB/mW', *Electronics Letters*, **26**, 1641, 1990.
20. R. J. Mears, L. Reekie, S. B. Poole and D. N. Payne; 'Low threshold tunable c.w. and Q-switched fibre laser operating at 1.55 μm', *Electronics Letters*, **22**, 159, 1986.
21. See for example W. Koechner; *Solid State Laser Engineering* Ch. 2, Springer-Verlag, Berlin, 1988.
22. See for example W. J. Miniscalo; 'Rare earth doped glass fibres: optical properties' in *Rare Earth Doped Fibre Lasers and Amplifiers*, Ed. M. J-F. Digonnet, Dekker, New York, 1993.
23. D. C. Hanna, R. M. Percival, R. G. Smart, J. E. Townsend and A. C. Tropper; 'Continuous-wave oscillation of holmium-doped silica fibre laser', *Electronics Letters*, **25**, 593, 1989.
24. See for example S. P. Craig-Ryan and B. J. Ainslie; 'Glass structure and fabrication techniques' in *Optical Fibre Lasers and Amplifiers*, Ed. P. W. France, Blackie, Glasgow, 1991.
25. S. R. Nagel; 'Fibre materials and fabrication methods', Ch. 4 in *Optical Fibre Telecommunications*, Vol. II, Eds. S. E. Miller and I. P. Kaminow, Academic Press, New York, 1988.
26. B. J. Ainslie, S. P. Craig, S. T. Davey and B. Wakefield; 'The fabrication and optical properties of Nd^{3+} in silica-based optical fibres', *Material Letters*, **5**, 143, 1987.

27. B. J. Ainslie, S. P. Craig, S. T. Davey and B. Wakefield; 'The fabrication assessment and optical properties of Nd^{3+} and Er^{3+}-doped silica-based fibres', *Material Letters*, **6**, 139, 1988.
28. W. J. Tomlinson; 'Passive and low-speed active optical components for fibre systems', Ch. 10 in *Optical Fibre Telecommunications*, Vol. II, Eds. S. E. Miller and I. P. Kaminow, Academic Press, New York, 1988.
29. D. B. Mortimore; 'Fibre loop reflectors', *Journal of Lightwave Technology*, **LT-6**, 1217, 1988.
30. K. O. Hill, Y. Fujii, D. C. Johnson and B. S. Kawasaki; 'Photosensitivity in optical fibre waveguides: applications to reflection filter fabrication', *Applied Physics Letters*, **32**, 647, 1978.
31. M. C. Brierley and P. W. France; 'Neodymium-doped fluorozirconate fibre laser', *Electronics Letters*, **23**, 815, 1987.
32. J. Stone and C. A. Burrus; 'Neodymium-doped silica lasers in end-pumped fibre geometry', *Applied Physics Letters*, **23**, 388, 1973.
33. I. P. Alcock, A. I. Ferguson, D. C. Hanna and A. C. Tropper; 'Tunable, continuous-wave neodymium doped monomode-fibre laser operating at $0.9-0.945$ and $1.070-1.135\,\mu m$', *Optics Letters*, **11**, 709, 1986.
34. M. Shimizu, H. Suda and M. Horiguchi; 'High efficiency Nd-doped fibre lasers using direct-coated dielectric mirrors', *Electronics Letters*, **23**, 768, 1987.
35. G. A. Ball and W. W. Morey; 'Efficient integrated Nd^{3+} fibre laser', *IEEE Photonics Technology Letters*, **3**, 1077, 1991.
36. I. D. Miller, D. B. Mortimore, P. Urqhart, B. J. Ainslie, S. P. Craig, V. A. Miller and D. N. Payne; 'A Nd^{3+}-doped cw fibre laser using all fibre reflectors', *Applied Optics*, **26**, 2197, 1987.
37. Y. Chaoyu, P. Jianngde and Z. Bingkun; 'Tunable Nd^{3+}-doped fibre ring laser', *Electronics Letters*, **25**, 101, 1989.
38. U. Ghera, N. Konforti and M. Tur; 'Wavelength tunability in a Nd doped fibre laser with an intracavity polariser', *IEEE Photonics Technology Letters*, **4**, 4, 1992.
39. U. Ghera, N. Friedman and M. Tur; 'A fibre laser with a comb-like spectrum', *IEEE Photonics Technology Letters*, **5**, 1159, 1993.
40. P. Mollier, V. Armbruster, H. Porte and J. P. Goedgebuer; 'Electrically tunable Nd^{3+} doped fibre laser using nematic liquid-crystals', *Electronics Letters*, **31**, 1248, 1995.
41. See for example P. Urqhart; 'Devices and configurations for fibre laser sources and amplifiers', Ch. 3 in *Rare Earth Doped Fibre Lasers and Amplifiers*, Ed. M. J-F. Digonnet, Dekker, New York, 1993.
42. E. Snitzer, H. Po, F. Hakimi, R. Tumminelli and B. C. McCollum; Paper PD5 'Double clad offset core Nd fibre laser' in *Digest of Conference on Optical Fibre Sensors*, 1988.
43. H. Po, E. Snitzer, R. Tumminelli, L. Zenteno, F. Hakimi, N. M. Cho and T. Haw; Paper PD7 'High-brightness Nd fibre laser pumped by a GaAlAs phased array' in the *Proceedings of Optical Fibre Commmunications*, OFC, 1989.
44. T. Weber, W. Luthy, H. P. Weber, V. Neuman, H. Berthou and G. Kotrotsios; 'Cladding-pumped fibre laser', *IEEE Journal of Quantum Electronics*, **QE-31**, 326, 1995.
45. T. Weber, W. Luthy, H. P. Weber, V. Neuman, H. Berthou and G. Kotrotsios; 'A longitudinal and side-pumped single transverse-mode double-clad fibre laser with a special silicone coating', *Optics Communications*, **115**, 99, 1995.

46. H. Zellmer, U. Wıllamowski, A. Tunnermann, H. Welling, S. Unger, V. Reichel, H. R. Muller, J. Hirchhof and P. Albers; 'High-power cw neodymium-doped fibre laser operating at 9.2 W with high beam quality', *Optics Letters*, **20**, 578, 1995.
47. R. Koch, Ugrienbner and R. Grunwald; 'High-average-power flashlamp-pumped Nd-glass fibre-bundle laser', *Applied Physics B*, **58**, 403, 1994.
48. L. Reekie, I. M. Jauncey, S. B. Poole and D. N. Payne; 'Diode-laser pumped Nd^{3+}-doped fibre laser operating at 938 nm', *Electronics Letters*, **23**, 884, 1987.
49. F. Hakimi, H. Po, R. Tumminelli, B. C. McCollum, L. Zenteno, N. M. Cho and E. Snitzer; 'Glass fibre laser at 1.36 μm from SiO_2', *Optics Letters*, **14**, 884, 1987.
50. W. J. Miniscalo, L. J. Andrews, B. A. Thompson, R. S. Quimby, L. J. B. Vacha and M. G. Drexhage; '1.3 μm fluoride fibre laser', *Electronics Letters*, **24**, 28, 1988.
51. D. S. Funk, J. W. Carlson and J. G. Eden; 'Room-temperature fluorozirconate glass-fibre laser in the violet (412 nm)', *Optics Letters*, **20**, 1474, 1995.
52. D. S. Funk, J. W. Carlson and J. G. Eden; 'Ultraviolet (381 nm) room-temperature laser in neodymium doped fluorozirconate fibre', *Electronics Letters*, **30**, 1859, 1994.
53. I. P. Alcock, A. I. Ferguson, D. C. Hanna and A. C. Tropper; 'Continuous-wave oscillation of a monomode neodymium-doped fibre laser 0.9 μm on the $^4F_{3/2}$ to $^4I_{9/2}$ transition', *Optics Communications*, **58**, 405, 1986.
54. M. C. Brierley and M. H. Hunt; 'Efficient semiconductor diode pumped fluoride fibre lasers', in the *Proceedings of SPIE*, **1171**, 157, 1990.
55. T. Komukai, Y. Fukasaku, T. Sugawa and Y. Miyajima; 'Highly efficient and tunable Nd^{3+} doped fluoride fibre laser operating in the 1.3 μm band', *Electronics Letters*, **29**, 755, 1993.
56. See for example Ch. 2 in *Optical Fibre Telecommunications*, Eds. S. E. Miller and A. G. Chynoweth, Academic Press, New York, 1979.
57. M. C. Brierley and P. W. France; 'Continuous wave lasing at 2.7 μm in an erbium doped fluorozirconate fibre', *Electronics Letters*, **25**, 935, 1988.
58. J. Y. Allain, M. Monerie and H. Poignant; 'Erbium doped fluorozirconate single mode fibre laser lasing at 2.71 μm', *Electronics Letters*, **26**, 28, 1989.
59. R. G. Smart, J. N. Carter, D. C. Hanna and A. C. Tropper; 'Erbium doped fluorozirconate fibre laser operating at 1.66 and 1.72 μm', *Electronics Letters*, **26**, 649, 1989.
60. J. Y. Allain, M. Monerie and H. Poignant; 'Lasing at 1.0 μm in erbium doped fluorozirconate fibres', *Electronics Letters*, **26**, 318, 1989.
61. C. A. Miller, M. C. Brierley, M. H. Hunt and S. F. Carter, 'Efficient up-conversion pumping at 800 nm of an erbium-doped fluoride fibre laser operating around 850 nm', *Electronics Letters*, **26**, 1218, 1990.
62. M. S. O'Sullivan, J. Chrostowski, E. Desurvire and J. R. Simpson; 'High power narrow-linewidth Er^{3+} doped fibre laser', *Optics Letters*, **14**, 438, 1989.
63. R. Wyatt, B. J. Ainslie and S. P. Craig; 'Efficient operation of an array-pumped Er^{3+} doped silica fibre laser at 1.55 μm', *Electronics Letters*, **24**, 1362, 1988.
64. C. G. Atkins, J. R. Armitage, R. Wyatt, B. J. Ainslie and S. P. Craig; 'Spectroscopic studies of erbium-doped single mode silica fibres', *Optics Communications*, **73**, 217, 1989.
65. R. I. Laming, S. B. Poole and J. E. Tarbox; 'Pump excited state absorption in erbium doped fibres', *Optics Letters*, **13**, 1084, 1989.
66. R. P. Davey, N. Langford and A. I. Ferguson; Paper CFE2 'Sub-picosecond pulse generation from an erbium doped fibre laser' in the *Proceedings of the Conference on Lasers and Electro-optics*, Baltimore, USA, 1991.

REFERENCES

67. K. Smith, J. R. Armitage, R. Wyatt, N. J. Doran and S. M. J. Kelly; 'The erbium fibre soliton laser', *Electronics Letters*, **26**, 1149, 1990.
68. V. P. Gasponstev, P. I. Sadovsky and I. E. Samartsev; Paper CPDP, in the *Proceedings of the Conference on Lasers and Electro-Optics*, Anaheim, USA, 1990.
69. W. L. Barnes, P. R. Morkel, L. Reekie and D. N. Payne; 'High-quantum-efficiency Er^{3+} fibrelasers pumped at 980 nm', *Optics Letters*, **14**, 1002, 1989.
70. R. Wyatt; 'High power broadly tunable erbium doped silica fibre laser', *Electronics Letters*, **25**, 1498, 1989.
71. J. L. Zyskind, C. R. Giles, E. Dersurvire and J. R. Simpson; 'Optimal pump wavelength in the $^4I_{15/2}-^4I_{13/2}$ absorption band for efficient Er^{3+} doped fibre amplifiers', *IEEE Photonics Technology Letters*, **1**, 428, 1989.
72. Y. Kimura, K. Susuki and M. Nakazawa; 'Laser diode pumped mirror free Er^{3+} doped fibre laser', *Optics Letters*, **14**, 999, 1989.
73. N. Langford; Unpublished results.
74. E. Dursurvire and J. R. Simpson; 'Evaluation of and Stark level energies in erbium-doped aluminosilicate fibres', *Optics Letters*, **15**, 547, 1990.
75. R. Kashyap, J. R. Armitage, R. Wyatt, S. T. Davey and D. L. Wllliams; 'All fibre narrowband reflection gratings at 1500 nm', *Electronics Letters*, **26**, 730, 1990.
76. G. A. Ball and W. H. Glenn; 'Design of a single mode linear cavity erbium fibre laser utilising Bragg reflectors', *Journal of Lightwave Technology*, **10**, 1338, 1992.
77. A. D. Kersey and W. W. Morey; 'Multielement Bragg-grating based fibre-laser strain sensor', *Electronics Letters*, **29**, 964, 1993.
78. G. A. Ball and W. W. Morey; 'Compression tuned single frequency Bragg grating fibre laser', *Optics Letters*, **19**, 1979, 1994.
79. J. J. Pan and Y. Shi; 'Tunable Er^{3+} doped fibre ring laser using fibre Bragg grating incorporated by optical circulator or fibre coupler', *Electronics Letters*, **31**, 1164, 1995.
80. C. M. Miller and F. J. Janniello; 'Passively temperature-compensated Fabry–Pérot filter and its applications in wavelength division multiple access computer network', *Electronics Letters*, **25**, 2122, 1990.
81. J. Minowa and Y. Fujii, 'Wide-bandwidth, sharp-cutoff bandpass filter for WDM transmission', *Electronics Letters*, **21**, 915, 1985.
82. K. McCallion, W. Johnstone and W. Fawcett; 'Tunable in-line fibreoptic bandpass filter', *Optics Letters*, **19**, 542, 1994.
83. N. Park, J. W. Dawson and K. J. Vahala; 'All fibre, low threshold, widely tunable single-frequency, erbium-doped fibre ring laser with a tandem Fabry–Pérot filter', *Applied Physicss Letter*, **59**, 2369, 1991.
84. S. Sanders, N. Park, J. W. Dawson and K. J. Vahala; 'Reduction of the intensity noise from an erbium-doped fibre laser to the standard quantum limit by intracavity spectral filtering', *Applied Physics Letters*, **61**, 1889, 1992.
85. J. W. Dawson, N. Park and K. J. Vahala; 'CO-lasing in an electrically tunable erbium-doped fibre laser', *Applied Physics Letters*, **60**, 3090, 1992.
86. G. W. Schinn, C. Y. Yue, J. Zhang, D. Yang, V. M. Paramonov and W. R. L. Clements; 'Recent developments in erbium fibre lasers and associated products at MPB Technologies Inc.', *Proceedings of OPTO 94*, Paris, France.
87. A. Gloag, K. MacCallion, W. Johnstone and N. Langford; 'A tunable erbium fibre laser containing a novel continuous fibre overlay filter', *Optics Letters*, **19**, 802, 1994.
88. P. L. Scrivener, E. J. Tarbox and P. D. Maton; 'Narrow linewidth tunable operation of Er^{3+}-doped single mode fibre laser', *Electronics Letters*, **25**, 549, 1989.

89. P. F. Wysocki, M. J. F. Digonnet and B. Y. Kim; 'Electronically tunable 1.55 μm erbium-doped fibre laser', *Optics Letters*, **15**, 273, 1990.
90. M. Y. Frankel, R. D. Esman and J. F. Weller; 'Rapid continuous tuning of a single-polarisation fibre ring laser', *IEEE Photonics Technology Letters*, **6**, 591, 1994.
91. M. W. Meada, J. S. Patel, D. A. Smith, C. L. Lin, M. A. Saifi and A. Vonlehman; 'An electronically tunable fibre laser with a liquid-crystal étalon filter as the wavelength tuning element', *IEEE Photonics Technology Letters*, **2**, 787, 1990.
92. H. Tobben; 'Room temperature cw fibre laser at 3.5 μm on Er^{3+} doped ZBLAN glass', *Electronics Letters*, **28**, 1361, 1992.
93. W. Hofle and H. Tobben; 'Analysis measurement and optimisation of threshold power of 3.5 μm ZBLAN-glass fibre lasers', *International Journal of Infrared and Millimeter Waves*, **14**, 1407, 1993.
94. R. Allen, L. Esterowitz and R. J. Ginther; 'Diode-pumped single mode fluorozirconate fibre laser from the $^4I_{11/2} - {}^4I_{13/2}$ transition in erbium', *Applied Physics Letters*, **56**, 1635, 1990.
95. H. Yangagita, I. Masuda, T. Yamashita and H. Toratani; 'Diode laser pumped Er^{3+} fibre laser operation between 2.7–2.8 μm', *Electronics Letters*, **26**, 1836, 1989.
96. L. Wetenkamp; 'Efficient cw operation of a 2.75 μm Er^{3+} doped fluorozirconate fibre laser pumped at 650 nm and 750 nm', *Journal of Electronics and Communications*, **5**, 328, 1991.
97. N. A. Swain and T. A. King; 'Analysis and performance of 2.7 μm Er^{3+} fibre laser amplifiers and oscillators', *Journal of Modern Optics*, **38**, 2023, 1991.
98. C. Frerichs; 'Efficient Er^{3+}-doped cw fluorozirconate fibre laser operating at 2.7 μm pumped at 980 nm', *International Journal of Infrared and Millimeter Waves*, **15**, 635, 1994.
99. J. Schneider; 'Midinfrared fluoride fibre laser in multiple cascade operation', *IEEE Photonics Technology Letters*, **7**, 354, 1995.
100. M. C. Brierley, C. A. Miller and P. W. France; 'Laser transitions in erbium doped fluoride fibres', Paper TUJ 22, *Proceedings of the Conference on Lasers and Electro-Optics*, 1989.
101. J. Y. Allain, M. Monerie and H. Poignant; 'Narrow line width tunable cw and Q-switched 0.98 μm operation of erbium-doped fluorozirconate fibre laser', *Electronics Letters*, **25**, 1082, 1989.
102. J. Y. Allain, M. Monerie and H. Poignant; 'Tunable green up-conversion erbium fibre laser', *Electronics Letters*, **28**, 111, 1992.
103. D. Piehler and D. Craven; '11.5 mW green InGaAs laser pumped erbium fibre laser', *Electronics Letters*, **30**, 1759, 1994.
104. J. Massicott, M. C. Brierley, R. Wyatt, S. T. Davey and D. Szebesta; 'Low threshold, diode pumped operation of a green Er^{3+} doped fluoride fibre laser', *Electronics Letters*, **29**, 2119, 1993.
105. D. C. Hanna, I. M. Jauncey, R. M. Percival, I. R. Perry, R. G. Smart, P. J. Suni, J. E. Townsend and A. C. Tropper; 'Continuous-wave oscillation of a monomode thulium-doped fibre laser', *Electronics Letters*, **24**, 1222, 1988.
106. D. C. Hanna, M. J. McCarthy, I. R. Perry and P. J. Suni; 'Efficient high power continuous wave operation of a monomode Tm-doped fibre laser at 2 μm pumped by Nd:YAG laser at 1.064 μm', *Electronics Letters*, **25**, 1365, 1989.
107. D. C. Hanna, R. M. Percival, R. G. Smart and A. Tropper; 'Efficient and tunable operation of a Tm-doped fibre laser', *Optics Communications*, **75**, 283, 1990.
108. W. L. Barnes and J. E. Townsend; 'Highly tunable and efficient diode pumped operation of Tm^{3+} doped fibre lasers', *Electronics Letters*, **26**, 746, 1990.

REFERENCES

109. R. G. Smart, J. N. Carter, A. C. Tropper and D. C. Hanna; 'Continuous-wave oscillation of Tm^{3+}-doped fluorozirconate fibre lasers at around 1.47 μm, 1.9 μm and 2.3 μm when pumped at 790 nm', *Optics Communications*, **82**, 563, 1991.
110. J. Y. Allain, M. Monerie and H. Poignant; 'Tunable cw lasing around 0.82, 1.48, 1.88 and 2.35 μm in thulium doped fluorozirconate fibre', *Electronics Letters*, **25**, 1660, 1989.
111. S. Boj, E. Delevaque, J. Y. Allain, J. F. Bayon, P. Niay and P. Bernage; 'High-efficiency diode pumped thulium-doped silica fibre lasers with intracore Bragg gratings in the 1.9–2.1 μm band', *Electronics Letters*, **30**, 1019, 1994.
112. D. C. Hanna, I. R. Perry, J. R. Lincoln and J. E. Townsend; 'A 1-watt thulium doped cw fibre laser operating at 2 μm', *Optics Communications*, **80**, 52, 1990.
113. R. M. Percival, D. Szebesta and S. T. Davey; 'Highly efficient and tunable operation of a 2 colour Tm-doped fluoride fibre laser', *Electronics Letters*, **28**, 671, 1992.
114. R. Allen and L. Esterowitz; 'Cw diode pumped 2.3 μm fibre laser', *Applied Physics Letters*, **55**, 721, 1989.
115. R. M. Percival, S. F. Carter, D. Szebesta, S. T. Davey and W. A. Stallard; 'Thulium doped monomode fluoride fibre laser broadly tunable from 2.25–2.5 μm', *Electronics Letters*, **27**, 1912, 1991.
116. T. Yamamoto, Y. Miyajima, T. Komukai and T. Sugawa; '1.9 μm Tm-doped fibre amplifier and laser pumped at 1.58 μm', *Electronics Letters*, **29**, 986, 1993.
117. R. M. Percival, D. Szebesta, C. P. Seltzer, S. D. Perrin, S. T. Davey and M. Louka; 'A 1.6 μm semiconductor diode pumped thulium doped fluoride fibre laser and amplifier of very high efficiency', *Electronics Letters*, **29**, 2110, 1993.
118. R. M. Percival, D. Szebesta, C. P. Seltzer, S. D. Perrin, S. T. Davey and M. Louka; 'A 1.6 μm pumped 1.9 μm thulium doped fluoride fibre laser and amplifier of very high efficiency', *IEEE Journal of Quantum Electronics*, **QE-31**, 489, 1995.
119. J. Y. Allain, M. Monerie and H. Poignant; 'Tunable cw lasing around 0.82, 1.48, 1.88 and 2.35 μm in thulium doped fluorozirconate fibre', *Electronics Letters*, **25**, 1660, 1989.
120. R. Allen, L. Esterowitz and I. Aggarwal; 'An efficient 1.46 μm thulium laser via a cascade process', *IEEE Journal of Quantum Electronics*, **29**, 303, 1993.
121. R. M. Percival, D. Szebesta and S. T. Davey; 'Highly efficient cw cascade operation of 1.47 and 1.82 μm transitions in Tm-doped fluoride fibre laser', *Electronics Letters*, **28**, 1866, 1992.
122. Y. Miyajima, T. Komukai and T. Sugawa; '1 W Tm-doped fibre laser at 1.47 μm', *Electronics Letters*, **29**, 660, 1993.
123. M. L. Dennis, J. W. Dixon and I. Aggarwal; 'High-power up-conversion lasing at 810 nm in Tm-ZBLAN fibre', *Electronics Letters*, **30**, 136, 1994.
124. J. Y. Allain, M. Monerie and H. Poignant; 'Blue up-conversion fluorozirconate fibre laser', *Electronics Letters*, **26**, 166, 1990.
125. S. G. Grubb, K. W. Bennett, R. S. Cannon and W. F. Humer; 'CW room temperature blue up-conversion fibre laser', *Electronics Letters*, **28**, 1243, 1992.
126. P. R. Barber, H. M. Pask, C. J. MacKechnie, D. C. Hanna, A. C. Tropper, J. Massicott, S. T. Davey and D. Szebesta; 'Improved performance of Tm^{3+} and Pr^{3+}-doped ZBLAN fibres', Paper CMF3, in *Proceedings of European Conference on Lasers and Electro-Optics ECLEO*, Amsterdam, 1994.
127. G. Tohmon, J. Ohya, H. Sato and T. Uno; 'Increased efficiency and decreased threshold in Tm-ZBLAN fibre laser co-pumped by 1.1 μm and 0.68 μm light', *IEEE Photonics Technology Letters*, **7**, 742, 1995.

128. D. C. Hanna, R. M. Percival, R. G. Smart, J. E. Townsend and A. C. Tropper; 'Continuous-wave oscillation of holmium-doped silica fibre laser', *Electronics Letters*, **25**, 593, 1989.
129. M. C. Brierley, P. W. France and C. A. Miller; 'Lasing at 2.08 µm and 1.38 µm in a holmium doped fluorozirconate fibre laser', *Electronics Letters*, **24**, 539, 1988.
130. R. M. Percival, D. Szebesta, S. T. Davey, N. A. Swain and T. A. King; 'High efficiency cw operation of 890 nm pumped holmium fluoride fibre laser', *Electronics Letters*, **28**, 2063, 1992.
131. L. Wetenkamp; 'Efficient cw operation of a 2.9 µm Ho^{3+}-doped fluorozirconate fibre laser pumped at 640 nm', *Electronics Letters*, **26**, 883, 1990.
132. J. Y. Allain, M. Monerie and H. Poignant; 'Room temperature cw tunable green upconversion holmium fibre laser', *Electronics Letters*, **26**, 261, 1990.
133. D. S. Funk, S. B. Stevens and J. G. Eden; 'Excitation-spectra of the green Ho-fluorozirconate glass-fibre laser', *IEEE Photonics Technology Letters*, **5**, 154, 1993.
134. J. Y. Allain, M. Monerie and H. Poignant; 'Tunable cw lasing around 610, 635, 695, 715, 885, and 910 nm in praseodymium-doped fluorozirconate fibre', *Electronics Letters*, **27**, 189, 1991.
135. R. G. Smart, D. C. Hanna, A. C. Tropper, S. T. Davey, S. F. Carter and D. Szebesta; 'Cw room-temperature upconversion lasing at blue, green and red wavelengths in infrared-pumped Pr^{3+} doped fluoride fibre', *Electronics Letters*, **27**, 1307, 1991.
136. R. G. Smart, J. N. Carter, A. C. Tropper, D. C. Hanna, S. T. Davey, S. F. Carter and D. Szebesta; 'Cw room-temperature operation of praseodymium-doped fluorozirconate glass-fibre lasers in the blue–green, green and red spectral regions', *Optics Communications*, **86**, 337, 1991.
137. A. C. Tropper, J. N. Carter, R. D. T. Lauder, D. C. Hanna, S. T. Davey and D. Szebesta; 'Analysis of blue and red laser performance of the infrared pumped praseodymium-doped fluoride fibre laser', *Journal of the Optical Society of America B*, **11**, 886, 1994.
138. Y. Shi, C. V. Poulsen, M. Sejka, M. Ibsen and O. Poulsen; 'Tunable Pr^{3+} doped silica-based fibre laser', *Electronics Letters*, **29**, 1426, 1993.
139. Y. Ohishi, T. Kanamori and S. Takahashi; 'Pr^{3+}-doped fluoride single mode fibre laser', *IEEE Photonics Technology Letters*, **3**, 688, 1991.
140. H. Doring, J. Peupelmann and F. Wenzel; 'Pr^{3+} doped 1.3 µm fibre laser using direct coated dichroic mirrors', *Electronics Letters*, **31**, 1068, 1995.
141. Y. Shi, C. V. Poulsen, M. Sejka, M. Ibsen and O. Poulsen', 'Tunable Pr^{3+} doped silica based fibre laser', *Electronics Letters*, **29**, 1426, 1993.
142. F. Sanchez, P. LeBoudec, P. L. François and G. M. Stephan; 'Self-pulsing in Er^{3+} doped fibre lasers: theory and experiment' Paper LDmoP141, in *Proceedings of European Quantum Electronics Conference*, Florence, 1993.
143. D. C. Hanna, R. M. Percival, I. R. Perry, R. G. Smart, P. J. Suni, J. E. Townsend and A. C. Tropper; 'Continuous-wave oscillation of a monomode ytterbium-doped fibre laser', *Electronics Letters*, **24**, 1111, 1988.
144. J. R. Armitage, R. Wyatt, B. J. Ainslie and S. P. Craig-Ryan; 'Highly efficient 980 nm operation of an Yb^{3+}-doped silica fibre laser', *Electronics Letters*, **25**, 299, 1989.
145. C. J. MacKechnie, W. L. Barnes, D. C. Hanna and J. E. Townsend; 'High power ytterbium (Yb^{3+})-doped fibre laser operating in the 1.12 µm region', *Electronics Letters*, **29**, 52, 1993.

REFERENCES

146. J. Y. Allain, J. F. Bayon, M. Monerie, P. Bernage and P. Niay; 'Ytterbium doped silica fibre laser with intracore Bragg gratings operating at 1.02 µm', *Electronics Letters*, **29**, 309, 1993.
147. J. M. Dawes, H. M. Pask, J. L. Archambault, J. E. Townsend, D. C. Hanna, L. Reekie and A. C. Tropper; 'Single frequency lasers and efficient cladding-pumped lasers using Yb^{3+}-doped silica fibre', Paper CHJ1, in *Proceedings of European Conference on Lasers and Electro-Optics ECLEO*, Amsterdam, 1994.
148. J. Y. Allain, M. Monerie and H. Poignant; 'Ytterbium-doped fluoride fibre laser operating at 1.02 µm', *Electronics Letters*, **28**, 988, 1992.
149. J. Y. Allain, M. Monerie and H. Poignant; 'High-efficiency ytterbium-doped fluoride fibre laser', *Journal of Non-Crystalline Solids*, **161**, 270, 1993.
150. M. E. Fermann, D. C. Hanna, D. P. Shepard, P. J. Suni and J. E. Townsend; 'Efficient operation of an Yb-sensitised Er fibre laser at 1.56 µm', *Electronics Letters*, **24**, 1135, 1988.
151. G. T. Maker and A. I. Ferguson; '1.56 µm Yb-sensitized Er fibre laser pumped by diode-pumped Nd–YAG and Nd–YLF lasers', *Electronics Letters*, **24**, 1160, 1988.
152. D. C. Hanna, R. M. Percival, I. R. Perry, R. G. Smart and A. C. Tropper; 'Efficient operation of an Yb-sensitized Er fibre laser pumped in 0.8-µm region', *Electronics Letters*, **24**, 1068, 1988.
153. J. D. Minelly, W. L. Barnes, R. I. Laming, P. R. Morkel, J. E. Townsend, S. G. Grubb and D. N. Payne; 'Diode-array pumping of Er^{3+}/Yb^{3+} co-doped fibre lasers and amplifiers', *IEEE Photonics Technology Letters*, **5**, 301, 1993.
154. J. Y. Allain, M. Monerie and H. Poignant; 'Red upconversion Yb-sensitised Pr fluoride fibre laser pumped in 0.8 µm region', *Electronics Letters*, **27**, 1156, 1991.
155. P. Xie and T. R. Gosnell; 'Room-temperature up-conversion fibre laser tunable in the red, orange green and blue spectral regions', *Optics Letters*, **20**, 1014, 1995.
156. A. Kermaoui, J. P. Denis, G. Ozen, P. Goldner, F. Pelle, B. Blanzat; 'Effect of Yb^{3+} on red to blue conversion fluorescence of Tm^{3+} in fluorozirconate glass', *Optics Communications*, **110**, 581, 1994.
157. J. Schneider, D. Hauschild, C. Frerichs and L. Wetenkamp; 'Highly efficient Er^{3+}/Pr^{3+}-codoped cw fluorozirconate fibre laser operating at 2.7 µm', *International Journal of Infrared and Millimeter Waves*, **15**, 1907, 1994.
158. L. Wetenkamp, G. F. West and H. Tobben; 'Co-doping effects in Er^{3+}-doped and Ho^{3+}-doped ZBLAN glasses', *Journal of Non-Crystalline Solids*, **140**, 25, 1992.
159. R. M. Percival, D. Szebesta, S. T. Davey, N. A. Swain and T. A. King; 'Thulium sensitised holmium-doped fibre laser of high efficiency', *Electronics Letters*, **28**, 2231, 1991.
160. J. X. Wang, M. Ahmad and T. A. King; 'Theoretical modelling of thulium-sensitised holmium continuous-wave fibre lasers', *Journal of Modern Optics*, **41**, 1457, 1994.
161. I. M. Jauncey, L. Reekie, R. J. Mears, D. N. Payne, C. J. Rowe, D. C. J. Ried, I. Bennion and C. Edge; 'Narrow linewidth fibre laser with integral fibre grating', *Electronics Letters*, **22**, 987, 1986.
162. I. M. Jauncey, L. Reekie, J. E. Townsend, D. N. Payne and C. J. Rowe; 'Single longitudinal mode operation of a Nd^{3+}-doped fibre laser', *Electronics Letters*, **24**, 24, 1988.
163. H. Sabert; 'Tunable narrow-band Nd^{3+} fibre laser', *Applied Physics Letters*, **59**, 2067, 1991.

164. H. Sabert, A, Koch and R. Ulrich; 'Reduction of spatial hole burning by single phase modulator in linear Nd^{3+} fibre laser', *Electronics Letters*, **27**, 2176, 1991.
165. H. Sabert and R. Ulrich; 'Spatial hole burning in Nd^{3+} fibre lasers suppressed by push pulse phase modulation', *Applied Physics Letters*, **58**, 2323, 1991.
166. H. Sabert; 'Continuous electronic tuning of a narrow-band Nd^{3+}-fibre laser', *Applied Physics Letters*, **62**, 452, 1993.
167. H. Sabert; 'Suppression of mode jumps in a single mode fibre laser', *Optics Letters*, **19**, 111, 1994.
168. C. R. O'Cochlain and R. J. Mears; 'Broad-band tunable single frequency diode-pumped erbium doped fibre laser', *Electronics Letters*, **28**, 124, 1992.
169. A. Gloag, L. Zhang, I. Bennion and N. Langford; 'Single-frequency travelling-wave erbium doped fibre laser incorporating a fibre Bragg grating', to be published in *Optics Communications*.
170. G. A. Ball, W. W. Morey and W. H. Glen; 'Standing wave monomode erbium fibre laser', *IEEE Photonics Technology Letters*, **3**, 613, 1991.
171. J. L. Zyskind, V. Mixrahi, D. J. DiGiovanni and J. W. Sulhoff; 'Short single frequency erbium doped fibre laser', *Electronics Letters*, **28**, 1385, 1992.
172. G. A. Ball and W. W. Morey; 'Continuously tunable single mode erbium fibre laser', *Optics Letters*, **17**, 420, 1992.
173. S. V. Chernikov, J. R. Taylor and R. Kashyap; 'Coupled-cavity erbium fibre lasers incorporating fibre grating reflectors', *Optics Letters*, **18**, 2023, 1993.
174. S. V. Chernikov, R. Kashyap, P. F. McKee and J. R. Taylor; 'Dual-frequency all fibre grating laser source', *Electronics Letters*, **29**, 1089, 1993.
175. G. A. Ball, G. Hullallen, G. Holten and W. W. Morey; 'Low noise single frequency linear fibre laser', *Electronics Letters*, **29**, 1623, 1993.
176. G. A. Ball and W. W. Morey; 'Compression tuned single frequency Bragg grating fibre laser', *Optics Letters*, **19**, 1979, 1994.
177. G. A. Ball, C. E. Holton, G. Hullallen and W. W. Morey; '60 mW 1.5 µm single frequency low noise fibre laser MOPA', *IEEE Photonics Technology Letters*, **6**, 192, 1994.
178. G. A. Ball, G. Hullallen and J. Livas; 'Frequency noise of a Bragg grating fibre laser', *Electronics Letters*, **30**, 1229, 1994.
179. V. Mizrahi, D. J. DiGiovanni, R. M. Atkins, S. G. Grubb, Y. K. Park and J. M. P. Delavaux; 'Stable single mode erbium fibre grating laser for digital communication', *Journal of Lightwave Technology*, **11**, 2021, 1993.
180. J. L. Wagener, P. F. Wysocki, M. J. F. Digonnet and H. J. Shaw; 'Modeling of ion-pairs in erbium-doped fibre amplifiers', *Optics Letters*, **19**, 347, 1994.
181. J. T. Kringlebotn, P. R. Morkel, L. Reekie, J. L. Archambault and D. N. Payne; 'Efficient diode pumped single frequency erbium-ytterbium fibre laser', *IEEE Photonics Technology Letters*, **5**, 1162, 1993.
182. J. T. Kringlebotn, J. L. Archambault, L. Reekie, J. E. Townsend, G. G. Vienne and D. N. Payne; 'Highly efficient low noise grating feedback Er^{3+} Yb^{3+} codoped fibre laser', *Electronics Letters*, **30**, 972, 1994.
183. M. Ibsen, B. J. Eggleton, M. G. Sceats and F. Ouellette; 'Broadly tunable DBR fibre laser using sampled fibre Bragg gratings', *Electronics Letters*, **31**, 37, 1995.
184. M. Sejka, P. Varming, J. Hubner and M. Kristensen; 'Distributed erbium doped fibre laser', *Electronics Letters*, **31**, 1445, 1995.
185. W. H. Loh and R. I. Laming; '1.55 µm phase shifted distributed feedback fibre laser', *Electronics Letters*, **31**, 1440, 1995.

186. A. Asseh, H. Storoy, J. T. Kringlebotn, W. Margulis, B. Sahlgren, R. Stubbe and G. Edwall; '10 cm Yb^{3+} DFB fibre laser with permanent phase shifted grating', *Electronics Letters*, **31**, 969, 1995.
187. K. Iwatsuki, H. Okamura and M. Saruwatari; 'Wavelength tunable single frequency and single polarisation Er-doped fibre ring laser with 1.4 kHz linewidth', *Electronics Letters*, **26**, 2033, 1990.
188. C. Y. Yue, G. W. Schinn, J. W. Y. Lit and Z. Zhang; 'Single mode EFL using an all-fibre subresonator', Paper WK7, in the *Proceedings of the OFC '94 Conference*, San Jose, 1994.
189. J. L. Zhang and J. W. Y. Lit; 'All-fibre compound ring resonator with a ring filter', *Journal of Lightwave Technology*, **12**, 1256, 1994.
190. A. Gloag, K. McCallion, W. Johnstone and N. Langford; 'Tunable single frequency erbium fibre laser using an overlay bandpass filter', *Applied Physics Letters*, **66**, 3263, 1995.
191. A. Gloag, K. McCallion, W. Johnstone and N. Langford; 'A continuously tunable single frequency erbium fibre laser', accepted for publication in the *Journal of The Optical Society of America B—Optical Physics*, **13**, 921, 1996.
192. A. D. Kersey and W. W. Morey; 'Multiplexed Bragg grating fibre laser strain-sensor system with mode-locked interrogation', *Electronics Letters*, **29**, 112, 1993.
193. See for example G. P. Agrawal; *Non-Linear Optical Fibres*, Ch. 2, Academic Press, New York, 1987.
194. F. Fontana, G. Bordogna, G. Grasso, M. Romagnoli, M. Midrio and P. Franco; 'Evaluation and measurement of the resonant group-velocity dispersion in erbium doped fibre lasers', *Optics Letters*, **18**, 2011, 1993.
195. See for example G. P. Agrawal; *Non-Linear Optical Fibres*, Ch. 5, Academic Press, New York, 1987.
196. See for example G. P. Agrawal; *Non-Linear Optical Fibres*, Ch. 7, Academic Press, New York, 1987.
197. H. G. Winful; 'Self-induced polarisation changes in birefringent optical fibres', *Applied Physics Letters*, **47**, 213, 1985.
198. See for example G. P. Agrawal; *Non-Linear Optical Fibres*, Ch. 5, Academic Press, New York, 1987.
199. N. J. Smith, K. J. Blow and I. Andonovic; 'Side-band generation through perturbations to the average soliton model', *Journal of Lightwave Technology*, **10**, 1329, 1992.
200. W. J. Tomlinson, R. H. Stolen and C. V. Shank; 'Compression of optical pulses chirped by self-phase modulation in fibres', *Journal of the Optical Society of America*, **B1**, 139, 1984.
201. E. B. Treacy; 'Optical pulse compression with diffraction gratings', *IEEE Journal of Quantum Electronics*, **QE-5**, 454, 1969.
202. O. E. Martinez, J. P. Gordon and R. L. Fork; 'Negative group velocity dispersion using refraction', *Journal of the Optical Society of America*, **B1**, 1003, 1984.
203. F. Ouellette; 'All fibre filters for efficient dispersion compensation', *Optics Letters*, **16**, 303, 1991.
204. See for example A. E. Siegman; *Lasers*, Ch. 27, University Science Books, 1988.
205. See for example A. E. Siegman; *Lasers*, Ch. 28, University Science Books, 1988.
206. E. J. Greer and K. Smith; 'All-optical FM mode-locking of fibre laser', *Electronics Letters*, **28**, 1741, 1993.
207. B. P. Nelson, K. Smith and K. J. Blow; 'Mode-locked erbium fibre laser using all-optical nonlinear loop modulator', *Electronics Letters*, **28**, 656, 1993.

208. N. J. Doran, D. S. Forester and B. K. Nayar; 'Experimental investigation of all-optical switching in fibre loop mirror device', *Electronics Letters*, **25**, 267, 1989.
209. K. J. Blow, N. J. Doran and B. K. Nayar; 'Experimental demonstration of optical soliton switching in an all-fibre nonlinear sagnac interferometer', *Optics Letters*, **14**, 754, 1989.
210. K. Smith, N. J. Doran and P. G. J. Wigley; 'Pulse shaping, compression, and pedestal suppression employing nonlinear-optical loop mirror', *Optics Letters*, **15**, 1232, 1990.
211. M. N. Islam, E. R. Sundermna, R. H. Stolen, W. Pleibel and J. R. Simpson; 'Soliton switching in a fibre loop mirror', *Optics Letters*, **14**, 811, 1989.
212. M. E. Fermann, F. Haberl, M. Hofer and H. Hochreiter; 'Nonlinear amplifying loop mirror', *Optics Letters*, **15**, 752, 1990.
213. D. J. Richardson, R. I. Lamming and D. N. Payne; 'Very low threshold Sagnac switch incorporating an erbium doped fibre amplifier', *Electronics Letters*, **26**, 1779, 1990.
214. A. W. O'Neill and R. P. Webb; 'All-optical loop mirror switch employing an asymmetric amplifier attenuator combination', *Electronics Letters*, **26**, 2009, 1990.
215. A. G. Bulushev, E. M. Dianov and O. G. Okhotnikov; 'Self-starting mode-locked laser with a non-linear ring resonator', *Optics Letters*, **15**, 968, 1990.
216. I. P. Alcock, A. I. Ferguson, D. C. Hanna and A. C. Tropper; 'Mode-locking of a neodymium-doped monomode fibre laser', *Electronics Letters*, **22**, 268, 1986.
217. M. W. Phillips, A. I. Ferguson and D. C. Hanna; *Optics Letters*, **14**, 21, 1989.
218. M. Hofer, M. E. Fermann, F. Haberl and J. E. Townsend; 'Active-mode-locking of a neodymium-doped fibre laser using intracavity pulse compression', *Optics Letters*, **15**, 1467, 1990.
219. M. E. Fermann, M. Hofer, F. Haberl, A. J. Schmidt and L. Turi; 'Additive-pulse-compression of a neodymium fibre laser', *Optics Letters*, **16**, 244, 1991.
220. M. Hofer, M. E. Fermann, G. Haberl, M. H. Ober and A. J. Schmidt; 'Mode-locking with cross-phase and self-phase modulation', *Optics Letters*, **16**, 502, 1991.
221. F. Haberl, M. H. Ober, M. Hofer, M. E. Fermann, E. Wintner and A. J. Schmidt; 'Low-noise operation modes of a passively mode-locked fibre laser', *IEEE Photonics Technology Letters*, **3**, 1071, 1991.
222. M. Hofer, M. H. Ober, F. Haberl and M. E. Fermann; 'Characterisation of ultrashort pulse formation in passively mode-locked fibre lasers', *IEEE Journal of Quantum Electronics*, **QE-28**, 720, 1992.
223. M. E. Fermann, M. J. Andrejco, Y. Silberberg and M. L. Stock; 'Generation of pulses shorter than 200 fs from a passively mode-locked Er fibre laser', *Optics Letters*, **18**, 48, 1993.
224. M. E. Fermann, M. J. Andrejco, Y. Silberberg and M. L. Stock; 'Passive mode-locking by using non-linear polarisation evolution in a polarisation-maintaining erbium-doped fibre', *Optics Letters*, **18**, 894, 1993.
225. M. W. Phillips, A. I. Ferguson and D. C. Hanna; 'Frequency-modulation mode-locking of a Nd^{3+}-doped fibre laser', *Optics Letters*, **14**, 219, 1989.
226. D. M. Pataca, M. L. Rocha, K. Smith, T. J. Whitley and R. Wyatt; 'Actively mode-locked Pr^{3+}-doped fluoride fibre laser', *Electronics Letters*, **30**, 964, 1994.
227. R. P. Davey, N. Langford and A. I. Ferguson; 'Sub-picosecond pulse generation from an erbium fibre laser', *Electronics Letters*, **27**, 729, 1991.
228. R. P. Davey, N. Langford and A. I. Ferguson; 'the role of polarisation rotation in the mode-locking of an erbium fibre laser', *Electronics Letters*, **29**, 729, 1993.

229. R. P. Davey, A. Gloag, N. Langford and A. I. Ferguson; 'The role of polarisation rotation in the mode-locking of an erbium fibre laser', Paper TuB20 in the *Proceedings of the Third International Conference on Non-Linear Optical Waveguides*, Cambridge, UK, 1993.
230. R. P. Davey, N. Langford and A. I. Ferguson; 'Interacting solitons in erbium fibre laser', *Electronics Letters*, **27**, 1251, 1991.
231. G. Geister and R. Ulrich; 'Neodymium-fibre laser with integrated-optic mode locker', *Optics Communications*, **68**, 187, 1988.
232. G. Geister and R. Ulrich; 'Integrated optical Q-switch/mode-locker for a Nd^{3+} fibre laser', *Applied Physics Letters*, **56**, 509, 1990.
233. D. B. Patterson, A. A. Godil, G. S. Kino and B. T. Khuri-Yakub; 'Detachable 400-mHz acoustooptic phase modulator for a single-mode optical fibre', *Optics Letters*, **14**, 248, 1989.
234. M. W. Phillips, A. I. Ferguson, G. S. Kino and D. B. Patterson; 'Mode-locked fibre laser with a fiber phase modulator', *Optics Letters*, **14**, 248, 1989.
235. M. W. Phillips, A. I. Ferguson and D. B. Patterson; 'Diode-pumped fm mode-locked fibre laser with coupled cavity bandwidth selection', *Optics Communications*, **75**, 33, 1990.
236. J. D. Kafka, T. Baer and D. W. Hall; 'Mode-locked erbium doped fibre laser with soliton pulse shaping', *Optics Letters*, **14**, 1269, 1989.
237. J. B. Schlager, Y. Yamubayashi, D. L. Franzen and R. I. Juneau; 'Mode-locked, long-cavity, erbium fibre lasers with subsequent soliton-like compression', *IEEE Photonics Technology Letters*, **1**, 264, 1989.
238. J. B. Schlager, P. D. Hale and D. L. Franzen; 'Subpicosecond pulse-compression and Raman generation using a mode-locked erbium-doped fibre laser-amplifier', *IEEE Photonics Technology Letters*, **2**, 562, 1990.
239. J. B. Schlager, S. Kawanishi and M. Saruwatari; 'Dual wavelength pulse generation using mode-locked erbium doped fibre ring laser', *Electronics Letters*, **27**, 2072, 1991.
240. H. Takara, S. Kawanishi, M. Surawatari and J. B. Schlager; 'Multiple wavelength birefringent cavity mode-locked fibre laser', *Electronics Letters*, **28**, 2274, 1992.
241. E. Greer, R. Wyatt, P. Wheatley, N. J. Doran and M. Lawrence; 'Totally integrated erbium fibre soliton laser pumped by a laser diode', *Electronics Letters*, **27**, 244, 1991.
242. A. Takada and H. Miyazawa; '30 GHz picosecond pulse generation from actively mode-locked erbium-doped fibre laser', *Electronics Letters*, **26**, 216, 1990.
243. A. Yoshida, Y. Kimura and M. Nakazawa; '20 GHz, 1.8 ps pulse generation from a regeneratively mode-locked erbium-doped fibre laser and its femtosecond pulse compression', *Electronics Letters*, **31**, 377, 1995.
244. R. P. Davey, K. Smith and A. McGuire; 'High speed, mode-locked tunable integrated erbium fibre laser', *Electronics Letters*, **28**, 482, 1992.
245. R. P. Davey, R. P. E. Fleming, K. Smith, R. Kayshap and J. R. Armitage; 'Mode-locked erbium fibre laser with wavelength selection by means of a fibre Bragg grating reflector', *Electronics Letters*, **27**, 2087, 1991.
246. E. J. Greer, Y. Kimura, E. Yoshida and M. Nakazawa; 'Generation of 1.2 ps, 10 GHz pulse train from all optically mode-locked erbium fibre ring laser with active non-linear polarisation rotation', *Electronics Letters*, **30**, 1764, 1994.
247. M. L. Stock, L. M. Yang, M. J. Andrejco and M. E. Fermann; 'Synchronous mode-locking using pump induced phase modulation', *Optics Letters*, **18**, 1529, 1993.

248. D. U. Noske, A. Boskovic, M. J. Guy and J. R. Taylor; 'Synchronously pumped picosecond ytterbium–erbium fibre laser', *Electronics Letters*, **29**, 1863, 1993.
249. D. M. Patrick; 'Mode-locked ring laser using nonlinearity in a semiconductor laser amplifier', *Electronics Letters*, **30**, 43, 1994.
250. M. Margalit, M. Orenstein and G. Eisenstein; 'High repetition-rate mode-locked Er-doped fibre laser by harmonic injection locking', *Optics Letters*, **20**, 1791, 1995.
251. D. M. Pataca, M. L. Rocha, R. Kashyap and K. Smith; 'Bright and dark pulse generation in an optically mode-locked fibre laser at 1.3 µm', *Electronics Letters*, **31**, 35, 1995.
252. M. H. Ober, M. Hofer and M. E. Fermann; '42 fs pulse generation from a mode-locked fibre laser started with a moving mirror', *Optics Letters*, **18**, 367, 1993.
253. N. Langford: unpublished data.
254. P. N. Kean, X. Zhu, D. W. Crust, R. S. Grant, N. Langford and W. Sibbett; 'Enhanced mode-locking of colour-centre lasers', *Optics Letters*, **14**, 39, 1989.
255. E. P. Ippen, H. A. Haus and L. Y. Lui; 'Additive pulse mode-locking', *Journal of the Optical Society of America*, **B6**, 1736, 1989.
256. G. Sargsan, U. Stamm, C. Unger, C. Zschocke and M. Ledig; 'Characteristics of a neodymium-doped fibre laser mode-locked with a linear external cavity', *Optics Communications*, **86**, 480, 1991.
257. C. Unger, G. Sargsan, U. Stamm and M. Muller; 'Coupled cavity mode-locking of a neodymium doped fibre laser', *Institute of Physics Conference Series*, **126**, 15, 1992.
258. Y. Shi, C. V. Poulsen, M. Sejka and O. Poulsen; 'Mode-locked Pr^{3+} doped silica fibre laser with an external cavity', *Journal of Lightwave Technology*, **12**, 749, 1994.
259. I. N. Duling; 'All-fibre ring soliton laser mode-locked with a non-linear mirror', *Optics Letters*, **17**,
260. M. Nakazawa, E. Yoshida and Y. Kimura; 'Low threshold, 290 fs erbium-doped fibre laser with a nonlinear amplifying loop mirror pumped by InGaAsP laser diodes', *Applied Physics Letters*, **59**, 2073, 1991.
261. D. J. Richardson, R. I. Laming, D. N. Payne, V. Matsas and M. W. Phillips; 'A self-starting passively mode-locked erbium fibre laser based on the amplifying Sagnac switch', *Electronics Letters*, **27**, 738, 1991.
262. D. J. Richardson, R. I. Laming, D. N. Payne, V. Matsas and M. W. Phillips; 'Pulse repetition rates in passive self-starting femtosecond soliton fibre laser', *Electronics Letters*, **27**, 1451, 1991.
263. D. J. Richardson, R. I. Laming, D. N. Payne, M. W. Phillips and V. Matsas; '320 fs soliton generation with passively mode-locked erbium fibre laser', *Electronics Letters*, **27**, 730, 1991.
264. S. J. Frisken, C. A. Telford, R. A. Betts and P. S. Atherton; 'Passively mode-locked erbium-doped fibre laser with non-linear fibre mirror', *Electronics Letters*, **27**, 887, 1991.
265. M. L. Dennis and I. N. Duling III; 'High repetition rate figure 8 laser with extra-cavity feedback', *Electronics Letters*, **28**, 1894, 1992.
266. E. Yoshida, Y. Kimura and M. Nakazawa; 'Laser diode pumped femtosecond erbium doped fibre laser with a sub-ring cavity for repetition rate control', *Applied Physics Letters*, **60**, 932, 1992.
267. A. B. Grudinin, D. J. Richardson and D. N. Payne; 'Energy quantisation in figure 8 fibre laser', *Electronics Letters*, **28**, 67, 1992.

268. M. J. Guy, D. U. Noske and J. R. Taylor; 'Generation of femtosecond soliton pulses by passive mode-locking of an ytterbium erbium figure 8 fibre laser', *Optics Letters*, **18**, 1447, 1993.
269. S. Wu, J. Strait, R. L. Fork and T. F. Morse; 'High power passively mode-locked Er-doped fibre laser with a non-linear optical loop mirror', *Optics Letters*, **18**, 1444, 1993.
270. E. Yoshida, Y. Kimura and M. Nakazawa; 'Femtosecond erbium doped fibre lasers and a soliton compression technique', *Japanese Journal of Applied Physics Part 1*, **32**, 3461, 1993.
271. M. L. Dennis and I. N. Duling III; 'Intracavity dispersion measurement in mode-locked fibre laser', *Electronics Letters*, **29**, 409, 1993.
272. M. Nakazawa, E. Yoshida and Y. Kimura; 'Generation of 98 fs pulses directly from an erbium doped fibre ring laser at 1.57 µm', *Electronics Letters*, **29**, 63, 1993.
273. M. L. Dennis and I. N. Duling; 'Role of dispersion in limiting pulse-width in fibre lasers', *Applied Physics Letters*, **62**, 2911, 1993.
274. D. U. Noske, N. Pandit and J. R. Taylor; 'Source of spectral and temporal instability in soliton fibre lasers', *Optics Letters*, **17**, 1515, 1992.
275. M. J. Guy and J. R. Taylor; 'Simultaneous dual-polarisation operation of a diode-pumped femtosecond fibre laser', *Electronics Letters*, **29**, 2044, 1993.
276. H. Lin, D. K. Donald and W. V. Sorin; 'Optimisation polarisation states in a figure 8 laser using a non-reciprocal phase shifter', *Journal of Lightwave Technology*, **12**, 1121, 1994.
277. M. L. Dennis and I. N. Duling; 'Experimental study of side-band generation in femtosecond fibre lasers', *IEEE Journal of Quantum Electronics*, **QE-30**, 1469, 1994.
278. G. Town, J. Chow and M. Romagnoli; 'Sliding frequency figure 8 optical fibre laser', *Electronic Letters*, **31**, 1452, 1995.
279. A. Boskovic, S. V. Chernikov and J. R. Taylor; 'Femtosecond figure of 8 Yb–Er fibre laser incorporating a dispersion decreasing fibre', *Electronics Letters*, **31**, 1446, 1995.
280. K. Tamura, H. A. Haus and E. P. Ippen; 'Self-starting additive pulse mode-locked erbium fibre ring laser', *Electronics Letters*, **28**, 2226, 1992.
281. D. U. Noske, N. Pandit and J. R. Taylor; 'Subpicosecond soliton pulse formation from self-mode-locked erbium ring fibre laser using intensity dependent polarisation rotation', *Electronics Letters*, **28**, 2185, 1992.
282. M. Nakazawa, E. Yoshida, T. Sugawa and Y. Kimura; 'Continuum suppressed uniformly repetitive 136 fs pulse generation from an erbium fibre laser with non-linear polarisation rotation', *Electronics Letters*, **29**, 1327, 1993.
283. V. J. Matsas, D. J. Richardson, T. P. Newson and D. N. Payne; 'Characterisation of a self-starting, passively mode-locked fibre ring laser that exploits non-linear polarisation evolution', *Optics Letters*, **18**, 358, 1993.
284. A. B. Grudinin, D. J. Richardson and D. N. Payne; 'Passive harmonic mode-locking of fibre soliton ring lasers', *Electronics Letters*, **29**, 1860, 1993.
285. W. H. Loh, A. B. Grudinin and D. N. Payne; 'Optically controlled wavelength adjustable passively mode-locked erbium doped fibre ring laser', *Electronics Letters*, **30**, 413, 1994.
286. H. A. Haus, E. P. Ippen and K. Tamura; 'Additive pulse mode-locking in fibre lasers', *IEEE Journal of Quantum Electronics*, **QE-30**, 200, 1994.
287. K. Tamura, K. Nelson, H. A. Haus and E. P. Ippen; 'Soliton versus non-soliton operation of fibre ring lasers', *Applied Physics Letters*, **64**, 149, 1994.

288. K. Tamura, Y. Kimura and M. Nakazawa; 'Femtosecond pulse generation over 82 nm wavelength span fron passively mode-locked erbium doped fibre laser', *Electronics Letters*, **31**, 1062, 1995.
289. T. Sugawa, E. Yoshida, Y. Miyajima and M. Nakazawa; '1.6 ps pulse generation from a 1.3 µm Pr^{3+}-doped fluoride fibre laser', *Electronics Letters*, **29**, 903, 1993.
290. M. J. Guy, D. U. Noske, A. Boskovic and J. R. Taylor; 'Femtosecond soliton generation in a praseodymium fluoride fibre laser', *Optics Letters*, **19**, 828, 1994.
291. L. E. Nelson, K. Tamura, E. P. Ippen and H. A. Haus; 'Additive-pulse mode-locked thulium-doped fibre ring laser', Paper CW14, in the *Proceedings of the Conference on Lasers and Electro-Optics CLEO 95*, Baltimore, USA, 1995.
292. M. Hofer, M. H. Ober, R. Hofer, M. E. Fermann, G. Sucha, D. Harter, K. Sugden, I. Bennion, C. A. C. Mendonca and T. H. Chiu; 'High power neodymium soliton fibre laser that uses a chirped fibre grating', *Optics Letters*, **20**, 1701, 1995.
293. P. N. Kean, J. W. D. Gray, I. Bennion and N. J. Doran; 'Dispersion modified actively mode-locked erbium fibre laser using a chirped fibre grating', *Electronics Letters*, **30**, 2133, 1994.
294. M. E. Fermann, K. Sugden and I. Bennion; 'High power soliton fibre laser-based on pulse width control with chirped fibre Bragg gratings', *Optics Letters*, **20**, 172, 1995.
295. W. H. Loh, D. Atkinson, P. R. Morkel, R. Grey, A. J. Seeds and D. N. Payne; 'Diode-pumped self-starting passively mode-locked neodymium doped fibre laser', *Electronics Letters*, **29**, 808, 1993.
296. M. H. Ober, M. Hofer, U. Keller and T. H. Chiu; 'Self-starting diode-pumped femtosecond Nd fibre laser', *Optics Letters*, **18**, 1532, 1993.
297. M. H. Ober, G. Sucha and M. E. Fermann; 'Controllable dual-wavelength operation of a femtosecond neodymium fibre laser', *Optics Letters*, **20**, 219, 1995.
298. W. H. Loh, D. Atkinson, P. R. Morkel, M. Hopkinson, A. Rivers, A. J. Seeds and D. N. Payne; 'All-solid-state passively mode-locked erbium doped fibre laser', *Applied Physics Letters*, **63**, 4, 1993.
299. W. H. Loh, D. Atkinson, P. R. Morkel, M. Hopkinson, A. Rivers, A. J. Seeds and D. N. Payne; 'Passively mode-locked Er^{3+} fibre laser using a semiconductor nonlinear mirror', *IEEE Photonics Technology Letters*, **5**, 35, 1993.
300. E. A. DeSouza, C. E. Soccolich, W. Pleibel, R. H. Stolen, J. R. Simpson and D. J. DiGiovanni; 'Saturable absorber mode-locked polarisation maintaining erbium doped fibre laser', *Electronics Letters*, **29**, 447, 1993.
301. O. G. Okhotnikov and J. R. Salcedo; 'Self-starting passively mode-locked fibre laser exploiting polarisation evolution in MQW Wave-guide', *Electronics Letters*, **30**, 1421, 1994.
302. O. G. Okhotnikov, F. M. Araujo and J. R. Salcedo; '1.48 µm pump-diode driven mode-locked Er fibre laser', *IEEE Photonics Technology Letters*, **6**, 933, 1994.
303. B. C. Barnett, L. Rahman, M. N. Islam, Y. C. Chen, P. Bhattacharya, W. Riha, K. V. Reddy, A. T. Howe, K. A. Stair, H. Iwamura, S. R. Friberg and T. Mukai; 'High power erbium doped fibre laser mode-locked by a semiconductor saturable absorber', *Optics Letters*, **20**, 471, 1995.
304. D. F. Voss and L. S. Goldberg; 'Simultaneous amplification and compression of continuous wave mode-locked Nd:YAG laser pulses', *Optics Letters*, **11**, 210, 1986.
305. M. Hofer, M. H. Ober, F. Haberl, M. E. Fermann, E. R. Taylor and K. P. Jedrzejewski; 'Regenerative Nd glass amplifier seeded with a Nd fibre laser', *Optics Letters*, **17**, 807, 1992.

306. M. L. Stock and G. Mourou; 'Chirped pulse amplification in an erbium-doped fibre oscillator fibre amplifier system', *Optics Communications*, **106**, 249, 1994.
307. M. E. Fermann, A. Galvanauskas and D. Harter; 'All-fibre source of 100 nJ subpicosecond pulses', *Applied Physics Letters*, **64**, 1315, 1994.
308. M. E. Fermann, A. Galvanauskas, D. Harter, J. D. Minelly, J. E. Caplen, Z. J. Chen and D. N. Payne; 'Cladding pumped Er^{3+} fibre amplifier generating femtosecond pulses with an average power of 0.26W', Post Deadline Paper CDP42. *Proceedings of Conference on Lasers and Electro-Optics*, Baltimore, USA, 1995.
309. See for example W. Koechner; *Solid State Laser Engineering* Ch. 5, Springer-Verlag, Berlin, 1988.
310. I. P. Alcock, A. C. Tropper, A. I. Ferguson and D. C. Hanna; 'Q-switched operation of a neodymium-doped monomode fibre laser', *Electronics Letters*, **22**, 84, 1986.
311. R. J. Mears, L. Reekie, S. B. Poole and D. N. Payne; 'Low threshold tunable cw and Q-switched fibre laser operating at 1.5 μm', *Electronics Letters*, **22**, 159, 1986.
312. I. M. Jauncey, J. T. Lin, L. Reekie and R. J. Mears; 'Efficient diode-pumped cw and Q-switched single-mode fibre laser', *Electronics Letters*, **22**, 198, 1986.
313. L. Reekie, I. M. Jauncey, S. B. Poole and D. N. Payne; 'Cw tunable and Q-switched operation at 939 nm of a diode laser pumped Nd^{3+} doped fibre laser', Paper THM46, in *Proceedings of Conference on Lasers and Electro-Optics*, Anaheim, USA, 1988.
314. L. A. Zenteno, H. Po and N. M. Cho; 'All solid state passively Q-switched modelocked Nd-doped fibre laser', *Optics Letters*, **15**, 115, 1990.
315. P. R. Morkel, K. P. Jedrzejewski, E. R. Taylor and D. N. Payne; 'Short-pulse highpower Q-switched fibre laser', *IEEE Photonics Technology Letters*, **4**, 545, 1992.
316. P. R. Morkel, K. P. Jedrzejewski and E. R. Taylor; 'Q-switched neodymium doped phosphate-glass fibre laser', *IEEE Journal of Quantum Electronics*, **29**, 2178, 1993.
317. I. Abdulhalim, C. N. Pannell, L. Reekie, K. P. Jedrzejewski, E. R. Taylor and D. N. Payne; 'High-power, short-pulse acoustooptically Q switched fibre laser', *Optics Communications*, **99**, 355, 1993.
318. P. Myslinski, J. Chrostowski, J. A. Koningstein and J. R. Simpson; 'High power Q-switched erbium doped fibre laser', *IEEE Journal of Quantum Electronics*, **28**, 371, 1992.
319. F. Seguin and T. Oleskevich; 'Diode pumped Q-switched fibre laser', *Optical Engineering*, **32**, 2036, 1993.
320. F. Chandonnet and G. Larose; 'High power Q-switched erbium fibre laser using an all fibre intensity modulator', *Optical Engineering*, **32**, 2031, 1993.
321. O. G. Okhotnikov and J. R. Salcedo; 'Dispersively Q-switched Er fibre laser with intracavity 1.48 μm laser diode as pumping source and non-linear modulator', *Electronics Letters*, **30**, 702, 1994.
322. P. Myslinski, J. Chrostowski, J. A. Koningstein and J. R. Simpson; 'Self-modelocking in a Q-switched erbium-doped fibre laser', *Applied Optics*, **32**, 286, 1993.
323. O. G. Okhotnikov, F. M. Araujo and J. R. Salcedo; 'Wavelength switching in pump diode modulated mode-locked and Q-switched Er fibre laser', *Applied Physics Letters*, **65**, 2910, 1994.
324. C. Frerichs and T. Tauermann; 'Q-switched operation of laser diode pumped erbium doped fluorozirconate fibre laser operating at 2.7 μm', *Electronics Letters*, **30**, 706, 1994.

325. T. Koumaki, T. Yamamoto, T. Sugawa and Y. Miyajima; 'Efficient up-conversion pumping at 1.064 μm of Tm^{3+} fluoride fibre laser operating around 1.47 μm', *Electronics Letters*, **28**, 830, 1992.
326. N. Kishi, J. N. Carter, R. Mottahedeh, P. R. Morkel, R. G. Smart, A. J. Seeds, J. S. Roberts, C. C. Button, D. N. Payne, A. C. Tropper and D. C. Hanna; 'Actively mode-locked and passively Q-switched operation of thulium doped fibre laser using multiquantum well asymmetric Fabry–Pérot modulator', *Electronics Letters*, **28**, 175, 1992.
327. P. Myslinski, X. Pan, C. Barnard, J. Chrostowski, B. T. Sullivan and J. F. Bayon; 'Q-switched thulium doped fibre laser', *Optical Engineering*, **32**, 2025, 1993.
328. S. M. Melle, A. T. Alavie, S. E. Karr, T. Coroy, K. Lui and R. M. Measures; 'A Bragg grating-tuned fibre laser strain sensor system', *IEEE Photonics Technology Letters*, **5**, 263, 1993.
329. G. A. Ball, W. W. Morey and P. K. Cheo; 'Single and multipoint fibre laser sensors', *IEEE Photonics Technology Letters*, **5**, 267, 1993.
330. A. Orthonos, S. Melle, A. T. Alavie, S. E. Karr and R. M. Measures; 'Fibre Bragg grating sensors', *Optical Engineering*, **32**, 2841, 1993.
331. A. T. Alavie, S. E. Karr, A. Orthonos and R. M. Measures; 'A multiplexed Bragg grating fibre laser sensor system', *IEEE Photonics Technology Letters*, **5**, 1112, 1993.
332. A. D. Kersey and W. W. Morey; 'Multielement Bragg grating based fibre-laser strain sensor', *Electronics Letters*, **29**, 964, 1993.
333. H. K. Kim, S. K. Kim, H. G. Park and B. Y. Kim; 'Polarimetric fibre laser sensors', *Optics Letters*, **18**, 317, 1993.
334. H. K. Kim, S. K. Kim and B. Y. Kim; 'Polarimetric fibre laser sensors', *Optics Letters*, **18**, 1465, 1993.
335. H. K. Kim, S. K. Kim and B. Y. Kim; 'Polarimetric fibre laser sensors using Er-doped fibre', *Optical and Quantum Electronics*, **27**, 281, 1995.

3
Fiber lasers in optical sensors

B. Y. Kim

3.1 INTRODUCTION

Rare earth-doped fiber amplifiers have been extensively developed for their applications to optical telecommunication systems (Desurvire, 1994). The technology is already mature for practical implementations. Fiber lasers using rare earth-doped fiber amplifiers have also been studied mainly for the possibility of using them as short pulse, wavelength tunable or soliton sources for optical communications (Duling, 1995). The fiber amplifier provides not only high gain but also a broad gain spectrum of tens of nanometers ideal for broadband wavelength division multiplexed (WDM) systems (Wysocki et al., 1994). This chapter develops the discussions of Chapter 3 and deals with another important application area of fiber amplifiers that takes advantage of their unique characteristics.

One of the original uses of fiber amplifier for sensors was with a Raman amplifier for a sustained recirculation of optical pulses in a so-called reentrant fiber gyroscope (Desurvire et al., 1988). The fiber Raman amplifiers, however, were soon replaced by rare earth-doped fiber counterparts for most of the applications. An amplified spontaneous emission source with a broad optical bandwidth turned out to be an ideal source for a fiber optic gyroscope with its low coherence, wavelength stability and high output power (Liu et al., 1988; Duling et al., 1990; Wysocki et al., 1991, 1994). The new fiber Bragg grating (FBG) technology made it possible to encode the change in effective grating spacing due to environmental perturbation in the lasing wavelength change (Kersey and Morey, 1993a; Kersey et al., 1994; Ball et al., 1993a; Alavie et al., 1993; Koo and Kersey, 1995). More recently, fiber lasers based on Er- or Nd-doped fiber amplifiers have been used to sense the change in the birefringence of the fiber cavity (H. K. Kim et al., 1993a, b, 1995; Ball et al., 1993b; H. Y. Kim et al., 1995a,b; J. S. Park et al., 1996a). This active polarimetric fiber laser sensor provides frequency readout rather than conventional intensity output. New forms of a fiber laser gyroscope (Jeon et al., 1993; K. H. Park et al., 1996; Lee et al., 1996; H. S. Kim and B. Y. Kim, 1996) also have been demonstrated that represent a drastic departure from the conventional interferometric fiber optic gyroscopes (IFOG), opening up a variety of new possibilities.

Optical Fiber Sensor Technology, Vol. 2. Edited by K. T. V. Grattan and B. T. Meggitt.
Published in 1998 by Chapman & Hall, London. ISBN 0 412 782 901

The potential advantages of fiber laser sensors compared to the conventional ones with external optical sources lie in the fact the optical phase change is directly translated into the change in the wavelength or optical frequency of the output. This form of output can be processed with much simpler electronic signal processors than those needed for conventional fiber sensors. In other words, the fiber laser sensors take advantage of inherent optical signal processing provided by the laser action in the fiber cavity. In the following sections, the principles and applications of fiber lasers to sensors will be described.

3.2 FIBER SOURCES FOR SENSORS

The rare earth-doped fiber amplifiers have broad gain spectra with high stability and high gain coefficients, which make them ideal sources for some sensor applications (Wysocki *et al.*, 1990, 1991, 1994; Duling *et al.*, 1990). One such application was demonstrated in an IFOG as depicted in Figure 3.1. The amplified spontaneous emission (ASE) source was found to provide most of the desired characteristics for the gyroscope, which has replaced the superluminescent diode (SLD) for high accuracy applications. The wavelength stability of a few ppm per °C and spectral line widths of between 5 and 15 nm have been successfully demonstrated (Liu *et al.*, 1988; Duling *et al.*, 1990; Wysocki *et al.*, 1990, 1991). A few milliwatts of optical power can easily be launched in single mode optical fibers. The Er-doped ASE source also provides an added advantage of being depolarized, which not only reduces the errors from polarization cross coupling but also enables the use of a conventional directional coupler for signal tapping. Significant progress has been made in the development of high performance fiber gyroscopes using the fiber ASE source at several laboratories (Fesler *et al.*, 1990; Moeller and Burns, 1991). One of the problems with the configuration in the figure is that the source property may be affected by optical feedback from the gyroscope (Wysocki *et al.*, 1994). It is not a serious problem for a closed loop gyroscope since the level of feedback is always kept constant. For an open loop gyroscope, this issue has to be solved. An obvious improvement with the IFOG using a fiber ASE source is shown in Figure 3.2. Here the fiber amplifier is used as a source and also as an amplifier since the amplified signal is detected at the backside of the fiber source/amplifier. Three orders of magnitude enhancement of the signal intensity have been

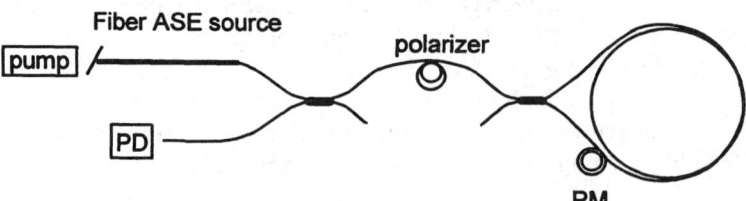

Fig. 3.1 IFOG using a fiber ASE source. PM = phase modulator, PD = photodetector.

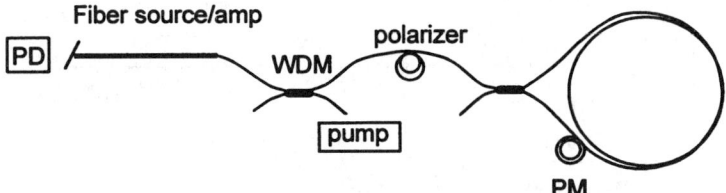

Fig. 3.2 IFOG using a fiber ASE source/amplifier. PM = phase modulator, PD = photodetector.

demonstrated in the range of a few milliwatts (Fesler et al., 1990). This configuration is useful for the suppression of electronic noise, making the electronic signal processing much simpler and inexpensive.

Other applications of fiber ASE sources include their use as a source for low coherence reflectometry where their high power capability is essential for increasing the sensitivity (Sorin and Baney, 1992, 1995; Baney and Sorin, 1993) (Figure 3.3). The source can also be used for wavelength multiplexed fiber grating sensors.

3.3 FIBER LASERS WITH BRAGG GRATINGS

The FBGs written directly in a fiber core using UV radiation were successfully utilized as a wavelength selective reflector in sensors (Morey et al., 1991; Measures, 1991; Kersey and Morey, 1993a, b; Kersey et al., 1993, 1994). The change in temperature or strain applied to the FBG modifies the reflection wavelength that can be monitored using dispersive optical elements. Initial attempts were made with an external broadband optical source and a demodulator based on filters with wavelength dependent optical transmission (Measures, 1991; Morey et al., 1991). Better signal to noise ratio can be achieved by replacing the external source–FBG combination with a fiber laser cavity formed with FBGs as reflectors (Alavie et al., 1992; Ball et al., 1993a; Kersey and Morey, 1993a). The wavelength of the stable and high power output signal is determined by the peak reflection wavelength of the FBG, λ which is a function of the grating period Λ as

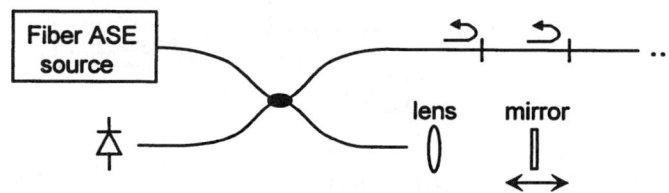

Fig. 3.3 Low coherence reflectometry.

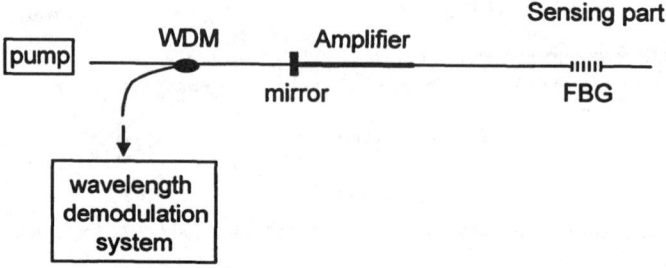

Fig. 3.4 Fiber laser sensor with Bragg gratings.

$$\lambda = 2n\Lambda \tag{3.1}$$

where n is the effective refractive index.

The fiber laser typically has a narrow optical line width but is not necessarily a laser with single longitudinal mode. The response of the output wavelength to the temperature and strain is (Ball and Morey, 1992)

$$\frac{\Delta\lambda_B}{\lambda_B} = (\alpha + \xi)\Delta T + (1 - \rho_e)\varepsilon \tag{3.2}$$

where α ($\sim 5.5 \times 10^{-7}$ °C^{-1} for silica glass) is the coefficient of thermal expansion and ξ ($\sim 8.3 \times 10^{-6}$ °C^{-1} for the Ge-doped silica core) is the thermooptic coefficient, ρ_e (~ 0.22 for silica glass fiber) is determined by photoelastic constants, and ε is the applied strain. Figure 3.4 illustrates a typical form of the fiber laser sensor based on FBGs, where one of the cavity mirrors is an FBG (Melle et al., 1993). Two different demodulation schemes are mainly studied, based on wavelength discriminating intensity filters (Melle et al., 1993) or interferometric demodulation (Koo and Kersey, 1994, 1995). The former is very simple and a sensitivity of a few micro strain has been demonstrated. The interferometric demodulator takes advantage of wavelength dispersion of an unbalanced Mach–Zehnder interferometer, as shown in Figure 3.5, that has an added

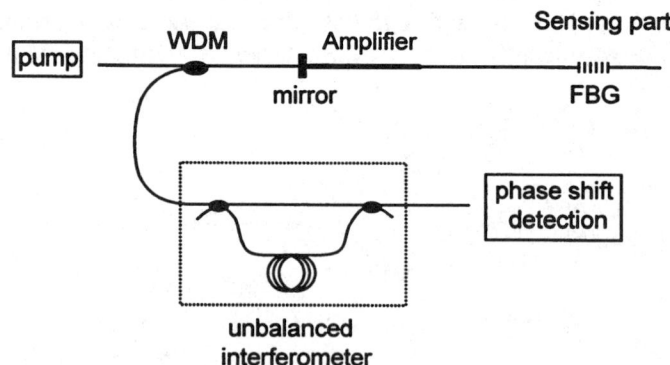

Fig. 3.5 Interferometric demodulation of an FBG laser sensor.

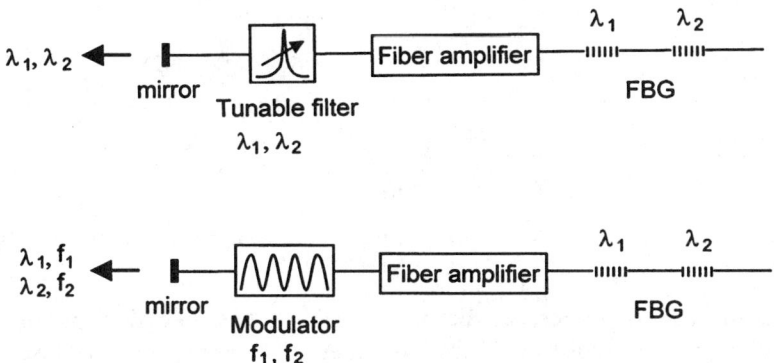

Fig. 3.7 Multiplexing of FBG laser sensors using mode locking.

advantage of sensitivity enhancement in proportion to the imbalance. The resulting output is the same as that from a conventional interferometer that should be electronically processed in order to retrieve the information of wavelength shift.

Since an FBG-based fiber laser sensor inherently operates within a narrow wavelength range, a number of sensors can easily be WDM as depicted in Figure 3.6 (Kersey and Morey, 1993a; Kersey et al., 1993, 1994; Alavie et al., 1993). Each sensing element located at different positions on a strand of single mode fiber operates at different wavelengths, the wavelength separation of which is designed to be greater than the dynamic range of each sensor. In this particular example, one of the mirrors that is common to all the laser cavities should have broad spectral response to cover the whole wavelength range of interest. It can be a simple dielectric mirror (Alavie et al., 1992; Melle et al., 1993) or a loop mirror (Kersey and Morey, 1993a; Kersey et al., 1993). Interrogation of each sensor can be done simultaneously using a WDM device or one at a time using a tunable wavelength filter. An alternative is to use mode locking as in Figure 3.7 in which identification of the sensors is done by the mode locking frequency that depends on the cavity length defined by a mirror and an FBG (Kersey and Morey, 1993b). Since the cavity lengths are known, one can monitor a particular sensor at a time by setting the modulation frequency applied to the mode locker at a predetermined value.

The FBG-based fiber laser sensors have great potential with simplicity in structure and also in signal readout in some cases. In particular, their application to strain/temperature monitoring of mechanical structures seems to be ideal and many research groups are actively developing the technology.

3.4 POLARIMETRIC FIBER LASER SENSORS

While the FBG-based fiber laser sensors rely on the macroscopic change of the wavelength determined by the FBG, the change in the optical frequency of

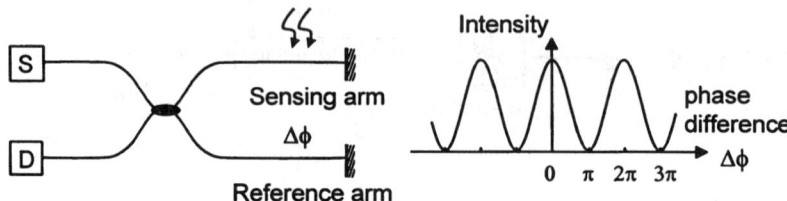

Fig. 3.8 Conventional interferometric sensor with intensity output. S = source, D = detector.

lasing longitudinal modes can be monitored for the detection of the cavity length change. The frequency of the Nth mode from a Fabry–Pérot laser cavity is

$$f_N = \frac{c}{2nL} N \qquad (3.3)$$

where L is the cavity length, n is the refractive index of the medium, and c is the speed of light in a vacuum. In this case the frequency spacing between the modes (free spectral range, FSR) is

$$f_{fsr} = f_{N+1} - f_N = \frac{c}{2nL} \qquad (3.4)$$

When the optical cavity length changes by $\lambda/2$, the laser mode frequency changes by one FSR. In other words, the optical length change is translated into an optical frequency change through laser action, which can be considered as optical signal processing. This contrasts with conventional interferometric fiber sensors where the optical length change in a fiber is measured by using interferometers that transform the optical phase shift to an intensity change as shown in Figure 3.8. An elaborate electronic signal processing scheme is required to recover the optical phase information from the interferometer output. The laser frequency shift, however, is linearly proportional to the phase shift and can be directly measured by the heterodyne technique as shown in Figure 3.9. In this example, two arms of an interferometer are replaced by two fiber lasers of approximately equal dimensions and their outputs are combined to produce a beat signal. With ideal single longitudinal mode lasers, the frequency of the beat signal is well defined and is the frequency difference between the two lasers. The reference laser is environmentally protected and the variation of the sensor laser is directly monitored by using a simple frequency counter.

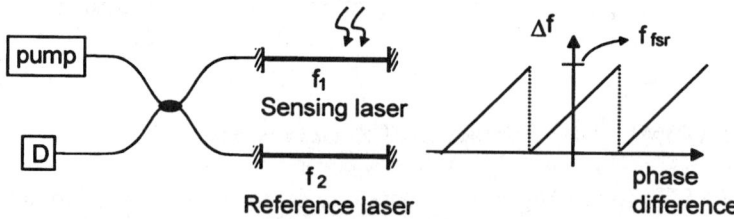

Fig. 3.9 Fiber laser sensor with beat-frequency output. f_{fsr} = free spectral range.

Direct implementation of the fiber laser sensor in Figure 3.9 may not be straightforward especially when a large number of longitudinal modes are lasing simultaneously. The most serious problem will be the difference in the laser mode spacing that will produce a very complicated r.f. frequency spectrum. A simple demonstration of the principle has been done using a polarimetric configuration as shown in Figure 3.10. In this case the two arms of the interferometer are the optical paths that the two eigenpolarization modes traverse. The optical path difference between them is determined by the amount of birefringence in the fiber cavity that can be made small enough for the purpose. The two polarization eigenmodes should experience the same optical loss in the cavity in order to have comparable intensities. A polarizer is used at the output of the dual polarization fiber laser to mix the two polarization components, producing a beat note, the frequency of which is determined by the total cavity birefringence. When external perturbations that alter the fiber birefringence are applied to the cavity, then the polarization beat frequency changes. The eigenpolarization states of a Fabry–Pérot (F–P) as appropriate cavity fiber laser with an arbitrary birefringence have a unique property. When the laser condition is satisfied in that the optical wave has to come back to the original polarization and phase after one complete round trip in the fiber cavity, it turns out that the eigenpolarization states at the position of the planar mirrors have to be linear polarizations (H. K. Kim *et al.*, 1993a; H. Y. Kim *et al.*, 1995a). It is true for arbitrary fiber birefringence, but the polarization directions vary as a function of the birefringence. At an arbitrary position in the fiber cavity, the Jones vector representation of states of polarization (SOPs) of the waves travelling in opposite directions are complex conjugates of each other in the laboratory reference frame (H. K. Kim, 1994). This means that the waves travelling in opposite directions will experience exactly the same optical paths. In order to stabilize the directions of the eigenpolarizations, a polarization maintaining fiber is used for the laser cavity, where a 90° splice at the cavity center was introduced to minimize the net birefringence (H. K. Kim *et al.*, 1993b, 1995). Strain and temperature were successfully measured using the fiber laser with a frequency readout. A polarimetric fiber laser sensor based on FBGs has also been demonstrated (Ball *et al.*, 1993b).

Another polarimetric fiber laser sensor configuration of interest is one with a Faraday rotating mirror (FRM) at one end of the cavity as shown in Figure 3.11. The FRM contains a 45° Faraday rotator and the reflected light always has orthogonal polarization states with respect to the input in the

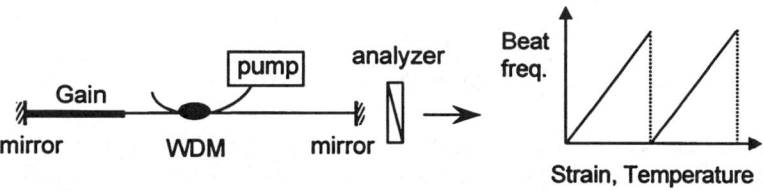

Fig. 3.10 Polarimetric fiber laser sensor.

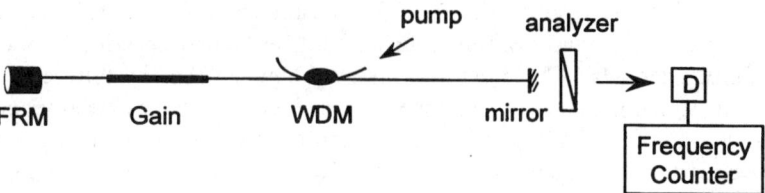

Fig. 3.11 Polarimetric fiber laser sensor with an FRM.

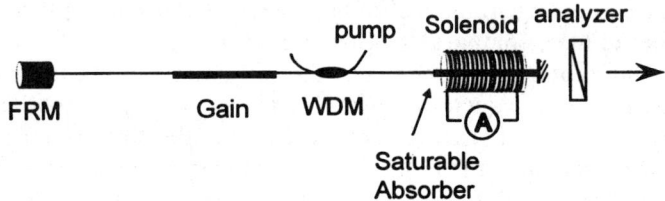

Fig. 3.12 Fiber laser current sensor.

propagating wave coordinate system. This configuration ensures that the two eigen-polarization states will experience the same loss, and also the polarization mode beat frequency is always the odd harmonics of half the FSR of the laser. Note that the beat frequency between the longitudinal modes in the same polarization state is the same as the FSR. Moreover, an interesting feature of the laser is that the eigenpolarization states at the location of the planar mirror are no longer linear but circular polarizations (J. S. Park *et al.*, 1996a, b). The PMB frequency and the eigenpolarization state are independent of the fiber birefringence as long as the cavity does not contain any nonreciprocal element other than the FRM. It makes the fiber laser with an FRM not useful for measuring any reciprocal perturbations. It is, however, ideal for the measurement of magnetic field or current through the Faraday effect. In Figure 3.12, a fiber laser with an FRM, a solenoid coil through which current is flowing, and a saturable absorber for the suppression of multilongitudinal modes is shown (J. S. Park *et al.*, 1996a, b). The magnetic field is applied near the planar mirror where the SOP is circular and therefore the sensitivity is at a maximum. Figure 3.13 shows the experimental results obtained for current sensing. One of the major issues with polarimetric fiber laser sensors is the stability of the lasing modes. Nevertheless, it demonstrated its great potential as a new class of fiber sensors with frequency readout.

3.5 FIBER LASER GYROSCOPES (FLAGS)

There have been considerable efforts to develop a ring laser gyroscope (RLG) in the fiber optic form to combine the advantages of the all solid state construction of an interferomatric fiber optic gyroscope (IFOG) and the straightforward

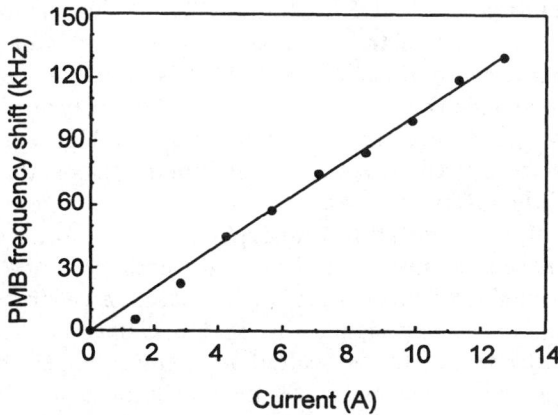

Fig. 3.13 Experimental results for current sensing. Total cavity length, 7.3 m; solenoid, 42 cm in length and 1530 turns.

signal processing of an RLG. For a successful RLG operation, the gain medium should have directional gain as in the case of a He–Ne RLG such that the counterpropagating laser beams do not compete for gain. Another important requirement for an RLG is that its cavity should have negligible backscattering so that the lock-in should not be significant. Solid state gain mediums in general cannot satisfy the necessary conditions for a good RLG. The most extensively studied approach is the one based on stimulated Brillouin scattering (SBS) as depicted in Figure 3.14 (Zarinetchi et al., 1991; Huang et al., 1993a, b; Nicati et al., 1993, 1994; Tanaka et al., 1996). The gain provided by SBS is in the opposite direction to the propagating direction of the pump. Therefore, signal laser beams circulating in the clockwise and the counterclockwise directions extract gain from pump beams circulating in the opposite directions. It means the signal laser beams do not share the pump and a stable bidirectional laser operation can be achieved at the same lasing wavelength that is needed for RLG operation. The backscattering-induced lock-in phenomena can be suppressed by a push–pull phase modulation technique that provides the same effect as

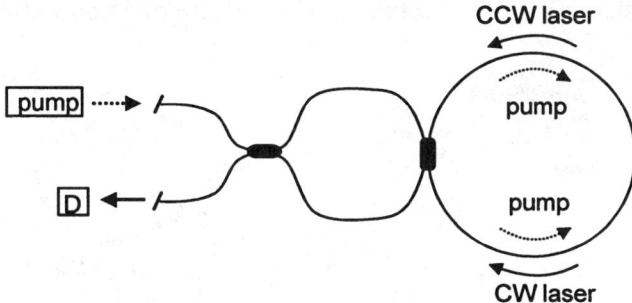

Fig. 3.14 Brillouin fiber laser gyroscope.

'dither' in the conventional RLG without an actual mechanical dither (Huang et al., 1993a, b). In spite of the successful demonstration of a Brillouin fiber optic RLG, a number of problems have to be solved before it can be considered practical. These include the issues of complexity due to many stabilization circuits, limited dynamic range and the nonlinear optical Kerr effect (Huang et al., 1993c; Nicati et al., 1994). However, the Brillouin RLG is the most successful fiber optic RLG demonstrated so far.

An observation of the rotation rate dependent beat note from an Er-doped fiber ring laser has led to investigations into the possibility of a fiber optic RLG (S. K. Kim et al., 1994; Kiyan et al., 1996). The laser had a large number of longitudinal modes, and the beat signal was small and unstable. At the same time, very complicated phenomena associated with gain competition and coupling between the counterpropagating waves have been observed. Nevertheless, the effort may be worthwhile considering the merits of this approach, including the facts that it does not require tight control of cavity length and birefringence and that its dynamic range is not limited as the Brillouin RLG. One of the important features of a fiber ring laser is that the eigenpolarization states for the two counterpropagating lasers at any particular position in the cavity are complex conjugates of each other regardless of the fiber birefringence in the ring cavity (S. K. Kim et al., 1994). This ensures the frequencies of the counterpropagating lasers are exactly the same without the presence of nonreciprocal effects, which is an essential feature for its application to gyroscopes.

Probably the most significant departure from conventional IFOG concepts can be found in fiber laser gyroscopes that are not in the form of an RLG (Jeon et al., 1993; H. S. Kim and B. Y. Kim, 1996). They make use of the advantageous features of both approaches in that the reciprocity is automatically obtained from the laser action and/or the readout is in the form that is much more convenient to process. One example is the mode-locked fiber laser gyroscope (ML-FLaG) as shown in Figure 3.15. In this configuration, the fiber laser is formed by a planar mirror at one end and a Sagnac loop mirror at the other. The amplifier can be a rare earth-doped fiber (Jeon et al., 1993) or a semiconductor amplifier (K. H. Park et al., 1996; Lee et al., 1996), of which the latter turns out to be more suitable for the application owing to its fast gain recovery. In the fiber loop mirror, a phase modulator is placed near one end of the loop to provide modulation of the reflectivity as in the case of a conventional IFOG.

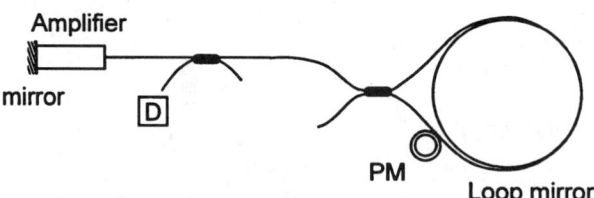

Fig. 3.15 Mode-locked fiber laser gyroscope (ML-FLaG). PM = phase modulator, D = detector.

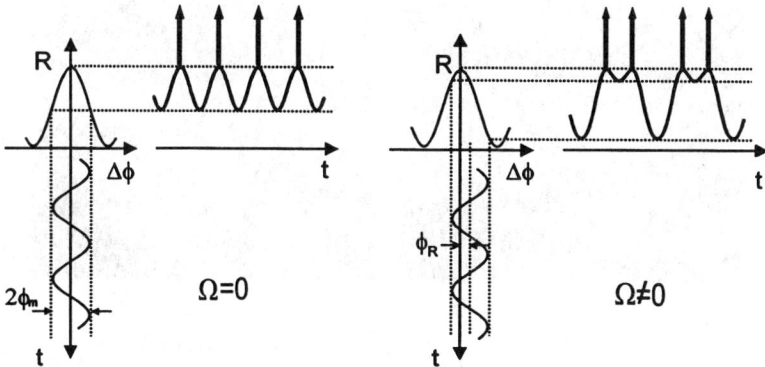

Fig. 3.16 Mode-locked pulses produced at the timing of peak reflectivity. $\Delta\phi$ = phase difference.

The modulator operates at the laser FSR (or longitudinal mode spacing) producing mode-locked pulses. When the fiber length outside the fiber loop is half that of the loop, the modulation becomes a pure amplitude modulation. For every cycle of the phase modulation signal, two mode-locked pulses will be produced at the time of peak reflectivity shown in Figure 3.16. The separation of the pulses in the time domain is exactly half of the phase modulation period without any rotation input. When the ML-FLaG is rotating at the rotation rate of Ω, the timings for the peak reflectivity of the loop mirror shifts and the time separation of the two adjacent pulses become

$$\Delta t = \frac{T_m}{2\pi}\sin^{-1}\left(\frac{\Delta\phi_R}{\phi_m}\right) \qquad \Delta\phi_R = \frac{8\pi AN}{\lambda c}\Omega \qquad (3.5)$$

where T_m is the modulation period, ϕ_m is the amplitude of phase modulation, $\Delta\phi_R$ is the rotation-induced nonreciprocal phase shift, A is the area of loop mirror, N is the number of fiber turns, and c is the speed of light. The experimental verification of the operating principle is shown in Figure 3.17. Since the output is the time interval between pulses that can be measured by using conventional electronic counters, signal processing can be much simpler than that for conventional IFOG. The scale factor in equation 3.5 depends on the phase modulation amplitude that has to be maintained at a constant value to achieve a stable scale factor. The polarization properties of the ML-FLaG are relatively complicated and are well analyzed in references (Jeong et al., 1996). In principle, the reciprocity of the ML-FLaG is automatically satisfied as in the cases of the F-P cavity and the fiber ring laser. Therefore high performance fiber components such as polarizers, polarization maintaining fiber, and directional couplers, needed for conventional IFOGs are not critically required. However, to obtain a stable reflection from the loop mirror, a polarization maintaining fiber with a modest performance is preferable for the fiber circuit. Figure 3.18 shows the performance of an ML-FLaG built with a diode amplifier and a polarization maintaining fiber. The short-term noise and long-term drift of a

FIBER LASERS IN OPTICAL SENSORS

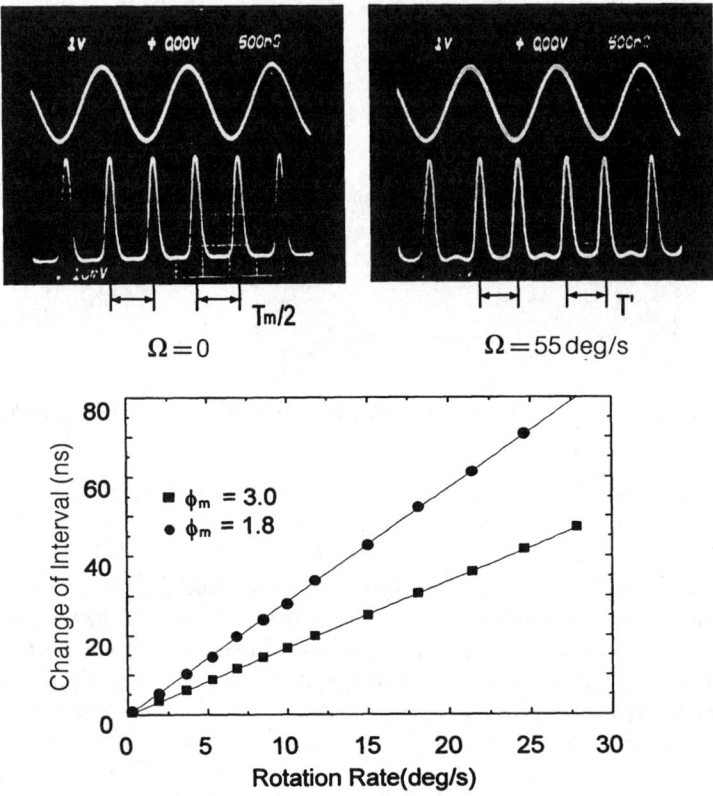

Fig. 3.17 Experimental results obtained from an ML-FLaG. Linear section length, 75 m; fiber loop, 150 m in length and 15.2 cm in diameter; modulation frequency, 644 kHz.

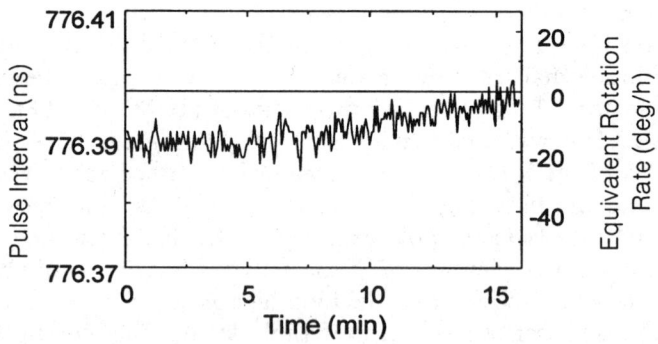

Fig. 3.18 Short-term noise of ML-FLaG. $\phi_m = 3.0$ rad; integration time, 1.55 s; other parameters are the same as those in Figure 3.17.

FIBER LASER GYROSCOPES (FLAGS)

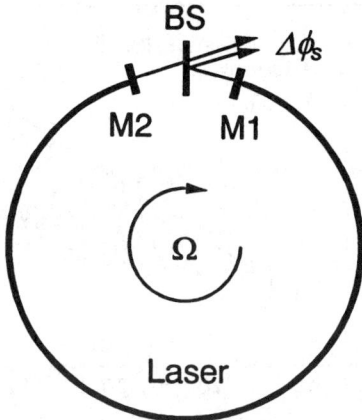

Fig. 3.19 Interferometric fiber laser gyroscope. BS = beam splitter; M1, M2 = mirror; $\Delta\phi_s$ = phase difference.

few degrees $h^{-1}(\sqrt{Hz})^{-1}$ and about a $100° \, h^{-1}$, respectively, are very promising for practical implementation of the ML-FLaG in the near future.

Another interesting development is a FLaG formed with aa F–P cavity as shown in Figure 3.19 (H. S. Kim and B. Y. Kim, 1996). The cavity is formed with two planar mirrors placed directly at the fiber end faces. The outputs from the two ends of the fiber laser are interfered to produce a fringe pattern as a function of their phase difference. The laser action makes sure that the phase accumulation during a complete round trip in the cavity is an integral multiple of 2π. The waves travelling in the opposite directions for a single transit through the cavity will experience the same phase shift. This means that the phase difference between the two interfering waves will be an integral multiple of π regardless of the cavity length and the fiber birefringence in the absence of a nonreciprocal element. As described earlier, the eigenpolarization states at the position of the mirrors are linear SOPs and they should be aligned to produce the maximum visibility. If the laser cavity formed into a multi-turn loop is rotating, then a Sagnac phase shift will be introduced and the interference signal changes. If the laser is operated in a single longitudinal mode, things are simple. For a multiple longitudinal mode, all other modes have to be suppressed to produce a nonvanishing interference signal, since the adjacent modes result in mutually complementary fringe patterns separated by π phase difference. A successful approach to mode control has been demonstrated by using a slow saturable absorber as a narrowband filter and the position control of the gain medium and the saturable absorber (H. S. Kim *et al.*, 1996). The initial results proving the principle are shown in Figure 3.20. The fiber laser gyroscope may have the simplest construction, and requires combining optics outside the laser similar to the case of an RLG. The combining optics will inevitably introduce phase errors that should be kept at low values for high performance operation

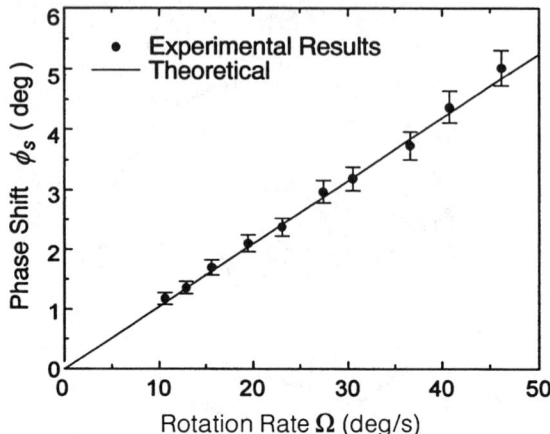

Fig. 3.20 Experimental results from an interferometric fiber laser gyroscope. Laser cavity length, 120 m; effective sensing area, 3.86 m^2; $\lambda = 1.53$ μm.

of the gyroscope. A number of signal processing techniques either similar to conventional ones or new schemes can be applied to the gyroscope.

3.6 CONCLUSIONS

Fiber lasers have great potential for their application to sensors, and it is one of the most active research areas. They are the integration of the optical source and the fiber sensing elements of the conventional passive fiber sensors with the addition of optical signal processing. The frequency readout will make the complicated electronic signal processing much simpler with reduction of cost. Additional advantages, that are hard to get otherwise, such as automatic reciprocity in the fiber laser gyroscopes are also derived. There seem to be many new fiber laser sensors yet to be discovered which might eventually replace conventional passive fiber sensors.

REFERENCES

Alavie, T., Othonos, A., Melle, S., Liu, K., Measures, R. (1992) Bragg fiber laser sensor. *Proc. SPIE Fiber Optic and Laser Sensors X*, **1795**, 194–203, Boston, Massachusetts.

Alavie, A. T., Karr, S. E., Othonos, A., Measures, R. M. (1993) A multiplexed Bragg grating fiber laser sensor system. *IEEE Photonics Technology Letters*, **5**, 1112–1114.

Ball, G. A., Morey, W. W. (1992) Continuously tunable single-mode erbium fiber laser. *Optics Letters*, **17**, 420–422.

Ball, G. A., Morey, W. W., Cheo, P. K. (1993a) Single- and multipoint fiber-laser sensors. *IEEE Photonics Technology Letters*, **5**, 267–270.

REFERENCES

Ball, G. A., Meltz, G., Morey, W. W. (1993b) Polarimetric heterodyning Bragg-grating fiber-laser sensor. *Optics Letters*, **18**, 1976–1978.

Baney, D. M., Sorin, W. V. (1993) Extended-range optical low-coherence reflectometry using a recirculating delay technique. *IEEE Photonics Technology Letters*, **5**, 1109–1112.

Desurvire, E. (1994) *Erbium-Doped Fiber Amplifiers*, John Wiley & Sons, Inc., New York.

Desurvire, E., Kim, B. Y., Fesler, K., Shaw, H. J. (1988) Reentrant fiber Raman gyroscope. *Journal of Lightwave Technology*, **6**, 481–491.

Duling III, I. N. (ed.) (1995) *Compact Sources of Ultrashort Pulses*, Cambridge University Press, Cambridge.

Duling III, I. N., Burns, W. K., Goldberg, L. (1990) High-power superfluorescent fiber source. *Optics Letters*, **15**, 33–35.

Fesler, K. A., Digonnet, M. J. F., Kim, B. Y., Shaw, H. J. (1990) Stable fiber-source gyroscopes. *Optics Letters*, **15**, 1321–1323.

Huang, S., Nicati, P.-A., Toyama, K., Kim, B. Y., Shaw, H. J. (1993a) Synthetic heterodyne detection in a fiber-optic ring-laser gyro. *Optics Letters*, **18**, 81–83.

Huang, S., Toyama, K., Kim, B. Y., Shaw, H. J. (1993b) Lock-in reduction technique for fiber-optic ring laser gyros. *Optics Letters*, **18**, 555–557.

Huang, S., Thévenaz, L., Toyama, K., Kim, B. Y., Shaw, H. J. (1993c) Optical Kerr-effect in fiber-optic Brillouin ring laser gyroscopes. *IEEE Photonics Technology Letters*, **5**, 365–367.

Jeon, M. Y., Jeong, H. J., Kim, B. Y. (1993) Mode-locked fiber laser gyroscope. *Optics Letters*, **18**, 320–322.

Jeong, H. J., Koh, Y. W., Lee, B. W., Jeon, M. Y., Kim, B. Y. (1996) Analysis of polarization properties of a mode-locked fiber laser gyroscope. *Applied Optics*, **35**, 2206–2210.

Kersey, A. D., Davis, M. A., Morey, W. W. (1993) Quasi-distributed Bragg-grating fiber-laser sensor. *Proc. OFS-9*, Postdeadline paper (Pd.5), Firenze, Italy.

Kersey, A. D., Morey, W. W. (1993a) Multi-element Bragg-grating based fiber-laser strain sensor. *Electronics Letters*, **29**, 964–966.

Kersey, A. D., Morey, W. W. (1993b) Multiplexed Bragg grating fiber-laser strain-sensor system with mode-locked interrogation. *Electronics Letters*, **29**, 112–114.

Kersey, A. D., Koo, K. P., Davis, M. A. (1994) Fiber optic Bragg grating laser sensors. *Proc. SPIE Fiber Optic and Laser Sensors XII*, **2292**, 102–112, San Diego, California.

Kim, H. K. (1994) Polarization and modal properties of the rare-earth doped fiber lasers and their applications. Ph.D. Thesis, Department of Physics, KAIST, Korea.

Kim, H. K., Kim, S. K., Park, H. G., Kim, B. Y. (1993a) Polarimetric fiber laser sensors. *Optics Letters*, **18**, 317–319.

Kim, H. K., Kim, S. K., Kim, B. Y. (1993b) Polarization control of polarimetric fiber-laser sensors. *Optics Letters*, **18**, 1465–1467.

Kim, H. K., Kim, S. K., Kim, B. Y. (1995) Polarimetric fiber laser sensors using Er-doped fibre. *Optical and Quantum Electronics*, **27**, 481–485.

Kim, H. S., Kim, B. Y. (1996) New fiber laser interferometer for rotation sensing. *Proc. OFS-11*, 76–79, Sapporo, Japan.

Kim, H. S., Kim, S. K., Kim, B. Y. (1996) Longitudinal mode control in few-mode erbium-doped fiber lasers. *Optics Letters*, **21**, 1144–1146.

Kim, H. Y., Kim, S. K., Jeong, H. J., Kim, B. Y. (1995a) Polarization properties of a twisted fiber laser. *Optics Letters*, **20**, 386–388.

Kim, H. Y., Kim, B. K., Yun, S. H., Kim, B. Y. (1995b) Response of fiber lasers to an axial magnetic field. *Optics Letters*, **20**, 1713–1715.

Kim, S. K., Kim, H. K., Kim, B. Y. (1994) Er^{3+}-doped fiber ring laser for gyroscope applications. *Optics Letters*, **19**, 1810–1812.

Kiyan, R. V., Kim, S. K., Kim, B. Y. (1996) Phase nonreciprocal bidirectional Er-doped fiber ring laser. *Proc. OFS-11*, 598–601, Sapporo, Japan.

Koo, K. P., Kersey, A. D. (1994) Fiber laser sensor system with interferometric read-out and wavelength multiplexing. *Proc. OFS-10*, 331–334, Glasgow, Scotland.

Koo, K. P., Kersey, A. D. (1995) Bragg grating-based laser sensors systems with interferometric interrogation and wavelength division multiplexing. *Journal of Lightwave Technology*, **13**, 1243–1249.

Lee, B. W., Jeong, H. J., Oh, M. S., Kim, B. Y. (1996) High resolution mode-locked fiber laser gyroscope using a semiconductor laser amplifier. *Proc. OFS-11*, 296–299, Sapporo, Japan.

Liu, K., Digonnet, M., Fesler, K., Kim, B. Y., Shaw, H. J. (1988) Superfluorescent single mode Nd:fiber source at 1060 nm. *Tech. Digest OFS-6*, **2**, 462–465, New Orleans, Louisiana.

Measures, R. M. (1991) Fiber optic sensor considerations and developments for smart structures. *Proc. SPIE Fiber Optic Smart Structure and Skins IV*, **1588**, p. 282.

Melle, S. M., Alavie, A. T., Karr, S., Coroy, T., Liu, K., Measures, R. M. (1993) A Bragg grating-tuned fiber laser strain sensor system. *IEEE Photonics Technology Letters*, **5**, 263–266.

Moeller, R. P., Burns, W. K. (1991) 1.06-µm all-fiber gyroscope with noise subtraction. *Optics Letters*, **16**, 1902–1904.

Morey, W. W., Dunphy, J. R., Meltz, G. (1991) Multiplexed fiber Bragg grating sensors. *Proc. SPIE Distributed and Multiplexed Fiber Optic Sensors*, **1586**, 216–224, Boston, Massachusetts.

Nicati, P.-A., Toyama, K., Huang, S., Shaw, H. J. (1993) Temperature effects in a Brillouin fiber ring laser. *Optics Letters*, **18**, 2123–2125.

Nicati, P.-A., Toyama, K., Huang, S., Shaw, H. J. (1994) Frequency pulling in a Brillouin fiber ring laser. *IEEE Photonics Technology Letters*, **6**, 801–803.

Park, J. S., Yun, S. H., Ahn, S. J., Kim, B. Y. (1996a) Novel fiber laser current sensor. *Tech. Digest OECC'96*, 56–57, Chiba, Japan.

Park, J. S., Yun, S. H., Ahn, S. J., Kim, B. Y. (1996b) Polarization- and frequency-stable fiber laser for magnetic-field sensing. *Optics Letters*, **21**, 1029–1031.

Park, K. H., Cho, H. S., Jang, D. H., Lee, B. W., Kim, B. Y. (1996) Mode-locked fiber laser gyroscope based on a distributed-feedback semiconductor laser amplifier. *Optics Letters*, **21**, 92–94.

Sorin, W. V., Baney, D. M. (1992) Measurement of Rayleigh backscattering at 1.55 µm with 32 µm spatial resolution. *IEEE Photonics Technology Letters*, **4**, 374–376.

Sorin, W. V., Baney, D. M. (1995) Multiplexed sensing using optical low-coherence reflectometry. *IEEE Photonics Technology Letters*, **7**, 917–919.

Tanaka, Y., Yamasaki, S., Hotate, K. (1996) Brillouin fiber optic gyro with directional sensitivity. *Proc. OFS-11*, 88–91, Sapporo, Japan.

Wysocki, P. F., Digonnet, M. J. F., Kim, B. Y. (1990) Broad-spectrum, wavelength-swept, erbium-doped fiber laser at 1.55 µm. *Optics Letters*, **15**, 879–881.

Wysocki, P. F., Digonnet, M. J. F., Kim, B. Y. (1991) Wavelength stability of a high-output, broadband, Er-doped superfluorescent fiber source pumped near 980 nm. *Optics Letters*, **16**, 961–963.

REFERENCES

Wysocki, P. F., Digonnet, M. J. F., Kim, B. Y., Shaw, H. J. (1994) Characteristics of erbium-doped superfluorescent fiber sources for interferometric sensor applications. *Journal of Lightwave Technology*, **12**, 550–567.

Zarinetchi, F., Smith, S. P., Ezekiel, S. (1991) Stimulated Brillouin fiber-optic laser gyroscope. *Optics Letters*, **16**, 299–301.

4
Multiplexing optical fiber sensors

J. D. C. Jones and R. McBride

4.1 INTRODUCTION

In this chapter we introduce the subject of the multiplexing of optical fiber sensors, explaining what is meant by multiplexing, and outlining the various techniques that are available for the implementation of multiplexing. We indicate the relative strengths and weaknesses of these techniques and provide some practical examples. We shall concentrate on multiplexing techniques for single-mode optical fiber systems, especially those based on interferometry, where multiplexing is arguably most highly developed.

An example of a stylized multiplexed fiber sensor system is shown in Figure 4.1. The system comprises an optical source or sources, an array of sensing elements connected by fiber downleads and upleads, and a set of optical detectors. The system will also often contain modulators and interconnections to preprocess the optical signals and means to optically and electronically demodulate and decode the signals from the individual sensing elements, thus returning the value of the measurand or measurands at each of the sensing element positions.

The measurand is defined as the quantity that is measured—for example temperature or strain. The position of a sensing element is defined only in so far as is necessary to identify the individual sensing element in the array, rather than to identify its absolute position coordinates. The term 'decoding' is used specifically to describe the process of deriving a signal corresponding to an individual sensing element.

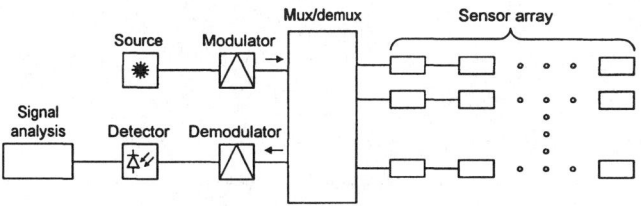

Fig. 4.1 Generalized multiplexed fiber sensor system.

Optical Fiber Sensor Technology, Vol. 2. Edited by K. T. V. Grattan and B. T. Meggitt.
Published in 1998 by Chapman & Hall, London. ISBN 0 412 782 901

A system is defined as multiplexed if the number of sources, downleads, or detectors is smaller than that which would be required if the same number of sensing elements, each detecting a single measurand, were assembled as individual measurement systems, without sharing sources, detectors and downleads.

More formally, we may say that the multiplexed system is capable of recovering signals $I_i(X_{1i}, \ldots, X_{Mi}; z_i)$ for each of the N sensing elements, where $i = 1, \ldots, N$, where the ith sensing element is designed to detect each of the measurands $X_{1i}, \ldots, X_{Mi}$, and where the ith sensing element has some unique location z_i in the array, as shown in Figure 4.2. The notation I_i is used to represent not only a single measurement of optical intensity but also sets of more general measured values and time-dependent quantities.

The exposition of the preceding paragraph is a more general description than is required for most practical multiplexed systems. Most systems are either location-multiplexed or measurand-multiplexed. In a location-multiplexed system, each sensor is normally designed to detect only one measurand, so that

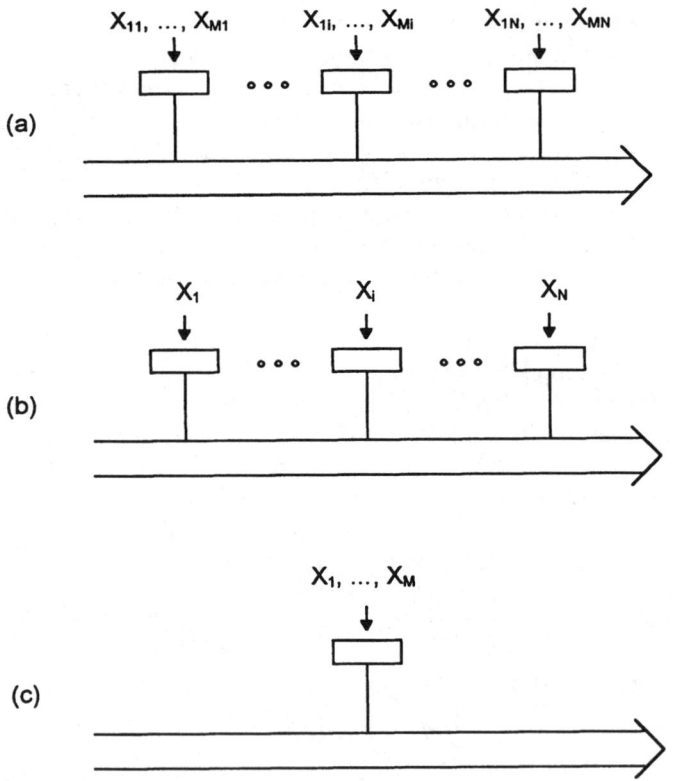

Fig. 4.2 Notation for multiplexed measurands and sensors: (a) general case; (b) location-multiplexed sensors; (c) measurand-multiplexed sensors.

INTRODUCTION

we require to recover $I_i(X_j, z_i)$. Most commonly, all sensors are designed to detect the same measurand, so we need only find $I_i(X, z_i)$. For example, in an array of temperature sensors, $X = T$, where T is temperature, so we need recover only $I_i(T, z_i)$.

In a measurand-multiplexed system, we are interested not so much in the value of a single measurand at different positions, but in the value of several measurands at a single position. Here, we wish to recover signals $I_j(X_j, z)$. One important example is in the simultaneous recovery of strain and temperature from a single sensing element. The requirement is of practical importance because interferometric and other types of optical sensors used to measure slowly varying strain are usually simultaneously responsive to temperature. Thus we seek to recover simultaneously $I_1(T)$ and $I_2(\sigma)$, where σ denotes strain.

Another example of measurand-multiplexing is in the detection of vector quantities. For example, fiber magnetometers have been designed to measure the three orthogonal components of magnetic field, B_x, B_y and B_z. Practical systems have been constructed in which the three components are measured in individual sensing elements spaced closely in comparison with field gradients in the measurement volume.

Whilst the concepts described above are most easily visualized in terms of arrays of point sensors, practical fiber sensors are incapable of making measurements at a geometrical point, and instead return a signal averaged over the length of the element, with no information yielded regarding the variation of the measurand over the length of the sensing element.

We must therefore distinguish between multiplexed sensors and distributed sensors, as illustrated in Figure 4.3. In the distributed sensor, there is only one sensing element, and the objective of the signal processing is to recover the measurand as a function of position along the sensing element. Conversely, the multiplexed system returns a value of the measurand for each of the sensing elements, usually averaged over the length of the individual sensor. It is possible to conceive of multiplexed distributed sensors, in which an array of distributed sensors are configured to share components. Measurand-multiplexed distributed sensors have been proposed for the simultaneous recovery of temperature and strain along a common distributed fiber sensor.

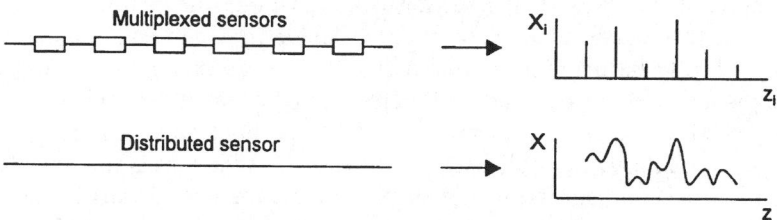

Fig. 4.3 Multiplexed and distributed sensors.

4.2 BASIC PRINCIPLES OF OPTICAL FIBER SENSORS

In this section we present an outline of the mode of operation of optical fiber sensors necessary to follow the remainder of the chapter. Extensive accounts exist elsewhere for the interested reader [1-3], in addition to those contained in the other chapters of this volume.

An optical fiber sensor may be defined as a device in which an optical signal is modulated in response to a measurand, in such a way that the measurand can be recovered by processing the optical signal. One way of classifying fiber sensors is thus in terms of that property of the optical signal that is modulated, such as its intensity, spectrum, phase, or state of polarization. We begin by following this method of classification.

4.2.1 Intensity modulation

In a sense, all optical sensors are intensity modulated, in that all optoelectronic detectors are capable of responding only to the time-averaged optical power. However, we shall here adopt the more restricted definition of an intensity-modulated sensor as one in which the intensity of a guided beam is modulated in the sensing element itself in response to the measurand.

We are here following the definition of intensity usually given in the optical fiber sensor literature as optical power per unit cross-sectional area of an optical beam, although a more formal term for that quantity is irradiance. We will also use the term 'intensity' when 'power' would be a more precise term, in order to avoid confusion between optical and other forms of power, such as mechanical power transmitted by the measurand. This usage is usually acceptable, since in a single-mode optical fiber the on-axis intensity is proportional to the transmitted power.

Many types of intensity-modulated sensor place no particular requirements on the spatial coherence of the beam, and hence may be constructed using multimode optical fiber. A simple example of an intensity-modulated sensor is the microbending sensor in which the curvature of a fiber is increased in response to a measurand, thus increasing the attenuation of the bend. For example, a sensor for transverse force can be made by holding a fiber between ridged plates, so that the bend loss increases as the plates are squeezed together. Alternatively, a fiber manufactured with built-in bends will experience an increase in transmission when stretched, providing a sensor for longitudinal strain [4].

The bend loss sensor is an example of a device in which fiber-guided modes are coupled to radiative modes [5]. Analogous devices exist in which the mode-coupling is induced not by bending the fiber, but by removing some of the cladding material to access the evanescent wave. Such devices typically involve applying a coating to the polished section of fiber to act as a superstrate. The transduction mechanism involves a change in the dimensions or the refractive index of the overlay, affecting the matching between guided modes in the fiber and modes in the superstrate, which are radiated. Such devices can be used, for example, as gas or chemical sensors [6] or for the detection of humidity [7]. A final example of the guided-to-radiative mode sensor is the long-period in-fiber-grating sensor, described in more detail in section 4.2.13 below.

Intensity-modulated sensors can involve more than one transduction mechanism. For example, a stress sensor can be constructed in which the primary sensing element is a piece of photoelastic material. The sensing element is illuminated with linearly polarized light, normally at 45° to the polarization eigenaxes of the sensing element to ensure maximum sensitivity. The state of polarization of the beam passing through the sensing element is modulated in response to the applied stress. The beam leaving the sensing element passes through a polarization analyzer, such that its intensity change represents the change in stress.

Many other types of intensity-based sensor have been developed for a very wide range of measurands. A helpful review is given in [3].

4.2.2 Wavelength-modulated sensors

Various techniques have been devised by which a measurand modifies the spectrum of a guided beam. Simple examples include mechanical devices in which the measurand moves a wavelength-selective element, such as a diffraction grating. More complex devices include sensors based on changing the optical path difference in an interferometer, which will be discussed in section 4.2.4.

In practice, all of the intensity-modulated sensors of the previous section are also wavelength-modulated sensors. For example, all processes of guided-to-radiative mode coupling show coupling coefficients that are functions of wavelength. Thus, when a fiber that is guiding light containing a range of wavelengths is bent, then the spectrum of the guided beam changes [8].

4.2.3 Frequency-modulated sensors

The distinction between wavelength-modulated and frequency-modulated sensors is arbitrary, in that the wavelength, λ, and the frequency, ν, of an optical beam are related by the elementary equation $\nu\lambda = v$, where v is the phase velocity of the beam. However, we shall use the following artificial but convenient distinction: we define frequency-modulated sensors as those in which the change in frequency is relatively small in comparison with the optical frequency (e.g. up to the GHz region, in comparison with the 300 THz optical frequency of radiation with a wavelength of 1 μm in free space), and where the frequency shift has been produced, for example, by the Doppler shift or by the action of an optoelectronic modulator.

There are two additional classes of frequency-modulated sensors: microresonator sensors, and subcarrier-modulated sensors. Microresonator devices are sensing elements in which the frequency of mechanical resonance is affected by the measurand [9]. Extensive work has been carried out in which silicon microresonators have been fabricated as sensing elements, are driven into oscillation optically, and interrogated optically in such a way that the intensity of the probe beam is modulated at the frequency of the resonator [10].

Subcarrier-modulated sensors are those in which the optical source is intensity modulated. The effect of the measurand on the sensing element is then to change either the frequency or the phase of the subcarrier. For example, the technique may be used to measure distance by comparing the phase of the subcarrier reflected from a remote target with the phase of the signal used to intensity-modulate the source [11, 12].

We shall consider the topic of frequency modulation again in our discussion of interferometry. Interferometry is the detection of the phase of an electromagnetic wave by mixing with a reference wave and observing the resultant intensity. Interferometry is widely used as a means of measuring optical path difference. In many situations it is convenient to frequency-shift the reference beam (localy oscillator) used in interferometry in order that the optical phase is recoverable as the electronic phase of the carrier at the offset frequency (heterodyne detection). Sensors can then be multiplexed by using different carrier frequencies for different sensing elements.

Further opportunities exist for the determination of optical path difference in interferometry by modulating the wavelength of the source in order to generate modulation frequencies at the output dependent on optical path difference.

4.2.4 Transfer function for two-beam interferometers

Figure 4.4(a) shows a prototypical optical fiber interferometer, of the Mach–Zehnder configuration. Light from an optical source is coupled into the downlead fiber, and is amplitude divided at a directional coupler into the signal and reference arms. The two arms recombine at a second coupler connected to the upleads, which terminate in photodetectors. At the photodetectors, optical interference is observed, such that the intensity observed at the lower detector is

$$I = \tfrac{1}{2}I_0(1 + V\cos\phi) \qquad (4.1)$$

and at the upper detector is

$$I = \tfrac{1}{2}I_0(1 - V\cos\phi) \qquad (4.2)$$

where I_0 is the total intensity falling on the two detectors (and is equal to the source intensity in an ideal lossless optical system), ϕ is the optical phase difference between the arms of the interferometer and V is the visibility of the interference. These equations assume a monochromatic source, identical lossless 50:50 couplers, and ignore effects due to birefringence.

The change in the sign of the modulation term between equations 4.1 and 4.2 arises from energy conservation, which determines the phase shifts occurring in the directional couplers. The 'reciprocal output' (one coupling and one transmission for each beam) takes the positive sign, and the other, 'nonreciprocal

BASIC PRINCIPLES OF OPTICAL FIBER SENSORS

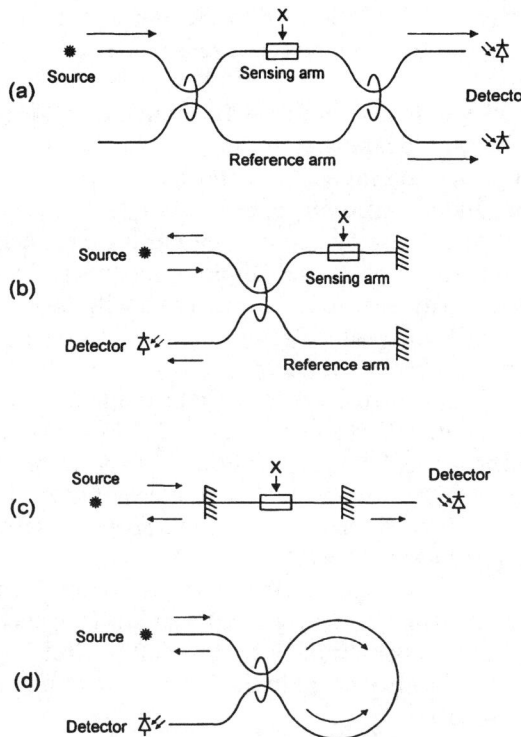

Fig. 4.4 Optical fiber interferometers: (a) Mach–Zehnder interferometer; (b) Michelson interferometer; (c) Fabry–Pérot interferometer; (d) Sagnac interferometer.

output', takes the negative sign. In this ideal case, the sum of the two outputs is constant, and they are thus said to be complementary. Power losses in real couplers affect these phase shifts, so that in practice the outputs are never perfectly complementary [13].

The optical phase difference is given by

$$\phi = \frac{2\pi L}{\lambda_0} \qquad (4.3)$$

where L is the optical path difference between the arms of the interferometer:

$$L = L_s - L_r \qquad (4.4)$$

where L_s and L_r are the optical path lengths of the signal and reference arms, which have physical lengths l_s and l_r and position-dependent effective refractive indices n_s and n_r, such that for a generalized situation for the ith interferometer:

$$L_i = \int_0^{l_i} n_i(z)\,dz \qquad (4.5)$$

The effective refractive index of the core of the fiber lies between the refractive index of the bulk material from which the core is made and the bulk index of the cladding material.

The visibility, shown as V in equation 4.1, is unity in an ideal interferometer, and is always positive. An ideal interferometer is one in which the recombining beams are of equal power and intensity, with wavefronts of the same shape and state of polarization, and are mutually coherent.

Equality of power and intensity depends on parameters such as the splitting ratios of the directional couplers and differential optical losses. Equality of shape of wavefronts of the beams is synonymous with saying that they have high mutual spatial coherence, which can be achieved by constructing the interferometer from single-mode optical fiber.

In many situations, unwanted birefringence in the fibers causes the states of polarization of the guided beams to be unequal. In cases where maximizing the visibility is important, active or passive control of the state of polarization is used [14]. Passive control techniques include the use of special birefringent elements, such as mirrors incorporating nonreciprocal polarization rotators [15], or by using special birefringent fiber.

The most common type of special fiber possesses high linear birefringence. Thus, power coupled into one of the eigenstates of birefringence remains in that state, even when the fiber is bent, inducing a little more birefringence. Fiber with high linear birefringence is also useful in the fabrication of polarimetric sensors, described in section 4.2.10.

4.2.5 Alternative interferometer configurations

The Mach–Zehnder has a transmissive configuration. For some applications it is desirable to have a reflective arrangement. An example is the Michelson configuration shown in Figure 4.4(b), where the signal and reference arms terminate in mirrors, which direct the guided beam back to recombine at the coupler originally used to split the input beam. One output is available at the otherwise unused coupler port, the other is directed back toward the source. The phase sensitivity of the Michelson is twice that of the Mach–Zehnder, because the guided beam traverses the sensing element twice.

Another interferometer often used in reflection is the Fabry–Pérot shown in Figure 4.4(c). This can be formed by using a reflective splice between the sensing element and downlead (the 'proximal' face), and a mirror on the other end of the sensing element (the 'distal' face). The sensing element is analogous to the signal arm in the Michelson interferometer, and interference occurs between light reflected at the distal face and the proximal face. For mirrors of high reflectivity, multiple reflections occur. However, in practice, most fiber Fabry–Pérots use low reflectivity mirrors, a configuration more properly called a Fizeau interferometer. It is then often sufficient to consider only the first-order reflections, and the two-beam transfer function of equation 4.1 adequately describes performance in reflection. The complementary output is directed forwards, but its visibility is low for the Fizeau type, because the interference arises between the strong transmitted beam and a weak beam that has been reflected by both the distal and proximal faces.

A configuration with special applications is the Sagnac interferometer shown in Figure 4.4(d). Here, the fiber is formed into a loop, so that interference is produced between the clockwise and anticlockwise-propagating beams. Most (so-called 'reciprocal') measurands produce equal phase changes in each beam, producing no phase shift at the detector. However, nonreciprocal measurands, such as angular velocity about an axis normal to the plane of the coil, produce a phase shift. The Sagnac interferometer is thus often applied as a gyroscope [16].

4.2.6 Interferometric fiber sensors

To see how the interferometer can act as a sensor, consider that the signal arm is exposed to a measurand whilst the reference arm is maintained in a constant environment. The measurand modulates the optical path length of the signal arm, L_s, and hence the optical path difference (OPD) of the interferometer, L. Thus, from equation 4.1, the effect of the measurand is to modulate the intensity output from the interferometer.

The change in L arises from a combination of a change in the effective refractive index of the guided mode and a change in the physical length of the fiber, such that in response to a change in measurand δX a corresponding change δL_s occurs, where

$$\delta L_s = \left(n \frac{\partial l_s}{\partial X} + l_s \frac{\partial n}{\partial X} \right) \delta X \qquad (4.6)$$

The change in effective refractive index is a function of the changes in the refractive indices of the bulk materials from which the fiber is made and the change in the fiber core radius induced by the measurand.

Consider the example of strain measurement. Axial strain makes the fiber physically longer, increasing l_s. Via Poisson's ratio, the axial extension reduces the core radius, causing a reduction in the effective index of the core effective index. The strain optic effect causes a further reduction in n_s. The physical extension term is the largest, and is positive. The effect of the reduced core radius is negligible. The strain optic effect is negative, and reduces the effect of the axial extension by typically 20%. Thus, at a wavelength of 633 nm, the optical phase change per unit axial extension $2\pi \delta l_s$ is about 6.5×10^7 rad m^{-1} [17].

Now consider the temperature sensitivity of the fiber. Here, the thermal expansion coefficient of fused silica is extremely small and the change in core bulk index is dominant, yielding (again at 633 nm) a sensitivity of about 100 rad K^{-1} in a meter of fiber [18].

4.2.7 The Doppler effect

One application of interferometers is in the measurement of velocity. Consider the arrangement of Figure 4.5 in which a fiber Michelson interferometer is adapted so that the signal arm terminates in a probe that directs light onto a remote target and captures light scattered from it. Suppose that the target is approaching the probe along its optical axis at some speed v_D. From equations 4.1 and 4.3, the interferometer phase changes continuously at a rate

$$\frac{\partial \phi}{\partial t} = \frac{4\pi v_D}{\lambda} \tag{4.7}$$

Thus the output intensity measured by the detector will be modulated at a frequency proportional to the target speed:

$$f = \frac{1}{2\pi} \frac{\partial \phi}{\partial t} = \frac{2v_D}{\lambda} \tag{4.8}$$

This frequency shift is equivalently described as the Doppler effect.

4.2.8 Spectral effects in interferometers

When light from a broad-band source is transmitted through an interferometer, its spectrum is modified. To understand why this should be so, we need to consider the relationship between phase in a two-beam interferometer and source wavelength. This is given in equation 4.3, and can be rewritten as

$$\phi = \frac{2\pi L \nu}{c} \tag{4.9}$$

for the Mach–Zehnder, where ν is the optical frequency and c is the speed of light *in vacuo*. Thus we see that the phase depends on frequency, and hence the intensity is also dependent on frequency.

Consider that we illuminate the interferometer with a source of uniform spectral density (a 'white-light' source). Suppose that at some frequency ν_0 we have $\phi = 2N\pi$, where N is integral, thus signal and reference beams are in phase and the output is a maximum at that frequency. At some frequency $\nu_0 + \delta\nu$, where

$$\delta \nu = c/L \tag{4.10}$$

we see that

$$\phi = 2\pi(N+1) \tag{4.11}$$

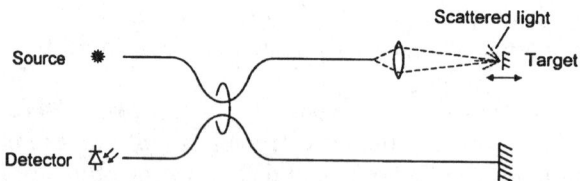

Fig. 4.5 Michelson interferometer as velocimeter.

BASIC PRINCIPLES OF OPTICAL FIBER SENSORS   127

and the output is again a maximum. However, for another frequency shift

$$\delta v' = c/2L \quad (4.12)$$

and we see that

$$\phi = 2\pi(N + \tfrac{1}{2}) \quad (4.13)$$

and the output is a minimum, zero for unit visibility. The output spectrum from the interferometer is shown in Figure 4.6.

The frequency interval δv is called the free spectral range (FSR). We see from equation 4.10 that a measurement of the FSR yields the OPD, L.

4.2.9 Signal processing

It should be evident from equation 4.1 that a single measurement of output intensity is insufficient to yield the interferometer phase, ϕ, and hence the measurand. There are two problems: how to identify the order of interference, and how to find the phase, modulo 2π.

We can write

$$\phi = 2N\pi + \phi' \quad (4.14)$$

where N is an integer defined as the fringe order and ϕ' is the phase modulo 2π. We shall first concentrate on methods to recover ϕ', which we shall denote simply as ϕ in the following.

Phase recovery depends on external control of the reference phase. Let us write

$$\phi = \phi_{\text{sig}} + \phi_{\text{amb}} - \phi_{\text{ref}} \quad (4.15)$$

where ϕ_{sig} is the measurand-induced phase that we wish to recover, ϕ_{amb} describes unwanted phase changes induced by the environment, and ϕ_{ref} is the externally controlled phase in the reference arm of the interferometer.

There are three basic ways in which ϕ_{sig} may be recovered: using homodyne, heterodyne and phase-stepped techniques. In the homodyne technique, an electronic servo is used to process the detector output to generate an error signal, which is in turn used to drive an actuator that controls ϕ_{ref} in order to maintain ϕ at some fixed value [19, 20]. The value chosen is normally

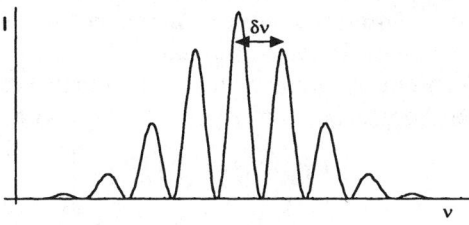

Fig. 4.6 Transmission spectrum of unbalanced two-beam interferometer.

$$\phi = 2\pi(N + \tfrac{1}{4}) \tag{4.16}$$

where $\partial I/\partial \phi$ has its greatest value—the so-called quadrature condition. If the phase change is restricted to a sufficiently small range that no electronic servo is required, ϕ may be recovered directly from the interferogram intensity. This is a restricted form of passive homodyne processing.

In a multiplexed system it is rarely convenient to provide each of the sensing elements with its own modulator, and heterodyne processing techniques are often preferred. In a heterodyne technique, a frequency offset, ω, is introduced between signal and reference beams, so that the intensity at one of the outputs becomes

$$I = \tfrac{1}{2} I_0 [1 + V \cos(\omega t + \phi)] \tag{4.17}$$

so that ϕ is readily recovered from electronic phase demodulation of the heterodyne carrier ωt, using conventional techniques [21].

Lastly, the technique of phase-stepping involves acquiring a set of intensity outputs from the interferometer, each separated by some reference phase difference [22]. As an example, consider a four-step technique with a phase step magnitude of $\pi/2$. Four values of intensity are then acquired, I_1, I_2, I_3 and I_4, with corresponding values of ϕ_{ref} of $0, \pi/2, \pi$ and $3\pi/2$.

The intensities can then be processed to find ϕ using

$$\tan \phi = \frac{I_4 - I_2}{I_1 - I_3} \tag{4.18}$$

4.2.10 Polarimetric sensors

The concept of fibers with high linear birefringence was mentioned above. We may consider them as guiding two orthogonal linear polarization eigenstates with effective refractive indices n_f and n_s for the modes with 'fast' and 'slow' phase velocities. In a polarimetric sensor, a single length of highly birefringent fiber serves as the sensing element, with the fast and slow modes analogous to the signal and reference arms [23]. For axisymmetric measurand fields, such as axial strain and temperature, the two modes experience the measurand equally, but exhibit different sensitivities.

To see how the polarimetric sensor operates, consider that a linearly polarized light source is coupled into the fiber with a polarization azimuth at 45° to the eigenaxes, equally populating the modes. At the exit from the fiber a polarization analyzer is placed with its azimuth again at 45°. The two modes are thus resolved equally and along the same azimuth such that they can interfere on the face of a detector, to produce an intensity (by analogy with equation 4.1) of

$$I = \tfrac{1}{2} I_0 \left[1 + V \cos\left(\frac{2\pi}{\lambda}\right) L \right] \tag{4.19}$$

Once again, we can define a measurand sensitivity by

$$\delta L = \left[l \frac{\partial (n_f - n_s)}{\partial X} - (n_f - n_s) \frac{\partial l}{\partial X} \right] \delta X \qquad (4.20)$$

The first term represents the effect that the measurand has on the physical length of the fiber, and the second the change in the effective refractive index.

A special case of the polarimetric sensor is the detection of magnetic fields via the Faraday effect. When a beam of linearly polarized light travels through a medium in a direction where there exists a nonzero component of magnetic field, then the polarization azimuth is rotated by an amount proportional to the field component. The polarization rotation is equivalent to a phase shift between left and right circularly polarized states. Thus the Faraday effect can be considered as magnetically induced circular birefringence [24].

4.2.11 Multiple wavelength and low-coherence interferometry

We now consider the problem of discovering the order of interference. The simple interferometer can yield ϕ only to mod(2π). Thus the unambiguous measurement range is restricted to an OPD change of λ.

We can extend the unambiguous range by multiple wavelength illumination [25]. Suppose that we illuminate the interferometer by light with wavelength λ_1, such that the corresponding phase is

$$\phi_1 = 2\pi L / \lambda_1 \qquad (4.21)$$

We then repeat the experiment at a new wavelength λ_2, where

$$\phi_2 = 2\pi L / \lambda_2 \qquad (4.22)$$

The unambiguous measurement range now corresponds to

$$\phi_1 - \phi_2 = 2\pi \qquad (4.23)$$

In other words, the unambiguous range expressed as an OPD change is

$$\lambda_{\text{eff}} = \lambda_1 \lambda_2 / (\lambda_1 - \lambda_2) \qquad (4.24)$$

Thus as the wavelength difference is reduced, the unambiguous range is increased concomitantly. A practical limit is set by the difficulty of stabilizing the wavelengths.

The range can be increased yet further by taking the logical extension of multiple wavelength interferometry to interferometry using a broad-band source [26]. To understand the usefulness of the technique, we must reconsider the question of optical temporal coherence and its effect on visibility.

To appreciate the effect of temporal coherence we shall make use of the helpful, although approximate, concept of coherence length. Consider an optical source with centre frequency v_0 and a smooth spectrum with spectral width δv, as shown in Figure 4.7. The coherence length is given approximately by [27].

$$L_c \sim c / \delta v \qquad (4.25)$$

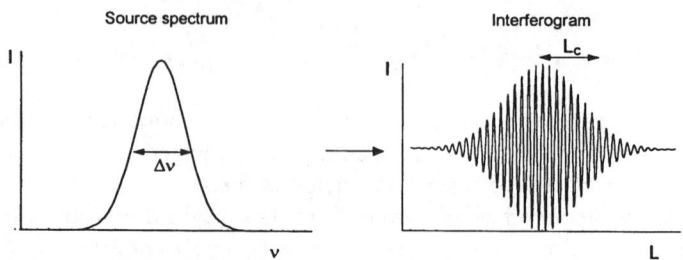

Fig. 4.7 Spectrum and coherence length.

If the OPD in an interferometer is much greater than the coherence length, $L \gg L_c$, then the visibility tends to zero. As L is reduced to below L_c the visibility improves, reaching a maximum at $L = 0$, when the fringe order is zero. Thus, when a broad-band source is used, the central (or zeroth-order) fringe can be identified because it is the one with the highest visibility.

Low-coherence interferometry is often carried out using the tandem arrangement shown in Figure 4.8. The system comprises two interferometers: a sensing interferometer (S) and a receiving interferometer (R). Each interferometer has two arms, a scanned arm (s) and a reference arm (r). The path differences of the individual interferometers are each too great to yield significant visibility, thus

$$L_i = L_{si} - L_{ri} \gg L_c \quad i = \text{S}, \text{R} \tag{4.26}$$

where L_i is the path length imbalance in the ith interferometer (sensing or receiving), L_{si} is the path length of the signal arm of the ith interferometer, and L_{ri} the length of the reference arm. Consider now that the path imbalance in the receiving interferometer is adjusted such that

$$|L_\text{S} - L_\text{R}| < L_c \tag{4.27}$$

There are now four possible paths through the interferometer, with path lengths given by

$$L_{ss} = L_{sS} + L_{sR} \tag{4.28a}$$

$$L_{sr} = L_{sS} + L_{rR} \tag{4.28b}$$

$$L_{rs} = L_{rS} + L_{sR} \tag{4.28c}$$

$$L_{rr} = L_{rS} + L_{rR} \tag{4.28d}$$

Each of these paths contributes to the mean intensity of the resultant interferogram, but only L_{sr} and L_{rs} are close enough in length to combine to produce interference with significant visibility:

$$L_{sr} - L_{rs} = L_{sS} + L_{rR} - L_{rS} - L_{sR} = L_\text{S} - L_\text{R} < L_c \tag{4.29}$$

thus fringes are observed, with maximum visibility when

$$|L_\text{S} - L_\text{R}| = 0 \tag{4.30}$$

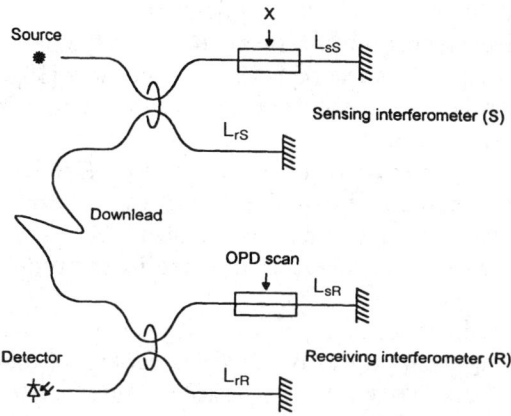

Fig. 4.8 Tandem interferometers.

superimposed on a 'pedestal' caused by the intensity of the transmission of all four paths. Since two of these paths generate no interference, the maximum fringe visibility is 0.5.

The above account of broad-band interferometry is intended to be sufficient to follow the arguments in the remainder of the chapter, but is very much simplified. The interested reader is referred to reference [26].

4.2.12 Components and modulators

We have seen that in order to recover the signal phase in an interferometer it is necessary to have control over the reference phase. The simplest of all possible fiber-optic phase modulators is the piezoelectric type, where part of the fiber is wrapped around a piezoelectric cylinder [28]. The cylinder expands in response to an applied voltage, straining the fiber and thus modulating the phase of the guided beam. Such modulators are appropriate for homodyne, heterodyne and phase-stepped schemes.

In heterodyne schemes, the frequency offset may be synthesized by periodic phase modulation. The case of serrodyne (saw-tooth) modulation is easiest to visualize. This generates a constant rate of change of phase, equivalent to a frequency shift [29].

There are practical difficulties in generating perfect serrodyne modulation, caused by the finite fly-back time, and the discontinuities in the temporal derivatives at the ends of the ramp. In consequence, most practical schemes instead use harmonic modulation, with the phase-modulated FM carrier synthesized from the first two harmonics of the frequency spectrum of the phase-modulated signal (i.e. the first two harmonics of the frequency of the modulating sine wave).

Piezoelectric elements are widely used for path-length scanning in broad-band interferometry, although often bulk-optic systems with moving mirrors are used instead. Alternatively, a full-field bulk-optic interferometer can be used as the receiver, incorporating shear such that the path length changes as a function of position across the detector field of view [30]. Thus an electronic array detector can be used to electronically scan the path imbalance: one such setup is described in section 4.7.

In many interferometer arrangements, the use of active piezoelectric phase modulators is unacceptable. As an alternative way of producing phase modulation, the source wavelength can be tuned instead. By equation 4.3, wavelength tuning combined with an unbalanced interferometer produces a phase modulation.

Wavelength modulation is especially attractive with diode laser sources, where the wavelength can be easily tuned via control of their injection current. A disadvantage of operating with unbalanced interferometers is that source frequency noise is transduced to phase noise that is indistinguishable from the signal [31].

An interesting feature of carrier generation by wavelength modulation is that the carrier frequency is dependent on the OPD in the interferometer. For example, suppose that the source laser is serrodyne modulated at a rate $\partial v/\partial t$. The generated carrier frequency is thus

$$f = (L/c)(dv/dt) \qquad (4.31)$$

The result offers two possibilities. Firstly, the same laser could be used to generate a set of different carrier frequencies in interferometers of unequal OPDs, with the opportunity for frequency-division multiplexing. Secondly, the carrier frequency f is directly proportional to L, so that L can be found without fringe-order ambiguity: such a frequency-chirped range measurement is known as a frequency-modulated continuous wave technique (FMCW) by analogy with radar [32].

Finally, various passive phase control techniques are available. For example, for phase stepping we require outputs separated in phase by angles other than pi rads. Directional couplers with three input and three output ports (3 × 3 couplers) give a $2\pi/3$ phase difference between their outputs and so are suitable phase-biasing elements [33]. Similarly, in polarimetric schemes a linear retarder (waveplate) generates a passive phase bias between outputs [34].

4.2.13 In-fiber grating sensors

A final class of single-mode fiber sensing elements are represented by in-fiber gratings, which are most usually regarded as wavelength-modulated sensors. The best-known type is the in-fiber Bragg grating (FBG). The grating is formed from periodic variations in effective refractive index in the axial direction along the core of the fiber. The grating period is designed to produce phase-matched coupling between the forward and backward-propagating fundamental guided mode, the LP_{01} mode [35].

The gratings are produced by exposing the fiber to short-wavelength visible or (more commonly) ultraviolet (UV) radiation. The most usual manufacturing technique is to expose the fiber through the side of the cladding using an excimer or other UV laser via a phase mask of suitable pitch.

The phase-matched coupling coefficient is wavelength-sensitive. If the fiber is used to guide broad-band light, then a range of wavelengths around the phase-matching wavelength are reflected. The grating can act as a sensor in the following way. The grating period is environmentally sensitive, in the same way as for an interferometric sensor, via changes in effective refractive index and length of the grating. The phase-matching wavelength, and hence the reflection spectrum of the grating, therefore shifts with temperature and strain [36].

More recently, other types of mode-coupling grating have been developed. One of these is the 'long period' grating (LPG), designed to couple the LP_{01} mode to radiative cladding modes, thus producing a wavelength-dependent loss. As for the Bragg gratings, the spectrum is modulated by stimuli such as temperature and strain [37].

Various techniques have been developed for the interrogation of grating sensors [35]. The most common is probably the use of a commercial optical spectrum analyzer, to detect changes in the center reflected wavelength when the grating is illuminated by a broad-band source. A simple alternative is to use a fixed spectral filter as a detector, so that the transmitted intensity is a function of the grating spectrum. The grating is thus interrogated as an intensity-modulated sensor. Other interrogation techniques include the use of electronically tunable filters or interferometers as spectrally selective elements.

4.3 MULTIPLEXING TOPOLOGIES

4.3.1 Parallel networks

One means of classifying multiplexing schemes is by their topology. Perhaps the simplest arrangement is the parallel scheme shown in Figure 4.9(a). Most often a single source, but occasionally more than one, is coupled into the fiber network, and the power distributed by a multiple-port directional coupler, or network of couplers, into a set of parallel downleads. Each downlead contains a single sensor. The sensors may be reflective or transmissive. For the moment, we shall assume them to be reflective. Power returns along the same downlead and is directed back to the detector array.

In the simple version of the technique shown in Figure 4.9(b), there is one detector for each sensor. If the couplers used to tap off power to the detectors have coupling ratios of 50:50 (the optimum value in terms of power budget), and the N-sensor network is otherwise lossless, then the detected power per sensor is

$$P_i = P_0/4N \tag{4.32}$$

where P_0 is the power coupled from the source into the network.

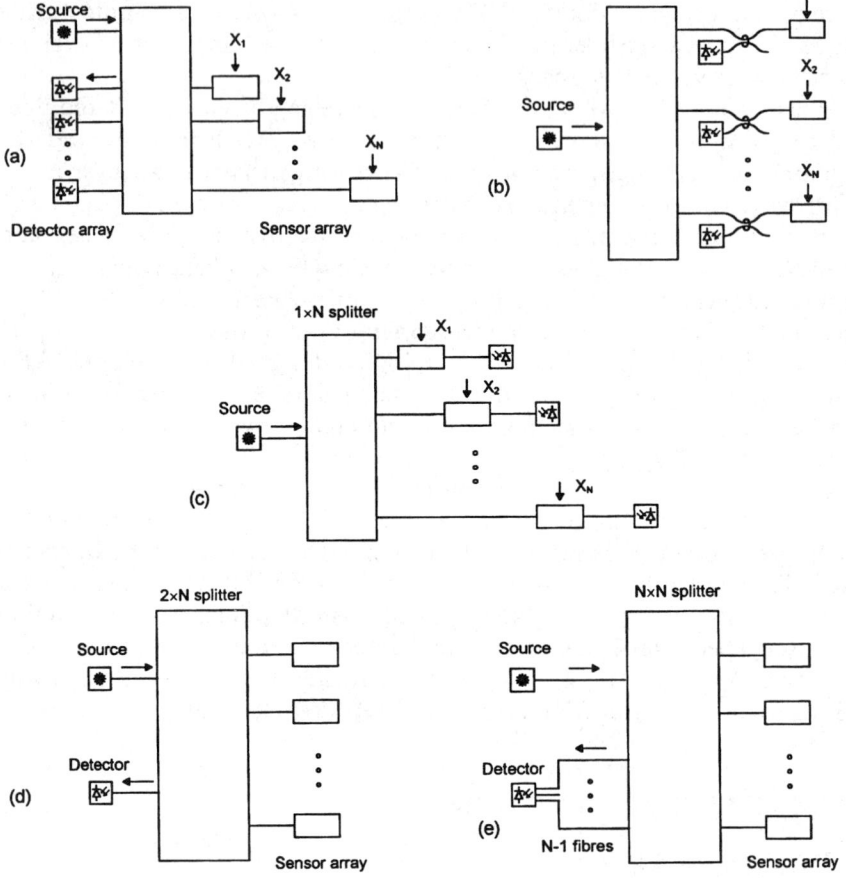

Fig. 4.9 Parallel multiplexed sensors: (a) reflective arrangement; (b) reflective arrangement, single detector per sensor; (c) transmissive arrangement; (d) single detector; (e) single detector, more efficient arrangement.

The very simple topology shown in the figure has a detector and downlead per sensor, so that the only gain in multiplexing is in sharing a single source between the sensors, at the cost of the additional couplers required for power distribution, and the reduced power per sensor. On the other hand, the detector outputs require no decoding to identify the sensors, so that cross-talk between the sensors is minimal.

The power budget is improved by using the transmissive arrangement indicated by the alternative detector positions in Figure 4.9(c). No tap couplers are now required, so that the power per sensor is

MULTIPLEXING TOPOLOGIES

$$P_i = P_0/N \tag{4.33}$$

However, in most practical arrangements it will be necessary to house the detectors with the source, so that the number of fibers is effectively doubled.

There is no requirement for there to be one detector per sensor if one of the decoding schemes described in later sections is used. For example, Figure 4.9(d) shows the simplest single-detector reflective arrangement, using a $1 \times N$ splitter. The power per sensor is now only

$$P_i = P_0/N^2 \tag{4.34}$$

thus quickly becoming unacceptably low as the number of sensors increases. However, using an $N \times N$ splitter as shown in Figure 4.9(e), the power from all $N - 1$ free return fibers can be combined onto a single detector, so that the power per sensor is then

$$P_i = P_0(N-1)/N^2 \approx P_0/N \tag{4.35}$$

where the approximation is valid for large N. In this case, the loss drops off far less rapidly with number of sensors. A similar arrangement can be used to combine n sources with $N - n$ downleads to detectors in order to increase the total received power. Alternative arrangements with some intermediate number of detectors are equally feasible.

4.3.2 Serial topologies

Figure 4.10(a) shows a serial arrangement. Here a single source and downlead are used to illuminate all of the sensors, disposed linearly along the fibre. For such an arrangement some decoding scheme is definitely required. The power budget for a single-detector arrangement is advantageous in comparison with the parallel arrangement. However, unless each sensor uses a different portion of the source spectrum, the power reaching sensors further away from the source will be diminished by the insertion loss or reflection occurring in the earlier sensors. In a reflective arrangement the insertion loss is experienced twice, once in each direction. Further loss is experienced in the tap coupler, assumed to have a 50:50 split. Thus, the detected power for the Nth sensor is

$$P_i = P_0 R T^{2(N-1)}/4 \tag{4.36}$$

where T is the transmission of a sensor, R is the power reflected by a sensor, and where the sensors are assumed similar. The 6 dB loss from the tap coupler can be reduced under some circumstances. For example, polarization techniques can be used to form an optical circulator [38]. Alternatively, certain sensors return an optical signal at a wavelength different from that of the source, and so can be separated using dispersive elements. Sensors that operate in transmission and can be serially multiplexed can benefit from the far more efficient arrangement shown in Figure 4.10(b).

Some sensors, such as Bragg gratings, are wavelength selective, and can have very low insertion loss or reflection outside their narrow range of operating wavelengths. Hence, in a wavelength-divided array, where each sensor has a different center wavelength, it is possible to multiplex a large number of sensors into a single array whilst retaining an acceptable power budget.

4.3.3 Ladders

An example of a ladder arrangement is shown in Figure 4.11(a). It comprises a source and downlead together with an uplead and detector, with sensors arranged like the rungs of a ladder. The sensors are transmissive, and could be, for example, Mach–Zehnder interferometers or intensiometric sensors. The ladder has much in common with the standard serial topology, in that it requires only a single source and detector.

The arrangement shown in Figure 4.11(a) is reflective, in the sense that the source and detector are at the same end of the array. In consequence, if the directional couplers used to tap light from the downlead into the sensors and from the sensors into the uplead are all of the same coupling ratio, then the power available per sensor falls with distance toward the distal end of the array.

The use of couplers with a fixed ratio is desirable for reasons of economy, where it is advantageous to purchase commercially standard components. Various techniques have been used to attempt to equalize the available power to each sensor. One solution involves the use of rare earth-doped fiber sections to serve as amplifiers to increase the power available at greater distances into the array. However, all such refinements add considerably to the complexity of the arrangement.

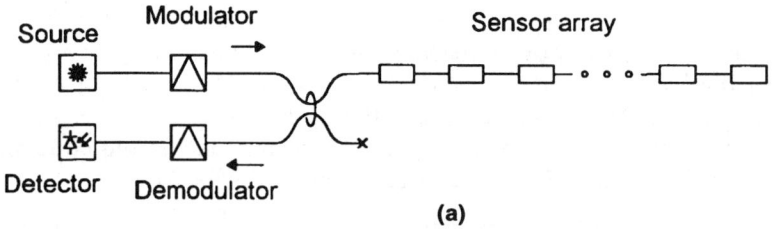

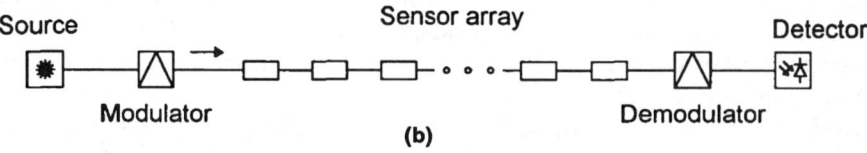

Fig. 4.10 Serially multiplexed sensors: (a) reflective arrangement; (b) transmissive arrangement.

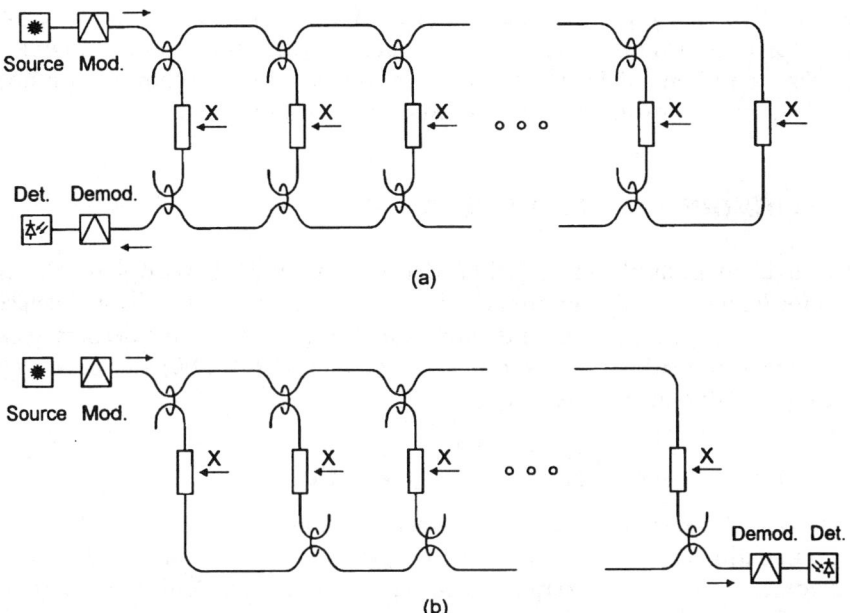

Fig. 4.11 Ladder networks: (a) reflective ladder network; (b) transmissive ladder network. Mod. = modulator, Demod. = demodulator, Det. = detector.

A simple approach to power equalization is provided by the transmissive ladder shown in Figure 4.11(b). Here, the basic arrangement is as before, except that the source and detector are now at opposite ends of the array. In consequence, light travelling from the source to the detector via any sensor traverses the same number of couplers, so that the power per sensor is easily equalized.

4.3.4 Nested arrays

For the ladder topology, any of the sensors can be replaced by a ladder subarray to form a nested array. An example is shown in Figure 4.12. Effectively, the downlead and uplead for an individual sensor in the simple array becomes the downlead and uplead for a ladder in the subarray.

4.3.5 Compound arrays

It is possible to combine more than one type of topology in a single system to form a compound array, thus increasing the total number of sensors. A simple example is to use a parallel arrangement to split light from a single source into a set of serial arrays, as shown in Figure 4.13. The sensors are divided into m groups, where each group is distinguishable in the decoding scheme (for example, by using a different heterodyne carrier frequency or different wavelength for each group). There are n sensors in each group, and n downloads, such that the ith member of each group is connected to the common ith downlead. Thus $m \times n$ sensors are multiplexed with only n downloads.

4.4 TIME DIVISION MULTIPLEXING

Time division multiplexing (TDM) identifies individual sensors by the time taken for light to travel from the source to the detector via the sensor. The use of different means of measuring this time delay leads to other time domain techniques, such as optical frequency domain reflectometry (OFDR) and polarization optical time domain reflectometry (POTDR).

4.4.1 Optical time domain reflectometry (OTDR)

OTDR was developed principally as a network diagnostic for optical fiber telecommunications systems, as a logical progression from electrical time domain reflectometry. The basic arrangement shown in Figure 4.14 is used to map attenuation as a function of length along a simple fiber.

The test fiber is illuminated with a short pulse of illumination of duration δt at time $t = 0$. As the light propagates through the fiber it experiences Rayleigh scattering, and a proportion of the scattered light is recaptured by the fiber core to be guided back toward the detector. Thus, the power recorded by the detector varies with time according to

$$P(t) = K \int_{z=0}^{t/2v_g} \exp[-2\alpha(z)] dz \qquad (4.37)$$

where α is the position-dependent fiber attenuation, v_g the group velocity of the pulse of light in the fiber, and K a constant determined by the launched power and the strength of Rayleigh backscatter. The attenuation coefficient of a

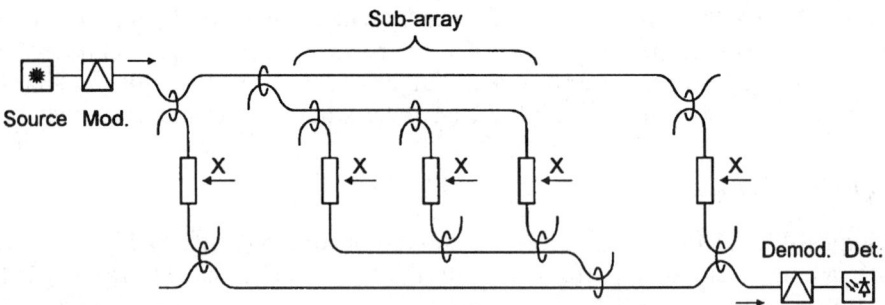

Fig. 4.12 Nested ladder network. Mod. = modulator, Demod. = demodulator, Det. = detector.

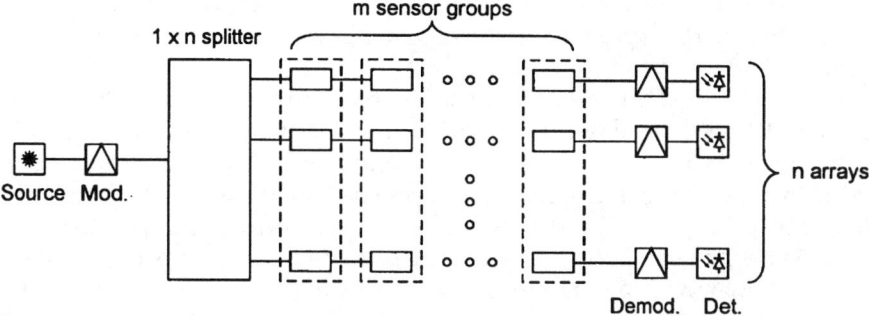

Fig. 4.13 Serial/parallel sensor array. Mod. = modulator, Demod. = demodulator, Det. = detector.

uniform region of the fiber is found from the gradient of the log-linear power versus distance curve, while discontinuities can be identified by reflection peaks and regions of high local attenuation. The backscattered signal is weak, so several pulses and signal averaging are usually required to obtain a good signal to noise ratio.

Consider now the more complicated situation where a set of sensors is deployed along the fiber in a serial array, where we shall assume that the effect of the measurand on a sensor is to modify the attenuation of the section of fiber within which it is incorporated. By using OTDR, the attenuation at each sensor is recovered, from which the measurand is derived [7].

One type of TDM array of current interest uses Bragg gratings as sensing elements. We saw in the previous section that a Bragg grating acts as a narrow-band reflector, where the center wavelength is measurand-dependent. In the array, a narrow-band source is used, tuned slightly off the reflectance peak. Consequently, the reflectance of an individual sensor is measurand-dependent. Thus, when deployed in a TDM array, the reflectance of each sensor, and hence the value of the measurand for each sensor region, is determined.

A complication with the simple TDM serial array using sensors based on intensity modulation is that the response of each sensor is modified by the

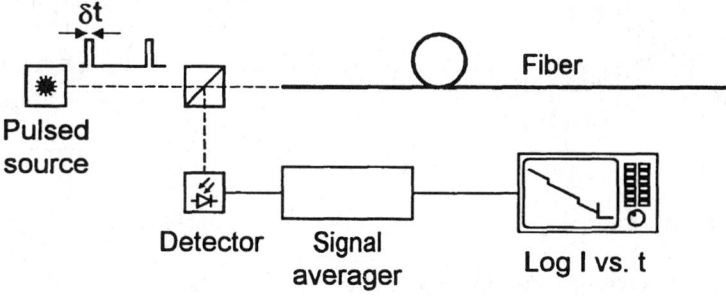

Fig. 4.14 Optical time domain reflectometry (OTDR).

response of earlier sensors in the same array. However, this difficulty can be avoided in sensors exploiting frequency or phase modulation (such as interferometers).

An example of TDM interferometers is shown in Figure 4.15. The arrangement was designed for application as a hydrophone array, where the sensing elements are fiber Fabry–Pérot interferometers [39]. The interferometers are formed by reflective splices uniformly disposed along the fiber. The array is illuminated by pulse pairs, f_1 and f_2, with a separation corresponding to the propagation time between a pair of splices. The second pulse of the pair is frequency shifted by an amount $\delta\nu$ relative to the first of frequency ν using an acousto-optic modulator. The ith sensor is defined by the ith and $(i+1)$th splice. The signal at frequency ν from the $(i+1)$th splice and $\nu + \delta\nu$ from the ith splice arrive at the detector together to give a heterodyne carrier at $\delta\nu$ phase modulated by path-length changes in the ith sensor.

Time division multiplexed systems need not be simple serial arrays. For example, in a reflective ladder arrangement, TDM may be used, provided that the time-of-flight difference between adjacent sensors is greater than the time resolution of the OTDR (usually set by the pulse duration of the illumination).

Many other possibilities exist, such as parallel arrays where the elements contain delay lines each of different length.

4.4.2 Optical frequency domain reflectometry (OFDR)

In OTDR measurements the spatial resolution (and hence the density with which sensors can be deployed) is set by the pulse duration, giving a practical minimum limit in the order of centimeters, but more often in the order of meters. A consequence of using short pulses is that the pulse energy is limited, raising the shot-noise-limited sensor resolution.

Alternative techniques using continuous wave frequency modulated lasers are available, yielding the OFDR technique. The basic principles can be understood

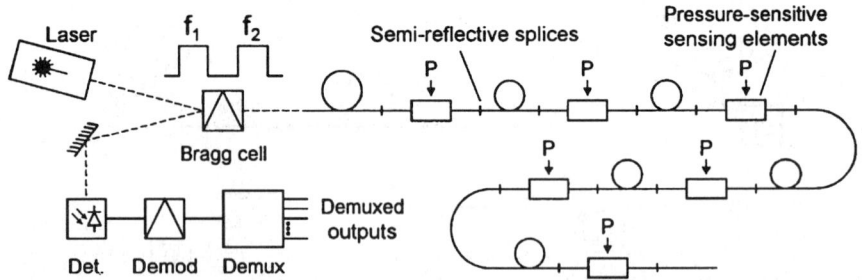

Fig. 4.15 Time division multiplexed interferometers. Det. = detector, Demod. = demodulator.

from the simple case in which the source optical frequency is serrodyne (sawtooth ramp) modulated, such that

$$v(t) = v_0 + \delta v \frac{t \bmod(\tau)}{\tau} \qquad (4.38)$$

where τ is the ramp period. Consider that the source is used to illuminate an interferometer of fixed path imbalance L. Hence, the phase of the interferometer is continuously modulated as a result of the frequency modulation, such that

$$\frac{d\phi}{dt} = \frac{2\pi}{c} L \frac{dv}{dt} \qquad (4.39)$$

so that the interferometer output is at a frequency

$$f = \frac{1}{2\pi} \frac{\partial \phi}{\partial t} = \frac{L}{c} \frac{\delta v}{\tau} \qquad (4.40)$$

Hence the frequency is proportional to the path imbalance.

The idea of using a frequency-modulated source to generate a beat frequency in an interferometer proportional to path imbalance is exploited in OFDR. Consider the arrangement shown in Figure 4.16, where a fiber containing an array of reflectors (defining sensing elements) is illuminated by a chirped (frequency-modulated) source. Light returned from the reflectors is mixed at the detectors with a reference wave. Thus the signals from each reflector are encoded with frequencies proportional to the distance of the reflector from the source.

Because of the practical difficulties in generating good approximations to serrodyne modulation, modifications of the technique using harmonic frequency modulation are more often used—the so-called **phase-generated carrier technique**, PGC [40, 41]. OFDR can, in principle, be used with the same array topologies as are appropriate for OTDR. For example, consider the ladder array shown in Figure 4.11(a), in which the sensing elements are intensiometric (i.e. the transmittance of a sensing element is a function of the measurand). If the first sensor is replaced by a reference fiber, the light transmitted by this reference will be mixed with the light returned from each of the sensors. The optical path from source to detector is different for each sensing element. Thus, when the source is chirped, the transmittance of each sensor is encoded as an amplitude modulation of a carrier whose frequency identifis the sensor.

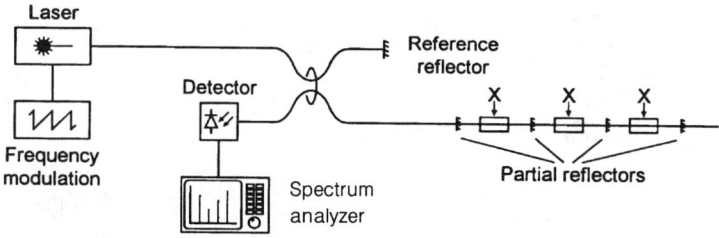

Fig. 4.16 Optical frequency domain reflectometry (OFDR).

In order to avoid crosstalk between sensors, the carrier and sidebands for each sensor must not overlap with those of an adjacent sensor. Hence, the OPD between sensors must be sufficiently large taking into account the chirp amplitude and the bandwidth of the measurand. Because the carriers are generated by a periodic frequency chirp, the carriers are distorted and their harmonics are also present. Thus the path imbalances must be chosen not only to prevent overlap of the fundamentals of their carriers and associated sidebands, but also of their harmonics. It is this effect that generally restricts the number of discrete sensors that can be multiplexed by OFDR [42].

The sensing elements may also be configured as part of an array of interferometers, in which the measurand modulates the optical path length of the sensing element. However, in a simple arrangement such as the ladder shown in Figure 4.11(a), the downlead and uplead form part of the interferometer, and the region over which the measurement is made cannot be defined. Instead, the principles of OFDR using a PGC are used extensively in the multiplexing of interferometers using an arrangement such as the one shown in Figure 4.17. The interferometers are arranged in a ladder array, where each interferometer has a different path imbalance, sufficient to allow the interferometers to be distinguished at the receiver by their different carrier frequencies. OFDR-based multiplexing of interferometers has been used with great success in hydrophone arrays [39]. Because the different sensors are identified by their carrier frequencies, the multiplexing of interferometers by PGC-based OFDR is usually classified as **frequency division** rather than time division multiplexing.

It is difficult to analyze the general case of distributed coupling of arbitrary strength in an OFDR system. Consequently, in practical implementation it is convenient to be able to isolate relatively small parts of the array using short illumination pulses, thus effectively using a combination of OTDR and OFDR.

4.4.3 Time domain techniques in birefringent fiber

The principles of OTDR and OFDR have a special application in birefringent fiber, where they may be used to map the strength and location of mode coupling points along the fiber. This technique is based on the fact that an optical path length imbalance ΔL accumulates between the polarization eigenmodes

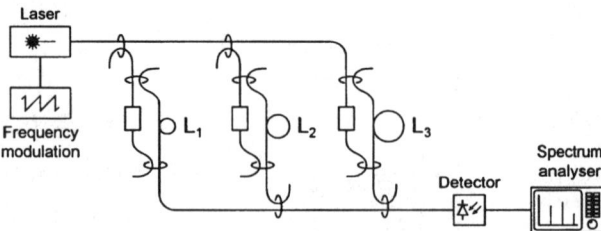

Fig. 4.17 OFDR with ladder of interferometers.

of the fiber, proportional to the physical distance travelled in the fiber l (assuming the birefringence to be uniform). We may define the refractive indices for the fast and slow eigenmodes in the fiber as n_f and n_s, so that

$$\Delta L = (n_f - n_s)l \tag{4.41}$$

The magnitude of the birefringence is often described in terms of a polarization beat length l_B, which is the physical propagation distance corresponding to a change in relative phase of the eigenmodes by 2π radians. Hence, ΔL increases by a wavelength for each beat length. A typical value of beat length for a commercially available highly birefringent fiber is in the order of 2 mm. Hence, at a wavelength of 780 nm, ΔL increases by about 0.39 mm per meter of fiber.

The principle of this type of measurement is illustrated in Figure 4.18. The source is coupled into the fast eigenmode of a highly birefringent fiber, which serves as both the downlead and the sensing element. Where the fiber experiences a strong localized transverse stress, light is coupled from the fast to the slow eigenmode. The power transfer is maximized if the direction of the stress is at an angle of $\pi/4$ to the birefringence eigenaxes. Light from the two eigenmodes is combined using a polarizer angled at 45° to the eigenaxes of the fiber so that interference between these modes is observed at the detector.

Consider the effect of a single stress point at a distance z along a fiber of total length l. The optical path length between the source and the coupling point is zn_f; the coupled light then travels an optical path distance $(l-z)n_s$ in the slow eigenmode to reach the detector, so that the total optical path travelled by the coupled light is

$$L_{\text{sig}} = zn_f + (l-z)n_s \tag{4.42}$$

whereas the uncoupled light travels an optical path length $L_{\text{ref}} = ln_f$. The optical path difference between the coupled and uncoupled beams is thus

$$\Delta L = (n_f - n_s)(l - z) \tag{4.43}$$

thereby providing a path imbalance dependent on the position of the point of stress.

ΔL is far smaller than l, and so the delay differences between the two paths are correspondingly small. The delay resolution of OTDR is therefore too poor

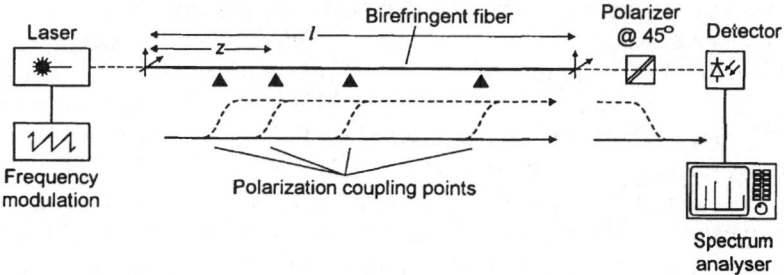

Fig. 4.18 Disturbance location in birefringent fiber.

to make this measurement on all but the longest fiber. OFDR, however, is capable of measuring delay on this scale. The path difference is used to generate a carrier frequency by chirping the source, exactly as in the standard OFDR technique. Thus the magnitude of the stress-induced mode coupling is revealed by the amplitude of the signal at a given carrier frequency, where the carrier frequency gives the position of the coupling point. This technique has been successfully used to multiplex sensors that are defined in birefringent fiber by introducing mode-coupling at specific points along the fiber.

In principle, we see how the concept may be extended for distributed measurement, where the stress varies continuously along the fiber. The stress distribution is revealed by the output intensity as a function of carrier frequency. Such a simple view, however, relies on the assumption that the coupling is sufficiently weak that we may assume only first-order coupling, i.e. we assume that once power is coupled from one mode to another at a particular coupling point, it is not coupled back to the original mode at some subsequent coupling point.

4.4.4 Optical coherence domain reflectometry: (OCDR)

In standard OTDR, the spatial resolution is set by the distance light travels during the illumination pulse—hence good resolution requires very short pulses. In comparison, in the OCDR technique, the 'effective pulse duration' is the **coherence time** of the optical source τ_c, which is inversely proportional to the spectral width, δv, such that $\tau_c \sim 1/\delta v$. Thus, the free-space spatial resolution of the OCDR technique is set by the source coherence length

$$l_c \sim c\tau_c = c/\delta v = \lambda^2/\delta\lambda \tag{4.44}$$

where $\delta\lambda$ is the wavelength spread of the source. As an example, a source with a spectral width of 100 nm at a mean wavelength of 1 μm thus has a coherence length of 10 μm. To achieve such spatial resolution with OTDR would require an impractical pulse duration of 33 fs.

The principles of OCDR are most easily understood by first considering the much simpler arrangement of illuminating two free-space interferometers in tandem with a source of short coherence length, as shown in Figure 4.19. Tandem interferometers have already been discussed in section 4.2.11. Assume that each of the interferometers possesses a path imbalance that is large in comparison with the source coherence length. Now consider that the path imbalance of the receiving interferometer is adjusted steadily. When the two interferometers come into balance, interference is observed, with maximum visibility when they are in exact balance. Therefore, provided that there exists a means to record the path imbalance of the receiving interferometer, the path imbalance of the sensing interferometer may be determined.

Now consider the extension of the tandem interferometer concept to the continuous measurement system illustrated in Figure 4.20. Light from the source is coupled into the test waveguide, which forms one arm of the sensing interferometer. Light backscattered in the waveguide is combined with the reflection

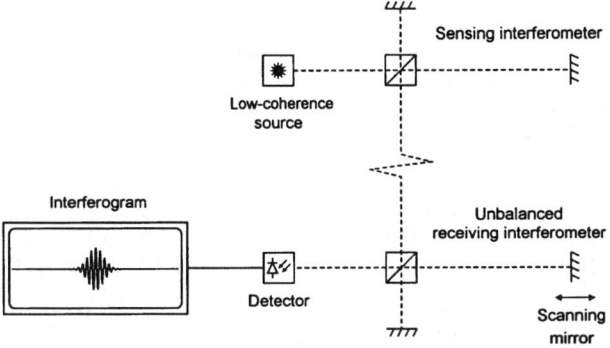

Fig. 4.19 Tandem interferometers for OCDR.

from the reference arm and is coupled into the receiving interferometer. The receiver comprises a scanning interferometer and a photodetector, such that the interference signal recorded at a particular path imbalance in the receiving interferometer is proportional to the reflected amplitude from the region of the test waveguide generating a similar path imbalance. In addition to measuring reflections and scatter in waveguides [43], this system has also been used to make measurements in multilayer bulk optic media [44], and in biological tissue [45].

By extension, we may consider the application of OCDR to the interrogation of a remote array of interferometers each containing a sensing element modulating either the optical path length or transmission of the interferometer. The interferometers all have different path-length imbalances, where the path difference between succeeding interferometers is large in comparison with the source coherence length. The array is interrogated with a scanning receiving interferometer, and thus each sensing interferometer is distinguished by the path imbalance of the receiving interferometer at which high-visibility interference is observed.

OCDR has great longitudinal resolution, and is used most often as a diagnostic technique in investigating integrated optic structures. Material and waveguide dispersion inevitably broaden the received interferogram, so that further signal processing is required in order to optimize spatial resolution [46]. OCDR is also used in sensor multiplexing, although it is successful only with small numbers of sensors. A difficulty is that even when the receiving interferometer is tuned to match the path imbalance of a particular sensing interferometer, optical signals from all the unmatched sensing interferometers are also present, generating a large incoherent background which constitutes an unwanted noise source. Coherence division multiplexing (CDM) is discussed in detail in section 4.8.

Whilst we have described OCDR in terms of the use of a scanning interferometer after the sensor, it is equally valid to place the scanning interferometer between the source and the system under test.

The distributed strain-sensing technique based on birefringent fiber described in section 4.4.3 can be adapted for use with OCDR. In this case, an

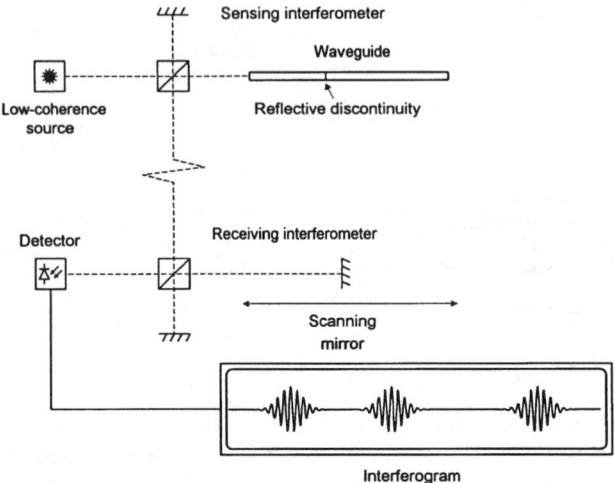

Fig. 4.20 OCDR for continuous measurement.

unmodulated broad-band source is used instead of a chirped narrow-band one, as shown in Figure 4.21. This allows extremely high resolution location of coupling points in birefringent fiber, which may be used for damage detection in structures.

OCDR derives its high resolution from the low source coherence length, using a precise OPD scan in the receiving interferometer to measure the autocorrelation function of the signal reflected from the detector. This autocorrelation function can be measured electronically, albeit with far lower delay resolution, by using an optically incoherent source the intensity of which is modulated by pseudo-random pulse sequences. Signals returned from the sensor array are electronically cross-correlated with a reference derived from a delayed version of the modulation pulse sequence, and show maximum correlation when the reference delay equals the propagation time from the source to the detector via the sensor. Thus the sensor is identified. The advantage of such a subcarrier technique is that the delay is controlled electrically, removing the need for an optical delay line and allowing arbitrarily large delays to be generated. The disadvantage is relatively poor spatial resolution, because of the long effective coherence times of the modulating pulse sequence.

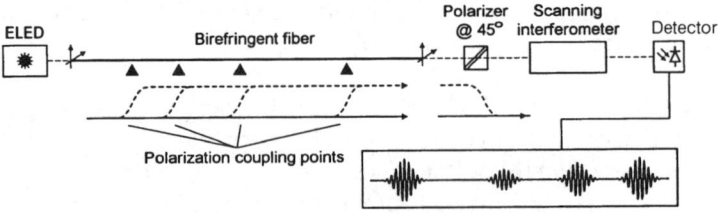

Fig. 4.21 Transmissive OCDR for disturbance detection.

4.4.5 Other time division techniques

A special case of TDM has been described for use with the Sagnac interferometer, shown in Figure 4.22. Consider the application of a harmonic phase modulation $\phi_{sig} = \phi_0 \sin \omega t$ at the mid-point of the Sagnac loop, with amplitude ϕ_0 and frequency ω. Because the optical arrangement is reciprocal, i.e. both clockwise and anticlockwise beams are affected equally, then no modulation is observed at the output. However, consider that the modulation point is now moved a distance l from the mid-point of the loop. The clockwise and anticlockwise beams in the loop now experience the modulation at times separated by the loop delay $\tau_l = 2l/v_g$, so that a phase modulation ϕ_{det} is now observable at the detector. Provided that $\omega/2\pi \ll 1/\tau_l$, then

$$\phi_{det} = \frac{d\phi_{sig}}{dt} \tau_l = \phi_0 \omega \frac{2l}{v_g} \cos \omega t \qquad (4.45)$$

so that the observed modulation amplitude is proportional to both l and ϕ_0. In one practical arrangement a separate Mach–Zehnder interferometer was used to make a position-independent measurement of ϕ_0, which when substituted into equation 4.45 yields the position l. A further refinement of the technique [47] uses a Bragg grating deployed at the center of the Sagnac loop to force a reflection there of the counter-propagating beams, which recombine at the detector as in a Michelson interferometer. Thus a Sagnac and Michelson interferometer are created in the same fiber system. These two superimposed interferometers are distinguished by wavelength division multiplexing, which is discussed in more detail in the next section.

4.5 WAVELENGTH DIVISION MULTIPLEXING

In wavelength division multiplexing (WDM), the signal from each sensor is encoded by the spectral range which it transmits or reflects. The optical source can be formed either as an array of narrow-band sources, chosen to match the properties of the sensors, or as a single broad-band source. Light-emitting

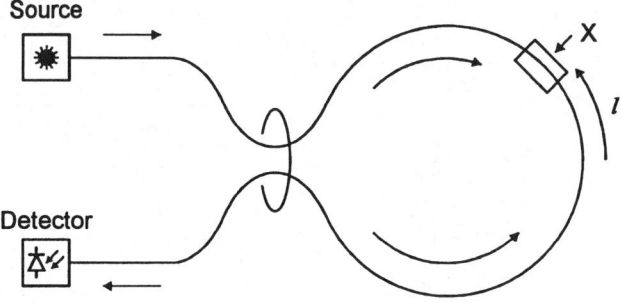

Fig. 4.22 Sagnac interferometer.

diodes are a popular choice for broad-band sources, although typical commercially available devices have bandwidths only in the order of tens of nanometers. For efficient coupling into single-mode fibers, edge-emitting or superluminescent diodes are preferred. Superluminescent diodes are effectively laser diodes where the reflections from the end facets have been suppressed to prevent laser action. Recently, superluminescent diodes with spectral widths in excess of 60 nm have become available, capable of coupling in excess of 1 mW of optical power into single-mode fibers [48].

WDM is a natural choice for use with sensors in which the transduction mechanism is the modulation of wavelength by the measurand. WDM has thus undergone a renaissance with the development of in-fiber Bragg grating sensors. A simple serial array can be constructed by concatenating a set of Bragg gratings along a downlead, where the center wavelengths of each sensor are chosen to be distinct, taking into account the maximum wavelength modulation likely to arise from the imposition of the measurand.

Various techniques have been implemented to decode the output from WDM arrays. The simplest involve the use of dispersive optics in the receiver, such as a scanning monochromator, thus the wavelengths corresponding to each of the sensors can be measured, the sensors distinguished, and the measurand recovered. The monochromator can be replaced by an electronically scanned filter, such as an acousto-optic device. The optical signal can be used more efficiently by using a Fourier transform spectrometer (FTS) to analyze the light returned from the array. The FTS is a two-beam interferometer, the path imbalance of which is scanned. The Fourier transform of the resultant interferogram yields the spectral distribution of the input light.

The sensing elements need not be Bragg gratings, or even wavelength modulated. Instead, wavelength-selective elements, such as Bragg gratings, can be deployed in the sensor array in order to encode the outputs of particular sensors with specific wavelengths. For example, Figure 4.23 shows a scheme using simple intensity-modulated sensors which are deployed as a parallel array. The signal from each sensor is returned by reflection from a Bragg grating, where the wavelength for each sensor is unique. The signal returned to the detector is monitored with an optical spectrum analyzer, and the intensity at each wavelength thus corresponds to the measurand at each sensor. Whilst a parallel

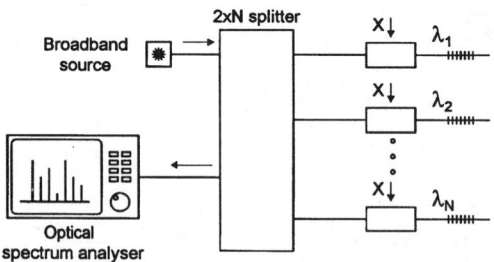

Fig. 4.23 WDM intensity-modulated sensors encoded using Bragg gratings.

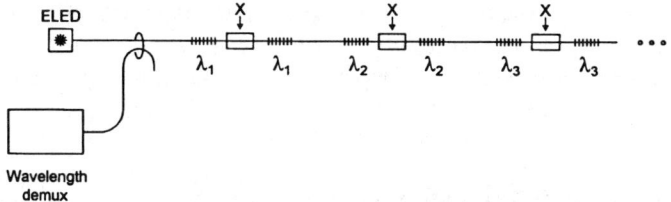

Fig. 4.24 Efficient WDM of interferometers using Bragg gratings.

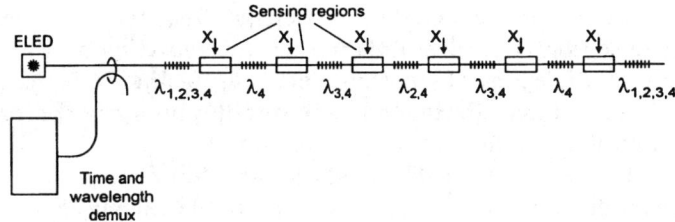

Fig. 4.25 Sensor array of controlled spatial resolution using Bragg gratings.

arrangement was described, serial arrays are equally feasible and make more efficient use of the downlead and upload fibers. A more sophisticated implementation is possible by replacing the intensity-modulated sensors by interferometers, where the interferometer mirrors are defined by Bragg gratings, as shown in Figure 4.24.

More complex arrangements can be created by writing gratings of different center wavelengths at the same point in the fiber. Figure 4.25 shows an example of an interferometer in which the gratings are deployed so that the two reflectors at the first wavelength define a sensing element that occupies the full length of the fiber; whereas those at other wavelengths define other sections of the fiber. By a combination of TDM and WDM, a sensor array of controlled spatial resolution is feasible [49].

Large numbers of sensors can be multiplexed by combining WDM with other multiplexing techniques. For example, Figure 4.26 shows a combination of WDM and TDM used to interrogate Bragg grating sensors, deployed as 5 sets of 12 in serial arrays. The system uses an optical switch to sequentially address each array. The application of this 60-sensor system was in strain

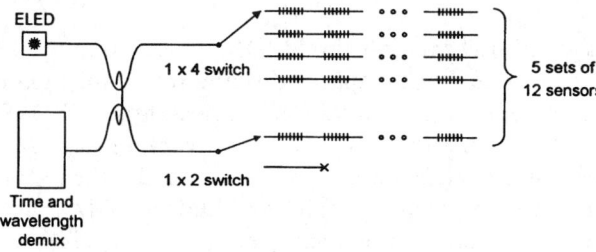

Fig. 4.26 A combination of WDM and TDM.

mapping, where one of the arrays was reserved to measure temperature, to compensate the effect of temperature on the strain sensors [50]. Similar numbers of sensors have been addressed as an all-serial array, but this approach introduces problems of crosstalk between sensors.

4.6 FREQUENCY DIVISION MULTIPLEXING

Frequency division multiplexing (FDM) can be achieved in sensor systems in which the measurand is encoded on a carrier, the frequency of which identifies the particular sensor. We have already seen the most important example of FDM in the multiplexing of interferometers using the PGC technique (see section 4.4.2). In that case, the individual sensor outputs are in the form of a heterodyne carrier phase modulated by the measurand.

Other FDM systems are feasible in which the heterodyne carrier is generated by an active modulator deployed in the vicinity of the sensing element. Such use of active modulators is not usually desirable in practical applications, but is essential in one important application—namely the interferometric magnetostrictive magnetometer [51].

The magnetostrictive magnetometer operates as follows. The sensor comprises a magnetostrictive element bonded to the sensing arm of a fiber interferometer. The magnetostrictive element changes its dimensions in response to an externally applied magnetic field, thus straining the fiber, and consequently producing an optical phase change. The magnetostrictive response is nonlinear, and is zero in the small-signal limit. Consequently it is usual to bias the sensing element with a DC field and a superimposed 'dither' field. The measurand is thus revealed as an amplitude modulation of the optical phase at a carrier frequency defined by the dither.

The opportunity for FDM is evident: the sensors may be deployed in an array and a different dither frequency used for each one. Hence, conventional frequency demultiplexing techniques may then be used to recover the outputs from each sensor. Such arrays have been applied, for example, in full-scale sea trials for tracking of surface vessels [52].

4.7 SPATIAL DIVISION MULTIPLEXING

The term 'spatial division multiplexing' (SDM) refers to multiplexing in which the decoding scheme relies on the spatial distribution of the light in the detector field. The most obvious, and essentially trivial, example of SDM is in parallel-topology arrays in which a detector is used for each sensor. Such an arrangement is often used in practical applications. Frequently, the technique is used in compound arrays, where a technique such as TDM or FDM is used to multiplex a set of sensors into a subarray, and then a set of subarrays is combined in a parallel arrangement, interrogated by SDM.

COHERENCE DIVISION MULTIPLEXING

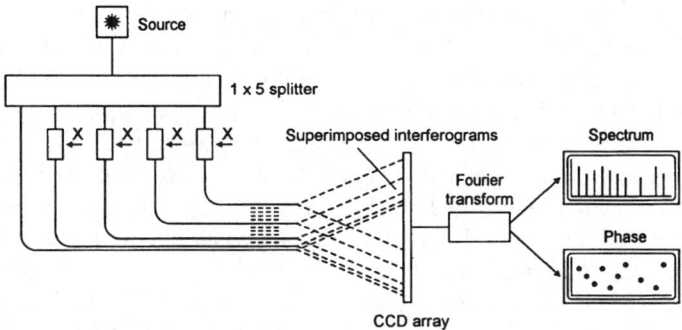

Fig. 4.27 Spatial division multiplexing with 2D interferogram.

A more sophisticated form of SDM has been reported where there is no direct one-to-one mapping between sensors and detectors. Chen and Hu [53] have described a version of SDM for interferometric sensors which operates as follows: one arm of the interferometer is used for sensing, and the output ends of multiple sensor fibers are disposed adjacently. The sensors share a reference arm in the form of a single fiber, also with its end adjacent to the end of the sensor fibers, as shown in Figure 4.27. The fibers are illuminated with a common source. Thus, light leaving the reference fiber and sensing fibers interferes to form a complex interference pattern in the far field. It is evident that the form of the interferogram contains the phase difference between the optical path through each of the sensor fibers relative to the reference fiber.

The interferogram is recorded using a one- or two-dimensional array detector, such as a CCD camera. The sensors are most easily demodulated and decoded by first taking the discrete Fourier transform of the sampled interferogram. Each combination of a sensing fiber and the reference fiber produces a pair of complex-valued peaks in the Fourier domain. The phase of either of these peak values gives the optical phase difference between the sensing and the reference fiber. Each combination of two sensing fibers produces a smaller parasitic peak, so the fiber spacings must be chosen carefully to minimize crosstalk.

This technique has been used to multiplex up to four sensing fibers [53, 54]. For practical exploitation a number of problems remain to be overcome, including the localization of a sensor along a particular sensor fiber, the stability of the reference path length and polarization-induced fading in the interferogram.

4.8 COHERENCE DIVISION MULTIPLEXING

4.8.1 Tandem interferometry and sensor multiplexing

In the tandem interferometer described in section 4.2.11, the single sensing interferometer can be replaced by an array of sensing interferometers, each with a different path-length imbalance, all interrogated by a single receiving interferometer with a long path scan. For most of the range of optical path imbalance of the receiving interferometer, no interference is observed, but when the receiver interferometer is tuned so that its path imbalance equals that of a particular sensing interferometer, then interference is observed. Thus, from the approximate value of the path imbalance necessary to produce interference, the particular sensor is identified. From the path imbalance corresponding to the observation of the 'central fringe' of highest visibility (approximately), the value of the measurand is derived.

This is the basis of coherence division multiplexing (CDM). CDM is based on time-of-flight, and so has a great deal in common with TDM, but CDM operates on far shorter timescales, and hence over far smaller OPDs.

An arrangement for a single sensor and receiving interferometer is shown in Figure 4.28(a), and in 4.28(b) is plotted the photodetector output as a function of path imbalance of the receiver. Following the argument of section 4.2.11, fringes are observed when the path imbalance of the two interferometers is matched (for both positive and negative values of the reference path imbalance) and when the reference path imbalance is zero. The last case gives the highest visibility fringes, but contains no measurand information. The measurand information is contained in the two sidelobe fringe sets, shown in Figure 4.28(b). For ideal interferometers (those for which the visibility would be unity for each interferometer when illuminated by a source of long coherence length), the maximum possible visibility of the sidelobes is 0.5.

Figure 4.29 shows two arrangements for using CDM to interrogate sensors in a serial array. Figure 4.29(a) shows Fizeau sensors interrogated in reflection, while 4.29(b) shows Mach–Zehnder sensors interrogated in transmission. Parallel topologies are also feasible.

These two arrangements involve quite different tradeoffs between crosstalk, received power, and the number of sensors multiplexed. For the case of serially addressed Fizeau sensors, reflectivity must be kept low in order to obtain sufficient power from the most distant sensors. Low reflectivities help prevent crosstalk between sensors due to multiple reflections, simplifying the choice of sensor lengths. A disadvantage of this CDM array is that the interferogram for one sensor will be superimposed on a pedestal due to the power reflected by all the sensors, thus multiplexing is achieved at the expense of signal to noise ratio.

For the case of serially addressed Mach–Zehnder sensors, received power drops off by a factor of two for each sensor, severely limiting the number of sensors that can be multiplexed. If all the directional couplers used have an equal 50:50 splitting ratio, then all received interferograms have a fringe visibility of unity, regardless of the number of sensors. This gives a far better signal to noise ratio than for the Fizeau case, and partially compensates for the additional power loss.

The number of Mach–Zehnder sensors may be severely restricted, however, by the need to avoid parasitic interferograms. A serial array of N Mach–Zehnder interferometers has 2^N separate optical paths, and hence 2^N pairs of

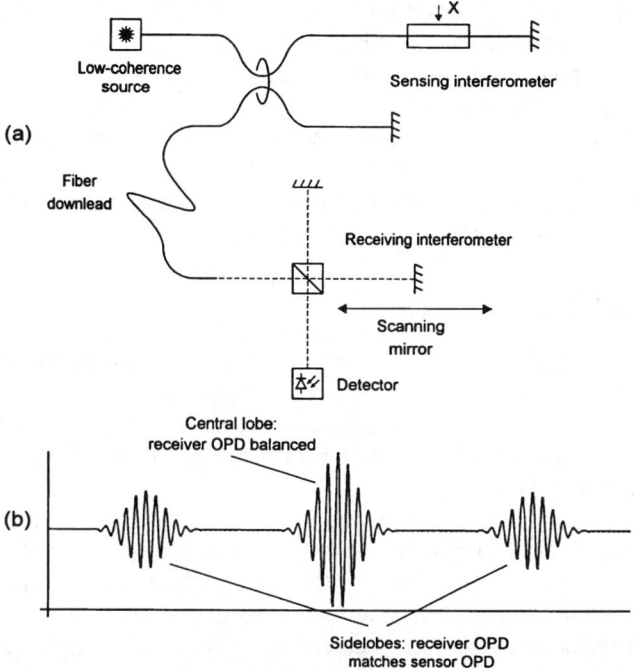

Fig. 4.28 Coherence division multiplexing, single sensor: (a) sensing and receiving interferometers; (b) detector output for receiver OPD scan.

sidelobes in a full scan of the receiving interferometer. Only N of these sidelobes pairs contain information about a single sensor. The others generate unwanted parasitic interferograms, producing crosstalk between sensors. Even when small numbers of sensors are used, their path differences must be carefully chosen to avoid crosstalk.

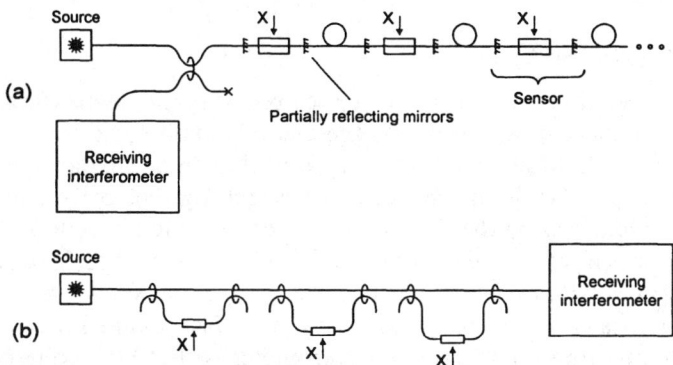

Fig. 4.29 Coherence division multiplexing, multiple sensors: (a) reflective Fizeau sensors; (b) transmissive Mach–Zehnder sensors.

CDM systems typically use either fiber optic or bulk optic interferometers to interrogate a single mode optical fiber sensor array. Arrangements using bulk optic interferometers as both sensors and receivers, linked by fibers, have also been reported [55]. The best performance is achieved using single mode fiber links, but multimode fibers can also be used successfully, with a penalty in signal to noise caused by mode-coupling in the link. However, this disadvantage is sometimes outweighed by the greater ease of coupling light from bulk optic sensors into multimode fibers.

4.8.2 Broad-band sources for CDM

An important design criterion in CDM systems is the coherence length of the source, where usually the shortest coherence length (and hence widest possible spectral bandwidth) is desirable. Often, there is a compromise between spectral width and power coupled into single-mode optical fiber, since the brightest sources are generally lasers.

The superluminescent light emitting diode (SLD) is the commonest choice, giving spectral widths of typically 30 nm with coupled powers of 0.1 mW into single-mode fibers [48]. Smoothness of the source spectrum is vital: poorly suppressed facet reflections lead to line structure in the spectrum and unwanted sidelobes in interferograms, hence power and spectral width alone are not sufficient parameters to specify an SLD for CDM.

An inexpensive alternative can be offered by using diode lasers at injection currents around threshold, giving a spectral width of around 8 nm, but with perceptible mode structure. The mode structure complicates the form of the low-coherence interferogram, makes recovery of the measurand phase more difficult, and can introduce crosstalk between sensors. Superfluorescent in-fiber sources have also been used. However, probably the most suitable practical sources available at present are very broad-band SLDs for which bandwidths of 60 nm and coupling powers into fibers of 2 mW at a wavelength of 800 nm are commercially available [48].

4.8.3 Signal processing

Many techniques have been reported for the recovery of the measurand phase. Perhaps the simplest to appreciate is the use of a bulk optic two-beam interferometer, such as a Michelson, with an adjustable path imbalance used as the receiver. The signal processing consists of first setting the receiver path imbalance to zero, identified by the position of the central high visibility fringe. The path imbalance is now scanned until the central fringe of the sidelobe is obtained where the path imbalance of the receiver and sensor are equal. Thus the sensor path imbalance is equal to the distance scanned. The scanned distance can be found by illuminating the receiver interferometer with a high-coherence beam of known wavelength and, for example, counting the high-coherence fringes as the receiver interferometer is scanned from the zero path imbalance to the central fringe of the sidelobe. Unless the interferogram is extremely narrow, and hence the source extremely broad-band, relatively small amounts of noise can prevent unambiguous identification of the central fringe. The center of the interferogram can then be found by a simple mathematical procedure, such as finding the centroid of the data around the balance point.

The situation is slightly more complicated when a fiber interferometer is used as the sensor. The fiber is a dispersive medium: dispersion broadens the low-coherence interferogram, reducing fringe visibility, and shifting the fringes so that the concept of a central fringe is no longer valid, as shown in Figure 4.30. Reducing the source coherence length then increases the width of the interferogram, degrading the signal still further. The centroid of the dispersed interferogram is also harder to estimate, due to increased interferogram width and lower fringe visibility.

Measurements of the interferogram centroid are in effect estimates of the group delay of the sensor. This group delay typically has a well-defined dependence on the measurand. The group delay can be derived from a broad-band interferogram in a more rigorous way, by taking the Fourier transform of the interferogram and calculating the first derivative of its unwrapped phase [56]. This gives an unambiguous value with good noise rejection, even in the presence of strong dispersion.

4.8.4 OPD scan and the receiving interferometer

A complication with CDM systems is that the path imbalances required can be fairly long, perhaps in the order of 0.1 m. There is thus a need for long path length scanning in the receiver interferometer, often conveniently achieved with bulk optics and motorized translation stages. Sensing bandwidths are consequently low.

There are practical advantages in all-fiber receiving interferometers. By using considerable lengths of fiber in standard coiled fiber piezoelectric phase modulators, optical path length tuning ranges of several centimeters have been achieved [57, 58]. However, the strains generated using these modulators are low, and the great length of fiber required can introduce problems due to

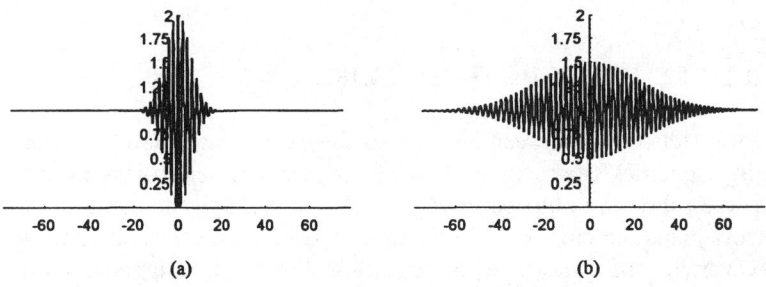

Fig. 4.30 Effect of dispersion on broadband interferogram: (a) undispersed interferogram; (b) dispersed interferogram.

dispersion and thermal drift. The birefringence of the fiber coil can reduce the visibility of the interference fringes and introduce errors in the phase demodulation. However, birefringence-compensated designs of fiber phase modulators have been described [59].

When long sensors must be used for reasons other than high resolution, e.g. in order to average the measurand over a long length, the requirements for long-range scanning can be reduced by using polarimetric sensing elements in the form of highly birefringent fibers, where we may see the 'fast' and 'slow' orthogonal eigenmodes as the two arms of the sensing interferometer, such that the effective delay between the modes is much less than the physical length of the sensing fiber. A polarizing receiving interferometer is used then to recover the delay between the eigenmodes of the sensor.

For general application in CDM, receiving interferometers requiring no mechanical path length scanning have been devised. Several designs employing bulk optic interferometers and linear array detectors have been demonstrated [30, 60]. The effective path imbalance varies linearly across the face of the detector array, producing a spatial, rather than a temporal, interferogram. This type of interferometer is said to offer electronic scanning of the path imbalance. The resultant scan is highly repeatable, but has a very limited range in comparison with mechanical scanning.

Other types of solid-state scanner have also been devised. One of these makes use of the different group velocities of the two lowest-order guided modes in an optical fiber, the LP_{01} and LP_{11} [61]. The technique is applied by choosing a fiber that will efficiently guide just these two modes. Power can be coupled from one mode to another by bending the fiber. Consider that a beam is launched into the LP_{01} mode at the input of the fiber and immediately encounters a bend that couples light to the LP_{11} mode. The beam then spends most of its time in the LP_{11} mode. Suppose now that the coupling point is moved away from the input end. The beam now spends progressively less time in the LP_{11} mode and more in the LP_{01} mode, thus reducing the group delay. Practical scanning devices have been constructed in which the moving coupling point is generated by coupling transverse flexure waves into the two-mode fiber using an ultrasonic piezoelectric transducer. To date, group delay scans of up to 1.6 ps have been achieved, equivalent to an OPD scan of 0.5 mm.

4.9 MULTIPLE MEASURAND SENSORS

The previous sections have been general to the extent that they have described multiplexing schemes that can be applied to sensors regardless of the measurand. These schemes could be used to measure multiple measurands using a set of sensors placed in close proximity to each other. They are more commonly used, however, to multiplex a set of sensors that are all designed for the same measurand (acoustic pressure, in the case of hydrophones, for example) at a set of different locations.

MULTIPLE MEASURAND SENSORS

In this section we shall consider the special case of multiplexed systems designed to measure a set of different measurands essentially at the same point.

4.9.1 Laser velocimetry

First consider the problem of determining the orthogonal components of a vector measurand. As an example, consider laser velocimeters designed with the objective of measuring the x, y and z components of velocity, typically in a fluid flow.

Many commercially available systems measure a single component of this velocity vector using the Doppler difference technique, illustrated in Figure 4.31. To measure a component transverse to the optical axis, two beams from the same laser are projected to intersect at their waists. A particle carried by the flow passing through the intersection point (which defines the measurement volume) scatters light from each beam to a detector. The scattered light components are Doppler shifted, and because the transmitted beams are in different directions the Doppler shifts seen at the detector are also different. They beat together at the detector to give a frequency proportional to the velocity components in the plane of the transmitted beams normal to the optical axis. The technique can be visualized by considering that the intersecting beams produce interference fringes (like Young's fringes), so that a particle passing through the fringe system scatters light that is intensity modulated at a frequency proportional to velocity. In order to ascertain the sign of the velocity component, it is common to apply a frequency offset to one of the transmitted beams, often using a Bragg cell. This has the effect of making the interference fringes move at a constant velocity in one direction.

The orthogonal velocity component can be determined by using a second set of transmitted beams to produce a fringe set perpendicular to the first, but retaining a single detector. Various means exist to distinguish signals arising from the two fringe sets, and hence to decode the two velocity components. The commonest, and often most satisfactory, multiplexing technique is to use two different wavelengths for the two fringe sets. Often an argon ion laser source is used as a common source, and dispersive optics are used to select the 514 nm line for one set of fringes and a 488 nm source for the other [62].

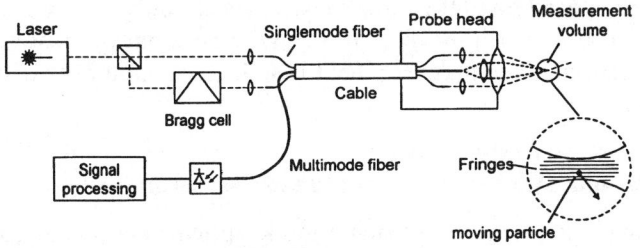

Fig. 4.31 Doppler difference technique.

An alternative method suitable for separating the two velocity components is to use orthogonal states of polarization for the two fringe sets. This method is particularly convenient for fiber optic arrangements where the properties of highly birefringent fibers can be exploited. However, a certain amount of crosstalk due to depolarization occurring when the light is scattered is inevitable [63].

A third multiplexing technique uses different offset frequencies for the two components. Such FDM is convenient when a wavelength-modulated source is used in conjunction with an unbalanced interferometer to generate the frequency offset. Here, a single source can be used for both velocity components, with the different frequency offsets generated by using interferometers with different path imbalances. Care must be taken to avoid crosstalk caused by the impure heterodyne carriers generated by the wavelength modulation technique, where harmonics of the two offset frequencies may overlap [62].

TDM has been also used in laser velocimetry, using pulsed laser sources and fiber delay lines. Inevitably the opportunity for truly simultaneous measurement of the different components is lost.

These various techniques can be extended to measure the third velocity component by adding two transmitted beams off axis and generating three orthogonal fringe sets. However, the off-axis angle required is large enough to make the transmitter optics bulky if a reasonable working distance is to be maintained. A smaller angle can be used, but now the detected velocity components are no longer orthogonal, and resolution in the direction of the optical axis is reduced [64].

The third (axial) velocity component is sometimes measured with a reference beam arrangement, in which a single on-axis beam illuminates the measurement volume via a confocal probe. Thus, the scattered light is collected by the same optics as were used to generate the transmitted beam. The scattered light then interferes with light derived from a static reference path, to generate a Doppler frequency proportional to the velocity component in the axial direction.

4.9.2 Magnetometry

In section 4.6 we discussed the optical fiber vector magnetometer as an example of simultaneous detection of the components of a vector measurand. We considered the special case of a magnetostrictive magnetometer, which is particularly well-suited to FDM. This sensor measured these three components at closely spaced points, rather than at a single point, but in most applications the spatial variation of the magnetic field is sufficiently small for this approach to be acceptable.

4.9.3 Simultaneous strain and temperature measurement

Amongst the many applications of single-mode optical fiber sensors, strain measurement is an important one. However, it is well known that although optical fibers show a strong phase response to axial strain, they also show considerable temperature sensitivity. Thus, it is difficult to measure a slowly varying strain whose phase modulation effect is indistinguishable from temperature. Consequently, an area of considerable current interest is the devising of sensors that can recover temperature and strain simultaneously, with the intention of providing temperature-independent strain measurement.

The overwhelming majority of reported methods can be called 'two-parameter' techniques; they operate as follows. Two different 'optical signals' S_1 and S_2 are derived from a single sensor. These signals are related to the temperature (T) and strain (σ) by the matrix equation

$$\begin{pmatrix} S_1 \\ S_2 \end{pmatrix} = \begin{pmatrix} a & b \\ c & d \end{pmatrix} \begin{pmatrix} T \\ \sigma \end{pmatrix} = M \begin{pmatrix} T \\ \sigma \end{pmatrix} \qquad (4.46)$$

which may be inverted to give

$$\begin{pmatrix} T \\ \sigma \end{pmatrix} = M^{-1} \begin{pmatrix} S_1 \\ S_2 \end{pmatrix} \qquad (4.47)$$

provided that

$$ad \neq bc \qquad (4.48)$$

The accuracy with which the inversion can be carried out is dependent on how well conditioned the linear relationships of equation 4.46 are. An extreme example of good conditioning would be the ideal situation in which S_1 is sensitive only to temperature, and S_2 only to strain. At present, no practical sensors approach the ideal, although emerging possibilities are discussed below.

The earliest method for temperature cross-sensitivity, and one still used today, is to use two physically distinct sensors, one of which is exposed to the measurand strain and (inevitably) ambient temperature, and the other to temperature only. Unfortunately, it is difficult in most situations to make the temperature reference sensors experience exactly the same temperature as the strain sensors, and so effort has been expended on the development of sensors in which the strain and temperature sensing regions are identical, or (in the case of compound sensors discussed below) are at least in contact.

Numerous possibilities exist and have been explored for dual parameter sensors for strain and temperature. The two signals S_1 and S_2 are often phase measurements, including: phase measurements made in a single-mode fiber sensing element at two different wavelengths; the phases for two orthogonal polarization modes in a highly birefringent sensing element [65]; and the phases of the LP_{01} and LP_{11} modes in a 'two-mode fiber'. Various combinations of these different phase terms have also been investigated, including the relative phase of the two polarization eigenmodes and the relative phase of the two spatial modes [23]. Effective laboratory demonstrations have been achieved. However, the conditioning of the transformation from the two phases to the two measurands is not ideal, and is critically dependent on the properties of the particular fibers used. Perhaps only limited future development may be expected.

A variation on the two-parameter technique is feasible with broad-band illumination in conjunction with the coherence methods of section 4.8. An example investigated by at least two groups is to extract from the broad-band interferogram the group delay and first-order dispersion of a fiber-sensing element, and then to apply a linear transformation of the form given in equation 4.47 to obtain strain and temperature. One group measured the group delay and first-order dispersion using Fourier transform methods [56], while the other based their measurements on three-wavelength interrogation [66]. A variation that has been explored involves measuring simultaneously the mean group delay in a highly birefringent sensing element and the differential group delay between the orthogonal polarization eigenmodes [67]. Whilst the conditioning of the transformation is not necessarily better than that of the 'two-phase' schemes, broad-band interferometry at least has the advantage that the delays are determined unambiguously, since, unlike for measurements of phase, there is no fringe order ambiguity in the measurements of group delay and dispersion. Thus such schemes show promise of being more robust.

In an effort to improve the conditioning of dual parameter sensors, compound sensing elements have been investigated. One example consists of using two Fizeau-type interferometers in series. One interferometer is formed from a short piece of hollow fiber, and the second from conventional fiber [68]. Since the majority of the temperature sensitivity of a conventional fiber arises from the thermo-optic coefficient (principally the variation of the refractive index of the core material with temperature), the hollow-core sensor shows very small temperature sensitivity in comparison with the conventional fiber. Hence the temperature and strain may be resolved with minimal cross-sensitivity.

Amongst the various types of fiber strain sensor, Bragg gratings are emerging as a preferred technology for many applications. Consequently, considerable attention is currently being devoted to simultaneous temperature and strain measurement for these sensors. In principle, the dual parameter techniques developed for conventional fiber interferometers may be adapted to Bragg gratings. We recall that in its simplest form the Bragg grating can be considered as a narrow-band reflector, where the central reflected wavelength shifts in response to an applied measurand. Just as for plain fiber, the wavelength is temperature sensitive through the thermal expansivity and the thermo-optic coefficient, and is strain sensitive through the physical extension and the strain-optic coefficient.

A single Bragg grating can be used to investigate simultaneous strain and temperature measurement by interrogating the grating at two wavelengths, one twice the other—the first and second harmonics of the grating. Reflection at the second harmonic is typically weak, and occurs when the index modulation of the grating is not perfectly sinusoidal. The two central wavelengths can be related to the two measurands through a linear transformation similar to that shown in equation 4.46.

This technique is limited by the weak reflection for the second harmonic and the practical difficulty in using the same fiber to accommodate two wavelengths an octave apart whilst requiring only a single mode to be guided without excessive losses. A more practical sensor can be made by writing two gratings with different first-order wavelengths at the same point in the fiber [69]. Just as for the two harmonic wavelengths, the two central wavelengths respond to temperature and strain via a non-singular linear relationship and the two measurands are hence extracted. The linear transformation is then less well conditioned than for the first and second harmonic case due to the two wavelengths being more closely spaced, but this is more than compensated by the improvement in the signal to noise ratio of wavelength measurement gained by the increase in returned power.

Alternative approaches are possible when more complex grating structures are used. In one of these, the grating is physically tapered. Thus, when the grating is placed in uniform tension, the thinner section experiences greater strain than the thicker part. In consequence, the uniformity of the grating spacing is affected—the grating becomes 'chirped'. In a chirped grating, the reflected spectrum is broader at the expense of maximum reflectivity. Thus, as the tension is increased, the degree of chirp (i.e. the nonuniformity of the grating spacing) is similarly increased, thus broadening the spectral width of the reflection. Conversely, temperature has minimal effect on the degree of chirp, and affects only the central wavelength of the reflection. Thus, by measuring simultaneously the bandwidth of the reflection and its central wavelength, the strain and temperature are recovered [70].

Compound interferometer and Bragg grating sensors have also been explored as a means of achieving a well-conditioned transformation to yield temperature and strain. One example uses a hollow-core fiber interferometer (with dominantly strain sensitivity) in series with a Bragg grating (with dominantly temperature sensitivity). The dual measurand determination is then made by interrogating the compound sensor at two wavelengths [71].

In all cases of dual strain and temperature measurement, the quality of conditioning of the transformation is set by the available sensing element measurement properties. Ideally, one would wish to tailor the sensing element properties to the requirements. There is evidence that this is possible in long-period in-fiber gratings (LPG). In a conventional Bragg grating, the grating spacing is chosen to satisfy the phase-matching condition to couple the forward-propagating lowest-order spatial mode (LP_{01}) to the backward-propagating one; in an LPG, the phase matching is arranged to couple the guided mode to a (lossy) cladding mode [37]. Thus, the effect of the LPG is to generate a spectrally dependent loss in the fiber. The phase-matching condition, and hence the loss spectrum, of the grating is controlled by the environment [72].

The strain and temperature sensitivities of LPGs are markedly different from those of FBGs, and can be controlled to some extent by fiber type. Patrick *et al.* reported an LPG with seven times the temperature sensitivity and one-half the strain sensitivity of a conventional FBG [72]. By combining this with a pair of conventional FBGs, they realized a sensing element from which they could recover two independent parameters the transformation of which to strain and temperature is very well-conditioned, allowing simultaneous strain and temperature measurement with high resolution. Alternatively, it may be possible to design LPGs in which the temperature sensitivity is intrinsically low, by controlling the constituent dopants of both the core and cladding.

4.10 CONCLUSIONS

Multiplexing techniques for fiber optic sensors are maturing. Perhaps the first substantial efforts in the field were for the FDM of hydrophones using the phase-generated carrier technique, and (to a lesser extent) magnetometers. Such developments by the NRL group mirrored their earlier pioneering work in the development of single fiber sensors for similar applications.

With the advent of in-fiber gratings, we see the possibility of the mass-production of single-mode fiber sensors, and the promise of their more extensive application in physical measurement. The wavelength-dependent transfer function of these devices makes them naturally well-suited to WDM arrays, and the fiber sensor community has been quick to exploit this opportunity.

Whilst the basic fiber sensing element may be becoming inexpensive, the same cannot be said for the techniques used to interrogate the sensors. Perhaps here, as in many other areas of the field of fiber sensors, progress will be made by adapting developments from that greater area of application of optical fiber technology: telecommunications. Thus advances in TDM may be foreseen, exploiting as they do the well-developed instrumentation of OTDR. OTDR-based interrogation of arrays of simple intensity-modulated sensors is now a commercial reality. With adaptation, OTDR is appropriate for use with arrays of fiber interferometers, and applications with grating-based sensors are emerging.

The development of multiplexing techniques is now a major driving force in optical fiber sensor technology; it is not only the key to new areas of application, such as strain-mapping in advanced materials, but perhaps the enabler for large-scale commercial exploitation, where the ability to address many sensors from a single sophisticated optical instrument promises to bring the cost per sensor in line with that for the established and competing conventional measurement technologies.

REFERENCES

1. D. A. Jackson. Recent progress in monomode fiber-optic sensors. *Meas. Sci. Technol.*, **5**, 621–638, 1994.
2. E. Udd. An overview of fiberoptic sensors. *Rev. Sci. Instrum.*, **66**(8), 4015–4030, 1995.

REFERENCES

3. G. Murtaza and J. M. Senior. Referenced intensity-based optical fibre sensors. *International Journal of Optoelectronics*, 9(4), 339–348, 1994.
4. Jonathan D. Weiss. Fiber-optic strain gauge. *J. Lightwave Technol.*, 7(9), 1308–1318, September 1989.
5. Martin Kull, Rye Widell, Georges Borak, and Leif Stensland. A strain sensor based on permanent bends in single mode fibers. In *Tenth International Conference on Optical Fibre Sensors*, Brian Culshaw and Julian D. C. Jones, editors, volume 2360 of *Proc. SPIE*, pp. 187–190. SPIE, October 1994.
6. F. A. Muhammad, G. Stewart, and W. Jin. Sensitivity enhancement of D-fiber methane gas sensor using high-index overlay. *IEE Proceedings-J. Optoelectronics*, 140(2), 115–118, 1993.
7. D. C. Bownass, J. S. Barton, and J. D. C. Jones. Detection of high humidity using serially addressed monomode fibre sensors for optical network monitoring. In *JSAP/IEICE/IEEJ/SICE Proc. 11th Optical Fiber Sensors Conference*, Sapporo, Japan, May 1996, pp. 590–593. Paper Th3-43.
8. A. J. Harris and P. F. Castle. Bend loss measurements on high numerical aperture single-mode fibers as a function of wavelength and bend radius. *J. Lightwave Tech.*, 4(1), 34–40, January 1986.
9. K. T. V. Grattan, A. W. Palmer, N. D. Samaan, and F. Abdullah. Mathematical analysis of optically powered quartz resonant structures in sensor applications. *J. Lightwave Technol.*, 7(1), 202–208, January 1989.
10. D. Auttamchandani, Z. N. Li, L. M. Zhang, and B. Culshaw. All-fiber optical-system for exciting and detecting the vibration of silicon microresonator sensors. *Optics and Lasers in Europe*, 16(2–3), 119–126, 1992.
11. Robert W. A. Ayre. Measurement of longitudinal strain in optical fiber cables during installation by cable ploughing. *J. Lightwave Technol.*, 4(1), 15, January 1986.
12. Arthur J. Barlow. Quadrature phase shift technique for measurement of strain, optical power transmission, and length in optical fibers. *J. Lightwave Technol.*, 7(8), 1264–1269, August 1989.
13. Simon Ramo. *Fields and Waves in Communication Electronics*, 2nd Edition. Wiley, New York, 1985, pp. 552–553.
14. H. C. Lefèvre. Single-mode fibre fractional wave devices and polarisation controllers. *Electron. Lett.*, 16(20), 778–780, 1980.
15. M. Martinelli. A universal compensator for polarisation changes induced by birefringence on a retracing beam. *Optics Comms.*, 72(6), 341–344, August 1989.
16. Hervé Lefèvre. *The Fiber-Optic Gyroscope*. Artech House, Boston, 1993.
17. Axel Bertholds and René Dändliker. Determination of the individual strain-optic coefficients in single-mode optical fibers. *J. Lightwave Technol.*, 6(1), 17–20, January 1988.
18. Gorachand Ghosh, Michiyuki Endo, and Takashi Iwasaki. Temperature-dependent Sellmeier coefficients and chromatic dispersions for some optical fiber glasses. *J. Lightwave Technol.*, 12(8), 1338–1341, August 1994.
19. K. Fritsch and G. Adamovsky. Simple circuit for feedback stabilisation of a single-mode optical fiber interferometer. *Rev. Sci. Instrum.*, 52(7), 996–1000, July 1981.
20. D. A. Jackson, R. Priest, A. Dandridge, and A. B. Tveten. Elimination of drift in a single-mode optical fiber interferometer using a piezoelectrically stretched coiled fiber. *Appl. Opt.*, 19(17), 2926–2929, September 1980.
21. D. Uttam, B. Culshaw, J. D. Ward, and D. Carter. Interferometric optical fibre strain measurement. *J. Phys. E: Sci. Instrum.*, 18, 290–293, 1985.

22. P. Hariharan. Digital phase-stepping interferometry: effects of multiply reflected beams. *Appl. Opt.*, **26**(13), 2506–2507, July 1987.
23. Wojtek J. Bock and Tinko A. Eftimov. Polarimetric and intermodal interference sensitivity to hydrostatic pressure, temperature, and strain of highly birefringent optical fibers. *Optics Letters*, **18**(22), 1979–1981, November 1993.
24. H. O. Edwards, K. P. Jedrzejewski, R. I. Laming, and D. N. Payne. Optimal design of optical fibres for electric current measurement. *Appl. Opt.*, **28**(11), 1977–1979, June 1989.
25. Peter J. de Groot. Extending the unambiguous range of two-colour interferometers. *Appl. Opt.*, **33**(25), 5948–5953, September 1994.
26. Yun-Jiang Rao and D. A. Jackson. Recent progress in fibre optic low-coherence interferometry. *Meas. Sci. Technol.*, **7**, 981–999, 1996.
27. Joseph W. Goodman. *Statistical Optics*. Wiley, New York, 1985, p. 168.
28. G. Martini. Analysis of a single-mode optical fibre piezoceramic phase modulator. *Optics & Q. El.*, **19**, 179–190, 1987.
29. J. P. Dakin, C. A. Wade, and C. Haji-Michael. A fibre optic serrodyne frequency translator based on a piezoelectrically-strained fibre phase shifter. *IEE Proc.*, **132**, Pt. J(5), 287–290, October 1985.
30. R. Dändliker, E. Zimmermann, and G. Frosio. Electronically scanned white-light interferometry: a novel noise-resistant signal processing. *Optics Letters*, **17**(9), 679–681, May 1992.
31. T. P. Newson, F. Farahi, J. D. C. Jones, and D. A. Jackson. Reduction of semiconductor laser diode phase and amplitude noise in interferometric fiber optic sensors. *Appl. Opt.*, **28**(19), 4210–4215, October 1989.
32. S. A. Al-Chalabi, B. Culshaw, E. N. Davies, P. Giles, and D. Uttam. Multiplexed optical fiber interferometers: an analysis based on radar systems. *IEE Proc. P+J*, **132**(2), 150–156, April 1985.
33. K. P. Koo, A. B. Tveten, and A. Dandridge. Passive stabilisation scheme for fiber interferometers using (3×3) directional couplers. *Appl. Phys. Lett.*, **41**(7), 616–618, October 1982.
34. D. A. Jackson, A. D. Kersey, and A. C. Lewin. Fibre gyroscope with passive quadrature detection. *Electron. Lett.*, **20**(10), 399–401, May 1984.
35. G. Meltz. Overview of fiber grating-based sensors. In *Distributed and Multiplexed Optical Fiber Sensors VI*, A. D. Kersey and J. P. Dakin, editors, volume 2838 of *Proc. SPIE*. SPIE, August 1996, pp. 2–22. Paper 2838-01.
36. Serge M. Melle, Kexing Liu, and Raymond M. Measures. Practical fiber-optic Bragg grating strain-gauge system. *Appl. Opt.*, **32**(19), 3601–3609, July 1993.
37. V. Bhatia and A. M. Vengsarkar. Optical fiber long-period grating sensors. *Optics Letters*, **21**(9), 692–694, May 1996.
38. Y. Fujii. High-isolation polarization-independent optical circulator coupled with single-mode fibers. *J. Lightwave Technol.*, **9**(4), 456–460, 1991.
39. P. Nash. Review of interferometric optical-fiber hydrophone technology. *IEE Proceedings—Radar Sonar and Navigation*, **143**(3), 204–209, 1996.
40. A. Dandridge, A. B. Tveten, and T. G. Giallorenzi. Homodyne demodulation scheme for fiber optic sensors using phase generated carrier. *IEEE Journal of Quantum Electronics*, **18**(10), 1647–1653, 1982.
41. A. Dandridge, A. B. Tveten, A. D. Kersey, and A. M. Yurek. Multiplexing of interferometric sensors using phase-generated carrier techniques. *Electronics Letters*, **23**(13), 665–666, 1987.

42. J. Mlodzianowski, D. Uttamchandani, and B. Culshaw. A simple frequency domain multiplexing system for optical point sensors. *J. Lightwave Technol.*, **5**(7), 1002–1007, July 1987.
43. E. Brinkmeyer and R. Ulrich. High-resolution OCDR in dispersive wave-guides. *Electronics Letters*, **26**(6), 413–414, 1990.
44. U. Schnell, R. Dändliker, and S. Gray. Dispersive white-light interferometry for absolute distance measurement with dielectric multilayer systems on the target. *Optics Letters*, **21**(7), 528–530, April 1996.
45. G. J. Tearney, B. E. Bourma, S. A. Boppart, B. Golubovic, E. A. Swanson, and J. G. Fujimoto. Rapid acquisition of in-vivo biological images by use of optical coherence tomography. *Optics Letters*, **21**(17), 1408–1410, 1996.
46. Andreas Kohlhaas, Carsten Frömchen, and Ernst Brinkmeyer. High resolution OCDR for testing integrated-optical waveguides: dispersion-corrupted experimental data corrected by a numerical control algorithm. *J. Lightwave Technol.*, **9**(11), 1493–1502, November 1991.
47. A. A. Chtcherbakov, P. L. Swart, and S. J. Spammer. Dual wavelength Sagnac–Michelson distributed optical fiber sensor. In *Distributed and Multiplexed Optical Fiber Sensors VI*, A. D. Kersey and J. P. Dakin, editors, volume 2838 of *Proc. SPIE*. SPIE, August 1996, pp. 301–307.
48. Data sheets for SLD-37 and SLD362. Superlum Ltd, P. O. Box 73, E-358 Moscow 111538, Russia.
49. A. D. Kersey and M. J. Marrone. Bragg grating based nested fiber interferometers. *Electronics Letters*, **32**(13), 1221–1223, 1996.
50. M. A. Davis, D. G. Bellemore, M. A. Putnam, and A. D. Kersey. Interrogation of 60-fiber Bragg grating sensors with microstrain resolution capability. *Electronics Letters*, **32**(15), 1393–1394, 1996.
51. A. D. Kersey, M. Corke, D. A. Jackson, and J. D. C. Jones. Detection of DC and low-frequency AC magnetic-fields using an all single-mode fiber magnetometer. *Electronics Letters*, **19**(13), 469–471, 1983.
52. F. Bucholtz, C. A. Villarruel, A. R. Davis, C. K. Kirkendall, D. M. Dagenais, J. A. McVicker, S. S. Patrick, K. P. Koo, G. Wang, H. Valo, T. Lund, A. G. Andersen, R. Gjessing, E. J. Eidem, and T. Knudsen. Multichannel fiberoptic magnetometer system for undersea measurements. *J. Lightwave Technol.*, **13**(7), 1385–1395, 1995.
53. S. Chen and Y. Hu. Recent progress in digital spatial domain multiplexing technique. In *JSAP/IEICE/IEEJ/SICE Proc. 11th Optical Fiber Sensors Conference*, Sapporo, Japan, May 1996, pp. 542–545. Paper Th3-31.
54. I. Yamaguchi and K. Hamano. Multiplexed optical fiber interferometer based on Fourier fringe analysis. *Optics Letters*, **20**(23), 2432–2434, December 1995.
55. Y. N. Ning, K. T. V. Grattan, A. W. Palmer, and K. Weir. Measurement of up-lead and down-lead fiber sensitivity caused by the lead in a multimode fiber in an interferometric system. *Appl. Opt.*, **33**(31), 7529–7535, 1994.
56. D. A. Flavin, R. McBride, J. G. Burnett, A. H. Greenaway, and J. D. C. Jones. Combined temperature and strain measurement with a dispersive optical fibre Fourier transform spectrometer. *Optics Letters*, **19**(24), 2167–2169, December 1994.
57. M. A. Page-Jones and J. K. A. Everard. Optical-fibre spectrum analyser. *Electronics Letters*, **26**(2), 117–118, January 1990.
58. M. A. Davis and A. D. Kersey. Application of a fiber Fourier transform spectrometer to the detection of wavelength-encoded signals from Bragg grating sensors. *J. Lightwave Technol.*, **13**(7), 1289–1295, July 1995.

59. D. G. Luke, R. McBride, J. G. Burnett, A. H. G. Greenaway, and J. D. C. Jones. Polarization maintaining single-mode fibre piezo-electric phase modulators. *Optics Communications*, **121**, 115–120, 1995.
60. M. J. Padgett, A. R. Harvey, A. J. Duncan, and W. Sibbet. Single-pulse Fourier-transform spectrometer having no moving parts. *Appl. Opt.*, **33**(25), 6035–6040, September 1994.
61. Pranay G. Sinha, Erling Kolltveit, and Kjell Bløtekjær. Two-mode fiber-optic time-delay scanner for white-light interferometry. *Optics Letters*, **20**(1), 94–96, January 1995.
62. D. A. Jackson and J. D. C. Jones. Extrinsic fibre optic sensors for remote measurement: Part two. *Optics & Laser Technology*, **18**(6), 299–307, December 1986.
63. C. N. Pannell, R. P. Tatam, J. D. C. Jones, and D. A. Jackson. Two-dimensional fibre-optic laser velocimetry using polarization state control. *J. Phys. E: Sci. Instrum.*, **21**(1), 103–107, 1988.
64. N. A. Ahmed, S. Hamid, R. L. Elder, C. P. Forster, R. P. Tatam, and J. D. C. Jones. Fiber optic laser anemometry for turbomachinery applications. *Optics and Lasers in Europe*, **16**(2–3), 193–205, 1992.
65. F. Farahi, D. J. Webb, J. D. C. Jones, and D. A. Jackson. Simultaneous measurement of temperature and strain: cross-sensitivity considerations. *J. Lightwave Technol.*, **8**(2), 138–142, February 1990.
66. Valeria Gusmeroli and Mario Martinelli. Nonincremental interferometric fiber-optic measurement method for simultaneous detection of temperature and strain. *Optics Letters*, **19**(24), 2164–2166, December 1994.
67. D. G. Luke, R. McBride, J. D. C. Jones, P. Lloyd, J. G. Burnett, and A. H. Greenaway. Composite-embedded highly-birefringent optical fibre strain gauge with zero thermal-apparent strain. In *Laser Interferometry VIII: Applications*, Ryszard J. Pryputniewicz, Gordon M. Brown, and Werner P. O. Jüptner, editors, volume 2861 of *Proc. SPIE*, pp. 26–31. SPIE, August 1996. Paper 2861-03.
68. J. Sirkis, T. A. Berkoff, R. T. Jones, H. Singh, A. D. Kersey, E. J. Friebele, and M. A. Putnam. In-line fiber etalon (ILFE) fiberoptic strain sensors. *J. Lightwave Technol.*, **13**(7), 1256–1263, 1995.
69. M. G. Xu, J.-L. Archambault, L. Reekie, and J. P. Dakin. Discrimination between strain and temperature effects using dual-wavelength fibre grating sensors. *Electronics Letters*, **30**(13), 1085–1087, 1994.
70. M. G. Xu, L. Dong, L. Reekie, J. A. Tucknott, and J. L. Cruz. Temperature-independent strain sensor using a chirped Bragg grating in a tapered optical-fiber. *Electronics Letters*, **31**(10), 823–825, 1995.
71. H. Singh and J. Sirkis. Simultaneous measurement of strain and temperature using optical fiber sensors: two novel configuration. In *JSAP/IEICE/IEEJ/SICE Proc. 11th Optical Fiber Sensors Conference*, Sapporo, Japan, May 1996, pp. 108–111. Paper Tu5-1.
72. H. Patrick, G. M. Williams, A. D. Kersey, J. R. Pedrazzani, and A. M. Vengsarkar. Hybrid fiber Bragg grating/long period fiber grating sensor for strain/temperature discrimination. *IEEE Photonics Technology Letters*, **8**(9), 1223–1225, 1996.

5
Progress in optical fiber interferometry

D. A. Jackson

5.1 INTRODUCTION

The implementation of many of the classic interferometers in an all fiber format has revitalized the field of interferometry; this has produced newtypes of interferometer such as the ring resonator which can be operated as an extremely high resolution optical spectrum analyzers, novel based fiber components and an entirely new generation of sensors offering many important measurement opportunities.

Single mode fiber optic interferometers were introduced in Chapter 7 of Volume 1 of this series. In this chapter the mode of operation of these devices is treated more formally. Initially the transfer function of the fiber optic directional coupler is derived as this is a key component in fiber interferometry. The Mach–Zehnder and Michelson interferometers are then analyzed using the Jones calculus. This leads naturally to the mode of operation of the recently introduced Faraday rotation mirror used to eliminate variable visibility due to random fluctuations in the fiber waveguide. The Fabry–Pérot, ring resonator, differential interferometers and the differentiating interferometers are also considered (except the Sagnac interferometer which is dealt with in Chapter 9). The next section deals with recently introduced processing methods, including low coherence interferometry, processed either using a second interferometer, or with a CCD array where the sensing interferometer can be considered as generating channel spectra. The use of the imbalanced interferometer as a processing unit for fiber Bragg grating sensors is briefly mentioned. The next section presents a brief review of new components used in fiber optic interferometry. The final section deals with selected applications for optical fiber interferometry:

1. Absolute sensors;
2. High bandwidth miniature temperature sensors;
3. Low coherence ranging systems for medical applications and flow measurements;
4. Polarization state generators for magnetic field studies and ellipseometry;

Optical Fiber Sensor Technology, Vol. 2. Edited by K. T. V. Grattan and B. T. Meggitt.
Published in 1998 by Chapman & Hall, London. ISBN 0 412 782 901

168   PROGRESS IN OPTICAL FIBER INTERFEROMETRY

5. Noncontact velocity measurement using differentiating interferometers for ultrasound NDT applications;
6. Very high resolution spectrometers;
7. Low coherence sensors using channel spectrum processing.

5.2 FIBER OPTIC INTERFEROMETERS

The new fiber optic interferometers (FOIs) are fabricated from monomode fiber optic waveguides and components such as directional couplers, as these components can all be represented using Jones matrix notation. It is instructive to use the Jones calculus to derive transfer functions of the most widely used FOIs.

5.2.1 Directional coupler

The directional coupler is possibly the 'key' component in all fiber optic communication and sensor systems as it enables the optical power transmitted in a single fiber to be coupled equally into two or more optical fibers [1–3]. A schematic diagram of the directional coupler is shown in Figure 5.1(a); essentially, two identical waveguides are in close proximity along the coupling region. The electric field amplitude in fiber (A) before the coupling region is shown in Figure 5.1(b). In the coupling region, Figure 5.1(c), it has been shown, using supermode

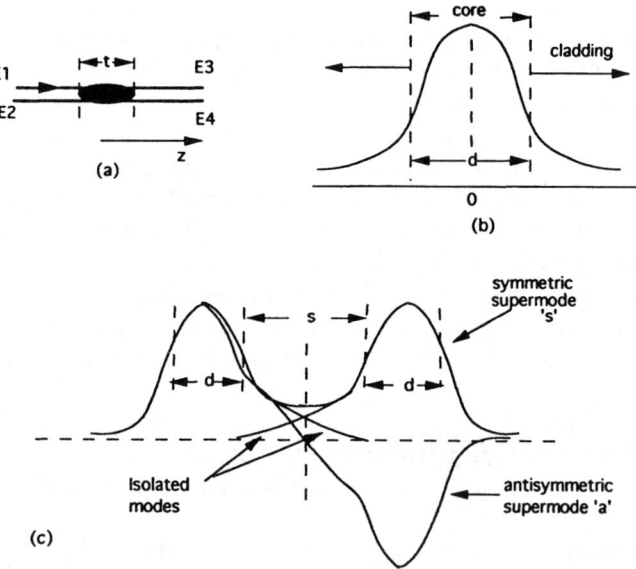

Fig. 5.1 Directional coupler: (a) schematic diagram showing the 3 dB fiber optic coupler; (b) schematic diagram showing the variation of the electric field in the lowest order (isolated) mode in monomode fiber; (c) electric field profiles in the coupling region for waveguides with core diameters of d separated by s. Coupling length $= t$.

FIBER OPTIC INTERFEROMETERS

theory, that the electric field in the structure can be represented by symmetric and antisymmetric modes. These modes are orthogonal and uncoupled with propagation constants K_s and K_a hence

$$\phi_{s,a} = K_{s,a} t \tag{5.1}$$

i.e. as the modes propagate in the coupler there is a relative phase shift where t is the length of the coupling region.

The amplitudes of the supermode fields at $z = t$ with respect to their values at $z = 0$

$$E_{s,a(z=t)} = E_{s,a(z=0)} e^{i\phi_{s,a}} \tag{5.2}$$

From the diagram we see that the input field amplitude

$$E_1 \propto (E_s(z=0) + E_a(z=0)) \equiv \text{input electric field}$$

$$E_2 \propto (E_s(z=0) - E_a(z=0))$$

If light is injected only into port 1, $E_{2(z=0)} = 0$

$$E_3 \propto (E_s(z=t) + E_a(z=t))$$

$$E_4 \propto (E_s(z=t) - E_a(z=t))$$

and

$$E_{s(z=0)} \propto \frac{E_1 + E_2}{2} \qquad E_{s(z=t)} \propto \frac{E_3 + E_4}{2} \tag{5.3}$$

Similar relations hold for $E_a(z)$.

Hence, the output electric fields E_3 and E_4 at $z = t$

$$E_3 = \left(\frac{e^{i\phi_s} + e^{-i\phi_a}}{2}\right) E_1 + \left(\frac{e^{i\phi_s} - e^{i\phi_a}}{2}\right) E_2 \tag{5.4}$$

$$E_4 = \left(\frac{e^{i\phi_s} + e^{-i\phi_a}}{2}\right) E_1 + \left(\frac{e^{i\phi_s} + e^{-i\phi_a}}{2}\right) E_2 \tag{5.5}$$

taking the mean phase shift over the coupling length as

$$\phi_m = \frac{\phi_s + \phi_a}{2}$$

and

$$\phi_s = \phi_m + \theta_c \qquad \phi_a = \phi_m - \theta_c$$

where θ_c is the coupling angle and

$$E_3 = [E_1 \cos\theta_c + iE_2 \sin\theta_c] e^{i\phi_m}$$
$$E_4 = [iE_1 \sin\theta_c + E_2 \cos\theta_c] e^{i\phi_m} \tag{5.6}$$
$$\theta_c = |k_c| t$$

where $|k_c|$ is the coupling coefficient.

For $E_2 = 0$

$$I_3(t) = I_{in} \cos^2 \theta_c$$
$$I_4(t) = I_{in} \sin^2 \theta_c \tag{5.7}$$

For a 3 dB coupler

$$\theta_c = \pi/4 \qquad I_3(t) = I_4(t) = I_{in/2}$$

and the phase shift across the coupler is $\pi/2$. It is convenient to write the matrix for the 3 dB coupler

$$C = \sqrt{\gamma} \begin{bmatrix} \sqrt{1-k} & i\sqrt{k} \\ i\sqrt{k} & \sqrt{1-k} \end{bmatrix} \tag{5.8}$$

where k is the intensity coupling factor $\sin^2 \theta_c$ and $\sqrt{\gamma}$ a normalizing factor to ensure energy is conserved.

5.2.2 Fiber delay

The matrix for a single delay line is

$$F = \begin{bmatrix} 1 & 0 \\ 0 & 1 \end{bmatrix}$$

and a differential delay τ or optical phase difference $\omega\tau$ between two modes of the same waveguide or between two fully symmetric waveguides

$$D = \begin{bmatrix} e^{i\omega\tau} & 0 \\ 0 & 1 \end{bmatrix} = \begin{bmatrix} e^{i\phi} & 0 \\ 0 & 1 \end{bmatrix} \tag{5.9}$$

where $\phi = \omega\tau$.

5.2.3 Fiber optic Mach–Zehnder interferometer

The all fiber optic Mach–Zehnder interferometer is shown in Figure 5.2, assuming lossless components and ideal 3 dB couplers, $k = 50\%$, then the

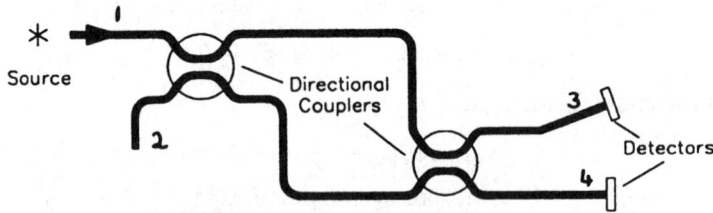

Fig. 5.2 All fiber Mach–Zehnder interferometer.

transfer function $T(\omega)_{mz}$ can be written, assuming the source coherence time $\tau_c \gg \tau$,

$$T(\omega)_{mz} = [C_2][D][C_1] = \frac{1}{\sqrt{2}}\begin{bmatrix} 1 & i \\ i & 1 \end{bmatrix}\begin{bmatrix} e^{i\omega\tau} & 0 \\ 0 & 1 \end{bmatrix}\frac{1}{\sqrt{2}}\begin{bmatrix} 1 & i \\ i & 1 \end{bmatrix} \quad (5.10)$$

$$= \frac{1}{2}\begin{bmatrix} -(1-e^{i\omega\tau}) & i(1+e^{i\omega\tau}) \\ i(1+e^{i\omega\tau}) & (1-e^{i\omega\tau}) \end{bmatrix} \quad (5.11)$$

$$\begin{bmatrix} E_3 \\ E_4 \end{bmatrix} = T(\omega)_{mz}\begin{bmatrix} E_1 \\ E_2 \end{bmatrix}$$

Hence, if $E_2 = 0$, i.e. light only injected into port 1, then the normalized input intensity $E_1 \cdot E_1^* = I_{in}$ and

$$E_3 = \frac{1}{2}E_1(e^{i\omega\tau} - 1)$$

$$E_4 = \frac{i}{2}E_1(1 + e^{i\omega\tau})$$

Hence, the optical power at ports E_3 and E_4

$$I_3 = \frac{I_{in}}{2}(1 - \cos \omega\tau) \qquad I_4 = \frac{I_{in}}{2}(1 + \cos \omega\tau) \quad (5.12)$$

i.e. the signal from the output ports vary in antiphase, at quadrature $I_3 = I_4$, a feature which has been exploited to stabilize the operating point of the Mach–Zehnder against environmentally induced phase shifts [4]. Another problem experienced with FOIs is that of polarization fading caused by environmentally induced birefringence—this results in random changes in the fringe visibility.

5.2.4 Fiber optic Michelson interferometer

Figure 5.3 shows a simple fiber Michelson interferometer where the distal fiber ends are directly coated with a highly reflective coating [5]. The transfer function $T(\omega)_m$ can be written, assuming $\tau_c \gg \tau$,

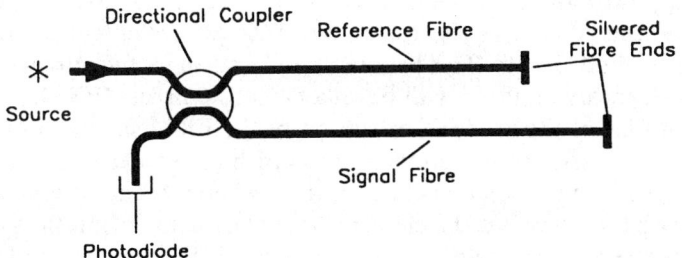

Fig. 5.3 All fiber Michelson interferometer.

$$T(\omega)_m = [C][\vec{D}][M][\overleftarrow{D}][C]$$

where $[\vec{D}]$ represents the Jones matrix for the light propagation in the fiber toward the mirror, $[M]$, and $[\overleftarrow{D}]$ represent the Jones matrix for light propagating in the same fiber toward the couplers [6].

Considering first only the fiber sensing elements which are represented by

$$[M][D]^t[M]^{-1}[M][D]$$

where $[D]^t$ is the transpose of $[D]$

$$= \begin{bmatrix} -1 & 0 \\ 0 & 1 \end{bmatrix} \begin{bmatrix} e^{i\phi} & 0 \\ 0 & 1 \end{bmatrix} \begin{bmatrix} e^{i\phi} & 0 \\ 0 & 1 \end{bmatrix} \quad (5.13)$$

$$= \begin{bmatrix} -e^{2i\phi} \\ 1 \end{bmatrix} \quad (5.14)$$

This vector must be multiplied by $[M]$ to take into account the change in the frame of reference on reflection,

$$= \begin{bmatrix} e^{2i\phi} \\ 1 \end{bmatrix}$$

Hence,

$$[E_0] = [C] \begin{bmatrix} e^{2i\phi} \\ 1 \end{bmatrix} [C][E_1] \quad (5.15)$$

(the matrices for the couplers are independent of the propagation direction of the light). Thus

$$I_0 = \tfrac{1}{2} I_{in}[1 + \cos 2\phi] \quad (5.16)$$

i.e. the variation of the output radiance as a function of phase is very similar to that of the Mach–Zehnder.

5.2.5 Faraday rotator mirror

In sensor applications, the birefringence of the fibers 1 and 2 varies due to environmental effects, this gives rise to random changes in the fringe visibility and hence the sensor sensitivity. This problem can be solved if the mirrors formed by coating the fibers are replaced with Faraday rotatot mirrors (FRMs) [7–9].

A brief outline of the mode of operation of the FRM shown in Figure 5.4 is given here. It consists of a section of glass of high Verdet constant, which is subject to a collinear magnetic field such that, when a beam of light with a specific wavelength transverses the element, its polarization azimuth is rotated by 45° (in the same manner as an optical isolator), the beam is then reflected from the mirror and retraces its path back through the Faraday rotator.

FIBER OPTIC INTERFEROMETERS

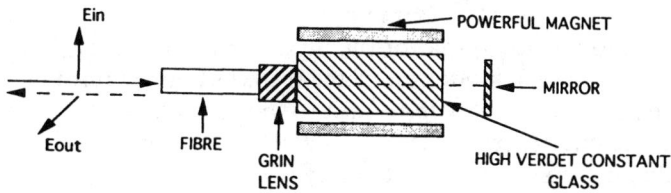

Fig. 5.4 Faraday rotator mirror.

Using the Jones matrix notation we may represent the action of the FRM by

$$[FRM] = [R(45°)]^t[M][R(45°)]$$

where $[R(45°)]$ and $[R(45°)]^t$ are the rotation and antirotation matrices and M is the matrix representing the reflection at the mirror [10, 11]. Hence

$$[FRM] = \frac{1}{\sqrt{2}}\begin{bmatrix} 1 & -1 \\ 1 & 1 \end{bmatrix}\begin{bmatrix} -1 & 0 \\ 0 & 1 \end{bmatrix}\frac{1}{\sqrt{2}}\begin{bmatrix} 1 & 1 \\ -1 & 1 \end{bmatrix}$$
$$= \begin{bmatrix} 0 & -1 \\ -1 & 0 \end{bmatrix} \quad (5.17)$$

We next consider the case where the FRM is placed at the distal end of a fiber where the fiber is illuminated with a state of arbitrary polarization represented by the vector

$$\begin{bmatrix} A \\ B\,e^{i\Delta} \end{bmatrix}$$

where A and B are real and Δ is the phase delay between A and B. Hence

$$[E_0] = ([M]^{-1}[M][D]^t[M]^{-1}[FRM][D]E_{in})^* \quad (5.18)$$

$$= \left(\begin{bmatrix} -1 & 0 \\ 0 & 1 \end{bmatrix}\begin{bmatrix} e^{i\phi} & 0 \\ 0 & 1 \end{bmatrix}\begin{bmatrix} 0 & -1 \\ -1 & 0 \end{bmatrix}\begin{bmatrix} e^{i\phi} & 0 \\ 0 & 1 \end{bmatrix}\begin{bmatrix} A \\ B\,e^{i\Delta} \end{bmatrix}\right)^* \quad (5.19)$$

$$= \begin{bmatrix} B\,e^{i\Delta} \\ -A \end{bmatrix}^*$$

where common phase terms have been dropped. The final multiplication of M^{-1} and the complex conjunction ensures that the output Jones vector is expressed in the same coordinate system as the input Jones vector. Then

$$E_0^\dagger E_m = (B\,e^{i\Delta} - A)\left(\frac{A}{B\,e^{i\Delta}}\right)$$

thus $[E_0]$ and $[E_{in}]$ are orthogonal.

Hence the polarization state of the light retracing its path back to the point of injection is orthogonal to that of the input state. Thus if FRMs are used as the mirrors in a Michelson interferometer the two beams returning toward the coupler are both orthogonal with respect to the input state, hence will strongly interfere, i.e. the interferometer will not suffer from polarization fading and the fringe visibility will always be close to unity.

5.2.6 Fiber Fabry–Pérot interferometer (FFP)

The fiber Fabry–Pérot interferometer has been implemented in two formats:

1. All fiber, where the cavity is formed in a length of monomode fiber by cleaving the fiber ends [12]; or
2. Air spaced, where the cavity is formed between the distal end of the fiber (or possibly a GRIN lens) and a reflecting surface [13].

The FFP can be operated in reflection or transmission; for sensor applications it is often used in reflection. The transfer function of the Fabry–Pérot is described by [14]

$$T(\omega)_{FP} = \frac{T^2}{(1-R)^2} \frac{1}{1 + F\sin^2(\phi/2)} \tag{5.20}$$

where T and R are the intensity transmission and reflectivity coefficients respectively, and

$$F = \frac{4R}{(1-R)^2}$$

known as the Airy function and ϕ is the phase delay between successive beams. When R is large, $T(\omega)_{FP}$ is a very sharply peaked function. In contrast when R is small, $\sim 5\%$, and the interferometer is used in reflection then

$$T(\omega)_{FP} = (R + R(1-R)^2 + 2R(1-R))\cos\phi \tag{5.21}$$

i.e. the transfer function is very similar to that of the two-beam interferometer and the output signal may be processed by the one of the several methods available for two-beam interferometers [15]. However, some caution is necessary; with mirror reflectives of only about 3%, phase errors will occur due to multipassing. Indeed, there is now a tendency to call this type of cavity a Fizeau rather than a Fabry–Pérot, in practice it is difficult to know whether multipassing occurs unless an exact fit to the measured transfer function of the cavity is made. This problem does not occur when low-finesse cavities are interrogated using low coherence techniques [16].

5.2.7 Ring resonator

The ring resonator [17] shown schematically in Figure 5.5 has a transfer function given by

$$T(\omega)_r = (1 - \gamma_0)\left(1 - \frac{(1 - K_r)^2}{1 + K_r^2 + 2K_r \sin \beta L_r}\right) \quad (5.22)$$

for the optimum amplitude resonance condition, where γ_0 is the insertion loss, K_r the resonance coupling coefficient, β the propagation constant of the recirculating mode and L_r the length of the ring.

The transfer function is very similar to a Fabry–Pérot interferometer operating in back-reflection. With very low coupler excess loss combined with low-loss fibers, a cavity finesse in excess of 1000 can be achieved [18]. At resonance, significant optical power is recirculating in the ring and nonlinear effects such as stimulated Brillouin scattering (SBS) can be observed for input powers of less than 100 µW [19].

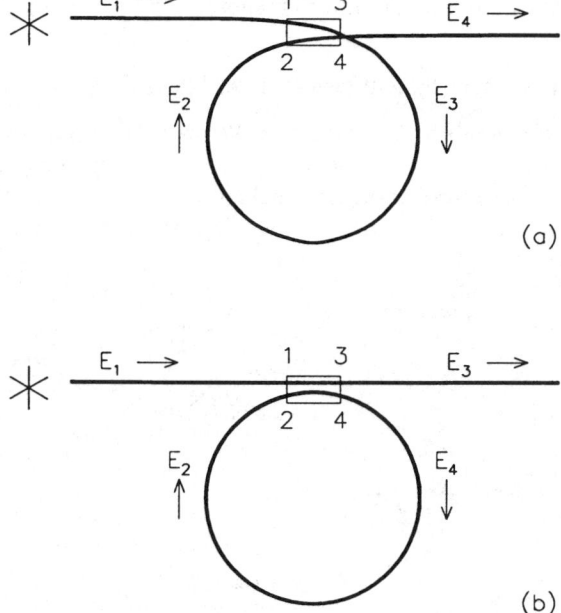

Fig. 5.5 Schematic of the all-fiber resonator geometries: (a) the cross-coupled ring resonator requires a coupling ratio, approaching unity; (b) the direct coupled ring resonator requires a coupling ratio approaching zero.

5.2.8 Differential (polarimetric) interferometers

Single mode fibres exhibiting a high level of birefringence can be used as differential or polarimetric interferometers. The basic interferometer is shown in Figure 5.6, where the transfer function $T(\omega)_{D,P} = A(\alpha)D$. $A(\alpha)$ is the Jones matrix for the analyzer where α is the angle its transmission axis makes with the eigenaxis of the fiber.

$$A = \begin{bmatrix} \cos^2 \alpha & \sin \alpha \cos \alpha \\ \sin \alpha \cos \alpha & \sin^2 \alpha \end{bmatrix} \quad (5.23)$$

and

$$D = \begin{bmatrix} e^{i(\phi_f)} & 0 \\ 0 & e^{i(\phi_s)} \end{bmatrix} \quad (5.24)$$

If $\alpha = 45°$ and the injected light equally populates both eigenmodes,

$$E_{out} = \frac{1}{2\sqrt{2}} \begin{bmatrix} 1 & 1 \\ 1 & 1 \end{bmatrix} \begin{bmatrix} e^{i\phi_f} & 0 \\ 0 & e^{i\phi_s} \end{bmatrix} \begin{bmatrix} 1 \\ 1 \end{bmatrix} \quad (5.25)$$

Hence the output irradiance is given by

$$I_0 = \tfrac{1}{2}(1 + \cos(\phi_f - \phi_s)) \quad (5.26)$$

which has the form of a two-beam interferometer, where $\tau_c \gg (\phi_f - \phi_s)/c$.

5.2.9 Differentiating fiber optic interferometer [20, 21]

The Sagnac interferometer (see Chapter 9) can be used to measure the time-dependent phase difference between the counterpropagating beams if the perturbation in the interferometer breaks the symmetry, i.e. the interaction occurs anywhere except at the center of the loop. In many Sagnac gyroscopes a

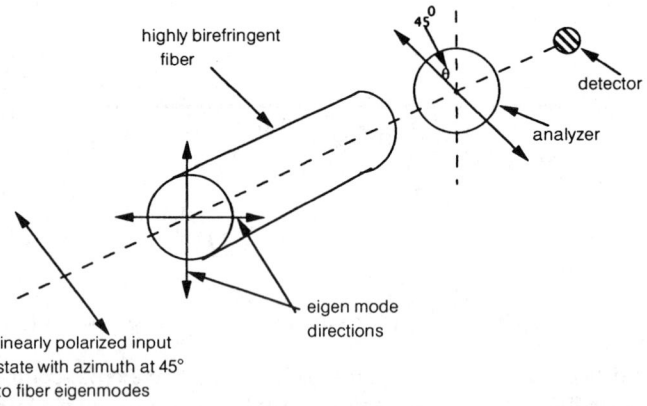

Fig. 5.6 Basic polarimetric fiber sensor.

fiber-wrapped nonreciprocal piezoelectric phase modulator is incorporated in one arm of the loop close to the coupler, the modulator is then driven at a frequency equal to $f_p = 1/2T$ where T is the transit time of the light in the loop [22]. Thus a dynamic phase shift bias is generated between the counter-propagating beams which allows the gyroscope to operate at its most sensitive operating point (i.e. quadrature).

In principle if a delay is incorporated into any interferometer such that the guided beams suffer the effects of a phase perturbation at different times the derivative of the induced phase modulation can be recovered. An advantage of this type of interferometer is that as the beams propagate along the same optical path (at different times) where the optical path difference between the beams is created by a periodic measurand, a low-coherence source can be used. The temporal response of the differential interferometer depends on the time delay between the beams, hence it tends to have the characteristics of a high pass filter.

Consider the case of the reflective ring resonator [23] shown in Figure 5.7. A nonreciprocal phase shift ϕ_{nr} is then generated between the two counter-propagating beams

$$\phi_{nr} = \phi\left(t - \frac{(nx)}{c}\right) - \phi\left(t - n\frac{(L_r + L_d - x)}{c}\right) \tag{5.27}$$

When n is the refractive index of the fiber, L_r the length of the fiber ring and L_d the delay. The phase perturbation $\phi(t)$ occurs at a distance x from the center of the ring. Assuming that $2nx/c$ is small then using Taylor's theorem,

$$\phi_{nr} = \left(\phi(t) - \frac{nx}{c}\frac{d\phi}{dt}\right) - \left(\phi(t) - \frac{n(L_r + L_d - x)}{c}\frac{d\phi}{dt}\right)$$

$$\phi_{nr} \propto \frac{n(L_r + L_d - 2x)}{c}\frac{d\phi}{dt} \tag{5.28}$$

If $L_r \ll L_d$ then

$$\phi_{nr} \propto \frac{nL_d}{c}\frac{d\phi}{dt}$$

i.e.

$$\phi_{nr} \propto T_d \frac{d\phi}{dt}$$

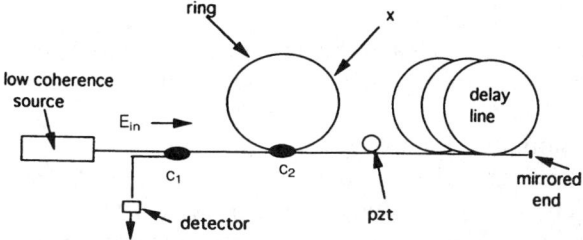

Fig. 5.7 Differentiating fiber optic interferometer based on a reflective fiber ring.

where T_d is the total delay caused by L_d, i.e. the output signal is proportional to the derivative of the phase.

5.3 COMPONENTS FOR FIBER OPTIC INTERFEROMETRY

Much of the recent progress in optical fiber sensors has occurred as a result of the development of new optical components.

5.3.1 Phase and frequency shifters for signal processing

In order to operate a fiber optic interferometric sensor (FOIS) at high resolution ($>10^{-3}$ rad $\sqrt{\text{Hz}}$) with constant sensitivity it is necessary to transpose the measurand-induced phase change to a modulation of an optical carrier which is detected at the output of the interferometer. Classically this is achieved with some form of acousto-optic or electro-optic modulator, incorporated in the interferometer itself [24]. As these devices are electrically driven, this approach cannot be used for electrically passive FOISs.

It is straightforward to show that the phase change $\Delta\phi$ induced in an unbalanced interferometer is given by

$$\Delta\phi = \frac{2\pi L \Delta v}{c} \qquad (5.29)$$

where Δv is the absolute optical frequency shift of the (laser) source, L is the optical path difference (OPD) in the interferometer and c is the velocity of light. Various schemes have been demonstrated to produce the carrier signals which rely upon serrodyne or sinusoidal modulation of the absolute frequency of the source [15]. The most successful scheme in terms of achievable phase resolution in a passive FOIS is the phase-generated carrier (PGC) technique described by [25] in conjunction with an electronic processing scheme often termed 'cross multiply and differentiate' (CMD).

This processing technique is linear over about five to six orders of magnitude with a noise floor of between 10^{-5} and 10^{-6} rad and a linear range of ~ 1 rad. Recently a digital version of the circuit has been developed to enable more sophisticated processing of the output signals from a network of sensors using time division multiplexing [26].

Although signal recovery based upon direct modulation of laser diode has proved very effective, problems have been encountered with:

1. Undesirable amplitude modulation;
2. Carrier frequencies restricted to ~ 100 kHz which limits the number of sensors that can be deployed in time division multiplexing networks;
3. The random fluctuations in the emission frequency of the laser are converted into intensity noise which increases linearly with the OPD.

In applications where it is necessary for the sensors to operate at large distances it is desirable to operate the sensor at longer optical wavelengths such as 1.3 or 1.5 μm where the fiber loss is significantly less. Studies of long wavelength laser diodes have shown them to be rather inferior in terms of power, phase noise and the maximum value of $\Delta v (\Delta v_{max})$ obtainable when compared with typical 800 nm lasers. Thus the use of such lasers would impose severe limits on the achievable performance of the FOIS. Comparable or superior performance to that demonstrated at 800 nm has been achieved by illuminating the FOIS with a laser diode pumped Nd:YAG laser operating at 1300 nm [27]. Unfortunately, these lasers have much lower values of Δv_{max} than typical laser diodes, hence signal processing based upon the PGC method requires a significantly larger OPD than for systems operating at 800 nm.

An alternative approach which still allows the sensor to be electrically passive is to use a phase modulator in conjunction with an unmodulated laser source as indicated in Figure 5.8. If the modulator is sinusoidally driven, at a frequency ω_p at an amplitude A, then the maximum amplitude of the induced phase shift in the interferometer is equal to $A\omega_p$. In the first demonstration of this method in an FOIS system, an in-line phase modulator, based upon a fiber wrapped under tension on a piezoelectric (PZT) cylinder, was used [28]. By driving the PZT at resonance (25.3 kHz) at a peak-to-peak phase modulation of $\sim 9 \times 10^3$ rad, a peak-to-peak frequency deviation of 228 MHz was achieved. With a large path imbalance in the interferometer (28 m) and selected gating of the output signal a carrier frequency of ~ 810 kHz was achieved. This method of processing has received relatively little attention as the gating process is very wasteful in power and the large values of OPD cause very high levels of intensity noise. A possible new approach to this problem is to use an integrated optic phase modulator, commercial devices can generate peak-to-peak phase deviations of ~ 10 rad at frequencies of up to 1 GHz. As the signal must be processed at the modulation frequency, one would probably choose a minimum value of ω_p of around 100 MHz, rather than 1 GHz, to minimize the speed requirements of the electronics. At this frequency, with a peak-to-peak phase deviation of 10 rad, the OPD required to enable the output of the interferometer to be swept over the 2.6 rad necessary for CMD processing is ~ 20 cm. Hence, the phase noise would be negligible for an FOIS illuminated with laser diode pumped Nd:YAG laser. Further benefits gained from this approach are: (1) there is no intensity noise associated with the method of signal processing as the laser is

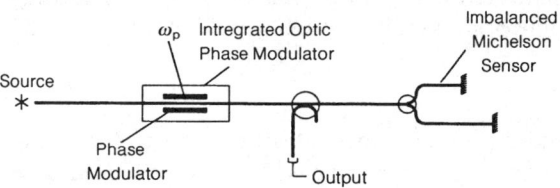

Fig. 5.8 Possible approach for remote interrogation of an unbalanced Michelson inter

operated without modulation and (2) a larger number of sensors could be incorporated into a multiplexing network.

A combined switch and phase modulator is commercially available [29] and could be used both to generate the phase-modulated carrier (phase modulator) and form part of a multiplexing system.

5.3.2 Optical sources

Several new types of optical source have been introduced which can be used in an FOIS; a brief summary of the important properties of new source types recently used with FOISs is presented below.

(a) Single frequency sources

In many applications for FOISs, the system must be illuminated with a single frequency source, laser diodes such as the Hitachi 1400 have proved particularly effective as it exhibits low phase noise compared with a value of $\Delta v_{max} \sim 40\,\text{GHz}$.

(b) Laser diode pumped Nd:YAG ring laser

Laser diode pumped Nd:YAG ring lasers are available commercially operating at 1064 or 1319 nm at power levels > 300 mW. Typically they exhibit extremely narrow optical linewidths $\sim 5\,\text{kHz}$ and very low frequency jitter (phase noise). A comparison of the normalized phase noise of a Nd:YAG ring laser and a Hitachi 1400 laser (commonly used as the optical source for 800 nm FOIS) [27] indicated that the ring laser phase noise is three orders of magnitude lower. Other advantages of these lasers when compared with typical laser diodes is that the output beam has low astigmatism, enabling higher coupling efficiencies into the fiber. As discussed previously in section 5.3.1, the main problem is that Δv_{max} is relatively small hence large OPDs are required in the interferometer to facilitate passive signal processing; significant reduction in the OPD is possible if an integrated optic phase modulator may be used.

(c) Fiber laser sources

Very powerful continuouswave (CW) and pulsed fiber laser sources have been developed in recent years which have as yet to find application in FOISs but are being used in distributed temperature sensors based on Raman backscattering [30] and optical time domain reflectometry (OTDR) systems [31].

(d) Low-coherence sources

Optical sources exhibiting extremely short coherence lengths are necessary for high performance fiber optic gyroscopes [32] and for FOISs based on white light interferometry (WLI). Unlike the development of highly coherent sources which have been mainly directed toward communications applications, most of the developments in low coherence sources have been for the gyroscope. The basic requirements for these sources are small spot size with a spatially coherent output beam to ensure efficient coupling into the fiber, high irradiance and very low temporal coherence with good stability of the mean wavelength. These requirements effectively rule out conventional optical sources and laser diodes.

(e) Superluminescent diodes

In order to achieve the desired source characteristics, superluminescent diodes have been developed from Fabry–Pérot type lasers. This involves depositing an antireflection coating on one of the laser facets to reduce its reflectivity and including an absorbing region at the other facet. As the optical gain of a typical semiconductor diode is high, optical powers of >1 mW can be obtained for single pass amplified spontaneous emission. The spatial coherence is the same as a typical laser diode, and the linewidth at 850 nm is of the order of 20 nm; invariably the broad emission spectrum has a superimposed modal structure associated with the basic Fabry–Pérot cavity as complete suppression of the lasing action is difficult to achieve [33]. It has recently been demonstrated that by using more sophisticated cavity structures, the modal structure can be suppressed [34]. This type of source is also available at longer wavelengths [35].

(f) Superfluorescent fiber sources

Unique superfluorescent sources have been developed based upon amplified spontaneous emission (ASE) in laser diode pumped doped monomode fibers. These sources are available at two wavelengths, ~ 1.06 μm for Nd:silica fibers and ~ 1.55 μm for Er^{3+}:silica fibers. The output powers, optical stability and linewidth of both sources are dependent on the wavelength of the pump, the fiber geometry and dopants, and the output power [36].

(g) Multimode laser diodes

An alternative approach to a truly low coherence source for WLI is a multimode laser. The emission spectrum of a multimode laser is composed of a set of oscillating longitudinal modes superimposed on some level of continuous, wideband, spontaneous emission spectrum. When multimode laser light propagates through a two-beam interferometer, each of the longitudinal modes generates its own interference pattern, given at the output [37]

$$P_{out} = A \sum_{i=-m}^{m} P_i \left(1 + \cos[(\omega_0 + i\Delta\omega)\tau] \exp -\frac{|\tau|}{\tau_c} \right) \quad (5.30)$$

where A is the optical attenuation (supposed equal for light that propagates in each arm), P is the optical power in mode i, $\omega_0 = 2\pi v_0$, $\Delta\omega = 2\pi\Delta v$, τ is the differential time delay, and τ_c is the coherence time associated with each longitudinal mode. The number of the modes is $2m+1$. Equation 5.30 can be rearranged to give the familiar expression

$$P_{\text{out}} = AP[1 + V\cos(\omega_i\tau)] \qquad (5.31)$$

where

$$P = \sum_{i=-m}^{m} P_i$$

is the total power emitted from the laser diode (spontaneous emission is neglected) and

$$V = \frac{1}{P}\left[P_0 + 2\sum_{i=1}^{m} P_i + \cos(i\Delta\omega\tau)\right]\exp-\frac{|\tau|}{\tau_c} \qquad (5.32)$$

where P_0 is the optical power in the central mode. Equation 5.31 shows that the fringe separation at the output of an interferometer illuminated by the multimode laser light is set by the central frequency ω_0. From equation 5.32 the visibility exhibits strong peaks when $\Delta\omega\tau$ is an integer multiple of 2π, corresponding to optical path imbalance ΔL for the interferometer given by

$$\Delta L = \Delta L_{\text{peak}} = 2pn_{\text{cav}}l_{\text{cav}} \qquad (5.33)$$

where p is an integer and n_{cav}, l_{cav} are the refractive index and length of the laser cavity, respectively. When this condition is satisfied, the interference signals generated by all the modes are in phase, giving a value for the visibility that is determined uniquely by the linewidth of the longitudinal modes. When $\Delta L \neq \Delta L_{\text{peak}}$ the longitudinal modes rapidly dephase, resulting in low visibility. Coherence sensing with such multimode sources is implemented by choosing the optical path imbalances of the sensing and receiving interferometers in such a way as to coincide with a region of low visibility in the source autocorrelation function (Figure 5.9) [38].

5.4 SIGNAL PROCESSING

Many of the signal processing schemes used to recover the measurand-induced phase change in the FOIS were discussed in Chapter 7 of the first volume of this series and hence only processing schemes recently developed are discussed here.

5.4.1 Low-coherence or white light interferometry

Possibly the most significant developments in fiber optic interferometry have been related to low-coherence interferometry [39] as it has enabled high precision absolute measurements to be made. In a typical low coherence sensing system two interferometers are combined to form a distributed interferometer, one acting as the sensor, the other the processing unit.

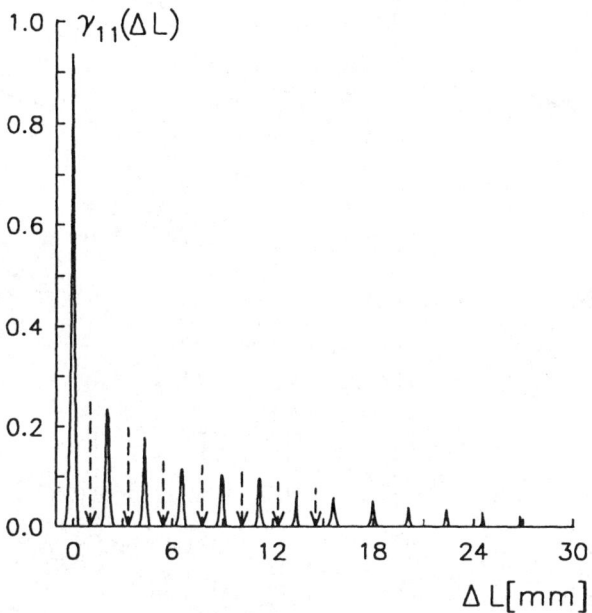

Fig. 5.9 Normalized autocorrelation function of a Mitsubishi 4406 laser; the broken arrows indicate possible operating point when used in white light systems.

The general advantages of low-coherence interferometry when applied to FOISs are:

1. Capability of determining the absolute phase of the sensor when the system is switched on, allowing the system to be used for quasi-static measurands such as temperature and pressure.
2. Sensor systems can be operated such that:
   (a) the measurement accuracy is virtually independent of the source stability; or
   (b) the effects of wavelength instabilities are greatly reduced.
3. The optical sensors can be fabricated with very short optical cavities $\sim 50\,\mu$ — which tends to greatly reduce the effects of measurand cross-sensitivity; this is particularly important for devices such as precision pressure sensors, for example.

5.4.2 Transfer function of tandem interferometers

For the two Mach–Zehnder interferometers operating in tandem shown in Figure 5.10, the transfer matrix is given by

$$T(\omega)_{TMZ} = [C_4][D_2][C_3][F][C_2][D_1][C_1] \qquad (5.34)$$

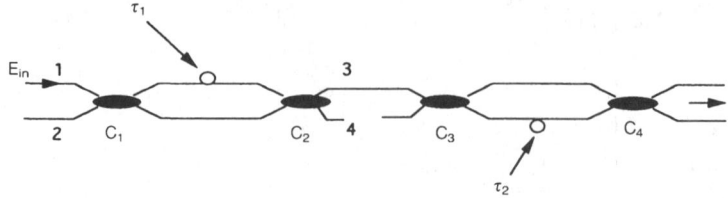

Fig. 5.10 Schematic diagram showing tandem Mach–Zehnder interferometers.

assuming all the couplers are perfect lossless devices with a 50/50 power splitting ratio. Then

$$[C_4] = [C_3] = [C_2] = [C_1]$$

$$T(\omega)_{TMZ} = \begin{bmatrix} 1 & i \\ i & 1 \end{bmatrix} \begin{bmatrix} e^{i\omega\tau_2} & 0 \\ 0 & 1 \end{bmatrix} \begin{bmatrix} 1 & i \\ i & 1 \end{bmatrix} \begin{bmatrix} 1 & 0 \\ 0 & 0 \end{bmatrix} \begin{bmatrix} 1 & i \\ i & 1 \end{bmatrix} \begin{bmatrix} e^{i\omega\tau_1} & 0 \\ 0 & 1 \end{bmatrix} \begin{bmatrix} 1 & i \\ i & 1 \end{bmatrix}$$

$$= \frac{1}{2}\begin{bmatrix} -(1-e^{i\omega\tau_2}) & i(1+e^{i\omega\tau_2}) \\ i(1+e^{i\omega\tau_2}) & (1-e^{i\omega\tau_2}) \end{bmatrix} \begin{bmatrix} 1 & 0 \\ 0 & 0 \end{bmatrix} \frac{1}{2}\begin{bmatrix} -(1-e^{i\omega\tau_1}) & i(1+e^{i\omega\tau_1}) \\ (1+e^{i\omega\tau_1}) & (1-e^{i\omega\tau_1}) \end{bmatrix}$$

$$= \frac{1}{4}\begin{bmatrix} -(1-e^{i\omega\tau_2})(1+e^{i\omega\tau_1}) - i(1-e^{i\omega\tau_2})(1+e^{i\omega\tau_1}) \\ i(1+e^{i\omega\tau_2})(1-e^{i\omega\tau_1}) - (1+e^{i\omega\tau_2})(1+e^{i\omega\tau_1}) \end{bmatrix}$$

(5.35)

When the interferometer is illuminated by a source with finite coherence we must also include the degree of coherence of the source defined by

$$\gamma(\tau_x - \tau_y) = \langle E_0(\tau_x)E_0^*(\tau_y)\rangle / I_0 \tag{5.36}$$

where τ_x and τ_y are the propagation delays to the detector, and I_0 is the intensity. The degree of coherence is related to the coherence time of the source

$$\gamma(t) = e^{-|\tau|/\tau_c} \tag{5.37}$$

There are several paths that the light can travel through the tandem interferometer. The corresponding delays are

$$\tau_1, \tau_2, \tau_1 - \tau_2, \tau_1 + \tau_2$$

In the matched condition $\tau_1 \approx \tau_2$, hence

$$\gamma|(\tau_1 - \tau_2)| \approx 1 \quad \text{and} \quad \gamma(\tau_1), \gamma(\tau_2) \text{ and } \gamma(\tau_1 + \tau_2) = 0$$

If light is only injected into the tandem interferometer at port 1 then

$$I_{01}(\omega) = \tfrac{1}{4}\left(1 + \tfrac{1}{2}\cos\omega(|\tau_2 - \tau_1|)\right)$$
$$I_{02}(\omega) = \tfrac{1}{4}\left(1 - \tfrac{1}{2}\cos\omega(|\tau_2 - \tau_1|)\right) \tag{5.38}$$

i.e. interference occurs for $\tau_1 \approx |\tau_2|$.

Interference effects are also observed if $\tau_1 = 0$ or $\tau_2 = 0$ corresponding to the OPD of either Mach–Zehnder being equal to zero. The fringe visibility for the condition $\tau_1 = |\tau_2| = \frac{1}{2}$.

5.5 PROCESSING FOR LOW-COHERENCE SYSTEMS

5.5.1 Coherence tracking [40, 41]

As indicated above, in order to find the coherence matched condition the OPD of one of the interferometers (number one say) has to be varied until $\tau_1 = |\tau_2|$. The accuracy of the measurement is determined by (a) the accuracy with which the point in the transfer function where $\tau_1 = |\tau_2|$ can be identified and (b) the accuracy of the tracking system (and its repeatability). Various techniques [42, 43] have been introduced to identify this fiducial point, which corresponds to the peak of the central fringe in the regions where $\tau_1 = |\tau_2|$.

A major difficulty with this technique is distinguishing between the central fringe and its nearest neighbors. The intensity difference is given by

$$\Delta I = [1 - e^{(2\lambda/l_c)^2}] \tag{5.39}$$

For typical low-coherence sources such as LEDs, laser diodes below the threshold and superluminescent light-emitting diodes (SLDs), the coherence lengths are ~ 20–$50\,\mu\text{m}$ and the resulting intensity variation between the fringes is extremely small. Methods to overcome this difficulty are described in detail in Chapter 9, Volume 1.

Recently, it has been shown that it is possible to modify the coherence function by adding the correlation functions from two sources with different wavelengths. As shown in Figure 5.11, the ease of identifying the central fringe is greatly enhanced by this technique. The method has been used for source pairs in the 700–800 nm [44, 45] and 1300–1500 nm regions. The technique is most effective at the longer wavelength regions as the differential fiber attenuation is relatively small allowing long lengths of fiber to be used [46].

Another problem with coherence tracking, particularly when used in multiplexed sensor networks, is that the finite time it takes to acquire and 'lock' to the desired point limits the bandwidth of the sensing system.

5.5.2 Two-wavelength low-coherence interferometry

By using a processing scheme based upon classical two-wavelength interferometry it is possible to solve the two major problems associated with coherence tracking.

If an interferometer is illuminated with separate sources of slightly different wavelengths, a source with an effective wavelength

$$\lambda_{\text{effective}} = \frac{\lambda_1 \lambda_2}{\lambda_1 - \lambda_2} \tag{5.40}$$

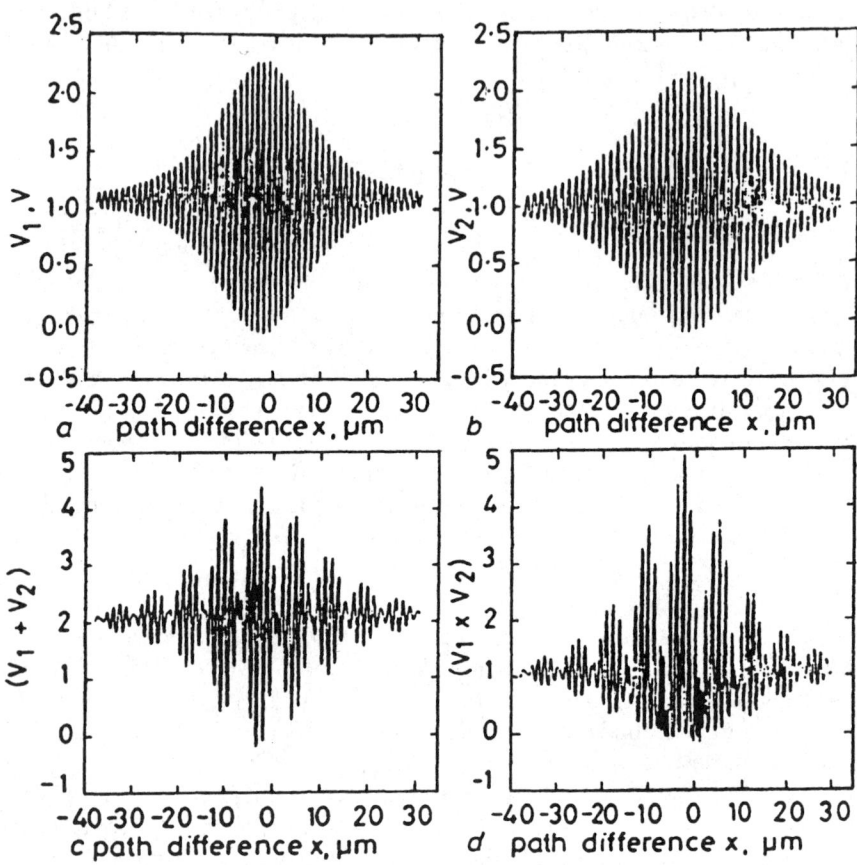

Fig. 5.11 Synthesized source for low coherence interferometry: (a) autocorrelation function of a 1.29 µm source; (b) autocorrelation function of a 1.566 µm source; (c) result of adding autocorrelation functions shown in (a) and (b); (d) result of multiplying autocorrelation functions shown in (a) and (b). The difficulty in identifying the central fringe is significantly reduced by using two low coherence sources and combining their autocorrelation functions by either adding or multiplying.

which is longer than either of the wavelengths of the individual sources λ_1 and λ_2. The unambiguous range of the system then becomes equal to $\lambda_{\text{effective}}$.

The basic system is shown in Figure 5.12. The output beams from the two low coherence sources are first spatially combined with a directional coupler and then injected into a transmitting interferometer (TMI) [47], either a Michelson or Mach–Zehnder (the Mach–Zehnder being preferred as it generates two outputs). This topology is ideal for sensor networks utilizing spatial multiplexing. The OPD of the TMI is matched to the sensor within a few microns. The signal is recovered by linearly ramping the OPD of the TMI over $\lambda n/2$ (where n is 1 or 2) such that transfer function of the sensor is scanned. The phase of the sensor is then determined modulo $2\pi m$, by comparing the

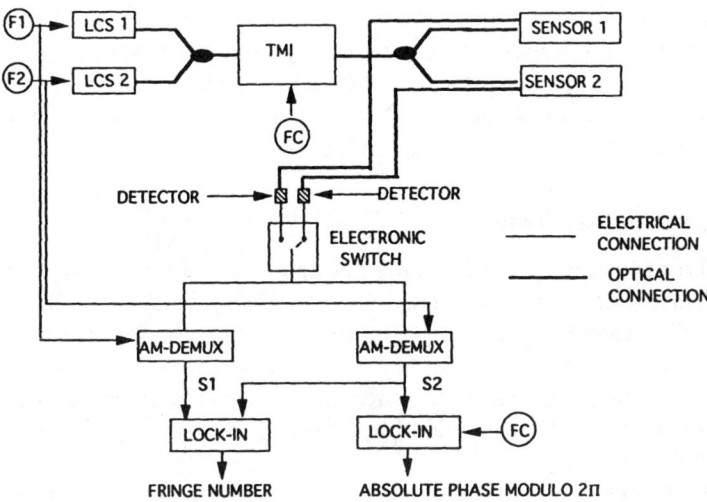

Fig. 5.12 Two-wavelength low coherence interferometry. LCS = low coherence source, TMI = transmitting interferometer.

fundamental frequency of the ramping signal against the output from the detector. The fringe number m and hence the absolute phase of the sensor is obtained by comparing the phase difference between the signals λ_1 and λ_2. The unambiguous range of the sensor is given by equation 5.40.

The major advantages of this approach are:

1. There is no need to identify the central fringe;
2. The TMI has to be linear only over $\lambda n/2$, rather than the full dynamic range of the sensor;
3. The bandwidth of the system can be as high as 1 kHz [48] even using piezoelectric transducers to drive the TMI.

The resolution of the system does depend on the stability of the sources, however, with an effective wavelength of $\sim 50\,\mu m$, a 1 nm resolution can be achieved with typically low-coherence sources by temperature stabilizing the sources to $\sim 0.1\,°C$.

5.5.3 Channel spectra processing [49, 50]

An alternative method for recovering the induced phase change in low coherence systems is to recover the signal in the spectral domain. The channel spectrum generated by illuminating an imbalanced interferometer with light from a broadband source is spectrally resolved. The OPD of the sensor determines its free spectral range equal to $c/2nd$, where d is the distance between the distal end of the fiber and the reflecting surface and n the refractive index. The channel spectrum fringe spacing varies as the target position changes. This technique provides an absolute measurement of displacement, over several millimeters, without recourse to a tracking stage as discussed in section 5.5.1. The bandwidth of the system is governed by the CCD readout time and can be much greater than that of the tracking approach. An example of the use of this technique to measure displacement is given in section 5.6.

5.5.4 Processing interferometers

Very recently imbalanced interferometers again illuminated with light from a low-coherence source have been used to interrogate multiplexed Bragg grating sensor (BGS) networks. The basic requirement is that the free spectral range of the interferometer should be equal to the anticipated measurand-induced wavelength shift in the BGS. In these studies the interferometer has been implemented in various formats, including integrated optics [51], all fiber and conventional bulk glass [52].

It has also been shown that an imbalanced interferometer can be used to simultaneously interrogate both BGS and FOIS provided the OPD of the FOIS equals the maximum induced wavelength change in the BGS [53].

(a) Processing interferometers based on highly birefringent fibers

Polarization state generators and controllers have been developed by exploiting the unique properties of highly birefringent fiber. In section 5.6 it is shown how this concept can be used (1) to recover the signal from Faraday current sensors [54] and (2) to realize a high speed polarimeter.

5.6 APPLICATIONS

Many applications for FOIS were discussed in the first volume of this series; in this section only a selection of the more recent applications for FOIS are presented.

5.6.1 Absolute sensors for quasi-static measurands

Absolute measurands with optical interferometers are difficult to determine due to the periodic nature of the transfer function which can lead to large ambiguities when the system is 'powered up'. As discussed in section 5.5.2, low coherence interferometry has the potential to solve this problem.

In many applications for sensors (conventional or optical) it is necessary to develop a sensor network which allows easy replacement in the event of damage or change in sensor configuration, without the requirement to re-calibrate the sensor. In addition, the deployment of the sensors must be such that if one of the fibers breaks the system can still function; this tends to mitigate against serial multiplexing topologies, favoring parallel topologies despite the fact that the multiplexing efficiency is lower (see Chapter 4).

(a) Sensor design

The fundamental working principle of any FOIS is that the measurand causes a change on the OPD of the sensor, hence measurands such as temperature, pressure, displacement, force refractive index can be directly detected. Thus if we wish to develop a sensing network capable of absolute measurement of multiparameters the sensor design ideally should have a common cavity structure to allow interchangeability.

The Fizeau/Fabry–Pérot structure offers this possibility, with the advantage that it can be very simply formed by the cleaved end of a single mode fiber and the surface of the sensing element (such as a coated diaphragm used for pressure measurement) essentially acting as a mirror. Other cavity structures are possible; for example, it can be fabricated entirely from fiber (ideal for temperature sensing). In order to make the sensors interchangeable the only requirement is that all the sensors must have the same nominal OPDs (to within a few microns). This type of sensor is reflective with the fiber acting as a transceiver link. Various sensors have been developed based on these concepts [48], for example Figure 5.13 shows a very high performance high pressure sensor [55]. The sensor itself is a hollow wall cylinder, which expands linearly with pressure. The Fabry–Pérot cavity is formed between the fiber and the end cap of the cylinder. The cavity length is $\sim 400\,\mu m$, for such a path length, which is much larger than the diameter of other fiber core; only the first-order reflected beam from the end of the single mode fiber and a small part of the first-order reflected beam from the end cap contribute to the interferometric signal. The normalized output signal is similar to that of a two-beam interferometer operating in reflection and has a sinusoidal transfer function modified by a visibility profile function resulting from the slight variation in intensity of the signal reflected from the cylinder as it moves in response to a pressure change and the coherence length of the source. A simple monomode fiber temperature sensor was mounted in the pressure sensor to enable errors in the pressure

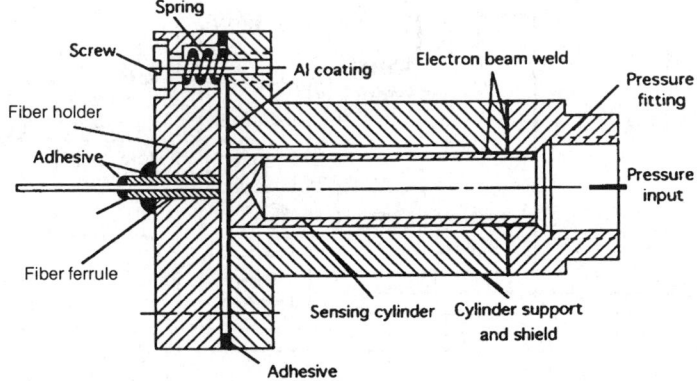

Fig. 5.13 Schematic of the ultra high pressure sensor.

measurement caused by thermally induced changes in the bulk modulus of the sensor to be corrected.

Sensors of this type have been used in a sensor network as shown in Figure 5.14, based upon a parallel multiplexing topology. The signal processing used is described in section 5.5.2. The range to resolution achieved for each sensor in the network was $10^4 : 1$ which is comparable with the best conventional electric sensors.

(b) Temperature

The very large thermo-optic coefficient of a typical optical fiber (phase change $°C^{-1}$) makes it an ideal thermometer or heat flux sensor when the induced phase change is measured interferometrically. Given a modest optical resolution of only 10^{-9} m, temperature changes of less than a millidegree can be determined using a 5 mm long fiber Fabry–Pérot sensor. Most applications for temperature sensors require that the output is always unambiguous, hence low coherence processing appears to be the optimum method for signal recovery. For low temperature applications, the short fiber sensing cavity can be bonded to the transceiver fiber with a suitable adhesive. However, a better approach is to fusion splice the sensing fiber to the transceiver fiber, in this case it is necessary that the ends of the cavity are coated to ensure reflection occurs at the interface. Metals cannot be used because the arc will 'track down' the coating causing it to evaporate; ordinary dielectric coatings are also unsuitable as they are damaged

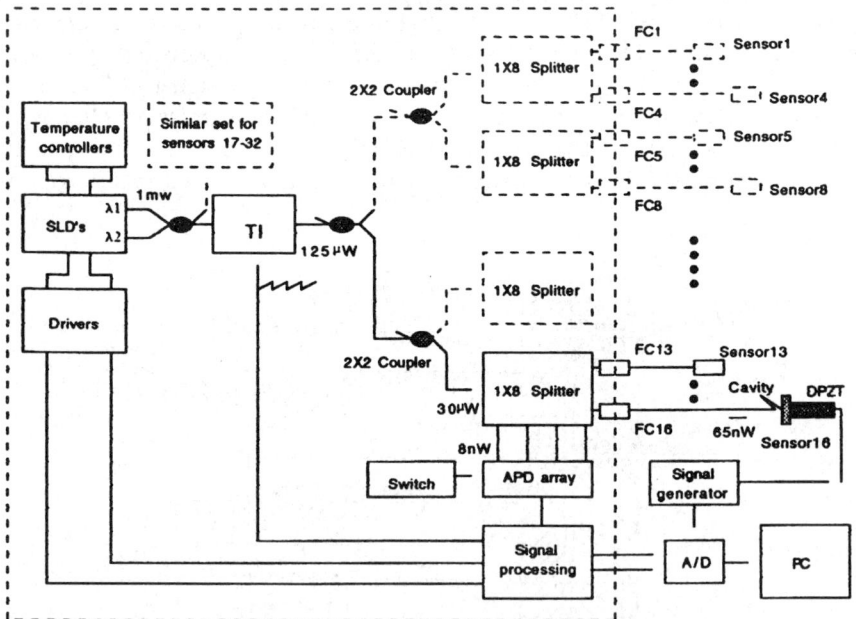

Fig. 5.14 Schematic diagram of the multiplexing system. SLDs = superluminescent diodes, TI = transmitting interferometer, FC1–16 = optical fiber connectors.

by the high temperature. Successful reflective splices have been reported for fibers coated with evaporated TiO_2 film [56, 57].

For applications where the temperature may exceed 1000 °C, the fiber cannot be used as the sensor. It is possible, using the type of probe reported by [58] using low-coherence processing, to make precise high temperature measurements. For measurements well above 1000 °C a sapphire sensing element is required [59].

One of the unique features of the miniature FFP is the very high bandwidth > 200 kHz which makes it ideally suited for the study of fast phenomena such as very fast chemical reactions and heat diffusion studies. In this type of application it is usually unnecessary to determine the absolute value of the induced phase change, and a single laser diode can be used to illuminate the sensors [60]. Sensors of this type have been used for heat diffusion studies in turbine blades [61].

5.6.2 Optical coherence domain reflectometry

Optical coherence domain reflectometry (which is based upon low-coherence interferometry) offers unique sensing opportunities. The technique was first described by Youngquist et al. [62] and can be used with a single interferometer or tandem interferometers. In the original paper the positions of very weak reflecting surfaces in a glass sample were located by using the sample as one of the mirrors of a Michelson interferometer, the other mirror driven at a frequency of f_r was mounted on a motor-driven translation stage. The interferometer was illuminated by a low coherence source. The positions of the reflecting surfaces were then located by varying the OPD of the Michelson whilst monitoring the output signal at f_r. The spatial resolution reported was ~ 10 µm with a dynamic range of 100 dB. Remote measurements are possible by incorporating optical fibers into the system; for example, the position of very weak reflecting sites in fibers have been located by this technique. A distributed laser Doppler velocimetry (LDV) system [63] has been developed based on this concept (Figure 5.15); here the fiber is placed in the flow and the position where the velocity is measured in the flow along the path of the probe beam, is governed by the OPD in the RI. A possible application for this system is the measurement of blood flows where it would be possible to move the measurement point away from the stagnation area around the fiber tip. A very large dynamic range is achieved with this system using differential detection. Similar instrumentation has been used to measure the reflectivity of biological tissues [64] and the length and positions of the lens and cornea of the eye [65, 66]. It is also possible to use this technique to build up images of objects buried in highly scattering materials [67] by scanning the sample in the x–y plane whilst measuring the position of reflecting surfaces in the z direction. At the present time, the application of the technique is limited by the relatively long scanning time; if this can be reduced, optical coherence reflectometry (or tomography) could replace X-ray techniques in the diagnosis of cancer in soft tissues.

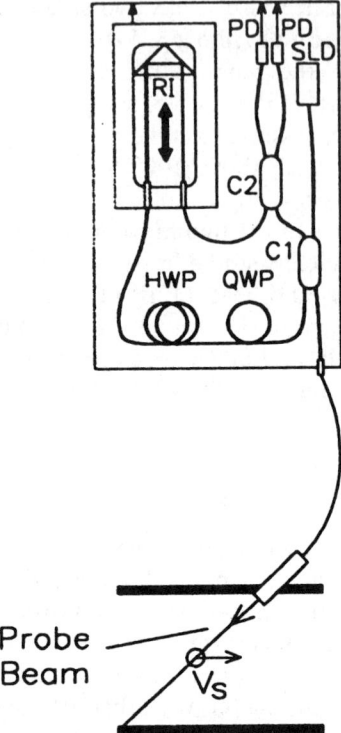

Fig. 5.15 Schematic diagram of a distributed fiber optic laser Doppler velocimeter, based upon low coherence interferometry. The sampling point in the flow V_S is controlled by the optical path imbalance of the receiving interferometer.

Another recent application of the technique is the high resolution study of the surface roughness. In this case a single Michelson interferometer is used, Figure 5.16(a), where the target acts as one of the mirrors and the position of the other mirror is swept. The output is detected with a CCD camera. This enables a 3D image of the surface to be reconstructed from the phase maps generated as a 'coherence plane' is swept across the target. In this application, the fiber is used essentially to produce and ensure that the light from the superfluorescent source is spatially filtered; an example of a surface damaged by a high velocity impact is shown in Figure 5.16(b).

5.6.3 Magnetometers

(a) Magnetostrictive devices

The optical configuration of the magnetostrictive magnetometer has remained virtually unchanged since ~1985. Typically, a fiber Mach–Zehnder or Michelson interferometer is used with the magnetostrictive element bonded to one of the arms of the interferometer. To avoid low frequency environmental noise the magnetostrictive element is subject to a dither field at ω_d to transpose the magnetically induced dimensional changes in the magnetostriction element to side bands of a carrier at ω_d [68]. A low frequency servo is used to maintain the interferometer at quadrature for maximum sensitivity. The use of the dither field also allows the magnetometer to be operated in a closed loop mode where a nulling field is used to maintain the sensor at a constant magnetic field. Most of the improvements in the performance of this sensor have occurred due to better designs of the magnetostrictive transducer or the magnetostrictive material itself. Excellent performance has been obtained with a fiber-wrapped cylindrical shell transducer [69]. In earlier work with these devices there was a tendency to dither the magnetostrictive element at a mechanical resonance; however, this has been shown to produce additional noise. The noise floor can be significantly reduced by selecting a dither frequency slightly away from the mechanical resonance. Resolutions of 10 pT Hz$^{1/2}$ at 1 Hz and 70 fT Hz$^{1/2}$ at 35 kHz close to a mechanical resonance of the transducer have been reported [70]. Gradient magnetometers and three axis magnetometers have also been demonstrated [71].

(b) Faraday effect magnetic probe sensors and polarization state rotators

Miniature high speed magnetic field probes have been developed based upon the Faraday effect. The Faraday effect is the rotation of the polarization azimuth ϕ_F which occurs when an optical beam propagates through a medium subject to a magnetic field H

$$\phi_F = \int_{L_F} V_{(\lambda\tau)} \bar{H} \, \bar{dl}_F \qquad (5.42)$$

where $V_{(\lambda\tau)}$ is the material-dependent Verdet constant, which is both dispersive and temperature dependent, and l_F is the interaction length. The basic concept of the magnetic probe, utilizing heterodyne signal processing, is shown in Figure 5.17. The probe is illuminated via a laser source oscillating at two different frequencies, ω_1 and ω_2, in orthogonally polarized modes. These two independent outputs from the laser are then transferred to the probe via a highly birefringent (hi-bi) fiber with the light at frequencies ω_1 and ω_2 being guided in different orthogonal modes of the fiber. At the probe head a reference frequency $(\omega_1 - \omega_2) h = \Delta\omega$ is generated by combining the two beams at polarizer (1) with its transmission axis at 45° to the eigenmodes of the hi-bi fibers. The remaining part of the input signal is coupled into the sensing element via a $\lambda/4$ plate where the two linear states are transformed into two orthogonal circular states. These states then transverse the sensor and are reflected at the mirror and retrace their paths back through the sensor. After passing through the $\lambda/4$ plate and polarizer (2), these states are converted to a linear state at $\Delta\omega$ where the phase depends on the instantaneous value of ϕ_f. If the applied magnetic field is periodic, then the amplitude of the phase deviation is proportional to the amplitude of the field and the rate of deviation, i.e. the resultant signal is a

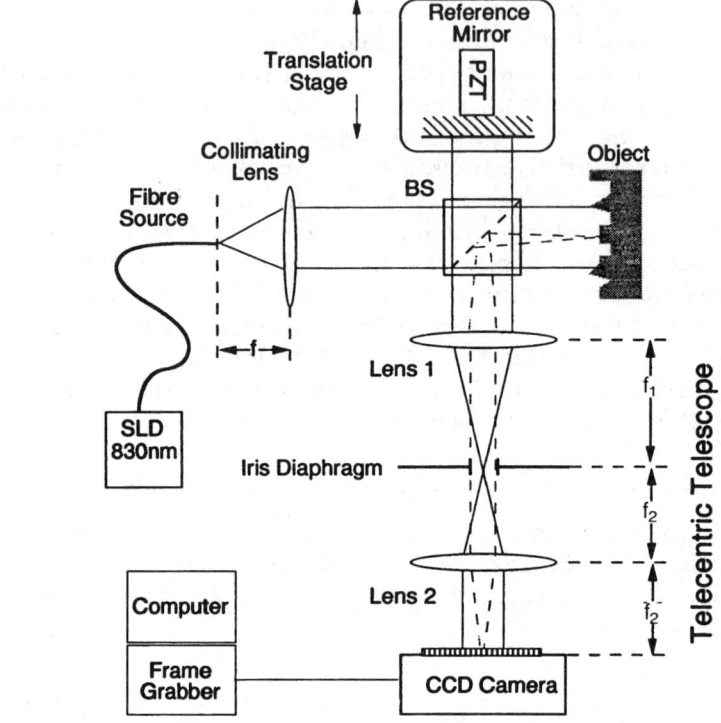

Fig. 5.16 (a) Basic system for coherence imaging based upon a Michelson interferometer with CCD readout; (b) plan view of crater caused by supersonic dust particle, cross-section of crater taken along dotted line.

phase-modulated carrier at $\Delta\omega$ and can be detected with a spectrum analyzer. If the hi-bi link is subject to environmental noise, the effects of this noise may be eliminated, to first order, by comparing the modulated signal with the reference signal using a phase meter. Using Hoya Fr-5 glass sensitivities of $50\,\mathrm{nT\,Hz}^{1/2}$ have been achieved. This heterodyne technique is particularly effective for measuring small amplitude high frequency magnetic fields [72] and could be used for probes based upon Ga:YIG films which have demonstrated $100\,\mathrm{pT\,Hz}^{1/2}$ sensitivities at 500 Hz with bandwidths up to 1 GHz [73].

The operation of the system may be described using the Jones matrix formalism; after the transfer through the $\lambda/4$ we have

$$E_1 = \frac{1}{\sqrt{2}}\begin{bmatrix}1\\i\end{bmatrix}e^{i(\omega_1 t+\phi_r)} \qquad E_2 = \frac{1}{\sqrt{2}}\begin{bmatrix}1\\-i\end{bmatrix}e^{i(\omega_2 t+\phi_s)}$$

APPLICATIONS

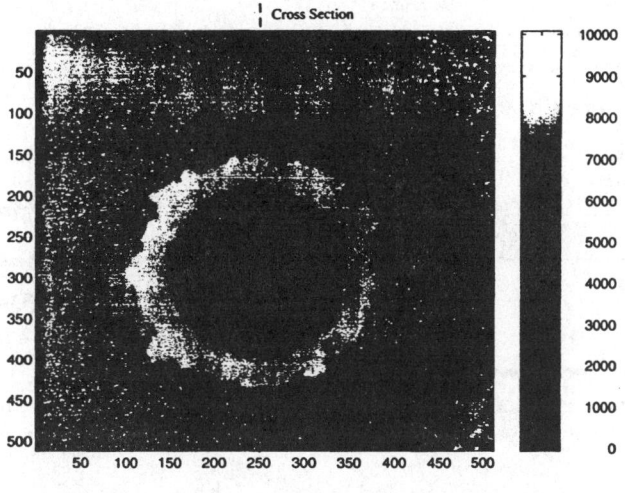

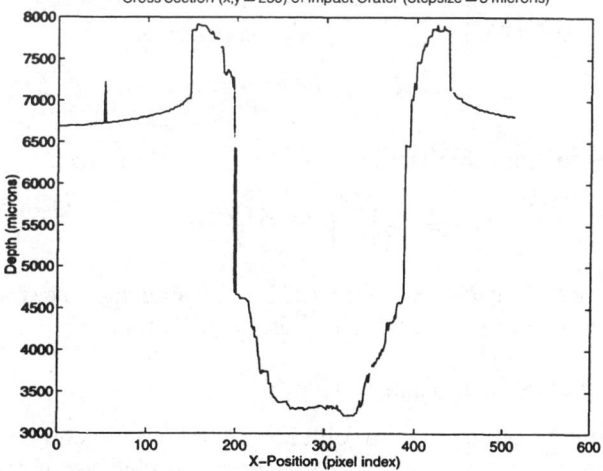

(b)

Fig. 5.16 *(continued)*

where ϕ_f and ϕ_s are the delays in the hi-bi fiber. The fields returning to the $\lambda/4$ plate are

$$E_{\lambda/4} = ([M]^{-1}[R(\theta)]^{-t}[M][R(\theta)][E_{in}])^*_{n=1,2} \qquad (5.43)$$

The complex conjugate is taken as the sensor is a nonreciprocal element. Hence

$$E_{\lambda/4} = \frac{1}{\sqrt{2}}\begin{bmatrix}1\\-i\end{bmatrix}e^{i(\omega_1 t+\phi_f+2\phi_F)} + \frac{1}{\sqrt{2}}\begin{bmatrix}1\\i\end{bmatrix}e^{i(\omega_2 t+\phi_s-2\phi_F)}$$

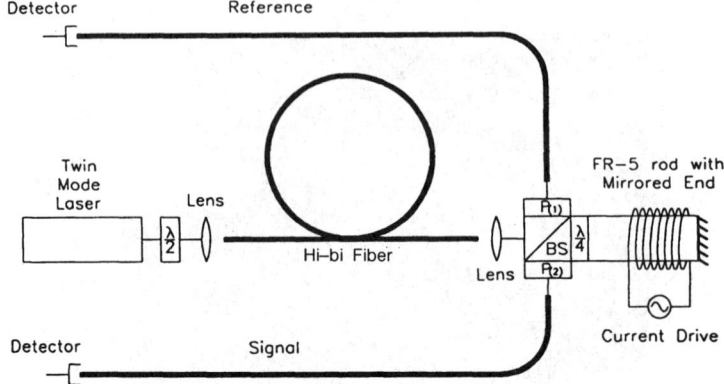

Fig. 5.17 A remote magnetic field probe based upon Faraday rotation and a common path optical fiber heterodyne interferometer.

which after passing through the $\lambda/4$ and the polarizer set at 45° to the axis of the input linear states yields

$$E_{\text{sig}} = \frac{1}{\sqrt{2}} \begin{bmatrix} 1 \\ 1 \end{bmatrix} e^{i((\omega_1 - \omega_2)t + (\phi_f - \phi_s) + 4\phi_F)} \quad (5.44)$$

The reference signal has a similar form:

$$E_{\text{ref}} = \frac{1}{\sqrt{2}} \begin{bmatrix} 1 \\ 1 \end{bmatrix} e^{i((\omega_1 - \omega_2)t + (\phi_f - \phi_s))} \quad (5.45)$$

Thus by comparing their phase difference the Faraday signal can be recovered free of any environmental noise picked up on the delivery fiber.

(c) Linear polarization state rotators

Another important application for differential interferometers is as a linear polarization state rotator, as they can form the basis of ellipsometers [74] which are used to measure the thickness of thin films. If we consider Figure 5.17 again, but remove the Faraday element, then we will have generated two linear rotating polarization states, one the reference as before and the second beam, which will propagate toward the test sample, where the state of polarization of the beam can be measured with suitable polarization components [75].

5.6.4 Nondestructive testing

Induced ultrasonic disturbances can be used to examine large structures for defects. Commonly high frequency contact piezoelectric transducers are used to study the propagation of the ultrasonic waves where peaks in the signal can be associated with surface cracks or flaws. In many applications it is desirable to use a noncontact method particularly where access is difficult. An example of such a system [76] is shown in Figure 5.18(a). The ultrasonic waves are generated by focusing the output power of a Q-switched Nd:YAG laser, 10 ns duration on the target. The motion of the surface is detected using a stabilized fiber optic Fizeau interferometer. The PZT is servo-controlled to maintain a constant distance between the fiber tip and the sample to keep the cavity in quadrature. Figure 5.18(b) shows a typical result for an ultrasonic waveform generated on an aluminum sample. The effectiveness of this technique can be clearly seen from the figure where nanometer displacements are determined in microseconds.

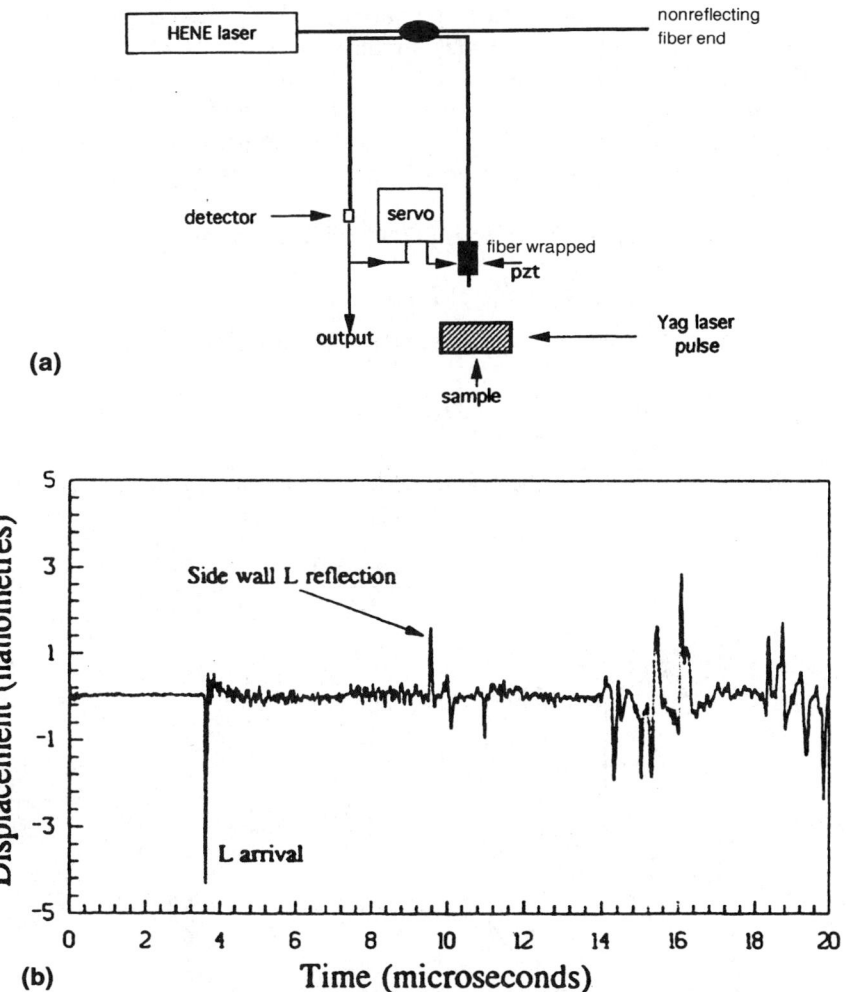

Fig. 5.18 (a) Fiber optic based Fizeau interferometer – servo is used to maintain distance between fiber tip and sample constant at low frequencies. Ultrasonic waves are generated by the YAG laser pulse; (b) ultrasonic wave detected on an aluminum sample.

(a) Nondestructive testing using differentiating interferometers

The differentiating fiber optic interferometer (see section 5.2.9) has great potential for high frequency, noncontact, nondestructive measurement in harsh industrial environments as it acts as a high bandpass filter, thus eliminating the effects of low frequency perturbations which are likely to totally swamp the nanometer displacements associated with ultrasound. The Sagnac interferometer [77] has recently been used for such studies. Figure 5.19 shows a simplified arrangement of the system used. The Sagnac ring is broken close to the input coupler such that the recombining counterpropagating beams arrive at the target at different times $t - t_1$ and $t - t_2$, a hi-bi fiber is used to take the signal from the Sagnac ring to the target. Polarization controllers are used in the Sagnac loop to ensure that when the counterpropagating beams reach the hi-bi fiber they are coupled into different modes of the fiber, the $\lambda/4$ plate at the end of the probe is used to ensure that back-reflected beams from the target propagate in different eigenmodes of the fiber as they return to the Sagnac ring — thereby ensuring they have followed a common path in opposite directions at different times.

Following the analysis of reference [77] the differential phase

$$\phi_D = \phi_2 - \phi_1$$
$$= 2k(x(t - t_1) - x(t - t_2)) \quad k = 2\pi n/\lambda$$
$$\phi_D = 2k \int_{t-t_2}^{t-t_1} v(t)\,dt = 2k\bar{v}\,\Delta t$$

where $\Delta t = t_2 - t_1$ and $\bar{v}$ is the mean velocity of the target between $t - t_1$ and $t - t_2$ and

$$\Delta t = \bar{n}\frac{(L_2 - L_1)}{C}$$

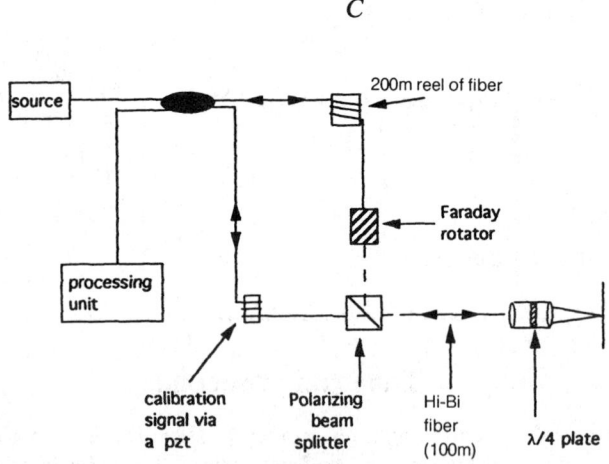

Fig. 5.19 Sagnac interferometer used in nondestructive testing.

where $\bar{n}$ is the mean refractive index of the fiber. If the target is subject to a periodic displacement of the form

$$x(t) = x_0 \sin \omega_s t$$

where x_0 is the amplitude and ω_s the angular frequency, it can be shown that

$$\phi_D = 2kV_p t_0 \mathrm{sinc}\left(\frac{\omega_s \Delta t}{2}\right) \cos \omega_s(t - t_0) \qquad (5.46)$$

where V_p is the peak velocity and $t_0 = (t_1 + t_2)/2$. Hence ϕ_D is directly proportional to the velocity of the target and the frequency response to the transit time of the light in the ring.

5.6.5 Ring resonator as a very high resolution optical spectrum analyzer

The availability of very low loss single mode fiber at wavelengths of 1.3 and 1.5 μm has made it possible to produce optical spectrum analyzers with unprecedented optical resolutions. High finesse Fabry–Pérots and ring resonators have been fabricated with absolute optical resolutions in excess of 20 kHz, i.e. ~ one part in 10^{10}. Previously this type of resolution was only achievable using heterodyne and homodyne techniques [78].

An example of a ring resonator operating as a high resolution spectrometer is described in [79]. The ring consists of very low loss polished coupler and a 40 m length of single mode fiber giving a free spectral range (FSR) of 5 MHz. A short section of the fiber is wrapped around a piezoelectric transducer in order to scan the transfer function. The fiber loss was 0.35 dB km^{-1} and the total loss of the device ~ 0.07 dB. The finesse was measured at 200 using a frequency tunable Nd:YAG laser diode pumped ring laser with a linewidth of 5 kHz, giving the resonator a linewidth of ~ 20 kHz [79]. This represents the highest optical resolution ever reported for a passive optical interferometer. The resonator finesse F is given by

$$F = \frac{\pi^4 \sqrt{kk_r}}{1 - \sqrt{kk_r}} \qquad (5.47)$$

where k is the intensity coupling coefficient and k_r the optimum resonance coupling coefficient which depends solely on the total optical loss.

Apart from being used as an optical spectrum analyzer, applications for such a high resolution device could include laser Doppler velocimetry and 'intensity fluctuation spectroscopy'. Due to the very long length of the ring and the number of recirculations the light makes in the ring, it is possible to analyze the spectrum of light where its frequency content varies in a time less than the transit time of the light in the ring. An experiment demonstrating the phenomenon is illustrated in Figure 5.20, where a phase modulator has been incorporated between the source and the ring [80]. The phase modulator was driven by a sine wave at frequency f_m such that the spectrum of the light will have components at f_{laser}, $f_L \pm f_m$, $f_L \pm 2f_m$, etc. where the amplitudes of these

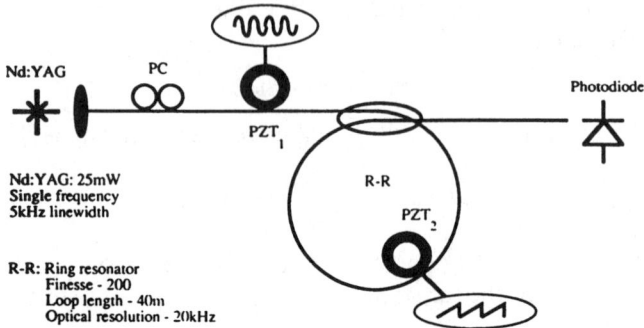

Fig. 5.20 The experimental configuration used to determine the resolution and dynamic response of the ring resonator. The piezoelectric transducer PZT_1 is used to generate a series of harmonics which are resolved using the scanning ring resonator spectrometer. PC = polarization controller.

components are proportional to $J_0(\phi) + J_1(\phi) + J_2(\phi)$ where J_0 etc. are Bessel functions. The observed spectrum is complex as shown in Figure 5.21(a) due to the phase of the input light being modulated at a rate comparable to the time of flight of the light in the ring.

A computer simulation gives good agreement as indicated in Figure 5.21(b).

5.6.6 Displacement sensor based upon channel spectra processing [81]

A system used to measure displacements exploiting channel spectra processing is shown in Figure 5.22(a). Light from a broadband source is transferred to the target via the fiber coupler, where a Fizeau cavity is formed.

Light reflected back from the FP cavity is transferred to a simple monochromator where the light diffracted from the grating is detected with a CCD array. The output from the CCD array can then be processed by a computer. This signal is the multiplication of the Gaussian source profile and the transfer function of the interferometer and is equal to

$$I(\lambda, \delta) = C \exp\left(\frac{-(\lambda - \lambda_0)^2}{\Delta \lambda^2}\right)\left(\frac{F \sin^2 \delta/2}{1 + F \sin^2(\delta/2)}\right) \quad (5.48)$$

where C is a constant, λ_0 the mean wavelength of the source, F a function of the cavity and δ a function of the wavelength and cavity width.

A typical channel spectrum obtained with this system is shown in Figure 5.22(b). Various methods for processing this signal are available using a software fast Fourier transform (FFT), combined with a Gaussian fit to the FFT. Resolutions of $\sim 0.1\,\mu m$ over a measurement range of $\sim 1.5\,mm$ have been demonstrated. Although this technique does not offer the nanometer resolution of the system described in section 5.4.1 the much greater working range makes it potentially useful for several industrial applications.

APPLICATIONS

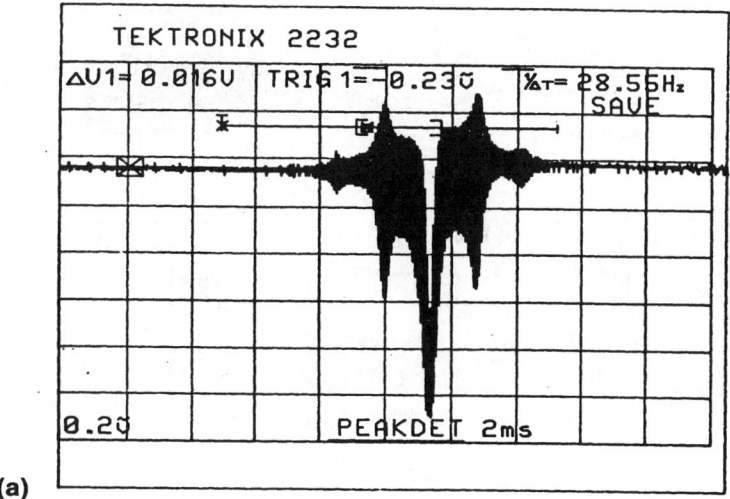

(a)

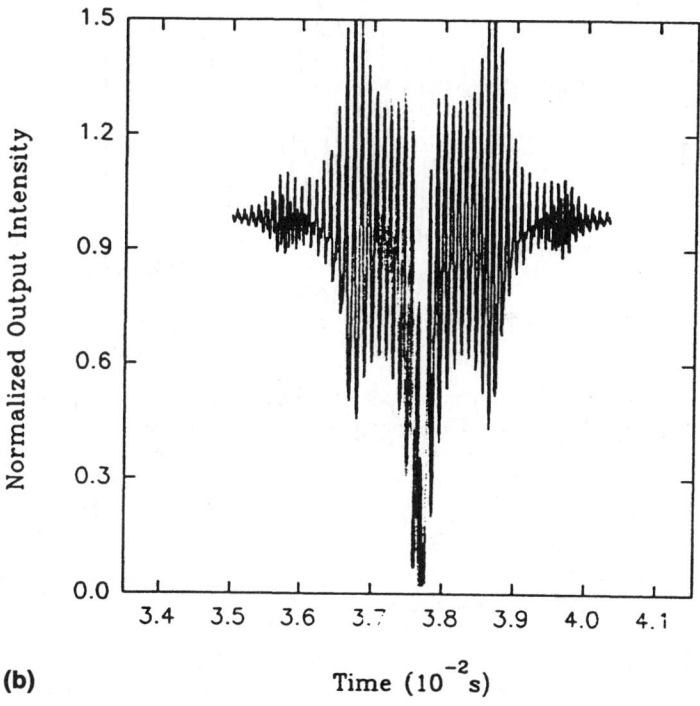

(b)

Fig. 5.21 (a) Spectrum observed with the ring resonator spectrometer when PZT_1 is modulated at 100 kHz for a modulation index of 0.7 rad; (b) theoretical curves showing the predicted temporal response for the experimental parameters used in (a).

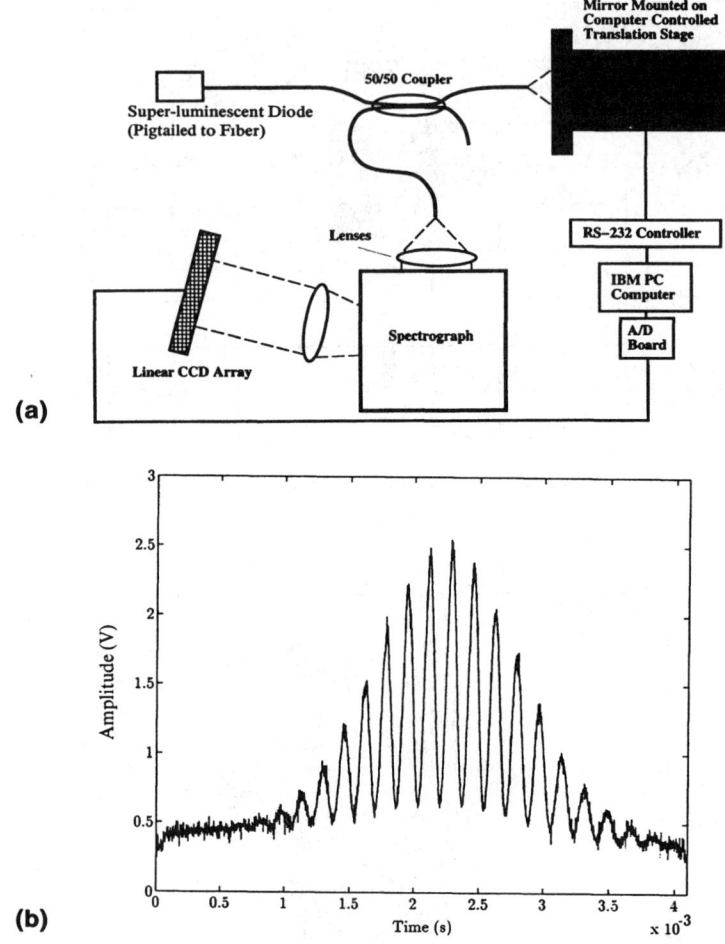

Fig. 5.22 (a) Experimental arrangement for channeled spectrum technique; (b) channeled spectrum output from linear CCD array for OPD ~ 0.1 mm.

5.7 SUMMARY

In this chapter a more formal approach to the derivation of the transfer functions of the common FOIs has been presented than in Volume 1. A summary of new components introduced into FOISs has been given together with several new applications for FOISs.

REFERENCES

1. McIntyre, P. D. and Synder, A. W., 1973, Power transfer between optical fibres, *Opt. Soc. Am.* No. 12, 1518–1527.
2. Buckman, A. B., 1992, *Guided Wave Photonics*, Saunders College Publishing, San Francisco.
3. Yariv, A., 1973, Coupled mode theory for guided-wave optics, *IEEE J. Quantum Electron.* **QE-9**: 919–933.
4. Jackson, D. A., Priest, R. P., Dandridge, A. and Tveten, A. B., 1980, Elimination of drift in a single-mode optical fibre interferometer using a piezo-electrical stretched coiled fibre. *Applied Optics*, **19**, 2926.
5. Kersey, A. D., Corke, M., Jackson, D. A. and Jones, J. C. D., 1983, All fibre Michelson interferometer, *Electron Letts.*, **19**, 471.
6. Yariv, A., 1987, Operator algebra for propagation problems involving phase conjunction and non reciprocal elements, *Applied Optics*, **26**, 4538.
7. Martinelli, M., 1989, A universal compensator for polarization changes induced by birefringence on a retracing beam, *Optics. Communs.* **72**, 341–344.
8. Kersey, A. D., Marrone, M. J. and Davis, M. A., 1991, Polarisation insensitive fibre optic Michelson interferometer, *Electron. Letts.* **27**, 518–519.
9. Marrone, M. J., Kersey, A. D. and Dandridge, A., 1992, Polarisation independent array configurations based on Michelson interferometer networks, *Distributed and Multiplexed Fibre Optic Sensors II*, Boston, A. D. Kersey and J. P. Dakin, Eds. SPIE, Vol. **1797**, 196–200.
10. Shurcliff, W. A., 1962, *Polarised Light*, Oxford Univ. Press, London.
11. Yariv, A., 1991, *Optical Electronics* (4th edn), Saunders College Pub. (Division of Holt, Rinehart and Winston), San Francisco.
12. Kersey, A. D., Jackson, D. A. and Corke, M., 1983, A simple fibre Fabry–Pérot sensor, *Opt. Communications*, **45**, 71.
13. Gerges, A. S., Newson, T. P., Farahi, F., Jones, J. C. D. and Jackson, D. A., 1988, A hemispherical air cavity fibre Fabry–Pérot sensor, *Optics Comm.*, **68**, 157–160.
14. Born, M. and Wolf, E., 1986, *Principles of Optics* (6th edn), Pergamon, Oxford.
15. Jackson, D. A., 1985, Monomode optical fibre interferometers for precision measurements, *J. Phys. E. Sci. Instrum. Instrument Science and Technology*, **18**, 981–1001.
16. Santos, J. L., Leite, A. P. and Jackson, D. A., 1992, Optical fibre sensing with a low-finesse Fabry–Pérot cavity, *Applied Optics*. **31**, 7361–7366.
17. Stokes, L. F., Chodorow, M. and Shaw, H. J., 1982, All single mode fibre resonators, *Opt. Letts.* **7**, 288–290.
18. Yve, C. Y., Peng, J. D., Liao, Y. B. and Zhou, B. K., 1988, Fibre ring resonator with finesse of 1260, *Elect. Lett.* **24**, 622–623.
19. Stokes, L. F., Chodorow, M. and Shaw, H. J., 1982, All fiber stimulated Brillouin ring laser with sub-milliwatt pump threshold, *Opt. Lett*, **7**, 509–511.
20. Dakin, J. P., Pearce, D. A., Wade, C. A. and Strong, A., 1987, A novel distributed optical fibre sensing system enabling location of disturbance in a Sagnac loop interferometer, *Proc. O.E. Fibre*, 1987, San Diego (*Proc. SPIE*, **838**) Paper 18.
21. Spammer, S. J. and Swart, P. L., 1995, Differentiating Mach–Zehnder interferometer, *Applied Optics*, **34**, 2350–23533.
22. Bergh, B. A., Lefèvre, H. C. and Shaw, H. J., 1981, All single-mode fibre optic gyroscope with long term stability, *Opt. Lett.* **6**, 502–504.
23. Booysen, A., 1994, Sensor applications of a reflective fibre optic ring interferometer, Ph.D. thesis, Rand Afrikaans University, Johannesburg, RSA.

24. Drain, L. E., 1980, *The Laser Doppler Technique*, John Wiley & Sons.
25. Dandridge, A., Tveten, A. B. and Giallorenzi T. G., 1982, Homodyne demodulation scheme for fibre optic sensors using phase generated carrier, *IEEE. J. Quantum Electron.* **QE-18**, 1647–1653.
26. Bush, I. J., Sherman, D. R. and Bostick, J. A., 1992, Time-division multiplexed interferometric demodulation technique with 5 million samples per second capability, *Distributed and Multiplexed Fibre Optic Sensors II*, J. P. Dakin and A. D. Kersey, Eds. *Proc. SPIE* **1797**, 242–248.
27. Kersey, A. D., Williams, K. J., Dandridge, A. and Weller, J. F., 1989, Characterisation of a diode laser-pumped Nd:YAG ring laser for fibre sensor applications, Springer-Verlag Proc. in Physics Vol. 44, 172–178, *Optical Fibre Sensors*, H. J. Arditty, J. P. Dakin and R. Th. Kersten, Eds Springer-Verlag, Berlin.
28. Sakai, I., Youngquist, R. C. and Party, G., 1987, Multiplexing of optical fibre sensors using a frequency-modulated source and gated output, *J. Lightwave Tech.*, **LT5**, 932–940.
29. New Focas Ltd, USA.
30. Dakin, J. P., 1992, Distributed optical fibre sensors, *Distributed and Multiplexed Fiber Optic Sensors II*, J. P. Dakin and A. D. Kersey, Eds. *Proc. SPIE* **1797**, Boston, 76–108.
31. France, P. W., 1991, *Optical Fibre Lasers and Amplifiers*, Blackie, Scotland.
32. Lefèvre H. C., 1993, *The Fibre Optic Gyroscope*, Artech House, Boston.
33. Kwong, N. S. K., Lau, K. Y., Bar-Chaim, N., Ury, I. and Lee, K. J., 1987, High power high efficiency window buried heterostructure GaAlAs superluminescent diode with an integrated absorber, *Appl. Phys. Letts*, **151**, 1879–1881.
34. Safin, S. A., Semenov, A. T., Shidlovski, V. R., Zuchov, N. A. and Kurnyavko, Y. V., 1992, High power 0.82 µm superluminescent diodes with extremely low Fabry–Pérot modulation depth, *Proc. 8th OFS Conf.*, Monterey, Calif., 78–80.
35. Kwong, N. S. K., Bar Chaim, N. and Chen, T., 1989, High-power 1.3mm superluminescent diode, *Appl. Phys. Letts*, **54**, 298–300.
36. Kim, B. Y., 1990, Broadband fibre sources for gyroscopes, *Proc. 7th OFS Conf.*, Sydney, 129–133.
37. Petermann, K. and Weidel, E., 1981, Semiconductor laser noise in an interferometer system, *IEEE. J. Quantum Electron.* **QE-17**, 1251–1256.
38. Gerges, A. S., Newson, T. P. and Jackson, D. A., 1990, Coherence tuned fiber-optic sensing system, with self-initialisation, based on a multimode laser diode, *Appl. Opt.* **29**, 4473–4480.
39. Bosselman, T. and Ulrich, R., 1984, High accuracy position-sensing with fibre coupled while light interferometers, *Proc. 2nd Int. Conf. on Optical Fibre Sensors*, Stuttgart (Berlin: UDE), 361–364.
40. Bosselmann, T., 1987, *Optical Fiber Sensors*, A. N. Chester, S. Msrtellucci and A. M. V. Scheggi, Eds, Martinus Nijhoff, 429–432.
41. Liu, T. Y., Cory, J. and Jackson, D. A., 1993, Partially multiplexing sensor network exploiting low coherence interferometry, *Applied Optics*, **32**(7), 1100–1103.
42. Gerges, A. S., Newson, T. P. and Jackson, D. A., 1990, A coherence tuned fiber-optic sensing system, with self-initialisation, based upon a multimode laser diode. *Applied Optics*, **29**(30), 4473–4480.
43. Chen, S., Palmer, A. W., Grattan, K. T. V. and Meggitt, B. T., 1992, Digital signal processing techniques for electronically scanned optical-fiber white-light interferometry. *Applied Optics*, **31**(28).
44. Rao, Y. J., Ning, Y. N. and Jackson, D. A., 1993, Synthesized source for white-light sensing systems, *Optics Letters*, **18**, 462–464.

45. Chen, S., Grattan, K. T. V., Meggitt, B. and Palmer, A. W., 1993, Instantaneous fringe order identification using dual broadband sources with widely spaced wavelength, *Electron. Lett.* **29**, 334–335.
46. Rao, Y. J. and Jackson, D. A., 1995, Long-distance fibre optic white-light displacement sensing system using a source-synthesizing technique, *Electron. Letts.*, **31**, 310–312.
47. Lobo Ribeiro, A. B., Rao, Y. J. and Jackson, D. A., 1994, Multiplexing interrogation of interferometric sensors using dual multimode laser diode sources and coherence reading. *Optics Communs.* **109**, 400–404.
48. Rao, Y. J., Jackson, D. A., Jones, R. and Shannon, C., 1994, Development of Prototype fibre-optic based Fizeau pressure sensors with temperature compensation and signal recovery by coherence reading, *Journal of Lightwave Technology*, **12**(9), 1685–1695.
49. Egorov, S. A., Ershov, Y. A., Likhachiev, I. G. and Marnaev, A. N. 1992, Spectrally encoded fibre optic sensors based on Fabry–Pérot interferometers, *SPIE*, Vol. **1972**, 8th Meeting on Optical Engineering in Israel, 362–368.
50. Podoleanu, A. Gh., Taplin, S. R., Webb, D. J. and Jackson, D. A., 1993, Channelled spectrum liquid refractometer, *Rev. Sci. Instrum*, **64**(10), Oct., 3028–3029.
51. Kersey, A. D., 1993, Interrogation and multiplexing techniques for fiber Bragg grating strain sensors, SPIE, Vol. **2071**, *Distributed and Multiplexed Fiber Optic Sensors III*, Boston, 30–48.
52. Rao, Y. J., Kalli, K., Brady, G., Webb, D. J., Jackson, D. A., Zhang, L. and Bennion, I., 1995, Spatially-multiplexed fibre-optic Bragg grating strain and temperature sensor system based on interferometric wavelength-shift detection, *Electron. Lett.*, **31**, 1009–1010.
53. Brady, G., Kalli, K., Webb, D. J., Jackson, D. A., Reekie, L. and Archambault, J. L., 1995, Simultaneous interrogation of interferometric and Bragg grating sensors, *Optics Letters*, **20**, 1340–1342.
54. Bartlett, S. C., Farahi, F. and Jackson, D. A., 1990, Current sensing using Faraday rotation and a common path optical fiber heterodyne interferometer, *Rev. Sci. Instrum.* **61**(9), Sept., 2433–2435.
55. Rao, Y. J. and Jackson, D. A., 1994, Prototype fibre-optic based ultrahigh pressure remote sensor with built-in temperature compensation. *Rev. Sci. Instrum.* **65**, 1695–1698.
56. Lee, C. E., Taylor, H. F., Markus, A. M. and Udd, E., 1989, Optical fiber Fabry–Pérot embedded sensor, *Opt. Letts.*, **14**, 1225–1227.
57. Inci, M. N., Kidd, S. R., Barton, J. S. and Jones, J. D. C., 1993, High temperature miniature fibre optic interferometric thermal sensors, *Meas. Science and Tech.* **4**, 382–387.
58. Gerges, A. S. and Jackson, D. A., 1991, A fibre-optic based high temperature probe illuminated by a multimode laser diode, *Optics Communs*, **80**, 210–214.
59. Wang, A., Collapudi, S., May, R. G., Murphy, K. A. and Claus, R. O., 1992, Advances in sapphire fibre based intrinsic interferometric sensors, *Opt. Letts.* **17**, 1544–1546.
60. Farahi, F., Jones, J. D. C. and Jackson, D. A., 1991, High speed fiber-optic temperature sensor, *Opt. Letts.* **16**, 1800–1802.
61. Kidd, S. R., Sinha, P. G., Barton, J. S. and Jones, J. D. C. 1992, Utilisation of fibre Fabry–Pérot interferometers in the determination of heat transfer transients in wind tunnels, *Proc. 8th OFS*, Monterey, Calif. 73–76.
62. Youngquist, R. C., Carr, S. and Davies, D. N., 1987, Optical coherence domain reflectometry; a new optical evaluation technique, *Opt. Letts*, **12**, 158–160.

63. Gusmeroli, V. and Martinelli, M., 1991, Distributed laser Doppler velocimeter, *Opt. Letts.* **16**, 1358–1360.
64. Chivaz, X., Marques-Weddle, F., Salathé, R. P., Novak, R. P. and Gilgen, H. H., 1992, High resolution reflectometry in biological tissues, *Opt. Letts*, **17**, 4–6.
65. Swanson, E. A., Huang, D., Hee, M. R., Fujimoto, J. G., Lin, G. P. and Puliafito, C. A. 1992, High speed optical coherence domain reflectometry, *Opt. Letts.* **17**, 151–153.
66. Chen, S., Wang, D. N., Grattan, K. T. V., Palmer, A. W. and Dick, A. L., 1993, A compact optical device for eye length measurement, *Photonic Tech. Letts.*
67. Hee, M. R., Izatt, J., Jacobson, M. J., Fujimoto, J. G. and Swanson, E. A., 1993, Femtosecond transillumination optical coherence tomography, *Opt. Letts*, **18**, 950–952.
68. Kersey, A. D., Jackson, D. A. and Cirke, M., 1985, Single-mode fibre optic magnetometer with DC bias field stabilisation, *J. Lightwave Tech.*, **LT-3**, 836–840.
69. Dagenais, D. M., Bucholtz, F., Koo, K. P. and Dandridge, A., 1989, Detection of low frequency magnetic signals in a magnetostrictive fibre-optic sensor with suppressed residual signal. *J. Lightwave Tech.*, **7**, 881–887.
70. Bucholtz, F., Dagenais, D. M. and Koo, K. P., 1989, Optic magnetometer with 70ft Hz resolution, *Electron. Letts*, **25**, 1719–1721.
71. Bucholtz, F., Dagenais, D. M., Koo, K. P. and Vohra, S., 1990, Recent developments in fibre optic magnetostrictive sensors, *Proc. SPIE*, Vol. **1367**, *Fiber Optic and Laser Sensors VIII*, Ramon P. DePaul, Eric Udd, Eds, 226–235.
72. Bartlett, S. C., Farahi, F. and Jackson, D. A., 1990, Current sensing using Faraday rotation and a common path optical fiber heterodyne interferometer, *Rev. Sci. Instrum.* **61**, 2433–2435.
73. Day, G. W., Deeter, M. N. and Rise, A. H., 1991, Faraday effect sensors: a review of recent progress, *Advances in Optical Fiber Sensors*, B. Culshaw, E. L. Moore and Z. Zhipend, Eds, SPIE Optical Engineering Press, Wuhan, China, 11–26.
74. Chitaree, R., Weir, K., Palmer, A. W. and Grattan, K. T. V. 1994, A highly birefringent fibre polarization modulation scheme for ellipsometry: system analysis and performance, *Measurement Science and Technology*, **5**, 1226–1232.
75. Jackson, D. A., Kersey, A. D., Akhavan Leilabady, P. and Jones, J. D. C., 1986, High frequency non-mechanical optical linear polarisation state rotator. *Phys. E. Sci. Instrum.* **19**, 146.
76. Pierce, S. G., Corbett, R. E. and Dewhurst, R. J. 1993, An actively-stabilised fibre-optic interferometer for laser-ultrasonic flow detection, *Review of Progress in Quantitative Non-destructive Evaluation*, **12**, 587–593.
77. Harvey, D., McBride, R., Barton, A. S. and Jones, J. D. C., 1992, A velocimeter based on the fibre optic Sagnac interferometer, *Measurement Science and Technology*, 1077–1083.
78. Cummins, H. Z. and Swinney, H. L., 1970, *Light Beating Spectroscopy*, Progress in Optics, VIII, 133–200.
79. Kalli, K. and Jackson, D. A., 1992, Ring resonator optical spectrum analyzer with 20 kHz resolution, *Optics Letters*, **17**(15), August, 1167–1169.
80. Kalli, K. and Jackson, D. A., 1993, Analysis of the dynamic response of a ring resonator to a time-varying input signal, *Optics Letters*, **18**(6), 465–467.
81. Taplin, S., Podoleanu, A. Gh., Webb, D. J. and Jackson, D. A., 1993, Displacement sensor using channelled spectrum dispersed on a linear CCD array, *Electronics Letters*, **29**(10), May, 896–897.

6
Optical fiber speckle interferometry

R. P. Tatam

6.1 INTRODUCTION

'Speckle interferometry' is a generic term for a range of optical configurations that utilize speckles, generated by the interference of light scattered from different points on an optically rough surface, to convey information about the object surface. This information can relate to the shape or surface profile of an object, the relative displacement under applied load, the vibrational characteristics or the surface strain.

In speckle interferometry the speckles are interferometrically encoded either by mixing the speckle pattern with a second speckle pattern or by mixing it with a smooth reference wavefront. Figure 6.1 shows an interferometrically encoded speckle pattern. The intensity of each individual speckle varies in a similar manner to a two-beam interferometer, that is, proportional to $(1 + V\cos\phi)$; V is the visibility and ϕ the phase. Due to the statistical nature of speckle formation [1] the phase relationship between neighboring speckles varies in a random manner. As a result the phase relationship across the object surface, represented by fringes, is obtained by correlating two separate speckle patterns. Correlation is usually achieved by subtraction as this removes the d.c. intensity and noise contributions from, for example, multiple reflections from optical surfaces and results in a higher signal to noise ratio. Correlation by addition is used for pulsed laser applications (section 6.4.2(c)) but generally results in lower quality fringes.

A detailed description of speckle formation and the conditions that need to be maintained to ensure that decorrelation between speckle fields does not occur can be found in reference [2].

Figure 6.2 summarizes schematically the optical configurations, the terminology and the geometry used in this chapter. Figure 6.2(a) represents an out-of-plane sensitive interferometer configuration where the speckle pattern is mixed with a smooth reference beam. Figure 6.2(b) depicts an in-plane sensitive configuration that mixes two speckle patterns generated from the same area on the object surface. Figures 6.2(a) and (b) are often generically called electronic speckle pattern interferometry (ESPI), and sometimes TV holography.

Optical Fiber Sensor Technology, Vol. 2. Edited by K. T. V. Grattan and B. T. Meggitt.
Published in 1998 by Chapman & Hall, London. ISBN 0 412 782 901

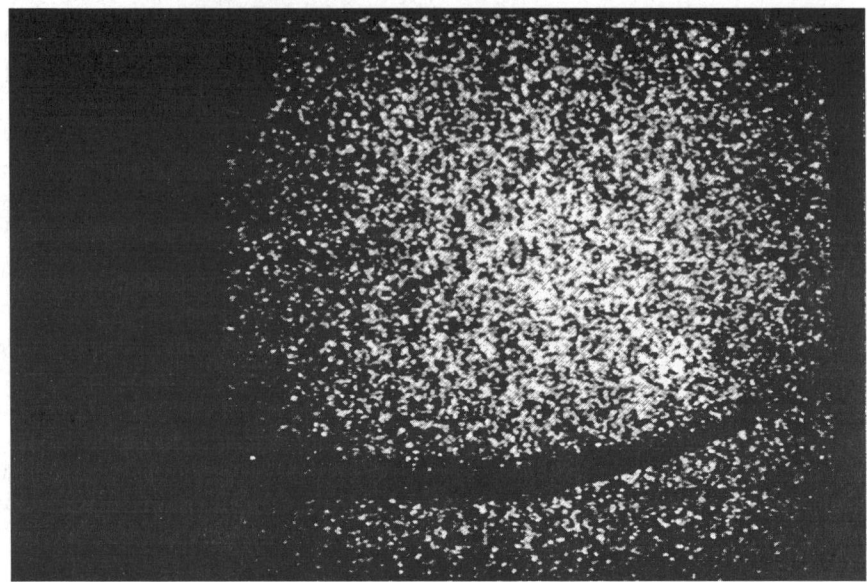

Fig. 6.1 Interferometrically encoded speckle pattern. The object is a gas turbine compressor blade (see Figure 6.7).

Figure 6.2(c) shows schematically a speckle shearing interferometer. In this arrangement the two speckle images of the object surface, produced by the Michelson interferometer arrangement, are mixed with one of the images sheared with respect to the other image. The resultant speckle field is due to interference of neighboring speckles on the object surface. For small shears this configuration is sensitive to gradient of displacement rather than displacement, that is, it provides optical differentiation of the displacement fringes. Also shown are speckle fringes obtained from the ESPI and shearography configurations. The correlation fringes were obtained by storing a reference speckle image of the object. The object, a flat plate, was then deformed by applying a load in the center of the back surface and a second speckle image stored. The two speckle fields were then subtracted to obtain the fringe patterns. In Figure 6.2(a) the fringes represent contours of out-of-plane displacement with each fringe equal to $\lambda/2$. In Figure 6.2(c) the fringes represent the out-of-plane displacement gradient.

Since the original ESPI papers published in the early 1970s [3–5] there has been extensive development and application of the techniques with several commercial systems available. Many of these have concentrated on out-of-plane sensitive vibration measurement which ESPI performs very well, providing full-field 'live' images of the vibrating surface on a TV monitor. However, these systems have typically resulted in rather large bulk optic assemblies with gas laser, usually HeNe, illumination. This restricted the use to the laboratory, often requiring an isolated table, and severely restricted the ease of use on more

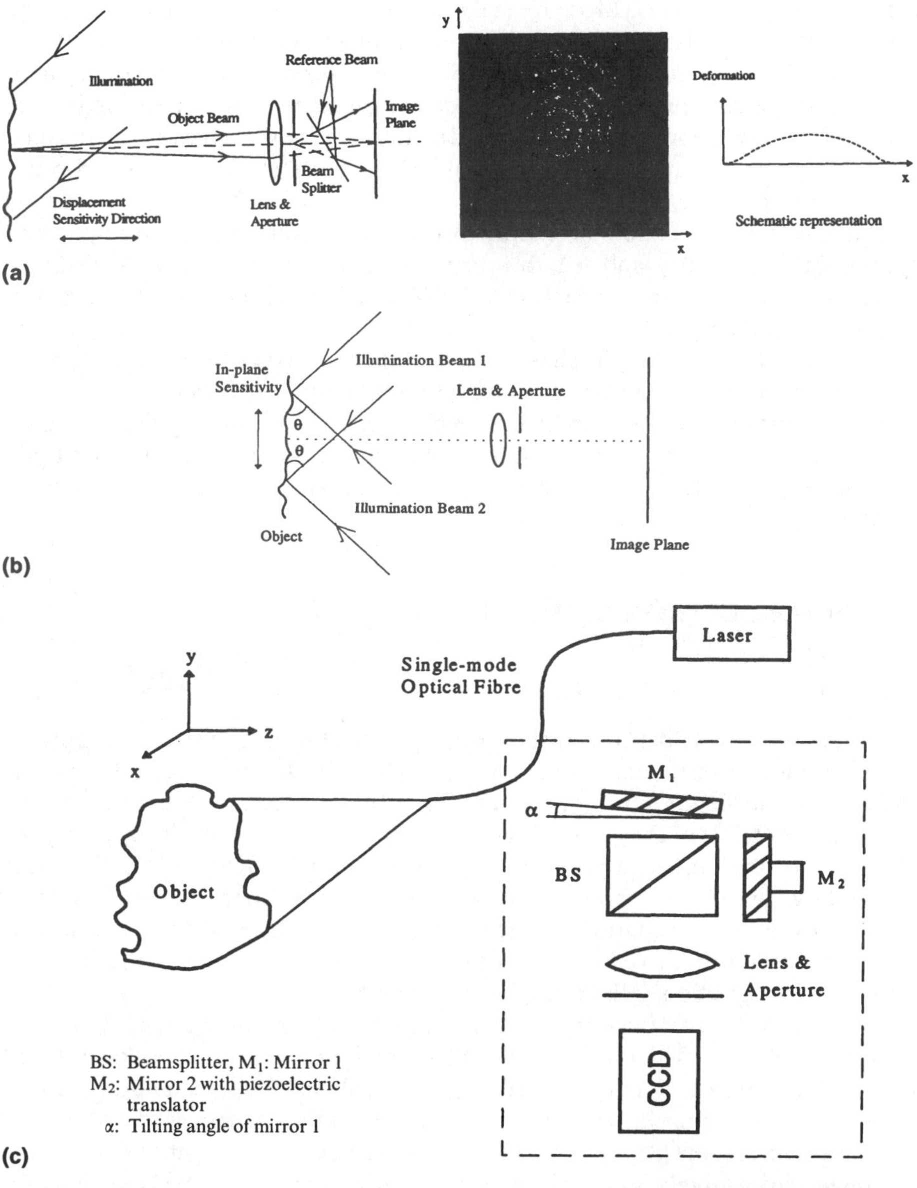

INTRODUCTION

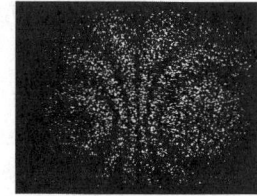

Fig. 6.2 Schematic of out-of-plane and in-plane ESPI and out-of-plane sensitive shearography configurations, with correlation fringes generated by shearography and ESPI. The test object is a composite carbon-fiber coupon clamped around the edges and mechanically loaded in the center of the back surface. (a) Out-of-plane sensitive ESPI arrangement; (b) in-plane sensitive ESPI arrangement; (c) out-of-plane sensitive shearography arrangement.

complex or large objects. More recently advances in optoelectronic technology such as solid state lasers and optical fibers combined with tremendous advances in CCD technology and computational processing power have resulted in small, robust systems capable of operating under a range of environmental conditions. This technology is combined with powerful analysis programmes which in some instances provides information compatible with finite element (FE) and other analysis packages.

This chapter will review how optical fiber technology has been applied to speckle interferometry and led to advances in the measurement capabilities. Image processing techniques will not be discussed in this chapter (for a review see, for example, [6]) although signal processing techniques implemented in the optical system, for example phase shifting, that are used to generate speckle interferograms to be utilized in subsequent data analysis are discussed.

A brief review of the optoelectronic technology is followed by describing its implementation in measurement of the following parameters: (i) surface profile and slope, (ii) static deformation/displacement, and (iii) dynamic displacement.

6.2 OPTOELECTRONIC TECHNOLOGY

6.2.1 Optical fiber implementations

Single mode optical fibers are able to propagate the lowest order, LP_{01} (approximately Gaussian intensity profile) mode and guide the input laser beam efficiently without degradation of its coherence properties. However the state of polarization (SOP) of the guided beam is modified due to the birefringence properties of the fiber, induced primarily by external perturbations of the fiber such as bending, twisting or thermally induced strain. For interferometric systems with two physically separated optical paths this results in unpredictable and environmentally sensitive SOP variations between the two beams, producing a concomitant variation in the resulting fringe visibility.

A technique to overcome this that is simple to implement is to use fiber optic SOP controllers [7] which utilize controlled bending and tension coiled induced linear birefringence to realize fiber optic waveplates. This technique is often used but does require adjustment to maintain optimum fringe visibility. An alternative, but more costly method, is to construct the entire fiber optic interferometer from highly birefringent optical fiber. Such fiber preserves linearly polarized light that is coupled into the fiber with the polarization vector aligned parallel to one of the eigenaxes of the optical fiber. Once aligned with the laser this implementation requires no further adjustment.

Optical fibers are useful at a number of levels. They may simply be used as flexible light guides to deliver the optical power to the object under test, or, as in early fiber optic ESPI systems, used to provide high quality reference beams [8]. Finally, by combining optical fibers with a fiber optic directional coupler a

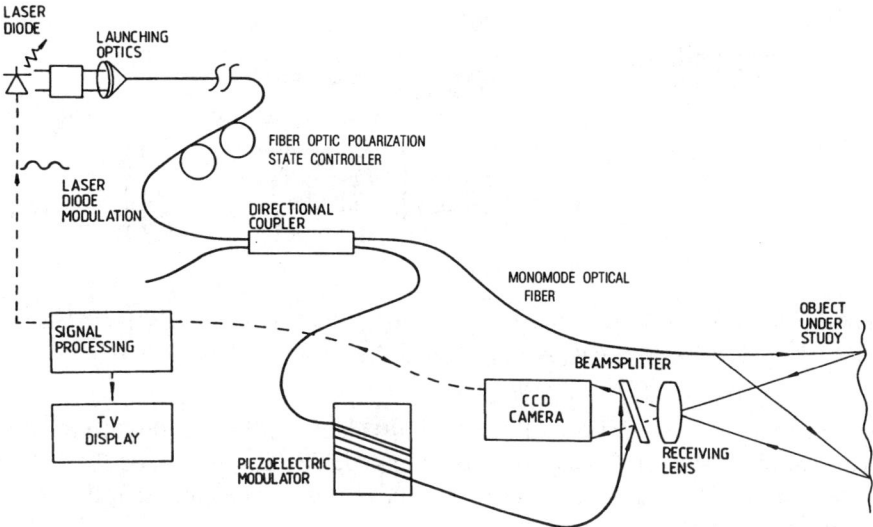

Fig. 6.3 Schematic configuration for a fiber-optic out-of-plane sensitive ESPI configuration [8–14].

complete ESPI system can be constructed. Figure 6.3 shows a schematic of an out-of-plane sensitive fiber optic ESPI arrangement similar to that reported by a number of authors [9–13] and used in a range of industrial research laboratories [14].

The directional coupler can have a fixed split ratio, typically 95:5 for out-of-plane ESPI and 50:50 for in-plane configurations. The most flexible approach that allows optimization of the signal to reference beam ratio for a range of collected scattered light powers is to use a variable split ratio directional coupler. Also shown in Figure 6.3 is a piezoelectric (PZT) cylinder tightly wrapped with optical fiber. Applying a voltage to the PZT changes its diameter which stretches or compresses the fiber. This changes the phase of the fiber guided beam primarily via a change in the refractive index of the fiber [15, 16]. This phase modulation capability is used in a number of signal processing techniques discussed in later sections.

Optical fibers have also been used in shearography systems either simply as delivery systems [17] or in a recent implementation [18] as an integral part of the shearing interferometer. The configuration is shown in Figure 6.4. Linearly polarized light is coupled into highly birefringent optical fiber with the polarization azimuth at 45° with respect to the fiber birefringence axes, thus producing equal intensities in the two orthogonal polarization eigenmodes. Light exiting the fiber illuminates the test object. The lens system and Wollaston prism produce two sheared images of orthogonal polarization. This is ensured by aligning the polarization axes of the Wollaston prism parallel to the birefringent fiber polarization axes. A linear polarizer in front of the CCD interferes the two

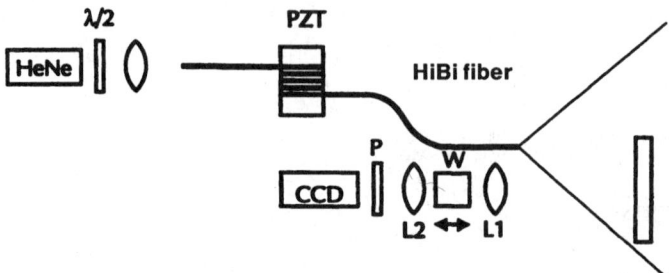

Fig. 6.4 Fiber-optic system for out-of-plane sensitive shearography (after [18]). $\lambda/2$ = half-wave plate, W = Wollaston prism, L1, L2 = lenses, P = polarizer.

sheared images. Translation of the Wollaston prism in the direction of the arrow enables the degree of shear to be changed. The fiber-wrapped PZT allows modulation of the phase between the two sheared images enabling optical signal processing techniques to be implemented.

A major disadvantage of optical fibers is the drift in the interferometer phase due to temperature. Since optical fiber has a typical thermal phase sensitivity of 100 rad K^{-1} m^{-1} [19] the thermal drift can lead to significant phase errors. A number of techniques have been used to overcome this problem. The simplest passive technique usually adopted is to thermally insulate the fibers and place as much as possible in a sealed box thus ensuring maximum common mode rejection to thermal variations. A second technique that can be used is to actively stabilize the interferometer [13, 20–25]. This has been achieved by monitoring the fringes formed from the fiber optic part of the system either in reflection [20] or transmission [13] and applying a feedback signal to the PZT. However, although this compensates for thermal drift in the fiber optic part of the interferometer it does not compensate for phase changes caused by drift between the optical system and the object surface. The system shown in Figure 6.5 [24] addresses this problem by combining a fiber optic laser velocimeter (LV) (point measurement) with the ESPI system. The object and reference beams for the ESPI system are derived from fibers B and G respectively. The object and reference beams for the LV are obtained from fibers D/E and C/F respectively. The output from the LV system, which measures the displacement at a particular point on the object surface, is used to provide a feedback signal to PZT1. Due to the number of different fiber paths and the lengths used the configuration is still sensitive to thermal drift but compensates well for high frequency operation. An alternative technique [22, 23] is to use the signal from a single pixel output from the image processing system. The pixel output can be chosen at any point on the object surface to correspond to a well-modulated speckle. This output can then be used as a feedback signal to a PZT or laser diode source (discussed further in the next section). The advantage of this technique is that it is optically very simple, requiring no further components. The disadvantage is that with conventional CCDs the bandwidth is Nyquist limited to ~12 Hz.

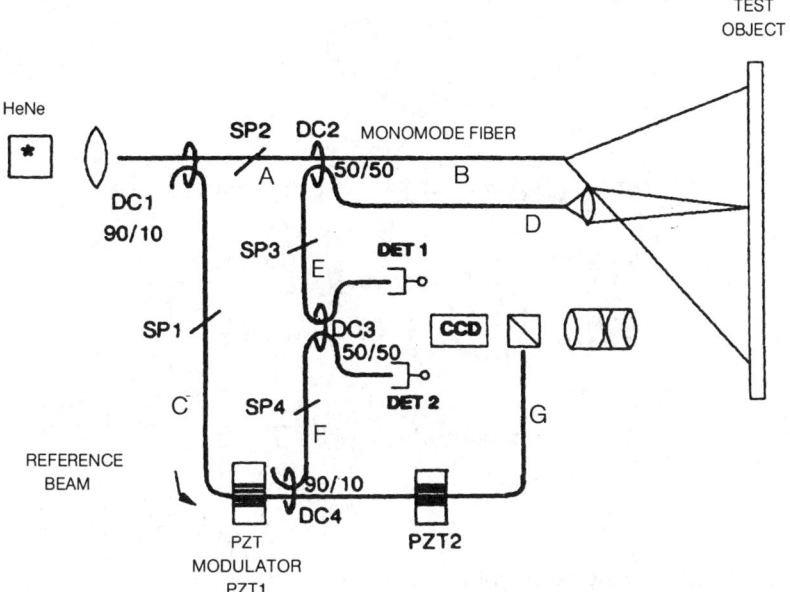

Fig. 6.5 Combined ESPI/LV system (after [24]). DC1–4 = directional couplers; SP1–4 = splices.

However, the use of a separate high readout speed CCD or a CID [26] (see section 6.2.3) detector would improve the bandwidth sufficiently to compensate for thermal drifts and rigid body motions of the test object. None of these techniques can compensate for tilt of the test object. Examples of optical measurements that have been implemented using these fiber optic configurations are presented in sections 6.3 and 6.4.

Imaging optical fiber bundles have been used in endoscopy applications to transmit speckle images to the CCD camera over a distance of up to 1 m [27]. The ESPI configuration shown in Figure 6.6 used a 1 m imaging fiber bundle composed of 80 000 fibers each of 10 μm diameter arranged in a square of $4 \times 4 \, \text{mm}^2$ with a fill ratio of $\sim 50\%$. This technique was also applied to endoscopic shearography.

6.2.2 Sources

Traditionally gas laser sources such as helium neon ($\lambda = 632.8$ nm) and argon-ion ($\lambda = 514.5$ nm) have been used. The argon-ion laser has high power, hundreds of milliwatts to several watts, that can be efficiently delivered by optical fiber enabling relatively large objects to be illuminated [8]. More recently solid state lasers such as frequency doubled YAG ($\lambda = 532$ nm) [14], and laser diodes have been utilized [10, 14, 17, 21, 28–34]. In particular the use of solid state laser diodes has resulted in small, compact, portable systems with the significant additional attribute that the emission wavelength of the laser is tunable by

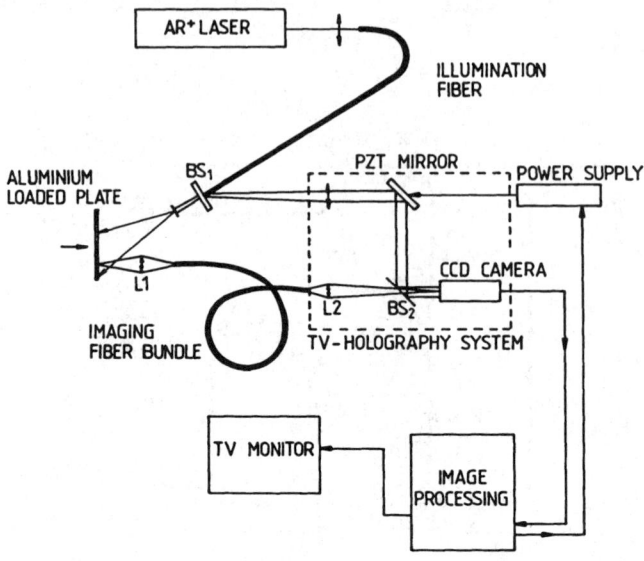

Fig. 6.6 Schematic of endoscopic ESPI system (after [27]).

modulation of the cavity effective index via temperature control (low frequency modulation capability) or the injection current (high frequency modulation capability).

Laser diodes are particularly attractive sources as they require very low power consumption, their emission wavelength corresponds to the peak wavelength sensitivity of CCD cameras, they have a linearly polarized output and they can be reasonably efficiently coupled into monomode optical fiber. The coherence length can easily be several meters although feedback into the lasing cavity can cause power fluctuation, mode hopping and multilongitudinal mode operation. One solution, although rather expensive, is to use a Faraday isolator. Alternatively, good results have been achieved by **pigtailing** fibers to diodes. This involves a miniature optical assembly with back reflections into the cavity reduced by angle cleaving the fiber and aligning the fiber at an angle to the center axis of the diode [35].

Commercially available single mode continuous wave laser diodes are limited to approximately 200 mW output power, restricted by limited heat dissipation of the absorbed optical power causing facet damage. However, in many situations this is sufficient light intensity. Very recently a device has been announced that claims 295 mW single mode operation at 680 nm. The increase in performance is obtained by growing a transparent window structure onto the laser end facet [36] that reduces the amount of optical power absorbed. When available the device will have exciting prospects in a number of extrinsic sensor applications. Unfortunately it does not yet appear to be available commercially.

The optical emission frequency of the laser diode can typically be continuously modulated by up to 50–100 GHz before mode hopping occurs. It is also possible to introduce injection current induced mode hops to produce much larger wavelength shifts of $\sim 0.3-1.5$ nm. Temperature tuning changes the emission wavelength by several nanometers at a typical rate of 0.3 nm °C^{-1}. The recent development of external cavity laser diodes has resulted in sources that have a long coherence length and can be wavelength tuned, whilst maintaining a reasonable power output, over 20 nm. However, reports of using this for speckle interferometry are limited.

Laser diodes are also capable of pulsed operation at rates up to GHz and with rise times <0.5 ns. This capability has also been utilized in the signal processing techniques detailed in section 6.5.

6.2.3 Cameras

Although original speckle interferometry used photographic techniques these have essentially been surpassed by CCD cameras.

CCD technology continues to develop at a rapid pace with pixel arrays up to 4k × 4k, analog and digital (up to 16 bit) outputs, high quantum efficiency in the visible and near infrared, readout speeds from hours to thousands of frames per second and intensifiers offering gains of $10^2 - 10^4$ available. The most typical employed for speckle interferometry are 512 × 512 analog outputs with frame rates of 25–30 Hz. The main reason for this is the compromise between adequate performance and cost particularly with the availability of inexpensive (and expensive!) frame stores to digitize the CCD output. A device that to the author's knowledge has not yet been used in speckle applications is the CID (charge injection device) [26], which offers the ability to address individual pixels at much higher rates (up to 20 kHz) as entire frames do not have to be read out to read the charge on an individual pixel. These may offer interesting signal processing options in the future; for example, the stabilization techniques discussed in section 6.2 could be implemented with high bandwidth.

6.3 SURFACE PROFILE/SLOPE MEASUREMENT

The measurement of surface profile and gradients of surface profile is a requirement for many engineering applications. For example, on-line quality control, solid modeling (the transfer of contour information from prototypes to CNC machines), biomedical applications, and the measurement of surface wear. In addition, for objects with nonplanar shapes, surface contours and slopes are required for a complete strain analysis [37] and have been used in conjunction with ESPI systems to obtain a complete surface strain analysis [38].

Various full-field optical methods have been developed for contour and slope measurement, for example moiré, holographic interferometry, fringe projection, ESPI and shearography. The techniques employed for speckle methods were originally developed from those of holographic interferometry and include refractive index change of the medium surrounding the object, tilting the object [39, 40], two-wavelength illumination [10, 33, 41–44] and two-source illumination [45–48] produced by shifting or tilting the illumination. All these techniques produce fringes that represent height contours on the object surface. In speckle interferometry essentially two methods that are applicable to engineering applications are employed. These are the two-wavelength and two-source techniques.

6.3.1 Two-wavelength technique

The intensity of light in the camera image plane, I_i, for the configuration of Figure 6.3, assuming normal incidence and detection and an object size small compared to the illumination and viewing distances, is given by

$$I_i(x, y) = I_0 \left[1 + V \cos\left(\frac{2\pi(2h(x, y) + \delta l)}{\lambda} + \phi_s\right) \right] \quad (6.1)$$

where I_0 is a constant, V the visibility, ϕ_s the randomly varying speckle phase, λ the illumination wavelength and h represents the height, in the z direction, of the object surface above some arbitrary reference plane (x–y plane). Equation 6.1 demonstrates that contours of equal intensity are obtained for height intervals of $\lambda/2$. However, these are too small for most practical purposes and the randomly varying speckle phase term tends to prevent contour fringes even for very small height variations. If the object is illuminated by two different wavelengths, λ_1 and λ_2, then the processed image, obtained by correlating two speckle fields by subtraction, is given by

$$\Delta I = KV \cos\left[\frac{2\pi}{\lambda_1}(2h + \delta l) + \phi_s\right] \cos\left[\frac{\pi}{\lambda_{\text{eff}}}(2h + \delta l)\right] \quad (6.2)$$

where K is a constant, and λ_{eff} the effective, or synthetic, wavelength described by

$$\lambda_{\text{eff}} = \frac{\lambda_1 \lambda_2}{|\lambda_1 - \lambda_2|} = \frac{c}{\Delta \nu} \quad (6.3)$$

where $\Delta \nu$ is the optical frequency difference between λ_1 and λ_2 and c the free space speed of light. Fringe intervals now represent height contours on the object surface of $\lambda_{\text{eff}}/2$. Laser diode wavelength modulation has been used to generate height contours in the range 30 μm to tens of millimeters [10, 29, 43, 49]. More recently the availability of continuously tunable laser diodes has increased the versatility of the technique. These are available in the 780–850 nm range with tuning ranges up to 30 nm, power outputs up to 40 mW and linewidths of < 100 kHz, equivalent to a coherence length > 1 km [50].

Contour fringes on a gas turbine compressor blade (Figure 6.7) produced by laser diode wavelength modulation are shown in Figure 6.8 using the optical configuration of Figure 6.3 [43]. The blade is 17 cm high, 10 cm wide and has an overall depth of ~4 cm. Figure 6.8(a) was obtained by modulating the laser diode injection current to produce an optical frequency shift of 18.4 GHz giving

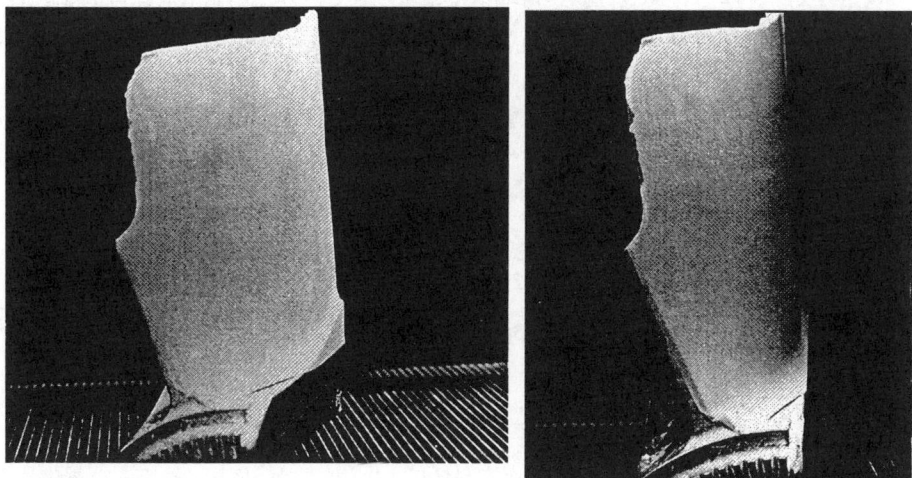

Fig. 6.7 Two views of the gas turbine compressor blade used as a test object. The blade is 17 cm high, 10 cm wide and 4 cm deep.

rise to 8 mm contour fringes. Figure 6.8(b) was obtained by making the laser diode mode hop by ~170 GHz to produce 0.9 mm contour fringes. These fringe patterns provide qualitative information of the object surface profile. Horizontal fringes represent the object surface tilted in the y–z plane (Figure 6.2) and the vertical fringes indicate the object surface oriented at an angle to the x–y plane. The absolute height, h, can be extracted using fringe analysis techniques. The range of techniques is extensive and are reviewed in [6]. One of the most versatile techniques is that of phase shifting/stepping [51]. This allows the absolute phase of the fringes to be evaluated, module 2π. Subsequent phase unwrapping removes the 2π ambiguity. As an example, if

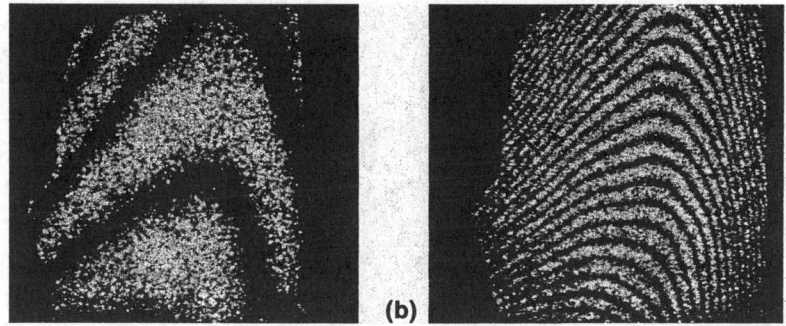

Fig. 6.8 Two-wavelength contour fringes of the turbine blade for contour intervals of (a) 8 mm ($\Delta\nu = 18.4$ GHz); (b) 0.9 mm ($\Delta\nu = 170$ GHz) (after [43]).

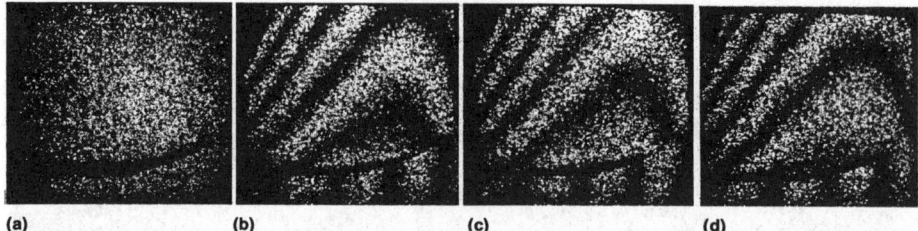

Fig. 6.9 (a) Speckle pattern obtained by illumination at one wavelength and (b), (c), (d) phase stepped contour fringes with a $2\pi/3$ phase shift between each image (after [43]).

three interferograms, $I_{0,1,2}$, are obtained with $2\pi/3$ phase shift between each of them, the phase ϕ can be evaluated from

$$\phi = \tan^{-1}\left[\frac{\sqrt{3}(I_0 - I_2)}{2I_1 - I_0 - I_2}\right] \qquad (6.4)$$

In the contouring case $\phi = 4\pi h/\lambda$ and hence the profile of the object surface can be evaluated. An example of this is shown in Figure 6.9 [11, 43]. Figure 6.9(a) is the speckle pattern when the compressor blade is illuminated at only one wavelength. Figures 6.9(b)–(d) are obtained by illuminating the blade with two wavelengths separated by $\sim 30\,\text{GHz}$ but with a $2\pi/3$ phase shift obtained, using the PZT fiber optic phase shifter, between successive images. The wrapped

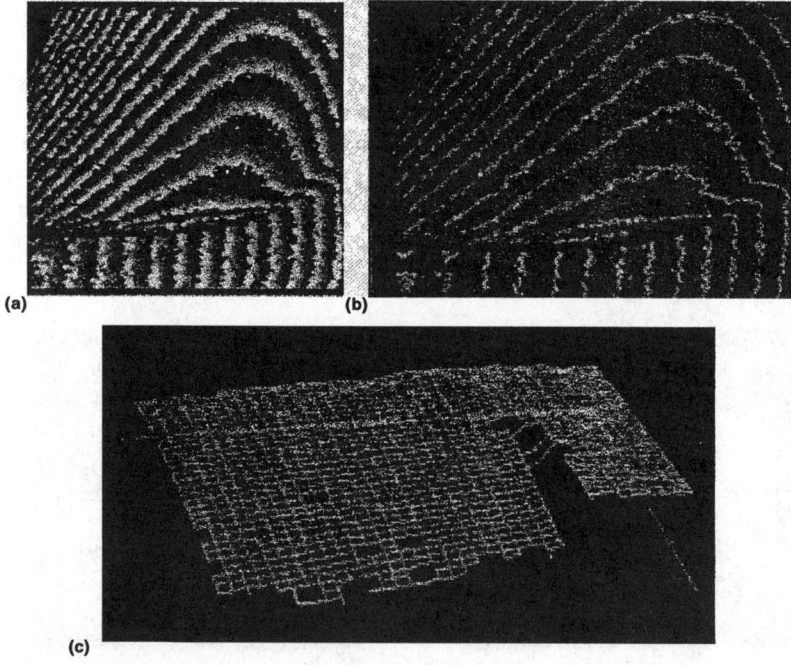

Fig. 6.10 (a) Wrapped, (b) unwrapped and (c) 3D wireframe representations of the phase variation across the turbine blade (after [43]).

phase map, equation 6.4, is shown in Figure 6.10(a) and the unwrapped map and 3D wireframe representation in Figures 6.10(b) and (c) respectively.

Figure 6.11 demonstrates that much smaller contours may be produced by increasing the wavelength difference. The contour interval on the UK 10-pence coin (~ 20 mm diameter) is 100 μm obtained using two separate laser diode sources separated by ~ 3 nm [52].

6.3.2 Two-source technique

Contour figures have also been obtained by shifting the illumination position. This was first reported in refs [45] and [46]. In this method the object illuminating beam of an out-of-plane ESPI system is laterally displaced, i.e. orthogonal to the illumination axis or rotated between image frames, and the subtraction process used to generate correlation fringes representing surface contours. The contour sensitivity is controlled by the movement of the illuminating beam. Implementing this technique using optical fibers considerably simplifies the optical arrangement and enables precise control on the translation of the illumination beam [52, 53]. An example of contour fringes obtained in this way are shown in Figure 6.12.

Although most of the contouring techniques can be extended to shearography for surface slope measurement, illumination shifting using optical fibers [54], and object tilting [55] need to record speckle interferograms on photographic film and use Fourier filtering techniques to generate slope fringes of reasonable visibility. This is due to speckle decorrelation caused when the illumination conditions are changed. As with contouring, the use of optical fibers simplifies the implementation [56]. Recently slope fringes have been generated using two-wavelength illumination from laser diodes [17, 56]. This technique was found to produce better fringe visibility and was implemented using a CCD camera.

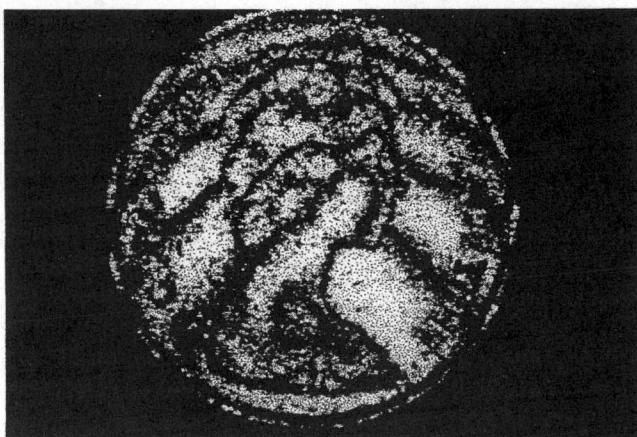

Fig. 6.11 Two-wavelength contour fringes produced by illuminating a UK 10-pence coin about 20 mm in diameter. The contour interval is 0.1 mm (after [52]).

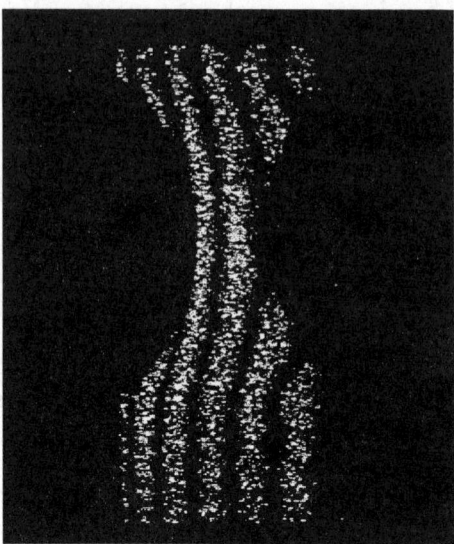

Fig. 6.12 Contour fringes on an engineering test component obtained by displacing the illumination beam by 100 μm in the x direction. The area imaged is approximately 10 cm high by 6 cm wide (after [52]).

In analogy with equation 6.2 the processed image for the slope fringes is given by

$$\Delta I(x, y) = K \sin\left[\frac{4\pi}{\lambda_1}\left(\frac{\partial z}{\partial x}\right)\delta x + \phi_s\right] \sin\left[\frac{2\pi}{\lambda_{\text{eff}}}\left(\frac{\delta z}{\delta x}\right)\delta x\right] \qquad (6.5)$$

where δx is the degree of shear in the x direction and the slope variation on the object surface is $\partial z/\partial x$. A similar expression can be written for shearing in the y direction. The slope fringes generated now depend on two parameters, that is, λ_{eff} and the degree of shear, both of which can be controlled to optimize the fringe visibility. An example of slope fringes generated using the two-wavelength method is shown in Figure 6.13(a) obtained using the optical configuration shown in Figure 6.2(c). The shear was $\delta x = 4$ mm and the wavelength difference $\Delta\lambda = 0.2$ nm. Phase-stepping techniques can be applied using the PZT mounted mirror. A wrapped phase map is shown in Figure 6.13(b) and an unwrapped 3D representation in Figure 6.13(c).

6.3.3 White light interferometric speckle

In the techniques discussed so far a high coherence source has been used to generate the interferometric speckles. In ref. [57] an alternative method using a very low coherence, sometimes called a 'white light', source is used. The experimental arrangement is shown in Figure 6.14(a). The configuration is similar to that of a Michelson interferometer with the object surface under test forming one of the 'mirrors'. Because of the short coherence length of the source, typically a few

SURFACE PROFILE/SLOPE MEASUREMENT

Fig. 6.13 (a) Slope fringes from a spherical object, 12.4 cm in diameter, generated using 4 mm of shear in the x direction; (b) wrapped and (c) unwrapped phase maps generated from phase-stepped slopes fringes (after [17]).

tens of microns, interferometric speckles are only observed when the path length from the object surface and the reference mirror are matched to within the coherence length of the source. By moving the reference mirror using the PZT translation stage, the object surface profile is reconstructed by locating the position at which interferometric speckles are observed; the intensity of interferometric speckle is greater than the noninterferometric speckles in the rest of the image. This can be seen in Figure 6.14(b) which shows the visibility variation of an individual speckle as the reference mirror is scanned. This is a technique that allows absolute measurement of surface profiles with a large dynamic range but does require a high stability between the test object and the optics for at least the duration of a scan, and a translation stage with reasonable resolution ($>$ μm). So far the technique has only been reported using an arc lamp as a source. Figure 6.14(c) demonstrates the potential of the method by contouring a German 1-pfennig coin.

222 OPTICAL FIBER SPECKLE INTERFEROMETRY

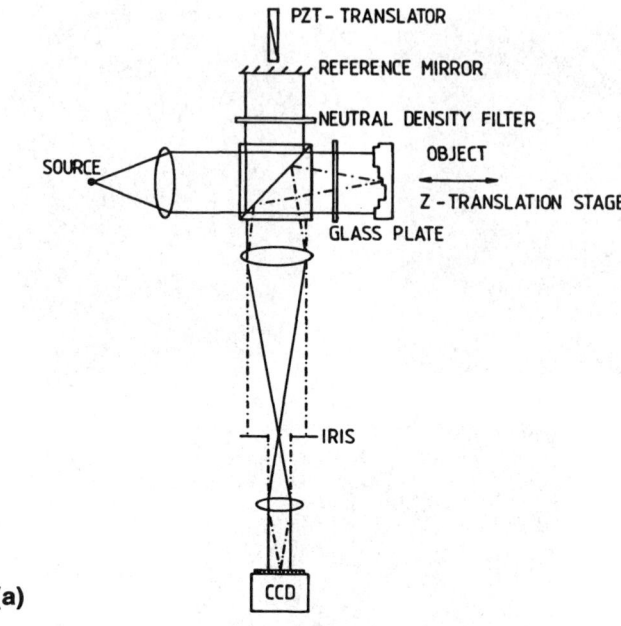

Fig. 6.14 (a) Experimental arrangement for 'white-light' speckle interferometry; (b) correlogram showing output of white-light interferometer for an individual speckle as the reference mirror is scanned along the z axis; (c) surface profile of a one pfennig coin with a diameter of 17 mm (after [57]). The depth resolution is 2 μm.

6.4 STATIC AND DYNAMIC DEFORMATION MEASUREMENT

6.4.1 Static measurements

ESPI has long been used to measure the displacement of objects under load, thermal or mechanical, to assess the effect of the loading conditions on structures [2] for example, deformation in an internal engine combustion chamber due to tightening the cylinder head bolts [58]. Figure 6.15 shows static displacement fringes obtained from a composite pressure vessel using a fiber optic ESPI system similar to Figure 6.3 and laser diode illumination. The vessel has been subjected to a thermal load between frames. The large fringes represent out-of-plane displacement. The smaller fringes at the top of the central black fringe are due to a defect in the composite vessel. Although this illustrates that ESPI can be used to detect defects, in many situations the gross deformation of the structure can mask this small detail. Shearography configurations do not measure the displacement but the gradient of displacement and are more sensitive to changes in the surface profile. An example of a defect detected using shearography is shown in Figure 6.16. This was obtained using the configuration of Figure 6.2(c) and laser diode illumination.

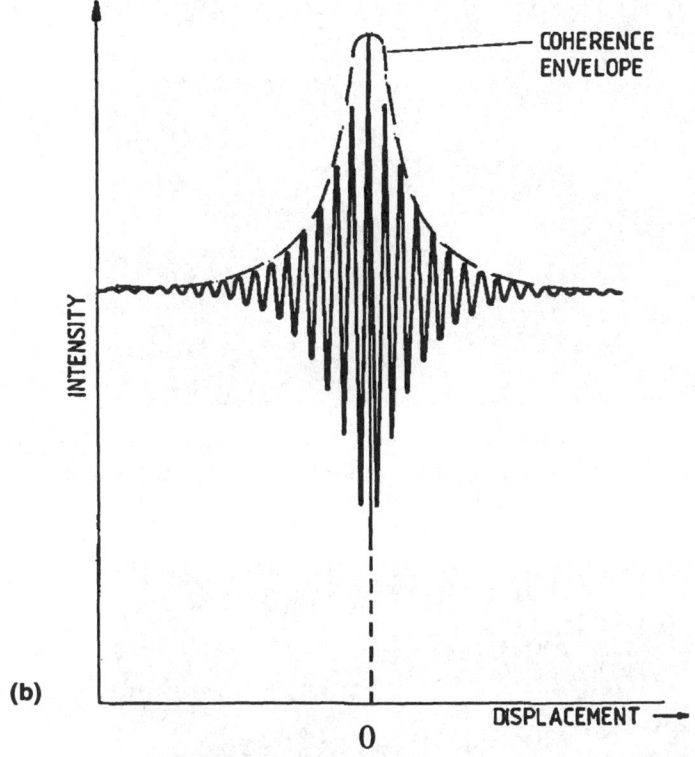

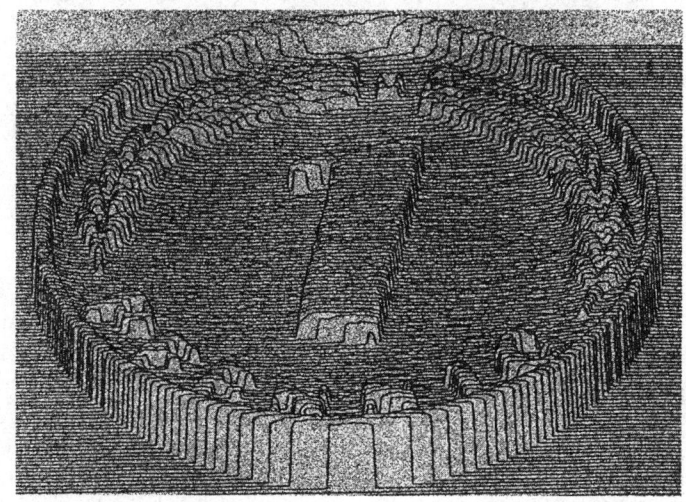

Fig. 6.14 cont'd.

224 OPTICAL FIBER SPECKLE INTERFEROMETRY

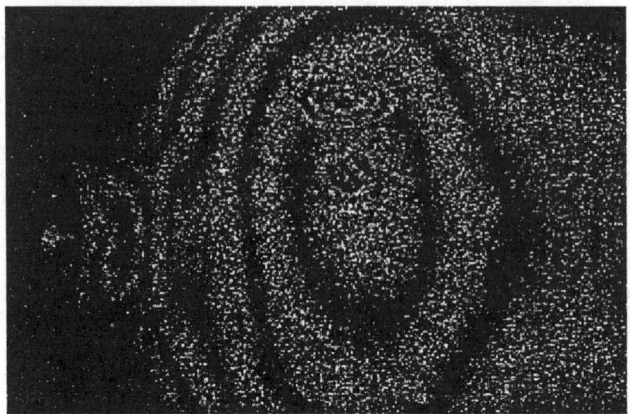

Fig. 6.15 Static ESPI displacement fringes from a thermally loaded composite pressure vessel, showing a defect towards the top center of the frame. The vessel is approximately 20 cm in diameter (Fig. courtesy of British Gas PLC).

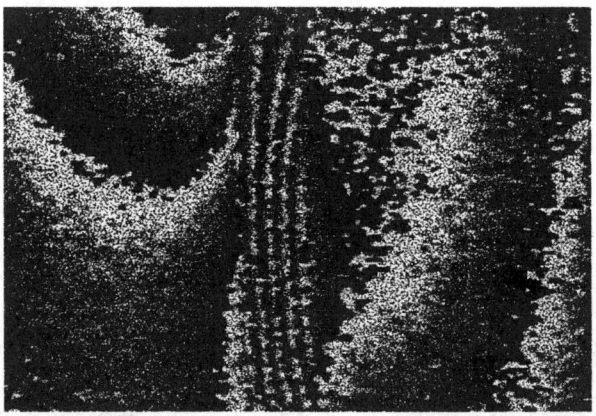

Fig. 6.16 Gradient-of-displacement shearography fringes for a pressurized gas filter, showing a line defect near the center of the frame. The area of view is approximately 15 cm across by 10 cm high.

As described in section 6.3, phase-shifting techniques can be used to determine the phase distribution in the interferograms. Phase shifting for static deformation measurements in ESPI and shearography can be implemented using the fiber wrapped PZT technique discussed in section 6.3 and using the configurations shown in Figures 6.3 and 6.4. Phase shifting using source wavelength modulation has also been demonstrated in path length imbalanced ESPI configurations [52]. The phase change induced by wavelength modulation for an optical path length imbalance of ΔL is given by

$$\Delta \phi = 2\pi \Delta L \, \Delta v / c \qquad (6.6)$$

This can be readily implemented in optical fiber based ESPI systems by incorporating extra fiber in one of the paths. It is generally more difficult to implement in shearography arrangements as unbalancing the interferometer results in unequal magnification and slight defocus of the two images. Recently a technique to overcome this has been presented [59] that uses the modified shearing interferometer shown in Figure 6.17. The high index glass block provides both compensation for image magnification and provides a path length imbalance. The wrapped and unwrapped phase maps obtained using this technique for a flat plate deformed centrally are shown in Figure 6.18.

6.4.2 Dynamic measurements

Dynamic displacement events arise due to object vibration or transient events such as impact. Vibrational mode shapes of objects, for example engine blocks, turbine blades and car body panels, have been investigated using ESPI [8, 60–63]. An important development in the technique was the introduction of **sequential-subtraction** processing of the images [8, 23]. In this technique a reference image is acquired, i_1. The next image, i_2, is acquired and correlation fringes obtained by subtraction, i.e. $i_2 - i_1$. The next image acquired is subtracted from i_2 and not i_1, as in the static case. This allows the reference frame to be updated at frame rate (25–30 Hz) reducing, but not altogether removing, the stability requirements usually associated with ESPI. The technique also allows repositioning of the interferometer relative to the object with automatic

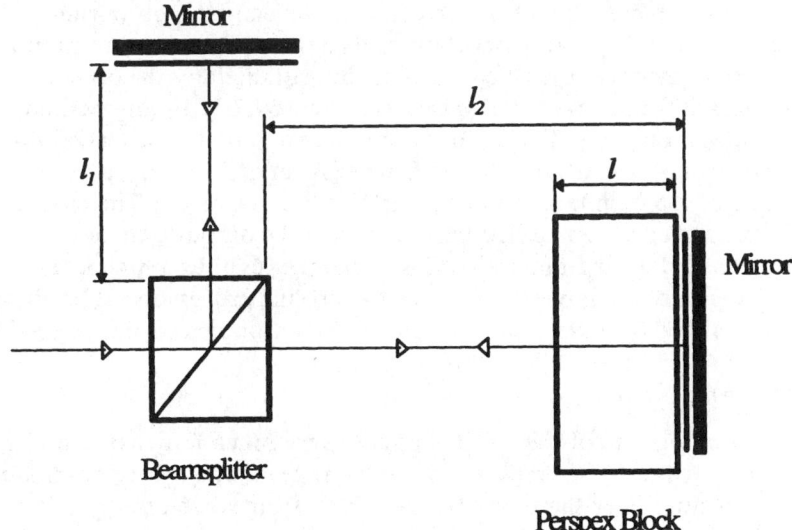

Fig. 6.17 Modified shearing interferometer with path-length imbalance ΔL, for use with wavelength modulation based phase-shifting techniques. The high-index Perspex block in one arm is used to compensate for the resulting image magnification mismatch (after [59]). $\Delta L = 2(1 + n_b)(l_2 - l_1)$ where n_b is the refractive index of the block.

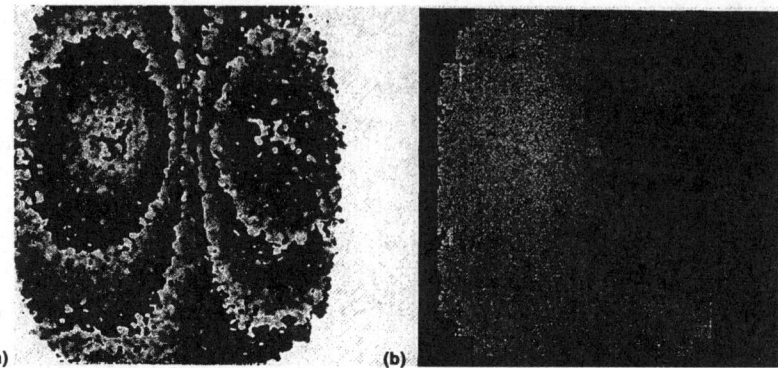

Fig. 6.18 (a) Wrapped and (b) unwrapped phase maps for a flat plate clamped around the edges and deformed by a force applied normal to the plate in the center of the back surface. The images were obtained using the shearography arrangement of Fig. 6.14, with a 6 mm shear in the horizontal direction. The area of the plate imaged is approximately 10×10 cm [59].

reacquisition of the vibration fringes. Assuming no movement between the object and the optical system subtraction of sequential frames results in a black image. Introducing a π phase shift between frames before subtraction restores the visibility of the fringes. This π phase is generally introduced using a fiber wrapped PZT or by laser diode modulation in conjunction with a path length imbalanced interferometer. This technique was originally developed for ESPI vibration measurement but has also been used in two-wavelength contouring [43] and shearographic slope [17] and vibration measurement [18]. An extension to this technique that improves the fringe visibility but still updates the observed correlation fringes at frame rate has also been reported [64, 65]. In this method, called electro-optic holography (EOH), four frames are required with a $\pi/2$ phase shift between frames. An image is obtained from each set of four sequential frames by subtracting alternate images, squaring and adding the result. The next, or fifth frame, is then used to replace the first frame and the output is recomputed, thus providing an update at frame rate. This technique requires a high level of computing power and was implemented using a pipeline processor but produces far less unmodulated speckles and therefore a much improved image quality.

(a) Time average vibration

Since the framing rate of the CCD is usually very much less than the vibration period each image frame represents the average of the vibration-modulated speckle intensities over the frame period. These fringes are usually called time-averaged and the intensity distribution in the final displayed image following sequential subtraction, or high pass filtering, and rectification, is given by

$$I(\underline{r}_d) = J_0^2\left[\frac{4\pi}{\lambda} a_0(\underline{r}_0)\right] \tag{6.7}$$

where $\underline{r}_d$ and $\underline{r}_0$ are conjugate points on the image plane and object surface respectively, and $a_0(\underline{r}_0)$ is the amplitude of the vibration motion. The amplitude of J_0^2, the square of the zero-order Bessel function of the first kind, becomes smaller for increasing displacement (Figure 6.19). The zero-order fringe has an intensity approximately six times larger than the first-order fringe intensity [2, 60] making it easy to visually identify the position of the nodal points. An example of a time-averaged ESPI vibration fringe map is shown in Figure 6.20(a) for the gas turbine compressor blade described previously. Although this provides useful qualitative information the averaging process removes the vibration phase (ϕ_0) information. This can also be seen from the absence of ϕ_0 in equation 6.7.

Shearography has also been applied to vibration analysis [18, 66–70]. However, the fringes now represent the square of a zero-order Bessel function whose argument is the derivative of the vibration amplitude. Time-average vibration fringes using shearography are shown in Figure 6.20(b). The test object is the same gas turbine blade as described previously vibrating at the same frequency of 5.2 kHz. The interferometer arrangement is as shown in Figure 6.2(c). A 100 mW laser diode was used as the source and the x-shear was 4 mm. Notice that the fringes exhibit more speckle noise than the ESPI fringes due primarily to the correlation interferogram being generated from two interfering speckle patterns and not from a speckle pattern and smooth reference beam. The bright ($J_0^2(0)$) fringes now occur not at the zero displacement point on the blade but at the points where the derivative of the displacement is zero.

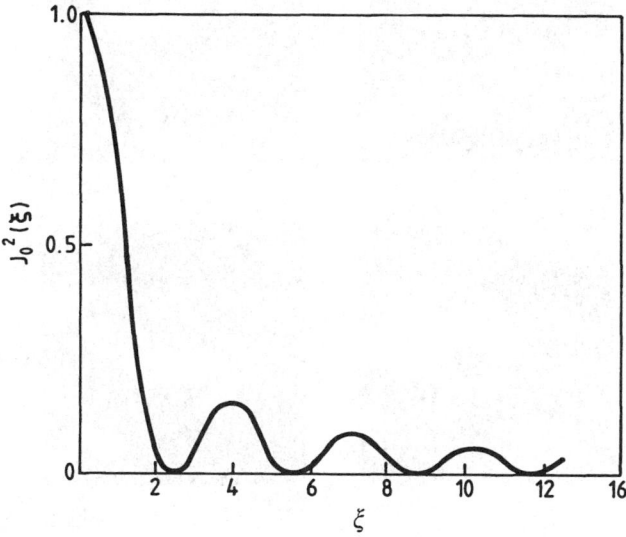

Fig. 6.19 $J_0^2(\xi)$ Bessel function distribution showing the decreasing intensity of the fringes at larger displacement amplitudes and the much higher intensity of the $J_0^2(0)$ fringe.

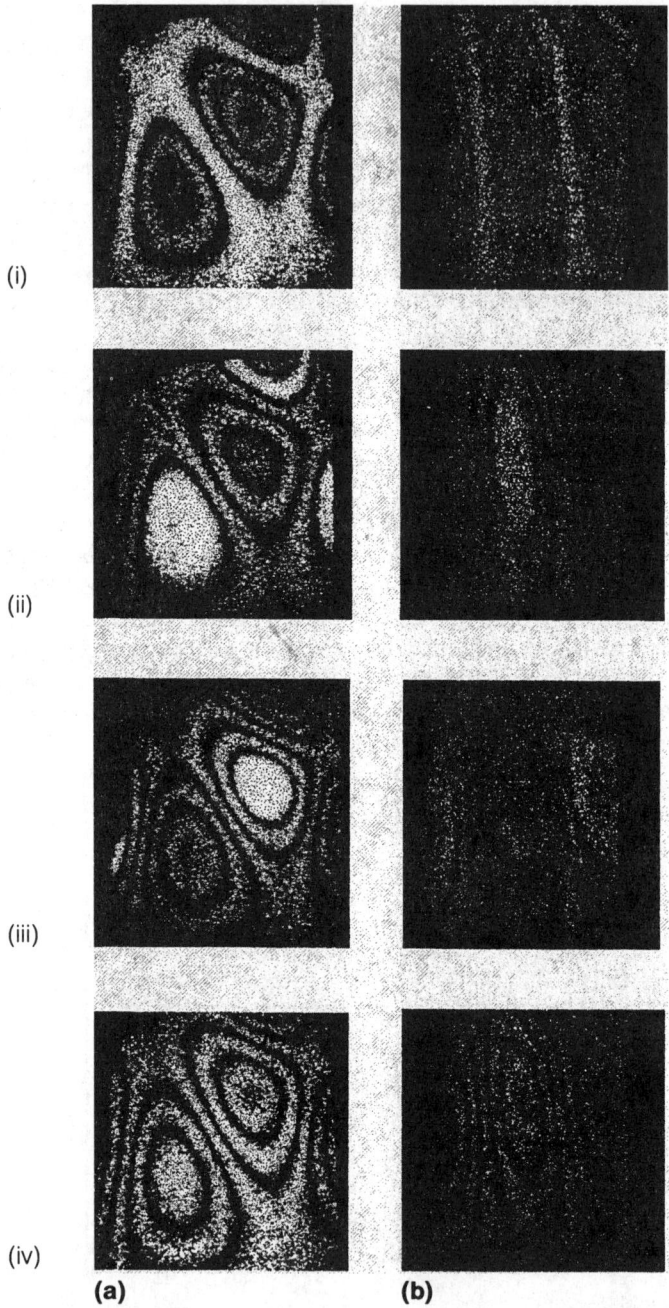

Fig. 6.20 (a) ESPI [30] and (b) shearographic [70] vibration fringes for the turbine blade at a frequency of 5.6 kHz: (i) time-averaged fringes; (ii) heterodyned fringes with a relative modulation phase of 0°; (iii) heterodyned fringes with a relative modulation phase of 180°; (iv) stroboscopic fringes.

(b) Heterodyning

One technique used to recover the phase information is called **heterodyning**. In this method the interferometer phase is modulated at the same frequency as the object vibration but with variable amplitude and phase. This phase modulation has been implemented using the fiber wrapped PZT [24, 58] and more recently using laser diode wavelength modulation [30]. For phase modulation of the reference beam, using the fiber wrapped PZT, for example, equation 6.7 becomes [71]

$$I(\underline{r}_d) = KJ_0^2\left[\frac{4\pi}{\lambda}\left(a_0^2(\underline{r}_0) + a_r^2 - 2a_0(\underline{r}_0)a_r \cos[\phi_0(\underline{r}_0) - \phi_r]\right)^{1/2}\right] \quad (6.8)$$

where a_r and ϕ_r represent the amplitude and phase of the reference beam modulation respectively.

The zero-order ($J_0^2(0)$) fringes will be observed when the amplitude and phase of the reference beam are the same as the object surface, that is,

$$a_0(\underline{r}_0) = a_r \qquad \phi_0(\underline{r}_0) = \phi_r \quad (6.9)$$

In contrast, heterodyning using wavelength modulation of the source affects the phase of both the object and reference beams. In this case equation 6.7 becomes [30]

$$I(\underline{r}_d) =$$
$$KJ_0^2\left[\left(\frac{4\pi v}{c}\right)\left(a_0^2(\underline{r}_0) + (\Delta L/2)^2(\Delta v/v)^2 - 2a_0(\underline{r}_0)(\Delta L/2)(\Delta v/v)\cos[\phi_0(\underline{r}_0) - \phi_v]\right)^{1/2}\right]$$
$$(6.10)$$

where ΔL is the path length imbalance. In an analogous manner to reference beam modulation the $J_0^2(0)$ fringes are observed wherever the phase and amplitude of the source frequency modulation are equal to the phase and amplitude of the object surface, that is,

$$\phi_0(r_0) - \phi_v = 0 \qquad a_0(\underline{r}_0) = (\Delta L/2)(\Delta v/v) \quad (6.11)$$

A significant difference to reference beam phase modulation is that the amplitude of the heterodyning is proportional to both the path length imbalance and the magnitude of the frequency modulation. An example of heterodyning using laser diode wavelength modulation and the configuration of Figure 6.3 is shown in Figures 6.20(ii) and (iii) [30]. The two heterodyned images have a relative phase difference of π rad, demonstrating the antiphase nature of the vibration of the object surface. The heterodyne technique has been demonstrated to extend the range of detectable vibration up to 8.4 µm [71] and down to 0.01 nm using photoelectric detection from a TV monitor or 1 nm visually [72].

In conventional bulk optic shearography systems heterodyning is generally difficult to implement since the modulation frequency is high, typically hundreds of Hz to tens of kHz, and the relatively large piezoelectrically mounted mirrors do not have adequate frequency response. The fiber optic arrangement shown in

Figure 6.4 has been successfully used to implement heterodyning by applying the phase modulation to the fiber wrapped PZT [18]. It can be shown that the intensity distribution $I(x_d, y_d)$ in the processed image for shear in the x direction, δx, is

$$I(x_d, y_d) = KJ_0^2 \left[\left(\frac{4\pi}{\lambda}\right) \left[\left(\frac{\partial a_0(x_0, y_0)}{\partial x} \delta x\right)^2 + a_r^2 - 2a_r \left(\frac{\partial a_0(x_0, y_0)}{\partial x} \delta x\right) \cos(\phi_0(x_0, y_0) - \phi_r) \right]^{1/2} \right]$$
(6.12)

Thus, in the shearography case the zero-order ($J_0^2(0)$) fringes will be displaced to positions where gradients of vibration displacement multiplied by the shearing distance is equal to the modulation amplitude and the vibration phase equals the modulation phase, i.e.

$$\frac{\partial a_0(x_0, y_0)}{\partial x} \delta x = a_r \qquad \phi_0(x_0, y_0) = \phi_r \qquad (6.13)$$

Source modulation can also be effectively used to perform the heterodyning [70] using the modified optical configuration described in section 6.4.1 (Figure 6.17). In this case the intensity distribution can be shown to be described by

$$I(x_d, y_d) = KJ_0^2 \left[\left(\frac{4\pi\nu}{c}\right) \left[\left(\frac{\partial a_0(x_0, y_0)\delta x}{\partial x}\right)^2 + \left(\frac{\Delta L}{2}\right)^2 \left(\frac{\Delta \nu}{\nu}\right)^2 \right.\right.$$

$$\left.\left. + 2\left(\frac{\partial a_0(x_0, y_0)\delta x}{\partial x}\right)\left(\frac{\Delta L}{2}\right)\left(\frac{\Delta \nu}{2}\right) \cos(\phi_0(x_0, y_0) - \phi_\nu) \right]^{1/2} \right] \qquad (6.14)$$

for the case where the optical path length difference, ΔL, in the interferometer is much greater than the path length difference due to height differences between the two neighboring points on the object surface that give rise to interference fringes. The position of the bright $J_0^2(0)$ fringe is obtained when the amplitude of the source frequency modulation is equal to the negative of gradient of displacement multiplied by the shearing distance and the phase of the source frequency equals the vibration phase, that is,

$$(\Delta L/2)(\Delta \nu/\nu) = -[(\partial a_0(x_0, y_0)/\partial x)\delta x] \qquad \phi_\nu = \phi_0(x_0, y_0) \qquad (6.15)$$

An example of the use of this technique is shown in Figures 6.20(b)(ii) and (iii) using the interferometer arrangement of Figure 6.17. A 100 mW laser diode was used as the source and the x-shear was 4 mm.

The heterodyned fringes with a π phase shift between them were generated by modulating the laser diode wavelength by 0.2 GHz. The $J_0^2(0)$ fringe has moved away from its original position to points where the time-varying optical phase difference between neighboring points on the object surface is varying with an amplitude and phase corresponding to the phase modulation between the interfered sheared images. Notice that for the two cases the bright fringes have moved in opposite directions. This is an inherent feature of shearography, since gradients of vibration displacement are symmetrically disposed (one positive, one negative) about peaks of vibration motion.

(c) Stroboscopic illumination

An alternative technique to heterodyning to recover the phase information is to illuminate, or **strobe**, the object at two points in its vibration cycle thus removing the time-average nature of the technique and consequently generating $\cos^2$ fringes. Alternatively, a reference image is obtained with the object stationary and a single pulse is used during the object vibration cycle. This latter method has the advantage that it is easier to phase step the images but has the disadvantage that high stability between the object and the optical system is required to avoid speckle decorrelation. In a recent paper [31] the reference speckle pattern was obtained by adjusting the phase of the stroboscopic illumination during the vibration cycle using a fiber wrapped PZT phase modulator, thus obtaining increased stability as the correlation fringes are renewed at frame rate. In addition the strobing was accomplished by injection current modulation of the laser diode output power rather than mechanical choppers or external modulators. To overcome the problem associated with wavelength chirp during the pulse an electronic equalization current was used to modify the pulse shape [73]. An alternative, but less flexible method, is to balance the path lengths in the interferometer [22, 52].

An example of ESPI fringes obtained by stroboscopic illumination is shown in Figures 6.20(a)(iv) and (b)(iv) for the blade vibrating at 5.2 kHz [52]. Notice the approximately equal intensity of the fringes compared to the time-averaged case. Shown in Figure 6.21(a) is an unwrapped phase map and 3D representation of the vibration deformation. The results show the antiphase vibration characterization of the blade at this frequency in agreement with the heterodyning case. Interestingly, stroboscopic techniques do not appear to have been applied to shearography. Similar techniques to ESPI are used but a particularly useful attribute is that the interferometer can be path length balanced so that, to first order, laser wavelength chirp is not observed on the correlation fringes and equalization circuits are not required. The only path length imbalance in the system is due to height differences between the interfering points on the object surface and the tilt of the mirror. However, these are generally very small. Examples of fringes obtained by shearing using stroboscopic illumination are shown in Figures 6.20(b)(iv) and 6.21(b) for the blade vibrating at 5.6 kHz [29, 70]. Also shown are wrapped and unwrapped phase maps obtained using the phase-stepping method described previously.

Stroboscopic illumination is straightforward to implement using laser diode illumination, and phase stepping can be used to recover the phase but the optical power received by the CCD is reduced by a factor equal to the duty cycle.

232 OPTICAL FIBER SPECKLE INTERFEROMETRY

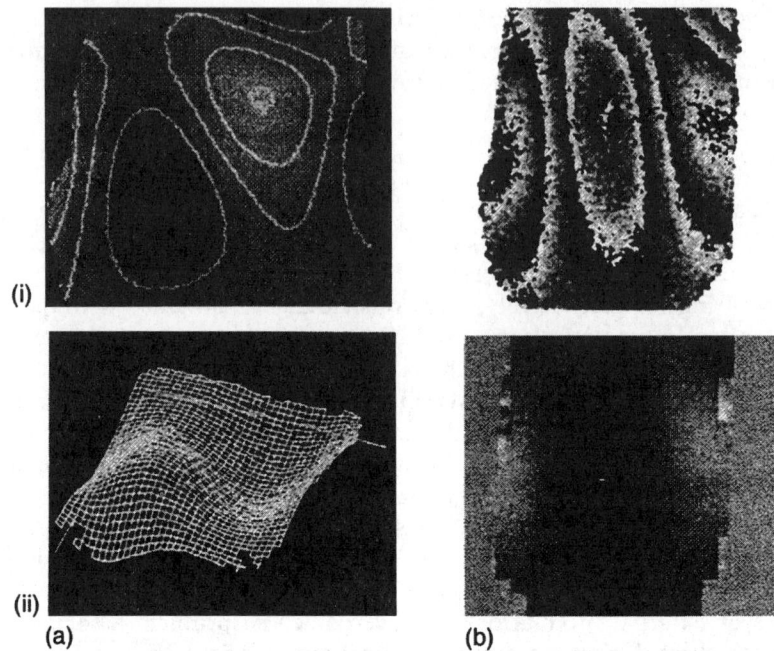

Fig. 6.21 (a) ESPI [30] and (b) shearographic [70] vibration fringes for the turbine blade at a frequency of 5.6 kHz: (i) ESPI image – unwrapped phase map; shearography image – wrapped phase map; (ii) ESPI image – wireframe representation of phase; shearography image – unwrapped phase map.

Finally, for transient events or vibration measurement on very unstable objects pulsed laser illumination is often used [74]. This uses two pulses from a ruby or more usually an injection-seeded frequency doubled YAG, typically of 10 ns duration and separated by tens of nanoseconds to tens of microseconds. This produces two speckle images on the same camera frame, that is correlation by addition. This produces noisier fringes than correlation by subtraction, although there has been recent progress in techniques to improve the fringe visibility [75].

6.5 CONCLUSION

This chapter has outlined how the use optical fiber technology and solid state laser sources can facilitate the use of speckle interferometers in engineering environments and in addition simplify and/or enhance the optical signal processing techniques. Examples of ESPI and shearography for surface profile/shape measurement, phase stepping and static and dynamic displacement have been presented.

It is apparent that the rapid advances in the enabling technology have led, and will lead, to further improvements and applications for the speckle technique.

ACKNOWLEDGEMENTS

The author would like to thank Helen Ford and Jen-Rong Huang for help in preparing the figures for this manuscript.

REFERENCES

1. Ennos, A. E. (1975) Speckle interferometry. *Laser Speckle and Related Phenomena* (Ed. Dainty, J. C.) Springer-Verlag, Berlin, pp. 203–253.
2. Jones, R. C. and Wykes, C. (1989) *Holographic and Speckle Interferometry*, Cambridge University Press, Cambridge, UK, 2nd edn.
3. Butters, J. N. and Leendertz, J. A. (1971) Holographic and video techniques applied to engineering measurements. *Journal of Measurement and Control*, 4, 349–354.
4. Macovski, A., Ramsey, D. and Shaefer, L. F. (1971) Time lapse interferometry and contouring using television systems. *Applied Optics*, 10, 2722–2727.
5. Schwomma, O. (1972) German Pat. 298830.
6. (1993) *Interferogram Analysis* (Eds Robinson, D. W. and Reid, G. T.) IOP Publishing Ltd, Bristol, UK.
7. Lefevre, H. C. (1980) Single-mode fibre fractional wave devices and polarisation controllers. *Electronics Letters*, 16(20), 778–780.
8. Davies, J. C. and Buckberry, C. H. (1986) Application of electronic speckle pattern interferometry in automotive product development. *VDI Berichte*, No. 617, 279–293.
9. Valera, J. D., Harvey, D. and Jones, J. D. C. (1991) Automatic heterodyning in fibre optic speckle pattern interferometry. *SPIE*, 1508, 170–179.
10. Tatam, R. P., Davies, J. C., Buckberry, C. H. and Jones, J. D. C. (1990) Holographic surface contouring using wavelength modulation of laser diodes. *Optics & Laser Technology*, 22(5), 317–321.
11. Atcha, H. and Tatam, R. P. (1991) The use of laser diodes and monomode optical fibre in electronic speckle pattern interferometry. *Proc. SPIE*, 1584, 221–232.
12. Paoletti, D. and Spagnolos, G. S. (1993) Application of fibre optic digital speckle interferometry to mural painting diagnostics. *Measurement Science and Technology*, 4(5), 614–618.
13. Mercer, C. R. and Beheim, G. B. (1989) Speckle interferometry using fiber optic phase stepping. *Proc. SPIE*, 1162, 180.
14. Atcha, H. British Gas, UK, private communication; Wiseall, S. S. Rolls Royce PLC, UK, private communication; Buckberry, C. H. Rover Group, UK, private communication.
15. Martini, G. (1987) Analysis of a single-mode optical fibre piezoceramic phase modulation. *Optics and Quantum Electronics*, 19, 179–190.
16. Tatam, R. P. (1995) Optical fibre modulation techniques for single mode fibre sensors in *Optical Fibre Sensor Technology* (Eds Grattan, K. T. V. and Meggitt, B. T.) Chapman & Hall, London, pp. 223–267.
17. Huang, J.-R., Ford, H. D. and Tatam, R. P. (1997) Slope measurement by two-wavelength electronic shearography. *Optics and Lasers in Engineering*, 27(3)
18. Valera, J. D. R. and Jones, J. D. C. (1995) Vibration analysis by modulated time-averaged speckle shearing interferometry. *Measurement Science and Technology*, 6, 97–202.

19. Hocker, G. B. (1979) Fiber-optic sensing of pressure and temperature. *Applied Optics*, **18**, 1445–1448.
20. Santos, J. L., Newson, T. P. and Jackson, D. A. (1990) Electronic speckle pattern interferometry using single-mode fibres and active fringe stabilisation. *Optics Letters*, **15**, 573–575.
21. Galannlis, K., Bunkas, T. and Ritter, R. (1995) Active stabilisation of ESPI systems for applications under rough conditions. *Proc. SPIE*, **2545**, 103–107.
22. Atcha, H. (1995) Optoelectronic speckle pattern interferometry, Ph.D. thesis, Cranfield University, Bedford, UK.
23. Moran, S. E., Law, R. L., Craig, P. N. and Goldberg, W. M. (1987) Optically phase locked electronic speckle pattern interferometer. *Applied Optics*, **26**, 475–491.
24. Valera, J. D., Doval, A. F. and Jones, J. D. C. (1993) Combined fibre optic laser velocimeter and electronic speckle pattern interferometer with a common reference beam. *Measurement Science and Technology*, **4**, 578–582.
25. Moran, S. E., Lagannani, R., Craig, P. N. and Goldberg, W. M. (1989) Optically phase locked electronic speckle pattern interferometry: system performance for vibration measurement in random displacement field. *Journal of the Optical Society of America*, **A6**, 252.
26. Ballantyne, R. L. (March 1996) Charge-injection devices create versatile sensors. *Laser Focus World*, S37–S38.
27. Smigielski, P., Albe, F. and Dischii, B. (1992) Progress in holographic and interferometric endoscopy. *Proc. SPIE*, **1732**, 481.
28. Wykes, C. and Flanagan, M. (1987) The use of a diode laser in a ESPI system. *Optics and Laser Technology*, **19**, 37–39.
29. Huang, J.-R. (1996) Optoelectronic speckle shearing interferometry, Ph.D. thesis, Cranfield University, Bedford, UK.
30. Atcha, H. and Tatam, R. P. (1994) Heterodyning of fibre optic electronic speckle pattern interferometers using laser diode wavelength modulation. *Measurement Science and Technology*, **5**, 704–709.
31. Anderson, D. J., Valera, A. F. and Jones, J. D. C. (1993) Electronic speckle pattern interferometry using diode laser stroboscopic illumination. *Measurement Science and Technology*, **4**, 982–987.
32. Ford, H. D., Atcha, H. and Tatam, R. P. (1993) Optical fibre technique for the measurement of small frequency separations: application to surface profile measurement using electronic speckle pattern interferometry. *Measurement Science and Technology*, **4**, 601.
33. Gulker, G. et al. (1992) Two-wavelength electronic speckle pattern interferometry for the analysis of discontinuous deformation fields. *Applied Optics*, **31**, 4519–4521.
34. Peng, X. et al. (1992) A simplified multi-wavelength ESPI contouring technique based on a diode laser system. *Optik*, **91**, 81–85.
35. De Groot, P. (1993) Fiber-coupled laser diode mount for interferometry. *Applied Optics*, Engineering and Laboratory Notes Supplement 1, Dec., 7122–7123.
36. Sharp Laboratories, Press Service, June 1994.
37. Hung, Y. Y., Turner, J. L., Tafralian, M., Hovanesian, J. D. and Taylor, C. E. (1978) Optical method for measuring contour slopes of an object. *Applied Optics*, **17**(1), 128–131.
38. Winther, S. (1988) 3D Strain measurements using ESPI. *Optics and Lasers in Engineering*, **8**, 45–57.
39. Jaisingh, G. K. and Chiang, F. P. (1981) Contouring by laser speckle. *Applied Optics*, **20**(19), 3385–3387.

REFERENCES

40. Ganesan, A. R. and Sirohi, R. S. (1988) New method of contouring using digital speckle pattern interferometry (DSPI). *SPIE*, **954**, 327–332.
41. Denby, D., Quintanilla, G. E. and Butters, J. N. (1975) *The Engineering Uses of Coherent Optics* (Ed. Robertson, E. R.) Cambridge University Press, Cambridge, 171–197.
42. Fercher, A. F., Hugh, Z. and Vry, U. (1985) Rough surface interferometry with a two-wavelength heterodyne speckle interferometer. *Applied Optics*, **24**, 2181–2188.
43. Atcha, H., Tatam, R. P., Buckberry, C. H., Davies, J. C. and Jones, J. D. C. (1991) Surface contouring using TV holography. *Proc. SPIE*, **1504**, 221–232.
44. Diao, H. Y., Peng, X., Zou, Y. L., Tıziani, H. J. and Chen, C. (1992) Contouring using two-wavelength electronic speckle pattern interferometry employing dual-beam illuminations. *Optik*, **91**(1), 19–23.
45. Wınther, S. and Slettemoen, G. A. (1984) An ESPI contouring technique in strain analysis. *Proc. SPIE*, **473**, 44–47.
46. Bergquist, B. D. and Montgomery, P. (1985) Contouring by electronic speckle pattern interferometry (ESPI). *Proc. SPIE*, **599**, 189–195.
47. Rodriguez-Vera, R., Kerr, D. and Mendoza-Santoyo, F. (1992) Electronic speckle contouring. *Journal of the Optical Society of America*, **A9**, 2000–2008.
48. Peng, X., Diao, H. Y., Zou, Y. L. and Tıziani, H. (1992) Contouring by modified dual-beam ESPI based on tilting illumination beams. *Optik*, **90**(2), 61–64.
49. Fercher, A. F., Vry, U. and Wermer, W. (1989) Two wavelength speckle interferometry on rough surfaces using a mode hopping diode laser. *Optics and Lasers in Engineering*, **11**, 271–279.
50. For example, Environmental Optical Sensors Inc., Boulder, Colo., USA.
51. Creath, K. (1993) Temporal phase measurement methods in *Interferogram Analysis* (Eds Robinson, D. W. and Reid, G. T.), IOP Publishing Ltd, Bristol, UK.
52. Atcha, H. and Tatam, R. P. (1993) Optoelectronic ESPI: applications to surface contouring and vibration measurements. *Proc. 9th Optical Fibre Sensors Conference* (IEEE/OSA), Fırenze, Italy, 337–340.
53. Quan, C. and Bryanston-Cross, P. J. (1990) Double source holographic contouring using fibre optics. *Optics and Laser Technology*, **22**,(4), 255–259.
54. Tay, C. J., Chau, F. S., Shang, H. N., Shim, V. P. W. and Toh, S. L. (1991) The measurement of slope using shearography. *Optics and Lasers in Engineering*, **14**, 13–24.
55. Rastogi, P. K. (1994) Measurement of the derivatives of curved surfaces using speckle interferometry. *Journal of Modern Optics*, **41**(4), 659–661.
56. Huang, J.-R. and Tatam, R. P. (1995) Optoelectronic shearography: two wavelength slope measurement. *Proc. SPIE*, **2544**, 300.
57. Dresel, T., Häusler, G. and Venzke, H. (1992) Three-dimensional sensing of rough surfaces by coherence radar. *Applied Optics*, **7**, 3919–3925.
58. Davies, J. C. and Buckberry, C .H. (1993) Television holography and its application in *Optical Methods in Engineering Metrology* (Ed. Wılliams, D. C.), Chapman & Hall, London.
59. Huang, J.-R., Ford, H. D. and Tatam, R. P. (1996) Phase-stepped speckle shearing interferometer by source wavelength modulation. *Optics Letters*, **21**(18), 1421–1423.
60. Løkberg, O. J. and Svenke, P. (1981) Design and use of an electronic speckle pattern interferometer for testing of turbine parts. *Optics and Lasers in Engineering*, **2**, 1–12.

61. Davies, J. C. (1985) The application of electronic speckle pattern interferometry to modal analysis. *Proc. Int. Symp. on Automotive Technol. and Automation*, Graz, Austria, 73–92.
62. Løkberg, O. J. (1988) Industrial applications of ESPI. *Proc. SPIE*, **952**, 208–217.
63. Buckberry, C. H. and Davies, J. C. (1990) The application of TV holography to engineering problems in the automotive industry. *Proc. Soc. for Exp. Mechanics on Hologram Interferometry and Speckle Metrology*, Bethel, Conn., USA, 268–278.
64. Stetson, K. A., Brokinsky, W. R., Wahid, J. and Bushman, T. (1989) An electro-optic holography system with real-time arithmetic processing. *Journal of Non Destructive Testing*, **8**, 69–76.
65. Pryputniewicz, R. J. and Stetson, K. A. (1989) Measurement of vibration patterns using electro-optic holography, *Proc. SPIE*, **1162**, 456–467.
66. Nakadate, S., Yatagai, T. and Saito, H. (1980) Digital speckle-pattern shearing interferometry. *Applied Optics*, **19**(24), 4241.
67. Toh, S. L., Sang, H. M., Chau, F. S. and Tay, C. J. (1991) Flaw detection in composites using time-averaged shearography. *Optics and Laser Technology*, **23**, 25–30.
68. Pryputniewicz, R. J. (1992) Electronic shearography and electronic holography working side by side. *Proc. SPIE*, **1821**, 27.
69. Mohan, N. K., Saldner, H. O. and Molin, N. E. (1994) Electronic shearography applied to static and vibrating objects. *Optics Communications*, **108**, 197–202.
70. Huang, J.-R., Ford, H. D. and Tatam, R. P. (1996) Heterodyning of speckle shearing interferometers by laser diode wavelength modulation. *Measurement Science and Technology* **7**, 1721–1727.
71. Løkberg, O. J. and Høgmoen, K. (1976) Use of modulated reference wave in electronic speckle pattern interferometry. *Journal of Physics E: Scientific Instruments*, **9**, 847–851.
72. Høgmoen, K. and Løkberg, O. J. (1977) Detection and measurement of small vibrations using electronic speckle pattern interferometry. *Applied Optics*, **16**(7), 1869–1875.
73. Anderson, D. J., Jones, J. D. C., Sinha, P. G., Kidd, S. P. and Barton, J. S. (1991) Scheme for extending the bandwidth of injection-current-induced laser diode optical frequency modulation. *Journal of Modern Optics*, **38**, 2459.
74. Tyrer, J. R. (1985) Application of pulsed holography and double pulsed electronic speckle pattern interferometry to large vibrating engineering structures. *SPIE*, **599**, 181.
75. Moore, A. J. and Pérez-López, C. (1996) Double-pulsed addition ESPI for harmonic vibration and transient deformation measurements. *Proceedings of the Applied Optics and Optoelectronics Conference* (Ed. K. T. V. Grattan) Institute of Physics Publishing, Bristol, UK.

7
Interferometric vibration measurement using optical fiber

K. Weir and B. T. Meggitt

7.1 INTRODUCTION

The measurement of vibration generally requires the determination of the displacement of a surface (from a mean position) as a function of time. The aim of the measurement is to determine the amplitude and frequency content of the vibration, or track the vibration. In many different systems the range over which these measurements are required can span several orders of magnitude. It is not unusual for a measurement system to be required to respond to vibrations over the range 1 Hz to 1 MHz, with amplitudes ranging from a few millimeters to a sub-ångström (1×10^{-10} m) making this a very demanding measurement.

Many of the standard means of monitoring vibration utilize accelerometers (e.g. piezoelectric transducers). However, these are generally contact devices, and may have their own mechanical resonances limiting bandwidth. The ideal vibration measurement instrument would be noncontact and nonloading so that the measurement procedure does not affect the actual measurement.

Interferometry, the coherent addition of two optical beams [1], has direct applications to the measurement of vibration. If the vibrating target can be incorporated into an interferometer this optical technique may be used as a noncontacting, nonloading technique to monitor vibration. The potential of such an optical technique was recognized at a very early stage in the history of interferometry [2]. It was not until the development of the laser (and its associated long coherence length) that the full potential of such an instrument was available. However, there were still problems as the surface to be monitored had to be incorporated into an interferometer, so direct line of sight is required imposing higher tolerances on alignment of the instrument. It is the potential of the combination of lasers and fiber optics that rekindled interest in this application. The fiber itself offers other advantages; in particular, the flexible nature of the medium gives the ability to measure 'round corners' — no direct line of sight is required — nor does the target require special preparation.

Optical Fiber Sensor Technology, Vol. 2. Edited by K. T. V. Grattan and B. T. Meggitt.
Published in 1998 by Chapman & Hall, London. ISBN 0 412 782 901

The small dimensions of the fiber also allow the instrument designer to define a small measurement region.

In this chapter vibration measurement will be discussed in terms of interferometric vibration sensing using optical fiber. However, it is worthy of note that there are other techniques to measure vibration, also incorporating fiber. Examples utilize intensity where the coupling efficiency between two fibers is modulated by the vibrating surface [3], polarization where the vibration modulates the fiber birefringence [4], frequency-modulated continuous wave [5], speckle interferometry and electronic speckle pattern interferometry (described elsewhere in this book). In addition many of the techniques of interferometric vibration measurement discussed here are directly applicable to other measurements, such as velocity (velocimetry), acceleration and flow (anemometry), discussed in Chapter 8.

7.2 INTERFEROMETRY

To introduce the use of fiber optics in interferometric vibration sensors it is convenient to pause to consider the response of bulk optic interferometers and how they respond to vibration input. Many of these considerations may then be carried across to optical fiber vibration measurement. There are many possible configurations of interferometer that may be used to measure vibration, the most common being a Mach–Zehnder, Michelson, Fabry–Pérot or Fizeau arrangement, the details of which can be found in many optics texts [1]. The fiber systems that will be discussed later are more comparable with a Michelson, two-beam interferometer, so this device is considered here. A simple schematic of this interferometer is shown in Figure 7.1.

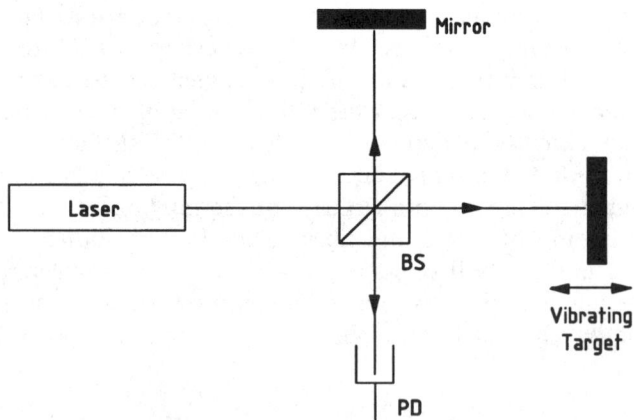

Fig. 7.1 A schematic of a bulk optic Michelson interferometer for vibration measurement. BS = beam splitter and PD = photodetector.

Considering the bulk optic Michelson interferometer of Figure 7.1, the fixed mirror provides a fixed reference path and the signal beam is reflected from the vibrating target. The intensity, I, detected at the photodector is given by [1, 6]

$$I = I_0[1 + \nu\cos(\phi)] \quad (7.1)$$

where I_0 is the total intensity due to both beams in the absence of interference, ν is the visibility of the interference pattern, which is dependent on the spectral characteristics of the light source, and ϕ is the phase difference between the two interfering beams. This may be written as

$$\phi = k2n\,\Delta L \quad (7.2)$$

where k is the wave number ($k = 2\pi/\lambda$ with λ the wavelength), n is the refractive index of the medium of propagation and ΔL is the path difference between the two beams. The factor of 2 accounts for the double pass of the Michelson interferometer. A graph showing the co-sinusoidal variation of detected intensity with phase difference is shown in Figure 7.2.

In the situation where the signal beam comes from a vibrating target, the path difference, ΔL, includes a constant average displacement plus the time-varying vibration. The vibration is then encoded in the time dependence of the phase of the detected intensity. Writing the path difference ΔL as a mean displacement L_m, plus a sinusoidal variation of angular frequency ω_s and amplitude L_s allows the equation for the detected intensity to be written as

$$I = I_0[1 + \nu\cos(2kn(L_m + L_s\sin(\omega_s t)))] \quad (7.3a)$$

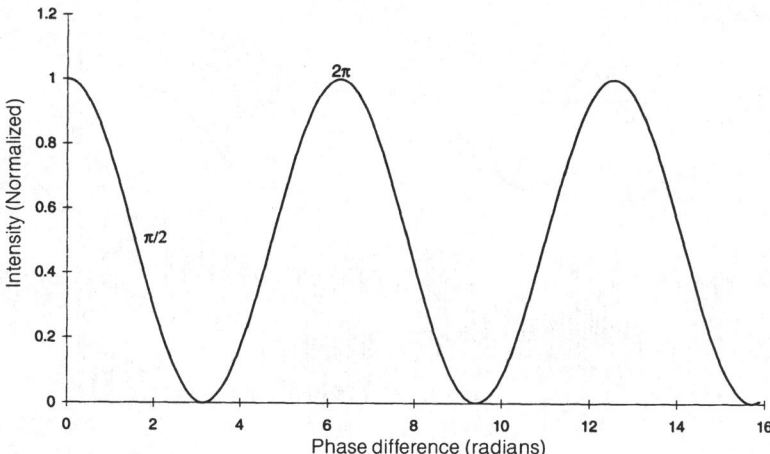

Fig. 7.2 The variation in output intensity of a Michelson type interferometer as a function of the phase difference between the interfering beams. Note the maximum slope occurs when the phase is an odd integer number of $\pi/2$.

or

$$I = I_0[1 + v\cos(\phi_m + \phi_s \sin(\omega_s t))] \quad (7.3b)$$

where ϕ_m is a bias phase due to the mean path difference and ϕ_s is the phase amplitude of the sinusoidal vibration. Figure 7.3 shows the variation in detected intensity for a sinusoidal vibration at frequency 10 Hz and amplitude 1 μm, assuming a laser source wavelength of 633 nm.

In principle the simplest interferometric fiber vibrometer consists of a single 2×2 coupler configured as shown in Figure 7.4. This is a two-beam interferometer with the reference beam derived from the light reflected from the distal end of the fiber, and the light recaptured by the fiber after reflection or scattering at the target forms the signal beam.

The analysis of this fiber equivalent of the Michelson is essentially the same. In this case the coupler (the fiber optic beam splitter) splits the light into the two fiber paths. Note that port 3 of the coupler is actually terminated in index-matching material to suppress unwanted reflections from this port. The interference signal is detected at the fourth port. In this fiber arrangement the detected intensity and phase are also described by equations 7.1 and 7.2, however, the visibility will in general be different as the intensities of the interfering beams will be different.

Thus, the output anticipated from a fiber optic interferometric vibration measurement system is the same as from a bulk optic system, and the problem that remains is extracting information on the vibration characteristics from the signals.

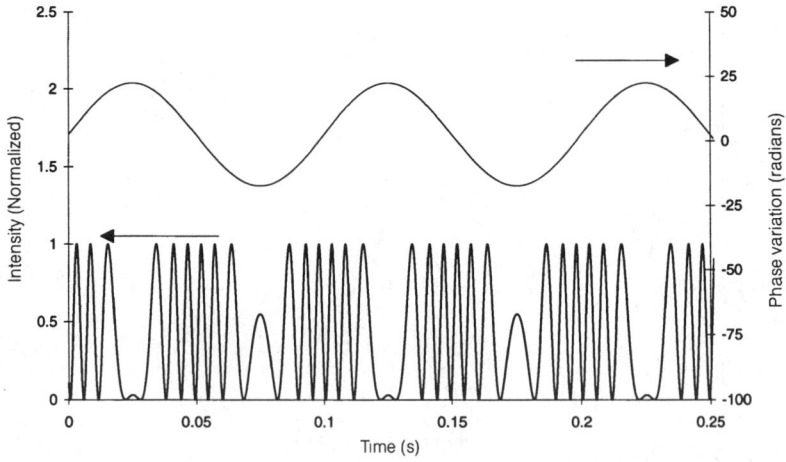

Fig. 7.3 The variation in phase and detected intensity in a Michelson interferometer as a function of time when one of the mirrors is vibrating. The frequency of the vibration is 10 Hz, the amplitude 1 μm and the wavelength of the laser source was taken to be 633 nm.

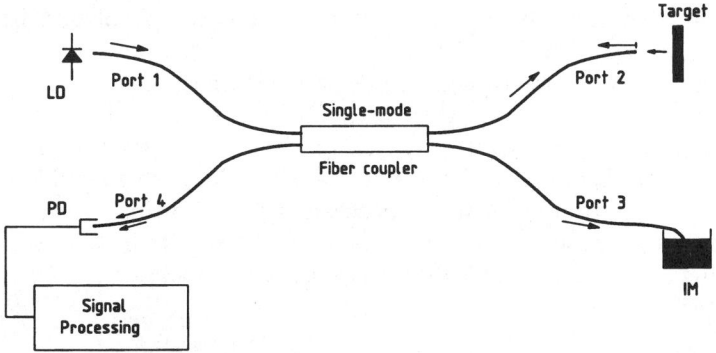

Fig. 7.4 The simple optical fiber interferometer for vibration measurement. LD = laser diode, IM = index matching and PD = photodetector.

7.3 SIGNAL PROCESSING

The measurement problem depends on extracting information about the amplitude and frequency content of the vibration from signals of the form shown in Figure 7.3. For large amplitudes a simple technique to determine its magnitude could be to count the number of interference fringes. As each fringe is separated from the next by a change in optical path difference (OPD) of $\lambda/2n$ (separation between maxima in equation 7.1), this provides information on the displacement as a function of time. However, this technique is not readily applicable to real-time signal processing, particularly of higher frequency signals, and other signal-processing techniques may be used, as described below.

In measuring small amplitude vibration the high sensitivity of interferometric techniques offers particular advantage. Here the problem is that the sensitivity of the change in detected intensity to phase is a function of the phase bias ϕ_m. Indeed if the phase bias is an integer number of 2π, the phase sensitivity is zero (see Figure 7.2). Maximum sensitivity is achieved for a phase bias of $(2m+1)\pi/2$ (where m is an integer), a condition known as quadrature. If this condition is satisfied the time-varying part of the detected intensity will give rise to an electrical signal, $V(t)$, which may be written as

$$V(t) = \Gamma I_0 \nu \cos(\pi/2 + \phi_s \sin(\omega_s t)) \tag{7.4}$$

where Γ incorporates all the constants of proportionality and amplifier gain, etc. In the small signal limit this may be written as

$$V(t) = \Gamma I_0 \nu \phi_s \sin(\omega_s t) \tag{7.5}$$

That is, the detected intensity follows the vibration pattern of the target [7]. However, this is only true for small vibrations and does not remain true if the interferometer drifts away from quadrature. By its very nature an interferometer is sensitive to small changes in path difference and environmental perturbation, therefore some form of dynamic control (feedback) is usually required.

The combined problems of controlling the bias phase in the interferometer and recovering the phase from the electrical signal derived from the detected intensity requires further signal processing. The most commonly used techniques in optical fiber sensor systems are based on homodyne and heterodyne techniques. These are used in signal processing for vibration-measuring interferometers and are described in detail elsewhere [8] but it will prove useful for the following discussion to summarize the main techniques here.

7.3.1 Passive homodyne

In this system two signals with a relative phase difference of $\pi/2$ are derived from the interferometer. These may then be combined by squaring and adding, or differentiating and cross-multiplying to give an output which is proportional to the phase of the interference signal [9]. These operations may easily be implemented using analog or digital signal-processing schemes to give real-time output.

A second homodyne technique requires a closer look at the output for the interferometer. The time-dependent output from the photodetector, $V(t)$, may be written (cf. equation 7.4):

$$V(t) = \Gamma I_0 \nu \cos(\phi_m + \phi_s \sin(\omega_s t)) \tag{7.6}$$

here ϕ_m is the phase bias, ϕ_s and ω_s represent the amplitude and angular frequency of the sinusoidal vibration which is to be measured. This expression for the output signal may be rewritten as a Fourier series with Bessel functions as the coefficients:

$$V(t) = \Gamma I_0 \nu \left[\cos(\phi_m) \left\{ J_0(\phi_s) + 2 \sum_{n=1}^{\infty} J_{2n}(\phi_s) \cos(2n\omega_s t) \right\} \right.$$

$$\left. - \sin(\phi_m) \left\{ 2 \sum_{n=0}^{\infty} J_{2n+1}(\phi_s) \sin((2n+1)\omega_s t) \right\} \right] \tag{7.7}$$

where J_n is the nth-order Bessel function of the first kind.

Thus if the spectrum of the output signal is examined, analysis of the amplitude components provides enough information to determine the amplitude of the phase variation. Most techniques are based on taking the ratio of even (or odd) harmonics as this ratio is independent of the phase bias. There are a range of variations on this type of processing which exploit the properties of the Bessel functions [10, 11].

7.3.2 Close loop (or active) homodyne

This technique involves the active control of the phase difference within the interferometer. The signals detected from the interferometer are processed to produce a control signal which is in turn used to modulate the OPD to maintain the quadrature position [12]. In a fiber system this may be achieved by stretching the fiber by winding it around a piezoelectric transducer [13]. The control signal is in direct proportion to the measurement-induced phase change in the interferometer and so provides the output signal.

7.3.3 Heterodyne

Heterodyne processing requires the frequency of one of the interfering beams to be shifted. This can be done with bulk optic components by using an acousto-optic modulator (Bragg cell) [14] which introduces a change in frequency of an optical beam, typically in the range 40–100 MHz. The resulting interference signal is a beat between the interfering beams and so consists of a sinusoidal variation in intensity at the frequency difference between the two beams. By using two Bragg cells, one in each arm of the interferometer, with slightly different frequencies (~ 100 kHz) the frequency of this carrier signal can be reduced [15]. The detected intensity may then be written

$$I = I_0[1 + \nu \cos(\omega_B t + \phi(t))] \tag{7.8}$$

where ω_B is the angular frequency shift introduced by the Bragg cell (or the frequency difference between the two beams) and $\phi(t)$ is the required phase variation due to the vibration. The required information is the phase term which may be demodulated using a phase tracker or phase-locked loop [15]. In addition, the use of the carrier frequency, ω_B, also removes the surface displacement (i.e. direction) ambiguity usually present in homodyne systems by providing an effective frequency offset.

Alternatively, the fact that the phase is changing with time can be considered to give rise to a change in the instantaneous frequency of the signal (that is the vibrating surface may be considered to have introduced a Doppler frequency shift). Thus, a frequency tracker may be used, the output of which is a signal in direct proportion to the velocity of the vibrating surface. A simple integrating circuit may then be used to recover the displacement pattern of the vibration.

The use of Bragg cells is the favored technique in bulk optic interferometers due to the high bandwidth available, but unfortunately there is not an efficient all fiber equivalent. This means that to use this technique the sensor must be carefully designed and will, in general, resort to a hybrid fiber and bulk optic system in order to incorporate a Bragg cell and fully exploit the advantages it offers.

7.3.4 Synthetic heterodyne

If the phase of the interferometer is modulated sinusoidally (by using a piezo-electric transducer in a fiber system) the time-dependent electrical output from the optical detector may be written as a voltage, $V(t)$, proportional to the cosine of the phase variation:

$$V(t) = \Gamma I_0 \nu \cos(\phi(t) + \phi_c \sin(\omega_c t)) \tag{7.9}$$

$\phi(t)$ contains the time-varying phase due to the vibration and ϕ_c and ω_c represent the amplitude and angular frequency of the induced phase modulation. As with equation 7.6, this expression for the output signal may be rewritten as a Fourier series with Bessel functions as the coefficients:

$$V(t) = \Gamma I_0 \nu \left[\cos(\phi(t)) \left\{ J_0(\phi_c) + 2 \sum_{n=1}^{\infty} J_{2n}(\phi_c) \cos(2n\omega_c t) \right\} \right.$$
$$\left. - \sin(\phi(t)) \left\{ 2 \sum_{n=0}^{\infty} J_{2n+1}(\phi_c) \sin((2n+1)\omega_c t) \right\} \right] \tag{7.10}$$

In this case the required information is encoded in $\phi(t)$. In a similar way to the spectral techniques used in homodyne processing, appropriate filtering and analysis of the amplitude of the harmonics allow the phase variation to be determined [16, 17].

7.3.5 Pseudo-heterodyne

This technique requires an unbalanced interferometer, and that the output frequency of the light source be modulated periodically [18]. The resultant intensity detected at the output of the interferometer is a periodic function, the period of which depends on the amplitude and frequency of the modulation, and the magnitude of the path imbalance.

If a linear ramp of the required amplitude (ideally with negligible flyback time) is used to modulate the frequency of the light source, with appropriate filtering the output of the interferometer is a sinusoidal variation in the intensity (cf. equation 7.8) which may be demodulated using standard phase recovery techniques. However, the limitations and noise implicit in this processing technique should be carefully considered [19].

Alternatively, if a sinusoidal modulation is applied, the output intensity may be expanded as a Fourier series where the coefficients of the terms are given by Bessel functions (in the same manner as equation 7.10). Techniques to recover the phase in this situation, like synthetic heterodyne, rely on appropriate filtering and exploitation of the properties of the Bessel functions.

The extent to which these, and any other, signal-processing schemes may be achieved in fiber-based systems is an important consideration. The signal-processing scheme adopted for each possible configuration is important in determining the overall system performance.

7.4 OPTICAL FIBER IN INTERFEROMETRIC VIBROMETERS

The addition of fibers to interferometric systems can facilitate the measurement procedure in a number of ways. The fiber can provide a 'light pipe' to carry the optical radiation to and from the measurement region. Alternatively, the optical fiber can take part in the measurement procedure. In the vibration measurement systems being considered here, where a remote target is to be addressed, it is clear that the fiber is acting largely as a flexible well-defined optical path for the light to and from the measurement region. However, this is not to say that the fiber does not play an integral part in the instrument as will be shown when specific configurations are considered. Another important classification of the use of optical fiber in interferometric vibrometers is the hybrid combination of both fiber and bulk optics, for example to allow the inclusion of a Bragg cell for heterodyne signal processing. This will be evident in some of the systems discussed.

7.4.1 Fibers to replace a path in an interferometer

Initial developments in interferometric vibrometers to utilize fiber used it to replace one of the arms of the interferometer. The earliest reports of fiber being used in vibration (and velocimetry) measurement systems were refs [20, 21]. Both of these configurations were hybrid configurations, designed to ensure that Bragg cells could be incorporated in the system. This then allowed variations in standard heterodyne signal-processing techniques to be applied.

An example of the configuration proposed by Nokes *et al.* [21] is shown in Figure 7.5. The system was designed to make vibration measurements of biological systems. This arrangement uses the zero- and first-order beams from the Bragg cell (driven at 40 MHz) to form the reference (the frequency shifted beam) and target beams. The fiber carried the light to and from the target. Care was taken to ensure the optical paths of the reference and target beam were approximately equal in order to reduce the phase noise arising from source wavelength fluctuations [22]. Light arising from reflections from the ends of the fiber probe, which would give rise to spurious signals, were minimized by using index-matching liquids and polishing the ends of the fiber at an angle. The signal processing utilized here was a variation on the standard heterodyne technique. The interference signal was mixed down from 40 MHz to 100 kHz using a local electrical oscillator at 39.9 MHz, which in turn was mixed with a reference oscillation of 100 kHz, also derived from the Bragg frequency and the local oscillator. Appropriate filtering provides an output in proportion to the phase of the interference signal for small signals. This puts a small signal limit on the maximum displacement of 25 nm, but the response is monotonic up to 93 nm providing greater range if appropriate signal processing is used. The calculated shot noise limit for this optical fiber vibration measurement system is of the order 0.005 nm $Hz^{-1/2}$.

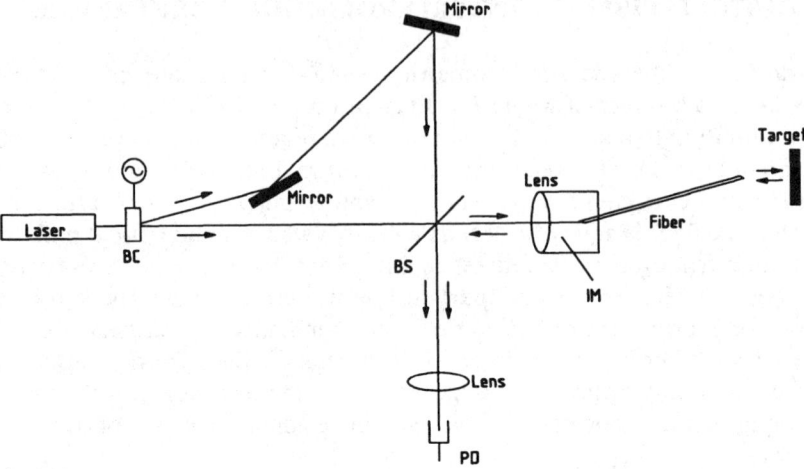

Fig. 7.5 A vibration measurement system that incorporates a fiber probe as one of the arms of the interferometer. BC = Bragg cell, BS = beam splitter, IM = index matching and PD = photodetector.

7.4.2 Single fiber interferometers

A simple all fiber vibrometer can be constructed from a single length of fiber, interference being provided between the light reflected from the end of the fiber and the vibrating target [23, 24]. However, as was discussed above, care must be taken to avoid unwanted reflections. The technique employed by Drake and Leiner [24] used polarization to isolate the required beams. The experimental arrangement is shown schematically in Figure 7.6. The polarizing beam splitter provides a linear input polarization to the fiber. A length of the single-mode fiber is coiled such that the induced birefringence leads to a quarter-wave device [25], the azimuth of which is controlled by adjustment of the azimuth of the fiber loop. Thus, by arranging the azimuth of the fiber quarter-wave device so that circularly polarized light arises from traversing the fiber once, the light which is reflected from the distal end of the fiber exits the fiber in a linear polarization state with the azimuth rotated by 90° with respect to the input polarization state. Similarly, light reflected from the vibrating target also traverses the fiber loop twice and exits the fiber in the same polarization state as the reference beam. The lens which launched the light into the fiber now recollimates the light and the polarizing beam splitter reflects the reference and target beam onto the photodetector. The polarizer thus eliminates unwanted polarizations arising from stray reflections. As the system is designed for the measurement of small vibrations, noise will ultimately limit the performance. Thus, efficient isolation of the reference and target beam is important and this will only hold true for high quality polarizing optics.

OPTICAL FIBER IN INTERFEROMETRIC VIBROMETERS

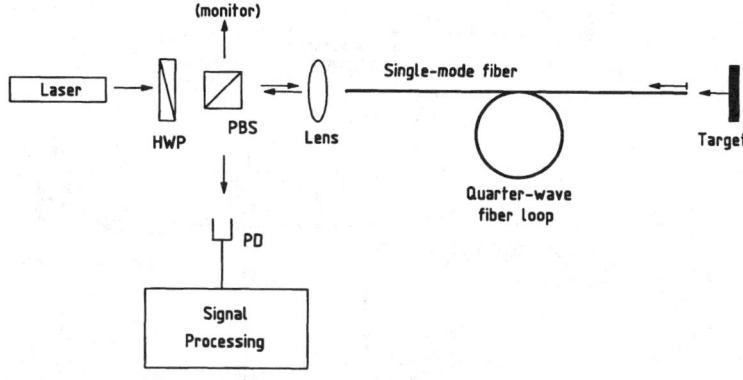

Fig. 7.6 A single fiber vibration measuring interferometer using polarization to control the interfering beams. HWP = half-wave plate, PBS = polarizing beam splitter and PD = photodetector.

This system was demonstrated addressing the tympanic membrane of a cricket, and determined the vibration amplitude when acoustically excited. One drawback of the system as described is that the signal processing is not entirely passive in that access to the target is required. A low frequency dither is applied to the target and a spectrum analyzer used to determine the frequency content. The dither was applied to the target via a piezoelectric transducer driven by a triangular waveform. The amplitude of the waveform was chosen to give a change in OPD of $\lambda/2$ thus ensuring that at some point in the period, maximum sensitivity is achieved. Thus, using the spectrum analyzer in peak detection mode, the response to a pure sinusoidal excitation could be determined. Sweeping the excitation frequency allowed the frequency response of the target to be determined.

7.4.3 Interferometers using fiber couplers

Another hybrid system was described by Lewin *et al.* [26]. This combined an optical fiber coupler and bulk optics in a modified Mach–Zehnder arrangement as shown in Figure 7.7. One arm of the fiber formed the reference path and the other carried light to the target probe optics. This configuration allowed the use of the full range of signal-processing techniques, with the active homodyne and synthetic heterodyne schemes utilizing fiber modulation techniques to achieve the required modulation. In addition a Bragg cell could have been placed in the air path to allow true heterodyne processing. This work showed the potential for the range of signal processing using fiber.

One of the problems of this and many of the systems discussed so far, is the sensitivity of the fiber to other environmental perturbations. In configurations where the two beams follow two separate paths, any perturbation which

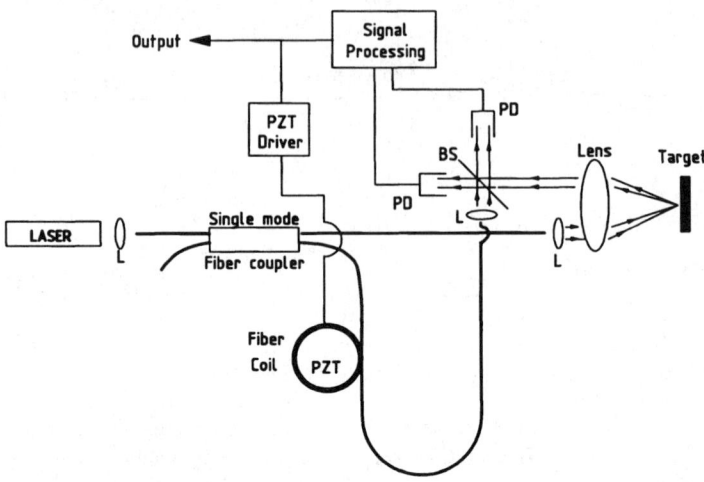

Fig. 7.7 A Mach–Zehnder interferometer using optical fiber and bulk optic components. L = lens, PZT = piezoelectric transducer, BS = beam splitter and PD = photodetector.

changes the optical path (i.e. the length of the fiber or its effective refractive index) will be interpreted as a change in the position of the target. A particular problem is that the fiber exhibits high sensitivity to strain and temperature [27]. Another potential source of noise in these configurations is the pickup of acoustic noise [28].

The simple coupler configuration, shown in Figure 7.4, reduces this problem. The beams have a common path within the fiber and so the output is not affected by any environmental perturbation that may be experienced by the fiber. However, signal processing does become more difficult as it is not possible to access one of the beams individually, except at the target. This, in general will not be possible as the target will be remote and the ideal system should be designed to be completely passive.

One signal-processing scheme which can be applied in this arrangement is pseudo-heterodyne as described by Laming *et al.* [29]. The experimental arrangement is shown in Figure 7.8. The pseudo-heterodyne technique applied here relies on a fiber probe to target distance in excess of 50 mm, in order that the modulation requirements can be met. Note that this means that some form of lens is required at the fiber probe output to ensure sufficient light is coupled back into the fiber. Sinusoidal modulation is applied to the drive current of the laser diode which in turn modulates the optical frequency of the laser diode output and hence the intensity detected at the output of the unbalanced interferometer. The second and fourth harmonics are filtered from the output (equation 7.10) and the amplitude of these harmonics (a second and fourth Bessel function of the amplitude of the phase modulation) are used in the feedback

OPTICAL FIBER IN INTERFEROMETRIC VIBROMETERS 249

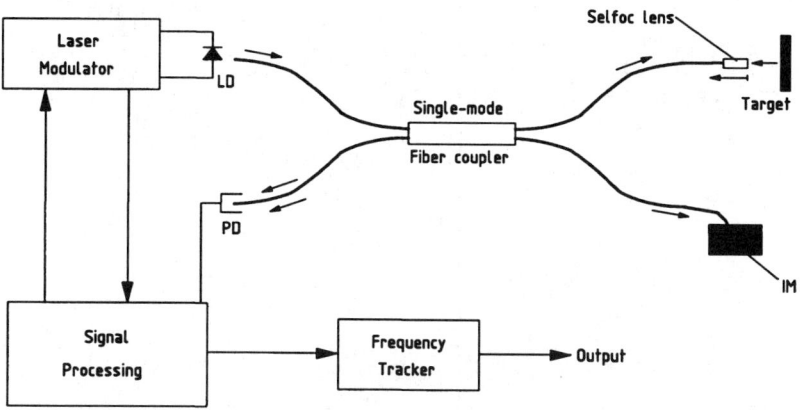

Fig. 7.8 A simple optical fiber vibration interferometer using pseudo-heterodyne demodulation. LD = laser diode, IM = index matching and PD = photodetector.

system to maintain the amplitude of the phase modulation at the optimum value. The phase is recovered using an analog switch to extract the second harmonic of the signal and a frequency tracker used to give an output in proportion to the target velocity. The system was demonstrated operating up to vibration frequencies of 20 kHz (limited by the frequency tracker).

This is an attractive simple configuration, but it does have some disadvantages for high performance applications. This is due to the fact that a large fiber probe to target distance is required. This has two consequences; first, it means that the noise level will be increased due to source phase noise. Second, the frequency of the fundamental optical signal is a function of path imbalance so slight, unpredictable variations will change this and the filtering is no longer optimized and the noise floor will increase.

A recently described fiber-based system for velocimetry is also directly applicable to vibration measurement [30]. This also uses polarization encoding of the signal and reference beams to allow the inclusion of a Bragg cell (operating at 60 MHz) in the system. This is achieved by using a highly birefringent fiber coupler to form the fiber probe. A schematic of the experimental arrangement is shown in Figure 7.9. The coupler performs the same function as a polarizing beam splitter and so by careful control of the polarization at the probe, the reference and signal beam are delivered to the second interferometer in orthogonal linear polarization states. This polarization discrimination is further utilized within the second interferometer to allow independent access to the reference and signal beam, ensuring that the carrier frequency from the Bragg cell is only imposed on one of the beams. A half-wave plate is also used to ensure efficient recombination of the interfering beams. This approach has the advantages that by controlling the input polarization state the ratio of the intensities of the reference and signal beam may be controlled to optimize the signal

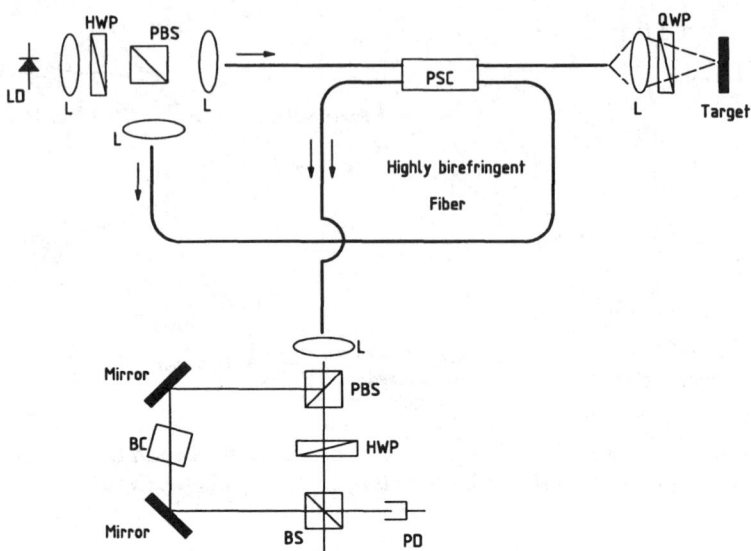

Fig. 7.9 An optical fiber interferometer utilizing highly birefringent fiber to enable the target and reference beam to be encoded in polarization. LD = laser diode, HWP = half-wave plate, PBS = polarizing beam splitter, L = lens, PSC = polarization selective coupler, QWP = quarter-wave plate, BC = Bragg cell, BS = beam splitter and PD = photodetector.

to noise characteristics and careful choice of fiber lengths removes the restriction on the working distance while maintaining low noise sensitivity to frequency fluctuation in the light source by operating close to zero OPD [22].

7.4.4 The Sagnac interferometer and vibration measurement

A novel variation on an all fiber interferometer is based on the Sagnac interferometer. The fiber Sagnac interferometer has received a lot of attention due to its application as a gyroscope [31], but is not so widely used in other applications. In a standard Sagnac interferometer, as shown schematically in Figure 7.10, the fiber loop is closed. Within the fiber loop two beams counterpropagate. In the absence of any asymmetry in the optical paths, the two beams recombine with zero OPD. However, when the asymmetry is broken, a phase shift is introduced between the recombining beams. The Sagnac effect, due to the rotation of the fiber in the plane of the loop, introduces such asymmetry and is utilized in the optical fiber gyroscope. Other origins of phase shift are birefringence and other nonuniform path length changes within the loop. For example, if a modulator is introduced close to one end of the fiber loop, the asymmetry is broken as the modulation is experienced by the counterpropagating beams at different times.

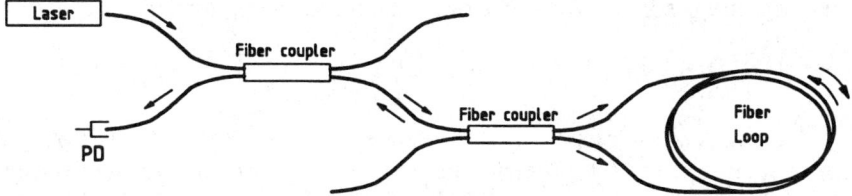

Fig. 7.10 A schematic of an all fiber Sagnac interferometer.

This has been exploited by Harvey et al. in constructing a velocimeter with potential applications in vibration measurement [32]. To construct the probe the fiber loop was broken close to one end, and the fibers arranged such that the light exiting from the two fibers is reflected from the target and coupled back into the other fiber to continue their counterpropagation. An example of this scheme is shown schematically in Figure 7.11. The phase difference between the beams at the output of the interferometer at a time t, is then given by the time average of the surface velocity, averaged over the time Δt which represents the time between the two counterpropagating beams encountering the target. This may be written as

$$\Delta t = (t - t_1) - (t - t_2) = \frac{(L_2 - L_1)n}{c} \qquad (7.11)$$

where t_1 and t_2 are the times at which the two beams encounter the target, L_1 and L_2 represent the length of fiber from the coupler to the probe for the two beams, n is the effective refractive index of the fiber and c is the speed of light. The phase difference, ϕ_D, for a target vibrating sinusoidally with a frequency of ω_s and amplitude x_0 may be written as

$$\phi_D = 2k_0 \Delta t \, \text{sinc}\left(\frac{\omega_s \Delta t}{2}\right) \cos(\omega_s(t - t_0)) \qquad (7.12)$$

where $v_0 = x_0 \omega_s$ and $t_0 = (t_1 + t_2)/2$.

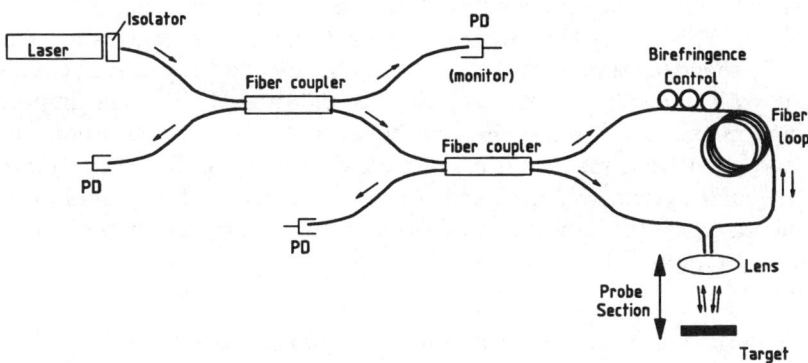

Fig. 7.11 A modification of the Sagnac interferometer for vibration measurement. PD = photodetector.

This response predicts a roll-off around the frequency f_r given by

$$f_r = \frac{1}{2\Delta t} \tag{7.13}$$

and indeed a zero in response at a frequency given by $1/\Delta t$. Clearly, the length of fiber and the asymmetric position of the probe dictate the averaging time and therefore the system bandwidth. In the system described, with a loop length of 220 m and an averaging time of 1.07 µs, the roll-off frequency was of order ~ 460 kHz.

As a Sagnac interferometer is used, the phase shift recorded is proportional to surface velocity as shown by equation 7.12. The output signal from the interferometer (for small signals) was derived by holding the interferometer in quadrature. This was achieved by birefringence control within the fiber loop, which was also used to control the depth of modulation. If required the inclusion of an integrating module would provide the vibration profile.

This approach to vibration measurement has distinct advantages over say a Michelson interferometer. For example, there is no requirement for a long coherence length light source; sensitivity to environmental perturbations are reduced; as it operates close to zero OPD sensitivity to source frequency noise is minimized. In addition to these it is also a completely passive instrument.

Recent work using this configuration has developed the probe further and uses highly birefringent fiber to allow a down-lead to the probe head of arbitrary length [33]. A polarizing beam splitter and a Faraday rotator (as a non-reciprocal element) are now required in the Sagnac loop. The two counter-propagating beams exit the fiber loop and the polarization of the beams is controlled such that they propagate along the down-lead of highly birefringent fiber in orthogonal modes to the probe head. The incorporation of a quarter-wave plate within the probe head ensures the light reflected from the target returns down the fiber down-lead in the orthogonal eigenmode. Thus, the use of highly birefringent fiber in the down-lead ensures that there is immunity to bending, and as both beams travel down both polarization modes of the fiber, only perturbations which are comparable with the time scale of the loop delay are detected. This falls in the low frequency range, and is a range which can be adequately shielded. Similar operating characteristics to the previous case were reported with a displacement resolution of 0.08 nm $Hz^{-1/2}$. The important advantage in this development is a probe with a remote probe head and no down-lead sensitivity, making it a practical device for in-field measurements. This particular system was designed for noncontact structural monitoring in power plants where the importance of an environmentally insensitive down-lead can readily be appreciated.

7.4.5 Vibration measurement using white light interferometry

As can be seen from the configurations discussed so far, much of the effort is directed at developing a system from which the required signal may be demodulated. It was stated that if a bulk optic configuration were to be used to perform vibration measurements, a Bragg cell would be used to introduce a carrier frequency and allow standard phase measurement techniques to be used, with a high measurement bandwidth. This is also evident from many of the systems which have been described which have been designed to incorporate a Bragg cell.

An alternative approach to gaining independent access to one of the interfering beams in optical fiber-based systems is to use the technique of white light interferometry. White light interferometry (WLI) is an alternative approach to interferometric sensors (intrinsic or extrinsic) which can be used to overcome some of the problems of standard interferometry to provide an unambiguous, absolute measurement [34, 35]. In addition to this it is directly applicable, in hybrid configurations, to interferometric vibration measurement using optical fiber. This then allows the standard techniques of signal processing, which would be applicable to bulk optic systems, to be incorporated within systems utilizing an optical fiber probe head.

In its simplest form the operation of a WLI sensor is summarized as follows. Light from a low coherence light source (a light source of broad spectral width such as a light emitting diode) is introduced into the first of two coupled interferometers. The coherence length of the light source, L_C, is necessarily short, so if the path difference in the interferometer is greater than the coherence length no interference would be observed at the output of this interferometer. The output from the first interferometer is then provided as input to the second interferometer. Interference will then be observed at the output of the second interferometer if the OPD of the second interferometer is zero, or if it matches that of the first interferometer (that is, matches to within the coherence length of the light source). Thus, if the two interferometers are adjusted such that the mean OPDs within the two interferometers are equal, any path variation in the first interferometer (due to the vibration of the target) will be observed in the output of the combined system. There is a concomitant reduction in visibility, and there may be a slight variation in the depth of modulation (depending on the vibration amplitude), but as the vibration is encoded in the phase of the signal this is not a serious problem, though there will be a reduction in the signal to noise ratio.

In a practical arrangement for vibration measurement, the first interferometer is the measurement interferometer and would be formed from a simple fiber probe (cf. Figure 7.4). The second recovery interferometer, which would be a bulk optic Michelson or Mach–Zehnder interferometer, allows independent access to the interfering beams. An advantage of such a scheme is that it always operates close to zero OPD and therefore noise arising from light source frequency variations is reduced.

The first exploitation of WLI utilized the technique to allow the introduction of a Bragg cell into the second interferometer [36, 37]. A schematic of the experimental arrangement is shown in Figure 7.12. Light from the optical fiber interferometer addressing the vibrating target is collimated and propagates through

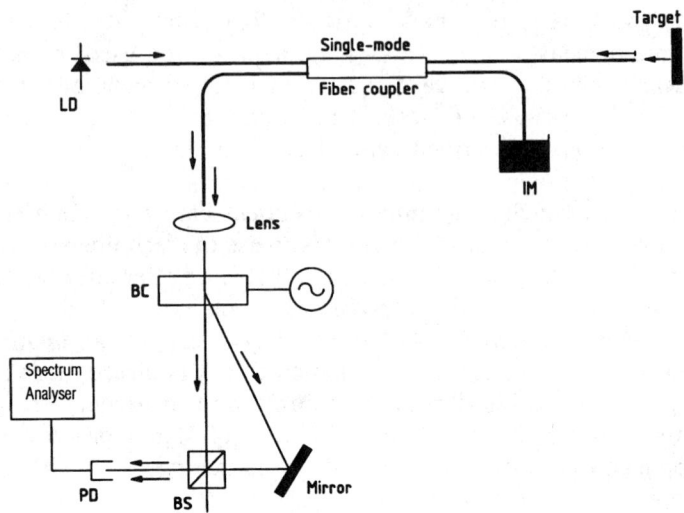

Fig. 7.12 A white light interferometer arrangement incorporating a Bragg cell for heterodyne demodulation. LD = laser diode, IM = index matching, BC = Bragg cell, BS = beam splitter and PD = photodetector.

the Bragg cell (operating at a frequency of 110 MHz). The OPD between the light reflected from the end face of the fiber and reflected from the target was arranged to be greater than the coherence length of the light source. In this case the light source was a multimode laser diode and an OPD of ~4.25 mm was used. In the arrangement described no lens was used at the end of the fiber probe, thus the amount of light recaptured by the fiber falls off quite dramatically with OPD. The addition of a lens, either to collimate the light or define the measurement region more specifically, should enhance the range of mean OPD over which the technique may be applied.

The recovery interferometer is formed by the beams which are output from the Bragg cell. The first-order diffracted beam and the undeviated beam were recombined, with appropriate optical path balancing to provide the interference signal. The combined system then offers all of the advantages of a true heterodyne system with the added convenience of an optical probe.

The second variation of this technique used the properties of the recovery interferometer to produce two outputs in quadrature, so providing the required signals for passive homodyne signal processing [38, 39]. A schematic of the apparatus is shown in Figure 7.13 and its operation is summarized as follows. The first interferometer, with an OPD greater than the coherence length of the light source, is the optical fiber measurement interferometer as in the previous system. The second interferometer is a bulk optic Michelson interferometer, with a mean path difference equal to that of the measurement interferometer. A lens is used to increase the beam diameter in the interferometer to the order

OPTICAL FIBER IN INTERFEROMETRIC VIBROMETERS 255

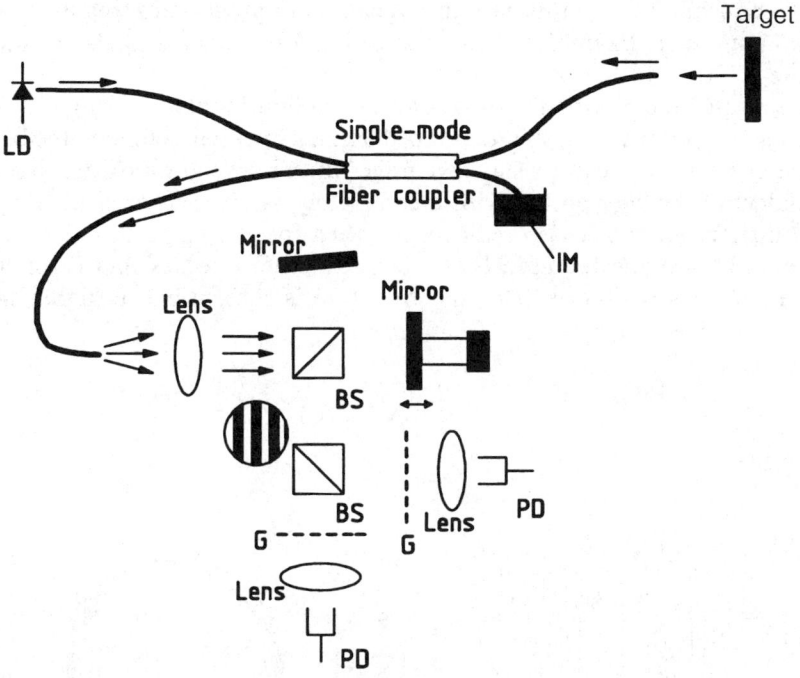

Fig. 7.13 A passive white light interferometer for the measurement of vibration. The operation of the interferometer arrangement is described in the text. LD = laser diode, IM = index matching, BS = beam splitter, G = transmission grating and PD = photodetector.

of ~10 mm. One of the mirrors is then tilted through a small angle to produce spatially distributed interference fringes across the expanded beam. If the target is vibrating, these spatial fringes within the beam profile 'vibrate' transversely in direct response to the vibration. The output of the recovery Michelson interferometer was further split at a second beam splitter and each of these beams was incident on a coarse transmission grating, with the transmitted light focused onto a photodetector. The period of the transmission grating was chosen to match that of the fringe pattern arising due to the tilted mirror. Thus, as the fringe pattern moved within the beam profile across the transmission grating, the detected intensity was modulated. The analysis showed that the detected signal at each detector is the same as would be measured with a standard interferometer (apart from a reduction in visibility) with the important addition that the initial phase of the signal may be controlled by adjusting the initial position of the transmission grating. As two signals are available, the relative phase between the signal may be adjusted by controlling the relative position of the two transmission gratings. Thus, it is possible to provide two outputs in quadrature which may be processed using passive homodyne techniques. In this case the signals were processed using an on-line personal computer. Examples of the detected and processed signals are shown in Figure 7.14.

These result were reported without any lens so limiting the working distance. There is no reason why one should not be included to either collimate the beam, of define the working region. The advantages of this configuration are that it is independent of the fiber probe giving a completely passive sensor head and processing interferometer. It also removes the need for a Bragg cell, reducing the expense and removing the need for the associated electronics that it requires. However, it does operate at DC and therefore is susceptible to higher noise levels.

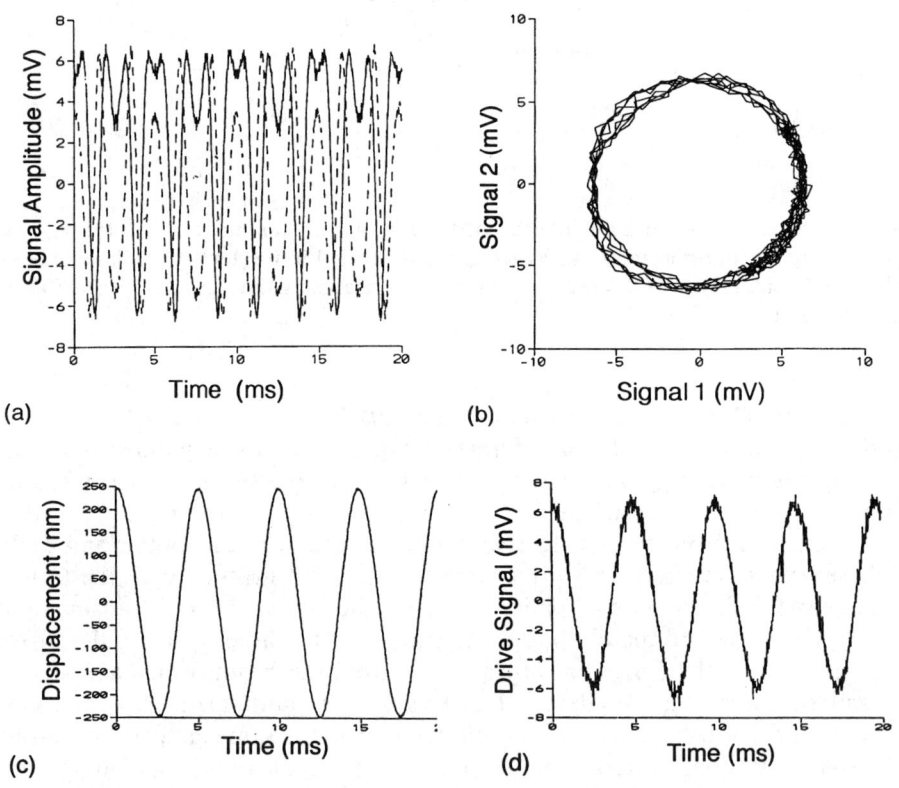

Fig. 7.14 Examples of the signals obtained from the system of Fig. 7.13: (a) the two interference patterns detected by the photodetectors; (b) a Lissajous' figure from the two detected signals showing the quadrature of the two signals; (c) the target displacement pattern reconstructed from the detected signals which can be compared with (d) the drive signal applied to the vibration target.

7.5 CONCLUSIONS

The use of optical fibre in interferometric vibration measurement has been an active area of research since the first examples were described in the late 1970s. An encouraging feature of this is that a search of the literature will uncover many examples of the actual applications of interferometric vibration measurement using optical fibers [40–44], many based on the techniques that have been described in this chapter. The advantages of optical fiber vibration measuring interferometers are recognized and they are being used.

The future is also bright. There is continued research into developing new optical components. The development of successful in-fiber frequency shifters [45] would open the way for many more new and innovative systems. New optical configurations have also been discussed which are directly applicable to vibration measurement and compatible with optical fibers [46–48]. Detection schemes too [49] continue to offer new schemes to simplify and improve the performance of vibration-measuring interferometers. With new measurement schemes, new components and advances in signal processing, there will continue to be developments and improvements in performance in interferometric vibration measurement using optical fiber.

REFERENCES

1. See for example Born, M. and Wolf, E. (1980) *Principles of Optics*, 6th edition, Pergamon Press, Oxford; or Hecht, E. and Zajac, A. (1987) *Optics*, 2nd edition, Addison-Wesley Publishing Co., Massachusetts.
2. Dyson, J. (1970) *Interferometry as a Measuring Tool*, Machinery Publishing Co., London.
3. Conforti, G., Brenci, M., Mencaglia, A., Mignani, A. G. and Scheggi, A. M. (1989) Multimode fibre-optic vibrometer. *Proceedings of 6th International Conference on Optical Fibre Sensors*, Paris, Springer-Verlag Proceedings in Physics, **44**, 194–199.
4. Pigeon, P., Pelissier, S., Mure-Ravaud, A., Gagnaire, H., Hosain, S. I. and Vıllas, C. (1993) A vibration sensor, using telecommunications grade monomode fiber, immune to temperature variations. *J. Phys. III France*, **3**, 1835–1838.
5. Waters, J. P. and Mottier, F. M. (1986) Fiber optic laser vibration sensor. *ISA Transactions*, **25**, 63–70.
6. Handerek, V. (1995) Single mode optical fiber sensors, in *Optical Fibers Sensor Technology* (eds K. T. V. Grattan and B. T. Meggitt) Chapman & Hall, London, 201.
7. Handerek, V. (1995) Single mode optical fiber sensors, in *Optical Fibers Sensor Technology* (eds K. T. V. Grattan and B. T. Meggitt) Chapman & Hall, London, 202.
8. Handerek, V. (1995) Single mode optical fiber sensors, in *Optical Fibers Sensor Technology* (eds K. T. V. Grattan and B. T. Meggitt) Chapman & Hall, London, 206–216.
9. Dandridge, A., Tveten, A. B. and Giallorenzi, T. G. (1982) Homodyne demodulation scheme for fibre-optic sensors using phase generated carrier. *IEEE J. Quantum Electron.* **QE-18**, 1647.
10. Sudarshanam, V. S. and Srinivasan, K. (1989) Linear readout of dynamic phase change in a fibre optic homodyne interferometer. *Opt. Lett.*, **14**, 140–142.

11. Jin, W., Uttamchandani, D. and Culshaw, B. (1992) Direct readout of dynamic phase changes in a fibre-optic homodyne interferometer. *Applied Optics*, **31**, 7253–7258.
12. Jackson, D. A., Priest, R., Dandridge, A. and Tveten, A. B. (1980) Elimination of drift in a single-mode optical fibre interferometer using a piezoelectrically stretched coiled fibre. *Applied Optics*, **19**, 2926–2929.
13. Davies, D. E. N. and Kingsley, S. A. (1974) Method of phase-modulating signals in optical fibres: applications to optical-telemetry systems. *Elecron. Lett.*, **10**, 21–22.
14. Tatam, R. P. (1995) Optical fiber modulation techniques for single mode fiber sensors, in *Optical Fibers Sensor Technology* (eds K. T. V. Grattan and B. T. Meggitt) Chapman & Hall, London, 239–241.
15. Oka, K., Tsukada, M. and Ohtsuka, Y. (1991) Real-time phase demodulator for optical heterodyne detection processes. *Meas. Sci. Technol.*, **2**, 106–110.
16. Cole, J. H., Danver, B. A. and Bucaro, J. A. (1982) Synthetic-heterodyne interferometric demodulation. *IEEE J. Quantum Electron.*, **QE-18**, 694–697.
17. Sudarshanam, V. S. (1992) New spectrum analysis techniques for interferometric vibration measurement. *Optics Communications*, **88**, 291–294.
18. Jackson, D. A., Kersey, A. D., Corke, M. and Jones, J. D. C. (1982) Pseudo-heterodyne detection scheme for optical interferometers. *Electron. Lett.*, **18**, 1081–1082.
19. Economou, G., Youngquist, R. C. and Davies, D. E. N. (1986) Limitations and noise in interferometric systems using frequency ramped single-mode diode lasers. *J. Lightwave Technol.*, **LT-4**, 1601–1608.
20. Cookson, R. A. and Bandyopadhyay, P. (1978) Mechanical vibration measurement using a fibre optic laser-Doppler probe. *Optics and Laser Technology*, 33–36.
21. Nokes, M. A., Hill, B. C. and Barelli, A. E. (1978) Fiber optic heterodyne interferometer for vibration measurement in biological systems. *Rev. Sci. Instrum.*, **49**, 722–728.
22. Dandridge, A., Tveten, A. B., Miles, R. O., Jackson, D. A. and Giallorenzi, T. G. (1981) Single-mode diode laser phase noise. *Appl. Phys. Lett.*, **38**, 77–78.
23. Kyuma, K., Tai, S., Hamanaka, K. and Nunoshita, M. (1981) Laser Doppler velocimeter with a novel optical fibre probe. *Applied Optics*, **20**, 2424–2427.
24. Drake, A. D. and Leiner, D. C. (1984) Fiber-optic interferometer for remote subangstrom vibration measurement. *Rev. Sci. Instrum.*, **55**, 162–165.
25. Lefèvre, H. C. (1980) Single-mode fibre fractional wave devices and polarisation controllers. *Electron. Lett.*, **16**, 778–779.
26. Lewin, A. C., Kersey, A. D. and Jackson, D. A. (1985) Non-contact surface vibration analysis using a monomode fibre optic interferometer incorporating an air path. *J. Phys. E: Sci. Instrum.*, **18**, 604–608.
27. Handerek, V. (1995) Single mode optical fiber sensors, in *Optical Fibers Sensor Technology* (eds K. T. V. Grattan and B. T. Meggitt) Chapman & Hall, London, 217–218.
28. Pannel, C. N., Jones, J. D. C. and Jackson, D. A. (1994) The effect of environmental acoustic noise on optical fibre based velocity and vibration sensor systems. *Meas. Sci. Technol.*, **5**, 412–417.
29. Laming, R. I., Gold, M. P., Payne, D. N. and Halliwell, N. A. (1986) Fibre-optic vibration probe. *Electron. Lett.*, **22**, 167–168.
30. James, S. W., Lockey, R. A., Egan, D. and Tatam, R. P. (1995) Fibre optic based reference beam laser Doppler velocimeter. *Optics Communications*, **119**, 460–464.
31. Lefèvre, H. C. (1993) *The Fiber-optic Gyroscope*, Artech House, Boston.
32. Harvey, D., McBride, R., Barton, J. S. and Jones, J. D. C. (1992) A velocimeter based on the fibre optic Sagnac interferometer. *Meas. Sci. Technol.*, **3**, 1077–1083.

REFERENCES

33. Carolan, T. A., Reuben, R. L., Barton, J. S., McBride, R. and Jones, J. D. C. (1995) Fibre optic Sagnac interferometer for non contact structural monitoring in power plant applications, in *Sensors and their Applications VII, Proceedings of the 7th Conference on Sensors and their Applications*, Dublin, Institute of Physics Publishing, Bristol, UK, 173–177.
34. Bosselmann, Th. and Ulrich, R. (1984) High-accuracy position sensing with fibre coupled white-light interferometry, *Proceedings of the 2nd International Conference on Optical Fiber Sensors*, Springer-Verlag Proceedings in Physics, 361–364.
35. Lefèvre, H. C. (1989) White light interferometry in optical fiber sensors. *Proceedings of the 7th Conference on Optical Fiber Sensors*, the Institution of Radio and Electronics Engineers, Australia, 345–351.
36. Meggitt, B. T., Boyle, W. J. O., Grattan, K. T. V., Baruch, A. E. and Palmer, A. W. (1991) Heterodyne processing scheme for low coherence interferometric sensor systems. *IEE Proceedings – Part J*, **138**, 393–395.
37. Ning, Y. N., Grattan, K. T. V., Palmer, A. W., Meggitt, B. T. and Weir, K. (1991) A novel optical heterodyne vibration sensor scheme, preserving directional information and using a short coherence length light source. *Optics Communications*, **85**, 10–16.
38. Weir, K., Boyle, W. J. O., Palmer, A. W., Grattan, K. T. V. and Meggitt, B. T. (1991) A novel optical processing scheme for interferometric vibration measurement using a low coherence source with a fibre optic probe. *OE/Fibers '91*, Boston USA, *Proc. SPIE*, **1584**, 220–225.
39. Weir, K., Boyle, W. J. O., Palmer, A. W., Grattan, K. T. V. and Meggitt, B. T. (1991) A low coherence interferometric fibre optic vibrometer using a novel optical signal processing scheme. *Electron. Letters*, **27**, 1658–1660.
40. Bower, J. E., Jungerman, R. L., Kuri-Yakub, B. T. and Kino, G. S. (1983) An all fiber-optic sensor for surface wave measurements. *J. Lightwave Technol.*, **1**, 429–436.
41. Powers, B. L. and Chuang, S. Y. (1983) Vibration monitoring of small lightweight components using fiber optic sensors. *Journal of Testing and Evaluation*, **19**, 493–496.
42. Ohki, M., Shima, N. and Shiosaki, T. (1993) Optical measurement of acoustic velocity and Poisson's ratio in disk piezoelectric transducers. *Jpn. J. Appl. Phys.*, **32**, 2463–2465.
43. McBride, R., Carolan, T. A., Barton, J. S., Wilcox, S. J., Borthwick, W. K. D. and Jones, J. D. C. (1993) Detection of acoustic emission in cutting processes by fiber optic interferometry. *Meas. Sci. Technol.*, **4**, 1122–1128.
44. Juang, P. A. and Wu, M. N. (1995) Active control using a fiber-optic interferometric sensor. *Smart Mater. Struct.*, **4**, 370–372.
45. Tatam, R. P. (1995) Optical fiber modulation techniques for single mode fiber sensors, in *Optical Fibers Sensor Technology* (eds K. T. V. Grattan and B. T. Meggitt) Chapman & Hall, London, 242–257.
46. Jentink, H. W., de Mul, F. F. M., Suichies, H. E., Aarnoudse, J. G. and Greve, J. (1988) Small Doppler velocimeter based on the self mixing effect in a diode laser. *Appl. Optics*, **27**, 379–385.
47. Wang, W. M., Boyle, W. J. O., Grattan, K. T. V. and Palmer, A. W. (1992) Fiber-optic velocimeter that incorporates active optical feedback from a laser diode. *Opt. Lett.*, **17**, 819–821.
48. Wang, W. M., Grattan, K. T. V., Palmer, A. W. and Boyle, W. J. O. (1994) Self-mixing interference inside a single-mode diode laser for optical sensing applications. *J. Lightwave Technol.*, **12**, 1577–1587.

49. Marshall, R. H., Sokolov, I. A., Ning, Y. N., Palmer, A. W. and Grattan, K. T. V. (1996) Photo-electromotive force crystals for interferometric measurement of vibration response. *Meas. Sci. Technol.*, **7**, 1683–1686.

8
Fiber optic laser anemometry

S. W. James and R. P. Tatam

8.1 INTRODUCTION

Laser anemometry (LA) is a technique that measures flow velocity distributions and higher moments such as turbulence levels, from light scattered by particles entrained in the flow [1-3]. LA was first demonstrated shortly after the invention of the laser, and quickly established itself as the instrument of choice for nonintrusive flow measurement. Early LA systems were generally large, bulky and difficult to use in inaccessible environments. The introduction of optical fibers led to the development of optical systems remotely linked to the laser source and detectors. Optical fibers are now used in a number of commercial laser anemometers, and other optical instrumentation. More recently solid state lasers and detectors have been combined with optical fiber technology to implement compact, robust multi-velocity component measurement instruments. This chapter reviews the principles of laser transit anemometry (LTA) and laser Doppler velocimetry (LDV), the optical fiber characteristics required to implement practical systems and optical signal-processing techniques employed. This is followed by a discussion of fiber optic implementations of LTA and LDV configurations, multiplexing techniques for multi-velocity component measurement and examples of applications in which they have been used.

8.2 LASER ANEMOMETRY PRINCIPLES

8.2.1 Laser Doppler velocimetry (LDV)

Laser Doppler velocimetry (LDV) is a well-established flow measurement technique, providing high quality, high spatial resolution data over a range of flow conditions. The basis of the technique is the measurement of the Doppler frequency shift imposed upon light scattered from an illuminating laser beam by particles entrained within the flow. The Doppler frequency shift is dependent upon the wavelength of the illuminating light, the orientation of the viewing direction with respect to the illuminating beam, and upon

Optical Fiber Sensor Technology, Vol. 2. Edited by K. T. V. Grattan and B. T. Meggitt.
Published in 1998 by Chapman & Hall, London. ISBN 0 412 782 901

the velocity component parallel to the bisector of the illuminating and viewing directions, as is shown in Figure 8.1.

Depending upon the optical configuration employed, the Doppler frequency shift typically lies in the range from hundreds of kHz (m s^{-1})$^{-1}$ to MHz (m s^{-1})$^{-1}$. LDV applications span a range of flow systems, ranging from blood flow (mm s^{-1}) to supersonic flows (> 1000 m s^{-1}). The Doppler frequency shift is thus very much smaller than the optical frequency, and, for the majority of flow systems, direct measurement is not practical. Measurement of the frequency shift is generally achieved using heterodyne detection, either by mixing the scattered light with a reference beam on a photodetector to yield a signal oscillating at the frequency difference between the beams, as in the reference beam anemometer [4], or mixing the light scattered from two illuminating beams, as in the Doppler difference technique [5, 6]. For high speed flows (in excess of 100 m s^{-1}), the frequency shift may be measured directly on a Fabry–Pérot interferometer [7], or using a wavelength filter, such as an iodine cell [8].

The Doppler frequency shift is given by

$$f_D = \frac{1}{2\pi}[\mathbf{k}_s + \mathbf{k}_0] \cdot \mathbf{v} \qquad (8.1)$$

where $\mathbf{k}_s$ and $\mathbf{k}_0$ are the wave vectors of the scattered and incident beams, respectively, and $\mathbf{v}$ is the velocity of the flow. For the geometry illustrated in Figure 8.1, this yields

$$f_D = \frac{2}{\lambda}\cos\left(\frac{\theta}{2}\right) v \cos\beta \qquad (8.2)$$

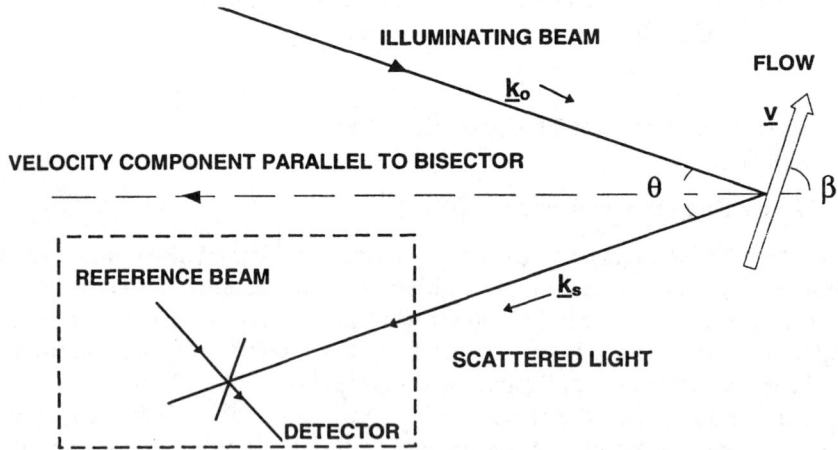

Fig. 8.1 Geometry for determining the Doppler shift imposed on light scattered from a moving particle. The components shown in the box are used for the reference beam anemometry configuration.

where λ is the wavelength of the illuminating light, θ is the angle between the illumination and collection directions, and $v\cos(\beta)$ the velocity component parallel to the bisector of the illumination and collection directions. In this section the principles of LDV will be established, and the operation, attributes and limitations of the predominant optical configurations outlined.

(a) Reference beam LDV

Reference beam LDV was the first reported laser anemometry configuration [4]. A single beam illuminates the flow, and light scattered by particles entrained within the flow is collected at an angle θ with respect to the illumination direction, as illustrated in Figure 8.1. A reference beam, derived from the same laser source, is mixed with the scattered light on the surface of a photodetector, yielding an electrical signal oscillating at the beat frequency. The beat frequency, given by equation 8.2, is then a measure of the velocity component of the flow parallel to the bisector of the illumination and collection directions. Working in backscatter allows the measurement of the velocity component along the axis of the instrument.

While the reference beam LDV configuration was used to measure flow velocities in real engineering applications, and formed the basis of the first three-dimensional LDV instrumentation [9], it suffers from a number of limitations which hinder its practical implementation, including the dependence of the measured frequency shift upon the viewing direction, poor signal noise to characteristics due to mismatch in the ratio of signal to reference beam powers, stringent alignment conditions, requiring alignment of the signal and reference beams to better than a few minutes of arc, and a limited collection aperture due to coherence considerations [10]. Section 8.5.2(a) will discuss ways in which the use of optical fibers allow many of these limitations to be overcome.

(b) Doppler difference LDV

The development of the Doppler difference technique [5, 6] led to the reference beam geometry being largely abandoned. In the Doppler difference technique two mutually coherent beams are crossed at their beam waists in the flow, as shown in Figure 8.2. Scattered light collected in any direction will thus contain components which have experienced different Doppler shifts by virtue of the different illumination directions of the two beams.

The Doppler frequency shifts, f_{D1} and f_{D2}, imposed on light scattered from the two beams, are given by

$$f_{D1} = \frac{1}{2\pi}[\mathbf{k}_{s1} + \mathbf{k}_{01}] \cdot \mathbf{v} \qquad (8.3)$$

$$f_{D2} = \frac{1}{2\pi}[\mathbf{k}_{s2} + \mathbf{k}_{02}] \cdot \mathbf{v} \qquad (8.4)$$

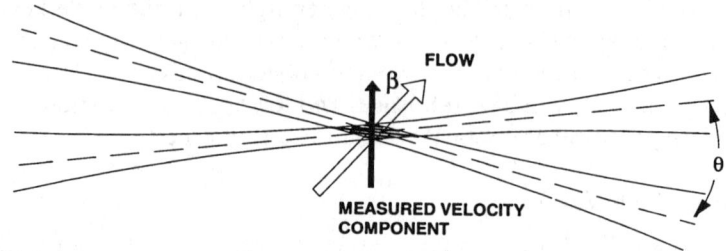

Fig. 8.2 The measurement volume of the Doppler difference laser anemometry technique.

Considering the superposition of these two waves on a photodetector, the beat frequency is given by

$$f_D^* = |f_{D1} - f_{D2}| = \left|\frac{\mathbf{v}}{2\pi}(\mathbf{k}_{s1} + \mathbf{k}_{01} - \mathbf{k}_{s2} - \mathbf{k}_{02})\right| \quad (8.5)$$

i.e. for the configuration illustrated in Figure 8.2, the measured frequency is given by

$$f_D^* = \frac{\mathbf{v}}{\pi}(\mathbf{k}_{01} - \mathbf{k}_{02}) = \frac{2v\cos(\beta)}{\lambda}\sin\left(\frac{\theta}{2}\right) \quad (8.6)$$

This may also be interpreted in terms of interference fringes formed at the intersection region. The crossing beams form an interference pattern, with fringe spacing, Λ, given by

$$\Lambda = \frac{\lambda}{2\sin(\theta/2)} \quad (8.7)$$

The light scattered by a particle traveling through the intersection region at angle β to the normal to the fringe planes will be amplitude modulated at frequency

$$f_D^* = \frac{v\cos(\beta)}{\Lambda} = \frac{2v\cos(\beta)}{\lambda}\sin\left(\frac{\theta}{2}\right) \quad (8.8)$$

This result illustrates a number of the key features of the Doppler difference technique: the measured frequency is independent of the observation direction, and is dependent only upon the geometry of the interfering beams. Alignment and beam ratio requirements are transferred to the illumination section of the system, where they may be more easily satisfied and optimized. Large collection apertures may be employed without broadening the spectrum of the signal. The scaling factor, governed by the illumination geometry, allows the measurement of high speed flows using conventional electronic signal processors.

The measurement region created in the Doppler difference system takes the form of an ellipsoid. The dimensions are normally defined in terms of the $1/e^2$ contours of the intersecting Gaussian beams. The length of the measurement volume is generally much less than that produced in the reference beam geometry, which is equal to the Rayleigh range of the focused beam.

A typical Doppler signal produced by a single particle transit through the measurement volume is shown in Figure 8.3. It consists of a cosinusoidal modulation at the Doppler frequency, sitting under a Gaussian envelope on a Gaussian pedestal.

It should be noted that the heterodyning process gives only the absolute frequency difference between the two waves being heterodyned, it does not indicate whether the Doppler shift is an upshift or downshift in comparison with the reference frequency. Thus the velocity inferred from a Doppler frequency shift contains a 180° directional ambiguity. For some flow systems the mean flow direction is known. However, for many engineering flows the direction of the flow is one of the unknown system parameters. The directional ambiguity may be overcome by imposing a frequency bias, such that the Doppler shift is no longer an even function of velocity [11]. The effect of the frequency bias is indicated by Figure 8.4. The frequency bias is imposed by shifting the optical frequency of one of the heterodyned beams. This may be achieved by passing one of the beams through an acousto-optic frequency shifter, or Bragg cell, via nonlinear optical processes such as stimulated Brillouin scattering, or using pseudo-heterodyning techniques borrowed from interferometric signal-processing schemes. Methods for producing a frequency bias will be discussed in section 8.4.

8.2.2 Laser transit anemometry (LTA)

The laser transit anemometer (LTA) [12, 13], or laser-dual focus anemometer (L2F), relies on the timing of the passage of a seeding particle between two

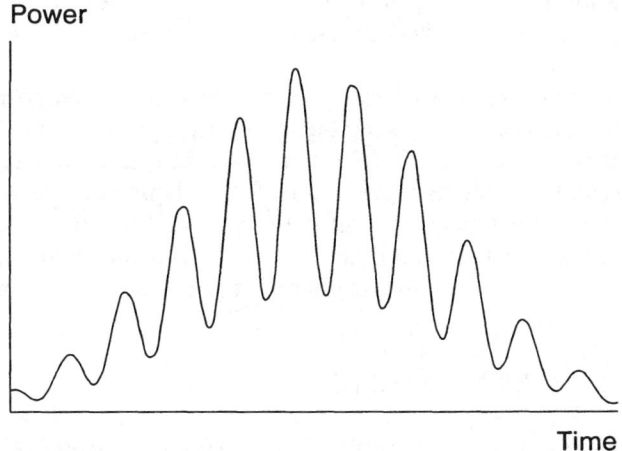

Fig. 8.3 A typical laser Doppler signal produced by a particle traveling through the measurement volume.

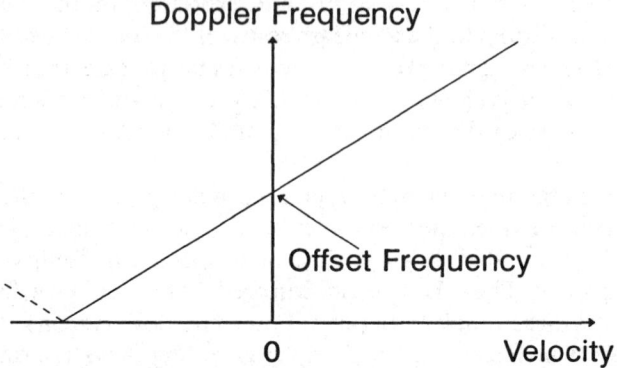

Fig. 8.4 This graph illustrates the effect of imposing an offset frequency on one of the beams in an LDV geometry to allow the determination of the direction of motion of the particle through the measurement volume.

discrete beams which are tightly focused into the flow region as illustrated in Figure 8.5. The beam diameter is typically of the order of 10 μm, and the spacing 0.5 mm. The passage of the particle produces two successive pulses of scattered light. Knowledge of the separation of the beams allows the velocity to be calculated from the time interval between pulses. To obtain the sequential pulses the plane defined by the two beams has to be rotated to lie in the direction of the flow, allowing the flow angle to be determined with resolution of $\pm 1°$. In turbulent flows sequential pulses may be obtained from two different particles passing through the beams. In general, the separation of such pulses is randomly distributed, with the peak in the distribution corresponding to the real measurement. Determining the direction of the flow relies upon knowledge of the order in which the beams were crossed. This may be achieved using two beams of differing polarization or wavelength to form the measurement volume, and two detectors with polarizers or wavelength filters to identify the source of the scattered light.

Since LTA does not rely upon heterodyning, there are no temporal coherence requirements for the source, allowing the use of high power, multi-longitudinal mode laser diodes. The use of tightly focused beams allows signals to be obtained from small seeding particles, which follow high speed flows more faithfully. The resultant short measurement volume allows measurements to be performed near surfaces, where the light scattered from the solid surface would introduce significant signal to noise problems for LDV configurations.

8.3 OPTICAL FIBER PROPERTIES

Single mode optical fibers are capable of preserving the coherence, phase, and, if highly birefringent, the state of polarization of the optical wave. These properties are required for all LDV systems based upon interferometric configurations.

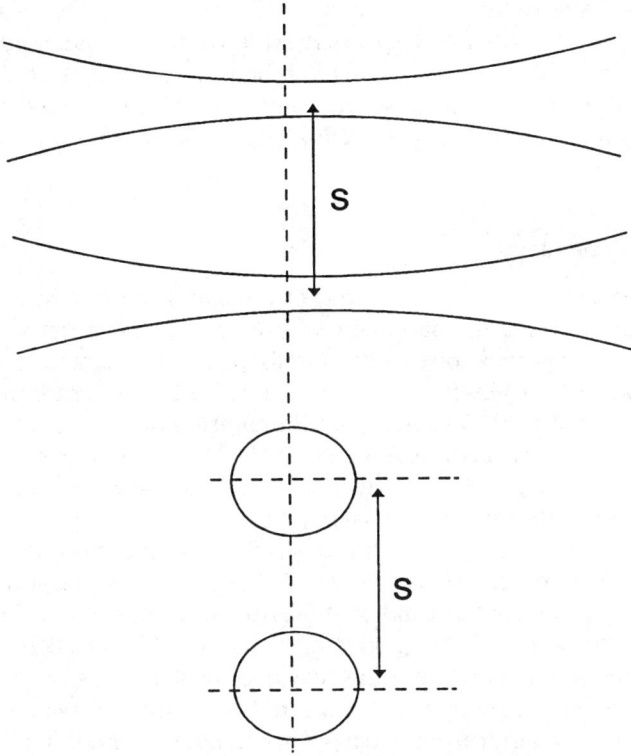

Fig. 8.5 Measurement volume of a laser transit anemometer; s is the beam separation.

In addition, for LTA, highly focused beams are required with well-defined (Gaussian), stable intensity distributions to form the measurement volume, again implying the use of single mode fibers, although polarization preservation is not necessarily required, unless polarization discrimination of the received light is being employed. Receiving fibers can be multimode for both LDV and LTA systems, since it is only intensity information that is required. One important exception is the reference beam anemometer (sections 8.2.1(a) and 8.5.2(a)) in which the received light is interferometrically combined with a reference beam. Hence, for this configuration, single-mode or highly birefringent single-mode fiber is employed.

Care in the choice of multimode fibers must be exercised, to ensure that dispersion effects in the fiber do not broaden the Doppler spectrum. In general, for the short lengths of fiber employed, typically 10 m, step-index fiber of 50–100 μm core diameter is suitable for the bandwidth of the signal processors (<150 MHz). Even for the pulsed laser techniques used in time-division-multiplexed multi-velocity component systems, discussed in section 8.5.3(b), the performance of 50 μm core diameter step index fiber is adequate [14].

Single-mode and multi-mode optical fibers are available for wavelengths in the range 200 nm to >10 μm. Light scattering from the small seeding particles (0.1–10 μm) used in LA applications is much more efficient at shorter wavelengths [15]. The combination of source, detector and fiber technology currently available results in most LA systems operating in the wavelength range 480 nm–1.3 μm.

8.3.1 Power handling

At short wavelengths, 488–532 nm, one of the major limitations on the delivery of optical power to the measurement volume by optical fibers is the slow deterioration in transmitted power when hundreds of milliwatts to several watts of continuous wave (c.w.) laser power are coupled into the fiber core. This effect can be partially reversed by subsequent illumination of the fiber but the original attenuation characteristics are not regained. This phenomenon is associated with Germania in the core, used to raise the effective refractive index of the fiber, interacting with the incident laser power to form color centers [16] and is thought to be due to absorption in the ~240–260 nm region, and is therefore associated with two photon processes at the pump wavelengths used. This ability to modify the refractive index of the fiber has led to a new class of in-fiber components, the so-called Bragg grating (Chapters 10–12). The effect can be severe, causing a reduction in transmitted optical powers. For example, an attenuation increase from 60 to 130 dB km^{-1} has been observed for ~100 m of single-mode highly birefringent fiber (York HB450) containing 5% germania [17]. Fibers are now available with very low levels of germania dopant, e.g. Fujikura Panda fiber has 1% germania content, allowing c.w. powers of several watts to be efficiently transmitted over short lengths of fiber (10–20 m). This photorefractive induced loss has not been reported for operation at longer wavelengths (600 nm–1.3 μm) and is not expected at the power levels used.

The second limitation of power transmission is imposed by the onset of non-linear stimulated scattering processes in the fiber; stimulated Raman scattering (SRS) and stimulated Brillouin scattering (SBS) (SBS is discussed further in section 8.4.2). The threshold for SRS is too high to be achieved with c.w. lasers, however the threshold for SBS is much lower, particularly when the linewidth of the source is less than the order of the SBS linewidth. This is the situation found for argon-ion lasers using an intra-cavity etalon to achieve single longitudinal mode operation (~3 MHz linewidth), and for high power semiconductor laser diodes (~10 MHz linewidth), and frequency-doubled diode-pumped YAG lasers (~1 kHz linewidth). This nonlinear interaction can be considered as the input optical beam creating a traveling acoustic wave, via electrostriction (high electric field intensity modulating the refractive index, altering the local fiber strain and thus changing the stress-optic coefficients), and subsequently scattering the pump light. In order to satisfy energy and momentum conservation the SBS wave propagates back along the fiber with an optical frequency lower than the pump and is often called a Stokes wave.

The threshold power, P_{th}, is usually defined as the input pump power that produces an SBS power at the fiber input equal to the pump. An approximate expression predicting P_{th} from fiber properties is given by [18]

$$P_{th} \approx \frac{21A}{L_{eff} g_B(\nu)} \qquad (8.9)$$

where A is the effective overlap between the electric field distribution of the pump and the SBS waves, and L_{eff} is the effective interaction length between the pump wave and the fiber core; $g_B(\nu)$ is related to the physical and optical properties of the fiber, the pump wavelength and the SBS linewidth [19]. For silica g_B is $\sim 5 \times 10^{-11}$ mW^{-1} [20] and has been found to vary from 2 to 4×10^{-11} mW^{-1} in fiber [17]. In practice P_{th} predicted by equation 8.9 agrees very well with the pump power required to experimentally observe the onset of SBS.

The effective interaction length, in kilometers, is given by

$$L_{eff} = \frac{10^{0.1aL} - 1}{10^{0.1aL} \ln(10^{0.1a})} \qquad (8.10)$$

where L is the actual fiber length (in kilometers) and a is the fiber loss (attenuation) in dB km^{-1}.

As an example [21], for an 800 m length of single-mode fiber and an attenuation of 27.4 dB km^{-1} at the pump wavelength of 514.5 nm ($L_{eff} = 160$ m), P_{th} was found to be 20 mW. At a launched power of 110 mW, $\sim 41\%$ conversion of the pump light into backward-propagating SBS was obtained. Although potentially useful for frequency shifter implementations (discussed in section 8.4.2) it is obviously a potential problem for power delivery. From equations 8.9 and 8.10, the P_{th} for shorter fiber lengths can be estimated. For a 25 m length of fiber ($L_{eff} = 23$ m), P_{th} is ~ 150 mW, while for a 10 m length ($L_{eff} = 9.97$ m), $P_{th} = 350$ mW. Many high power fiber delivered LDV systems currently use argon-ion lasers without intra-cavity etalons, resulting in multi-longitudinal mode operation. For this case the laser produces higher output power, but each line generally, contains insufficient power to achieve the SBS threshold. However, if sufficient power is available, several lines can produce SBS, which can then combine to produce multiple frequency components via the parametric process of four-wave mixing [22].

The onset of SBS can also be prevented by decreasing the gain coefficient and by increasing the fiber core size, hence reducing the power density. The Panda fiber that has reduced Germania concentration, to suppress color center formation, has a lower gain and significantly higher SBS threshold than many other single-mode fibers operating in the region 480–520 nm [17].

8.4 DIRECTIONAL DISCRIMINATION

To obtain complete flow information, particularly in complex flows with time-varying velocities, requires that the laser anemometer is able to distinguish the flow direction. This is straightforward to achieve for a laser transit configuration as the order in which the signals are received at the detectors is known. However, for Doppler systems only a single detector is used for each velocity component. With reference to Figure 8.2, it is clear that the direction a particle has traveled through the fringe volume is not obtained from a single detector; particles traveling either top to bottom or bottom to top produce the same frequency modulation at the detector. However, a constant frequency offset can be achieved by making the fringes move at constant velocity thus providing a means to determine the flow direction; particles of velocity v traveling in the same direction as the fringes produce a lower frequency signal than particles of velocity v traveling in the opposite direction. In addition to enabling direction discrimination this frequency-shifting technique allows higher resolution at small Doppler frequencies, by shifting the measurement from d.c. to a higher frequency.

8.4.1 Bragg cell

A conventional optical component that is used to achieve the frequency shifting is called a Bragg cell. In this device (Figure 8.6) a traveling acoustic wave sets up alternate regions of high and low density in the crystal causing a concomitant periodicity in the refractive index, via mechanical strain (electrostriction) [23]. This periodic index modulation acts as a diffraction grating which moves through the crystal at the acoustic wave velocity. The moving grating produces the frequency shift on the diffracted optical beam. Bragg cells produce frequency shifts from ten to hundreds of MHz. Lower frequencies can be achieved by using two Bragg cells with a slight frequency difference. For example, two Bragg cells operating at 80 and 81 MHz will result in a carrier at 1 MHz.

Bragg cells are still used extensively in laser Doppler systems as they can operate with high power lasers, e.g. argon-ion, and over a range of optical

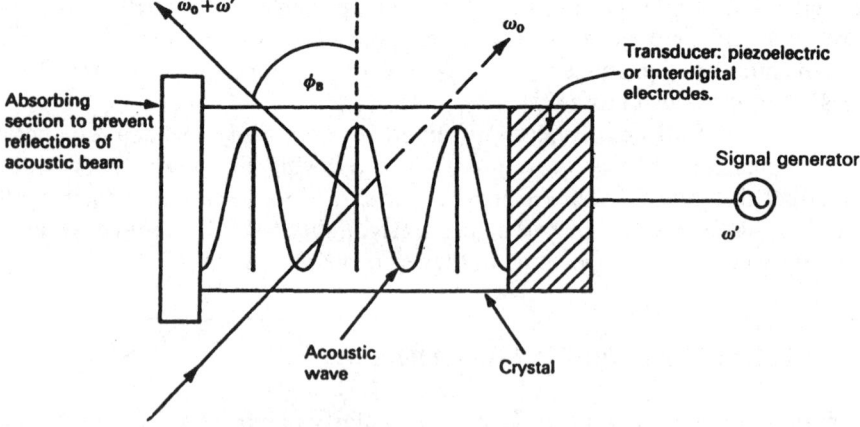

Fig. 8.6 Schematic of Bragg cell.

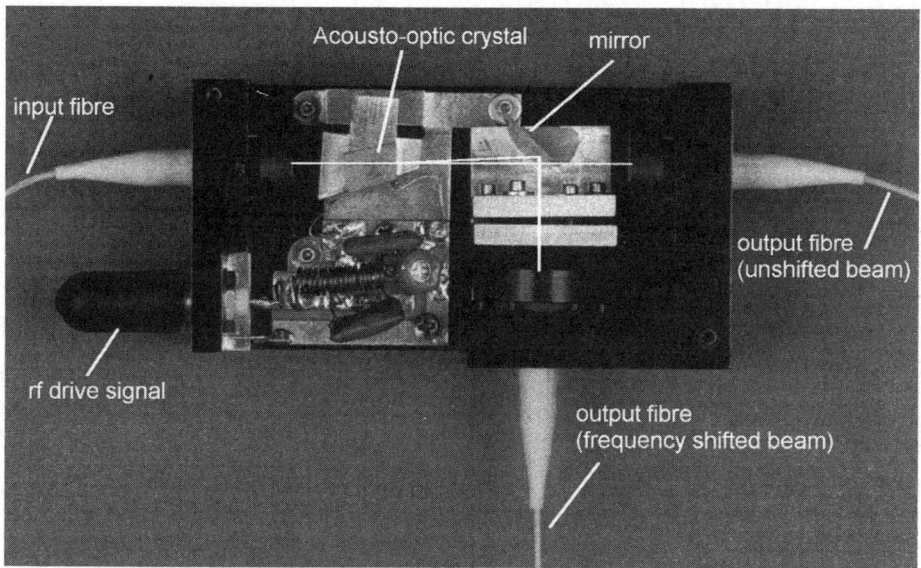

Fig. 8.7 Pigtailed Bragg cell (device courtesy of Gooch and Housego, Dorset, UK).

wavelengths (~400 nm – 1.5 µm). The devices generally require r.f. powers of several watts. The major limitation when combined with single-mode optical fiber is the very high optomechanical stability required between the fiber and the device. Recently this problem has been partially alleviated by incorporating optical fiber, micro-optics and the Bragg cell into one unit, often called pigtailed (Figure 8.7). The linearly polarized output from the input birefringent fiber is collimated and coupled into the Bragg cell. This frequency shifter, a shear wave device, produces the shifted and unshifted beams in orthogonal linear polarization states. The two output beams are coupled into two separate birefringent fibers, using a polarizing beam splitter and micro lenses, aligned such that an eigenaxis in each fiber is parallel to the linear output polarization states. The coupling efficiency is ~60%. This is effectively an 'engineering' solution to improve the stability and practicality of the system. However, in order to further decrease the losses and reduce the optomechanical stability and power requirements, a number of alternative methods, discussed in the following section, have been investigated.

8.4.2 Optical fiber frequency shifters

Extensive research effort has been aimed at developing an in-fiber frequency shifter. These devices can be divided into two classes, **extrinsic** in which the traveling acoustic wave is generated externally to the fiber and coupled to it by a transducer and **intrinsic**, in which the traveling acoustic waves are generated internally within the optical fiber. These two classes of device are discussed in detail in ref. [24] and therefore only recent developments that could be or have been applied to laser velocimetry will be discussed here.

(a) Extrinsic devices

Extrinsic fiber frequency shifters utilize the optical frequency shift that occurs when light is coupled from one propagation mode to an orthogonal mode by a traveling acoustic wave. These modes can be polarization [25, 26], spatial [27, 28] or the modes of fibers with dual cores [29]. At present only one of the devices has been commercialized [30] based on coupling between the two lowest order spatial modes ($LP_{01} \Longleftrightarrow LP_{11}$) of a two-mode fiber [27]. However, although the principle of operation of the device is the same at all wavelengths, the availability of fibers with appropriate characteristics combined with much larger market potential in communication systems has restricted the availability to 633 nm, 800 nm, 1.3 μm and 1.55 μm. The device has a quoted frequency shift of 3–5 MHz. This is suitable for LDV applications where velocities of up to a few tens of meters per second are to be measured. Lower frequency shifts can be obtained using two frequency shifters with slightly different center frequencies. However, many LDV systems employ 40 or 80 MHz or even higher frequency shifts and to date an extrinsic fiber frequency shifter has not been demonstrated at these higher frequencies.

A recently presented fiber frequency shifter technique [31] has the potential to achieve higher frequency operation (hundreds of MHz). In this device two fibers with different propagation constants are used to form a fused tapered coupler, that is, the two fibers are laid parallel to each other heated and stretched. The geometry is shown in Figure 8.8. Since the fibers are so mismatched none of the optical power input in one fiber is coupled into the other and all the power emerges from the same fiber as the input power. However, a traveling flexural acoustic wave can cause resonant coupling between the fundamental and second mode of the coupler waist resulting in the light being coupled out of one fiber into the other, with a concomitant frequency shift on the

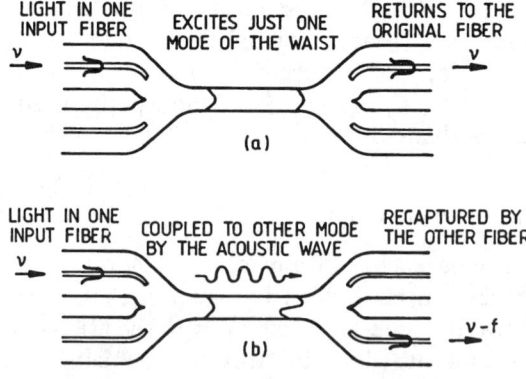

Fig. 8.8 Fiber frequency shifter based on a null coupler (after [31]).

coupled beam. If the beat length of the two modes matches the acoustic wavelength then a frequency shift occurs on the coupled beam. A device has been fabricated for use at 633 nm and a frequency shift of ~1.8 MHz obtained. The device is potentially attractive for fiber optic LDV as no mode converters or filters are required, power requirements are low (<1 mW) and the magnitude of the frequency shift can in principle be substantially increased by appropriate design of the coupler waist.

(b) Intrinsic devices

Intrinsic optical fiber frequency shifters are based on the generation of stimulated Brillouin scattering (SBS) within the optical fiber [24]. The SBS frequency shift in an optical fiber is given by [32]

$$\nu_{SBS} = \frac{2nV_a}{\lambda_p} \qquad (8.11)$$

where λ_p is the input (pump) optical wavelength, V_a is the acoustic velocity and n the effective refractive index of the fiber core. Table 8.1 shows the typical SBS frequency shift as a function of pump wavelength.

The linewidth varies inversely with pump wavelength and is typically 100–150 MHz for a 514.5 nm pump. The frequency shift achieved (~10–35 GHz) is generally too high to be used directly in LDV applications where signal processors typically have maximum bandwidths of 100–150 MHz. A low frequency carrier can be produced by mixing two SBS signals of slightly different frequency. The two SBS signals are generated using different effective refractive indices for each signal. This has been achieved by using separate fibers with different refractive indices [36] such that the beat frequency can be tuned by changing the temperature, and hence the refractive index difference, between the two fibers. For this system using two 500 m reels of fiber a beat frequency of ~750 MHz and linewidth of <1 MHz was achieved for an argon-ion laser pump (514.5 nm). A temperature sensitivity of ~4 MHz °C^{-1} was measured. The two polarization eigenmodes of 800 m of a single birefringent fiber have been used to reduce the beat signal to ~10 MHz [21]. Optical fiber ring resonators [37, 38] have been used to reduce the lengths of fiber required and reduce the power requirements to produce SBS from tens to hundreds of mW to tens to hundreds of μW. However, ring resonator systems do not produce sufficient power for many LDV applications, being limited to a few tens of mW.

Table 8.1 Typical SBS frequency shifts, ν_{SBS}, as a function of pump wavelength, λ_p

λ_p (nm)	ν_{SBS} (GHz)	Ref.
1.3	13	33
0.8	22	34
0.6328	27	35
0.5145	34	22

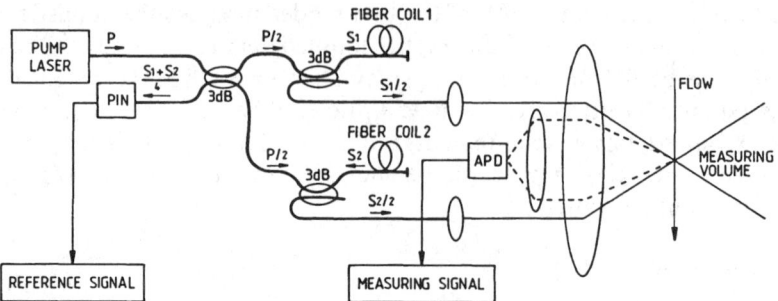

Fig. 8.9 Schematic of LDV using SBS frequency shifters (after [39]). P = pump power, S = SBS power.

SBS has recently been demonstrated as a technique to generate an optical heterodyne carrier for LDV [39] (Figure 8.9). The pump laser is an Nd:YAG MISER (monolilthic isolated single-frequency end-pumped ring laser), operating at 1319 nm, with a linewidth of < 100 Hz [40]. The laser output (100 mW) is coupled into two fiber coils each approximately 20 km long, via a Faraday isolator (used to prevent feedback into the laser cavity). The two different fibers used generated 10 mW of SBS with a frequency difference of ~18 MHz. This beat frequency shows stochastic jitter with a bandwidth of ~10 MHz. To remove this jitter from the demodulated signal requires that part of the SBS signal is used synchronously as a reference signal. This reference signal is divided and used to produce quadrature components (sine and cosine) which are both mixed with the measurement signal. This method eliminates the effect of large SBS frequency fluctuations and enables directional discrimination.

In principle this method could be used at lower wavelengths using high power single-mode frequency-doubled YAG lasers (532 nm) which have output powers up to 5 W [41]. However, the fiber photosensitivity at this wavelength would need to be investigated, as this would still limit the maximum effective interaction length possible.

8.4.3 Pseudo-heterodyne processing

In the previous sections the heterodyne carriers were produced by mixing two slightly different optical frequencies. An alternative technique is to synthesize a heterodyne carrier by changing the phase of the interferometer in a controlled manner. This can be achieved using a piezoelectric modulator (PZM) tightly wrapped with optical fiber. Applying a voltage to the PZM causes it to expand or contract with a concomitant change in the optical path length in the fiber [25, 42]. Figure 8.10(a) demonstrates the method with a serrodyne (sawtooth) waveform applied to the PZM. The resulting interferometric fringes are shown in Figure 8.10(b). Notice the flyback at the end of the ramp which causes the fringes to move in the opposite direction very rapidly. This is where the term

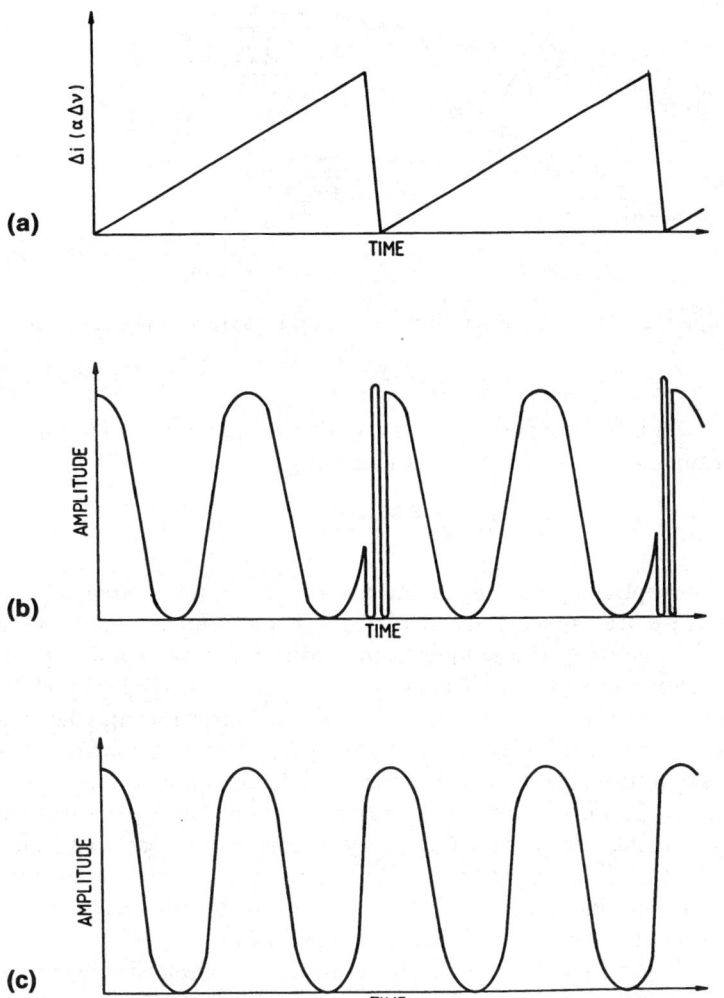

Fig. 8.10 Pseudo-heterodyne processing using either phase modulation or wavelength modulation: (a) serrodyne modulation; (b) interferometric fringes; (c) bandpass filtered version of (b).

'pseudo-heterodyne' comes from, that is, there is not a continuous change in the phase. This carrier can be bandpass filtered to remove the higher harmonics associated with the ramp flyback (Figure 8.10(c)). This method has been employed in LDV [43] but is limited to low velocities since the frequency shift achievable is generally <200 kHz.

A variation of this technique that allows higher frequency modulation is to combine the wavelength modulation capabilities of laser diodes with a path

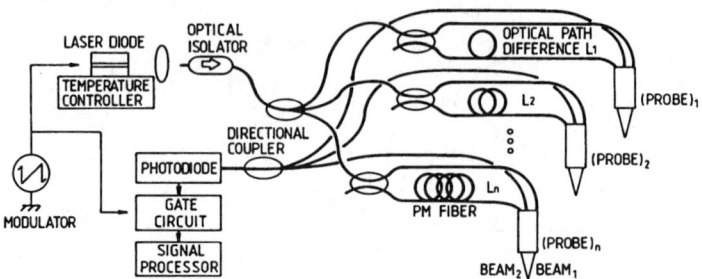

Fig. 8.11 Multiplexed LDV using laser diode wavelength modulation (after [48]).

length imbalanced interferometer. The phase difference between the two paths in the interferometer, $\Delta\phi$, varies linearly according to

$$\Delta\phi = \frac{(2\pi \Delta L \Delta v)}{c} \quad (8.12)$$

where ΔL is the optical path length imbalance, Δv is the optical frequency change and c is the free space velocity of light. The output wavelength can be varied slowly by changing the temperature of the diode or rapidly by modulating the injection current. Application of a ramp modulation signal to the injection current produces a concomitant linear variation in the optical frequency [44, 45]. The applied modulation and output waveforms are shown in Figure 8.10. Limitations of this technique include the concomitant intensity modulation that accompanies injection current modulation. This effect can be minimized by keeping the optical frequency change small, $<5-10\,\text{GHz}$, but this then implies a path length imbalance of several centimeters to achieve a 2π phase change which increases the phase noise in the interferometer. For example, to obtain a 2π phase shift for a 5 GHz optical frequency modulation requires a path length imbalance of $\sim 6\,\text{cm}$. This requires a relatively small injection current modulation, typically 1 mA, and therefore the intensity change is only a few percent. This technique has been demonstrated for LDV measurements [46, 47] with a multiplexed system recently reported [48]. In this system (Figure 8.11) the multiplexing was achieved by producing a different carrier frequency for each channel by using a different ΔL for each measurement channel. Laser diodes can be modulated at much higher frequencies than PZMs, with serrodyne frequencies $>100\,\text{MHz}$ reported.

8.5 OPTICAL FIBER IMPLEMENTATIONS

Fiber optic links are incorporated in laser anemometry instrumentation for a number of reasons. They may be used simply as a light pipes, to deliver the illuminating light to the flow, and collect the scattered light and deliver it to the

OPTICAL FIBER IMPLEMENTATIONS

detectors. This allows the laser source and detectors to be remote from the often harsh environment surrounding the flow region and allows the construction of a compact, robust and lightweight probe head, which is easy to traverse to map out the flow field. The fibers perform a spatial filtering operation, producing high quality beams in the measurement volume. The use of fiber amplifiers has been proposed to increase the power in the measurement volume, and to amplify the collected scattered radiation [49]. The fibers may also be used to perform some of the signal processing and demultiplexing for multi-component configurations. This section will describe how optical fibers have been utilized in LA instrumentation.

8.5.1 Laser transit anemometry

Figure 8.12 shows a fiber-based laser transit anemometer [50]. The distal ends of two lengths of polarization-maintaining (PM) fiber are positioned side by side and rotated to align their eigenaxes. The optical system produces a measurement volume consisting of two 14 µm beam waists separated by 320 µm. The scattered light is imaged onto the cleaved ends of two 50 µm core multi-mode fibers. The configuration allows the use of polarization or wavelength discrimination by the inclusion of appropriate optics and sources.

A multi-component fiber optic laser transit anemometer was developed by Schodl [51], and commercialized by Polytech GmbH. The optical configuration for a two-dimensional (2D) system is shown in Figure 8.13. The output from a multi-line argon-ion laser is launched into a single-mode fiber. The output from the fiber is collimated and incident upon a dispersing prism to separate the colors. The five resultant beams are focused to produce a set of parallel beams with differing spacing. Backscattered light is collimated and passed back

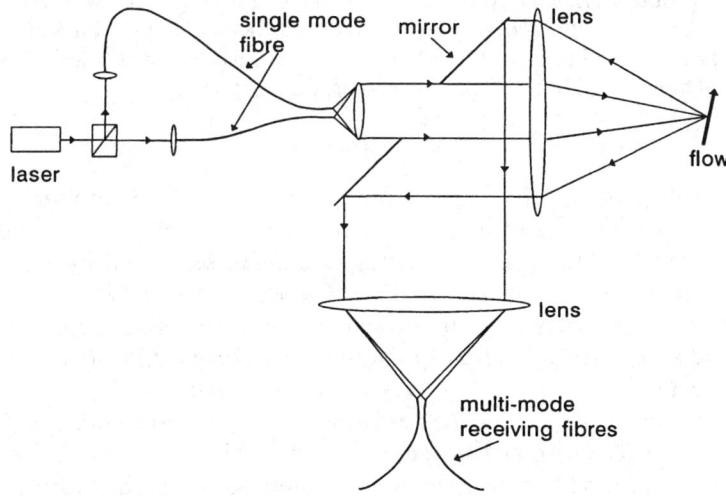

Fig. 8.12 A fiber optic laser transit anemometer (after [50]).

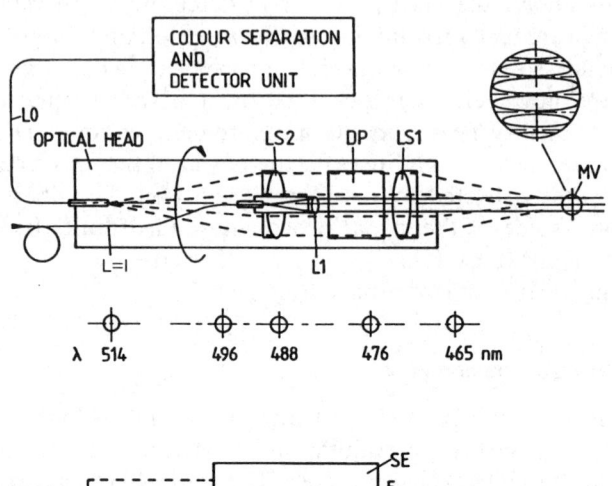

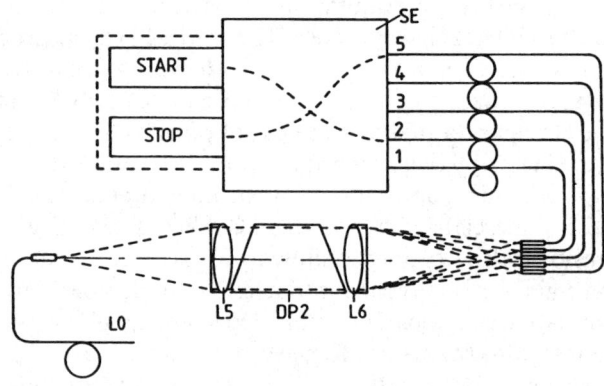

Fig. 8.13 Two-dimensional laser transit anemometer (after [51]). The lower diagram shows the details of the colour separation and detector unit llustrated above. MV = measurement volume, LS1, LS2, L1, L5 and L6 = lenses, L0 and LI = single-mode optical fiber, DP, DP2 = dispersive prisms, SE = switching unit.

through the dispersing prism to recombine the colors. The resultant beam is launched into a multi-mode fiber, which delivers the scattered light to a detector unit, as shown in Figure 8.13. The colors are again separated by a dispersing prism, and launched into a set of optical fibers which guide the light to a switching unit. The switching unit allows two arbitrarily chosen colors to act as the start and stop signals. Each color combination corresponds to a well-defined beam separation. The use of variable beam separation in LTA systems has been shown to be necessary to allow the performance of measurements in highly turbulent flows [51]. The compact system was mounted inside the shaft of a stepper motor, which was used to rotate the whole system to adjust the beam's plane to be parallel to the flow direction.

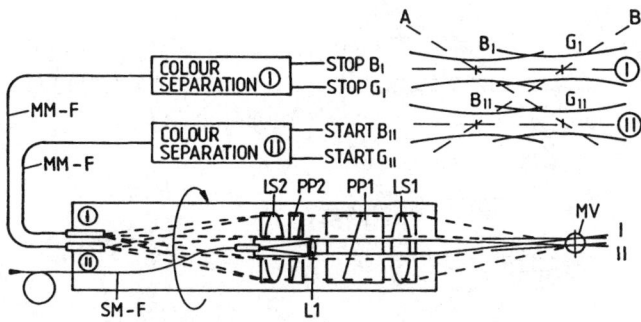

Fig. 8.14 Three-dimensional laser transit anemometer (after [51]). LS1, LS2, L1 = lenses, MM-F = multi-mode fiber, SM-F = single-mode fiber, PP1, PP2 = Wollaston prisms, MV = measurement volume.

A 3D version was described, in which the focusing lens on the probe has a degree of chromatic aberration. The effect of the chromatic aberration on the measurement volume is illustrated in Figure 8.14. For a given pair of beams, chromatic aberration produces an axial displacement of the focal position. When the flow is parallel to the connecting line of the focal points a maximum in the data acquisition frequency is obtained. The 3D information is obtained from two measurements, made with a 180° rotation of the beam plane. The system shown in Figure 8.14 allows simultaneous measurements by using a combination of wavelength and polarization discrimination. A Wollaston prism is employed as a polarizing beam splitter to produce the beam configuration. A second Wollaston prism is used to separate the orthogonally polarized elements of the scattered light, and the two beams launched into multi-mode fibers. Two color separation units then produce the four beams corresponding to the four focal regions in the measurement volume, allowing the 180° separated measurements to be made simultaneously. A major application of this instrument is the measurement of flows in gas turbines.

8.5.2 Fiber optic laser Doppler velocimetry

(a) Fiber reference beam system

The development of the first fiber optic based LDV configuration was driven by the requirement for making blood flow measurements *in vivo* [52, 53], fibers offering the capability for flexible delivery and collection of light in otherwise inaccessible regions. The experimental configuration is shown in Figure 8.15. The output from an He–Ne laser passes through a hole in a mirror and is launched into a multi-mode optical fiber. The output from the multi-mode fiber illuminates the flow, and light scattered backward from seeding particles is collected by the fiber. The scattered light populates all of the modes in the multi-mode fiber, producing a large diameter beam on collimation. The reference

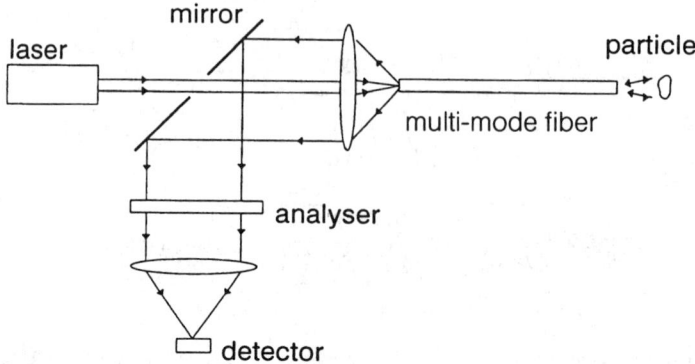

Fig. 8.15 Fiber optic reference beam anemometer (after [53]).

beam is derived from light reflected from the cleaved end of the fiber, automatically aligning the signal and reference beams for optimum heterodyning. By virtue of the numerical aperture (NA) of the fiber, and attenuation of the scattered light by absorption and secondary scattering, the fiber collects light scattered from particles within a few tens of core diameters from the fiber end.

The configuration of Figure 8.15 has the advantage of down-lead insensitivity, since the optical cavity is formed between the end of the fiber and the particle in the flow. In addition, the small fiber diameter, $\sim 125 \mu m$, does not significantly obstruct the flow. However, there is a stagnation region of several fiber diameters around the fiber end in which particles are traveling much slower than in the free stream flow. With highly coherent sources the origin of the returned signal is from particles < 1 mm from the fiber end. The configuration shown in Figure 8.16 [54] uses a low coherence length (l_c) source, typically with $l_c < 200 \mu m$, to illuminate the flow. The source could be an ELED, SLD or multi-mode laser diode. A receiving interferometer with a path length imbalance equal to the distance from the fiber end to the desired measurement volume in the flow is used ($2L$ in Figure 8.16). In this example a Bragg cell is used as a beam splitter and a frequency shifter to implement heterodyne processing and thus recover the flow direction.

Extending the range of operation of such configurations was achieved by terminating the end of the optical fiber with a graded-index rod lens [55], which collimates the output from the fiber to illuminate the flow, and collects and launches the backscattered light down the fiber. The reference beam is again derived from the 4% reflection from the fiber end.

The light efficiency of the design was improved by the use of PM single-mode fiber, as illustrated in Figure 8.17, allowing the use of a polarizing beam splitter in place of the trepanned mirror [56]. The output from a semiconductor laser diode is launched into the PM fiber along an eigenaxis. The tip of the fiber is terminated with a GRIN rod lens and quarter-waveplate. Light backscattered from the flow launched into the fiber having undergone a double pass through

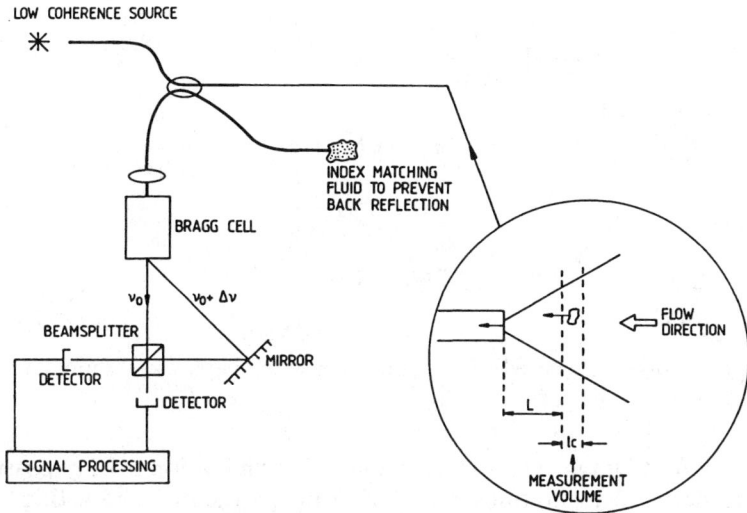

Fig. 8.16 Coherence-based reference beam LDV (after [54]).

the quarter-waveplate propagates along the orthogonal eigenaxis. Likewise, the polarization of light reflected from the outer surface of the plate, which acts as the reference beam, is rotated by 90°. The reference and signal beams then propagate back along the fiber and are reflected by the polarizing beam splitter and monitored on a photodetector.

Derivation of the reference beam from the 4% reflection from the cleaved fiber end considerably eases alignment issues, and provides common mode rejection; however, it does suffer from significant limitations. The working distance for such devices is limited to half the coherence length of the source, the

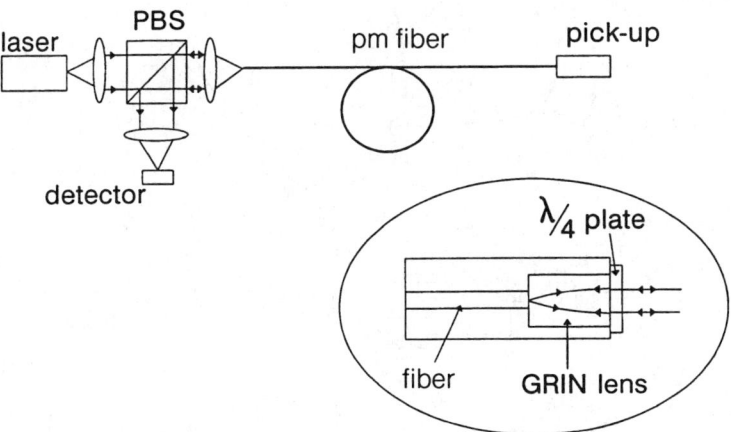

Fig. 8.17 Polarization-based fiber optic reference beam anemometer. PBS = polarizing beam splitter, pm = polarization-maintaining fiber (after [56]).

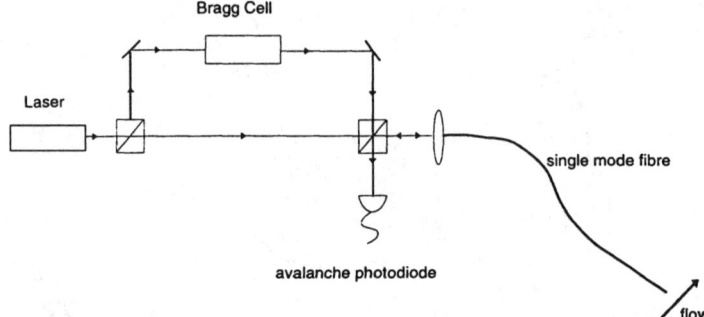

Fig. 8.18 Fiber optic reference beam anemometer with directional discrimination (after [57]).

incorporation of Bragg cells for directional discrimination is only possible for low coherence implementations, and there is no provision for adjusting the ratio of signal to reference beam powers for optimization of the heterodyne signal.

In Figure 8.18, the reference beam is derived using the beam splitter and is passed through a Bragg cell to facilitate directional discrimination. The signal beam is launched into single-mode optical fiber and the scattered light returned through the same fiber to be mixed with the reference beam on an avalanche photodiode [57].

Figure 8.19 shows a polarization-based fiber optic reference beam LDV configuration [58]. The output from a semiconductor laser diode is split at a polarizing

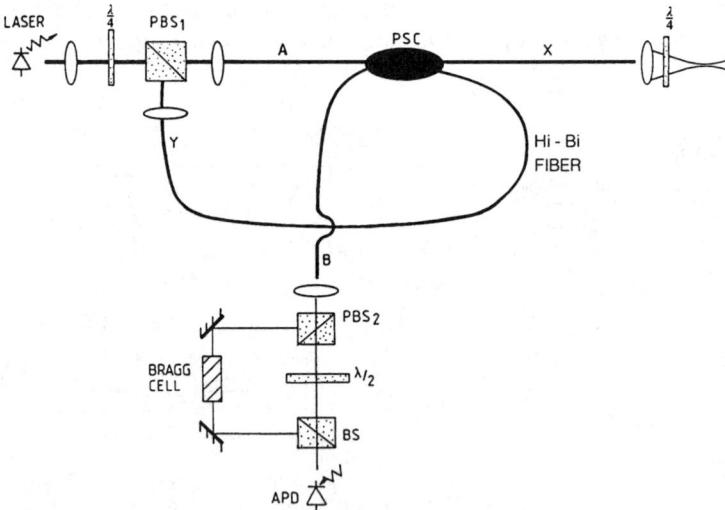

Fig. 8.19 Polarization-based reference beam laser Doppler anemometer (after [58]). PSC = polarization selective coupler, PBS = polarizing beam splitter, $\lambda/2$ = half-wave plate, $\lambda/4$ = quarter-wave plate, BS = beamsplitter.

beam splitter, PBS_1, to form the signal and reference beams. A half-waveplate is used to control the split ratio. The signal beam is launched into arm A of a polarization-splitting coupler, PSC (the fiber optic analog of a polarizing beam splitter) along the TE axis. The beam propagates through the PSC into arm X. Light scattered from the measurement volume, having undergone a double pass through the quarter-waveplate, is relaunched down the TM axis of the device, to be coupled across into arm B. The reference beam, launched along the TE axis of arm Y, propagates through the PSC to arm B. The output from arm B, consisting of the orthogonally polarized signal and reference beams, is collimated and split at PBS_2, the reference beam propagating through a Bragg cell, the signal beam through a half-waveplate to rotate its plane of polarization to coincide with that of the reference. The two beams are recombined at the beam splitter, and are heterodyned on an avalanche photodiode (APD). Removal of the optical signal processing to the receiving Mach–Zehnder interferometer improves the optomechanical stability of the system [59].

Alignment requirements for such a configuration could be completely removed via the use of all fiber components in the Mach–Zehnder interferometer, or by using an integrated optical circuit [60].

Environmental perturbations of the fiber propagation constants introduce phase noise to the signal, which has been shown to be a source of spectral broadening in fiber-based LDV configurations that lack common mode rejection. However, this environmental contribution to the broadening of the signal is found to be significant only at low mean velocities, below 1 m s^{-1} for a system with fiber leads 10 m long in an environment with an acoustic noise level of 100 dB. Transit time broadening becomes the dominant contribution as the mean velocity increases [61].

(b) Fiber Doppler difference systems

A number of fiber optic Doppler difference configurations have been reported using both single-mode fiber [62] and PM fiber [63]. A single-mode fiber may be used to deliver the illuminating beams to the probe head, where the two beams are derived and conditioned using bulk optic components, and a Bragg cell imposes a frequency shift on one of the beams for directional discrimination [64] (Figure 8.20). Alternatively, two lengths of single-mode fiber may be used to deliver the two beams to the probe head, removing the requirement for bulk optic components in the probe. Frequency shifting for these geometries may be achieved by employing pseudo-heterodyne techniques as illustrated in Figures 8.21 and 8.22, using a piezoelectric stretcher [65], or a path length imbalance and frequency-modulated laser diode [66]. The use of a PM fiber link ensures that the polarization of the interfering beams is maintained for maximum fringe visibility, and can produce a compact device, offering frequency shifting for directional discrimination and minimal down-lead sensitivity. Two linearly polarized beams are derived from the output from a laser, one of which is frequency shifted using a Bragg cell. The two beams are launched down orthogonal

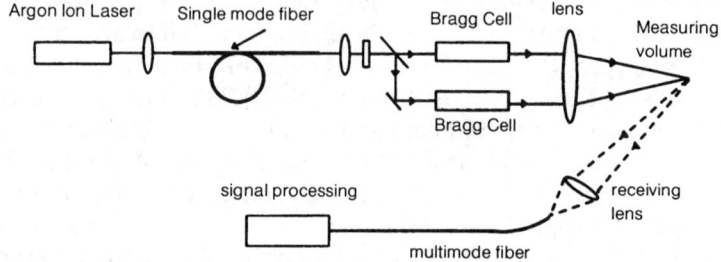

Figure 8.20 High power Doppler difference laser anemometer (after [64]).

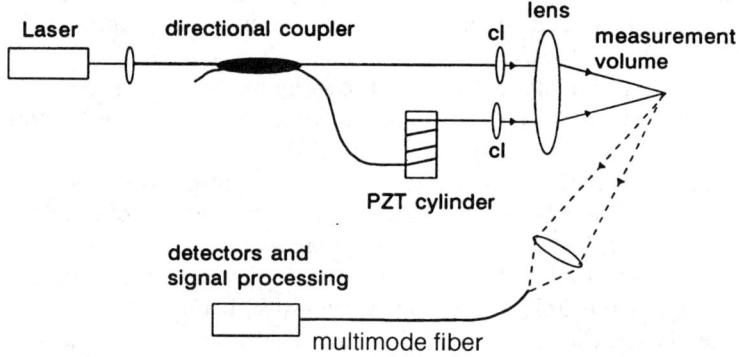

Fig. 8.21 Fiber optic Doppler difference laser anemometer using a fiber wrapped PZM to perform pseudo-heterodyning to permit determination of the direction of the flow. cl = collimating lens (after [66]).

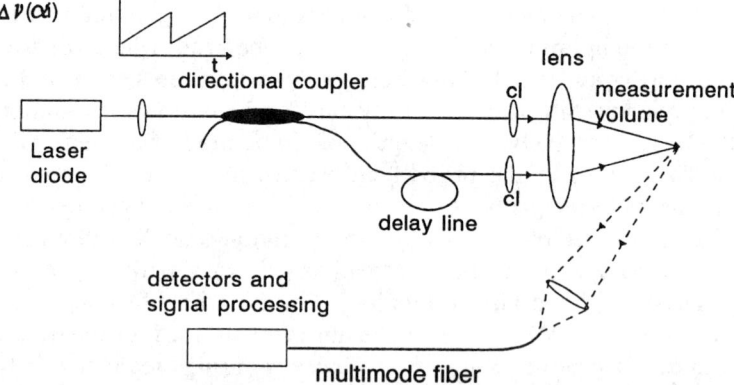

Fig. 8.22 Fiber optic Doppler difference laser anemometer using a frequency chirped laser diode and path length imbalanced interferometer to perform pseudo-heterodyning to permit determination of the direction of the flow. cl = collimating lens (after [66]).

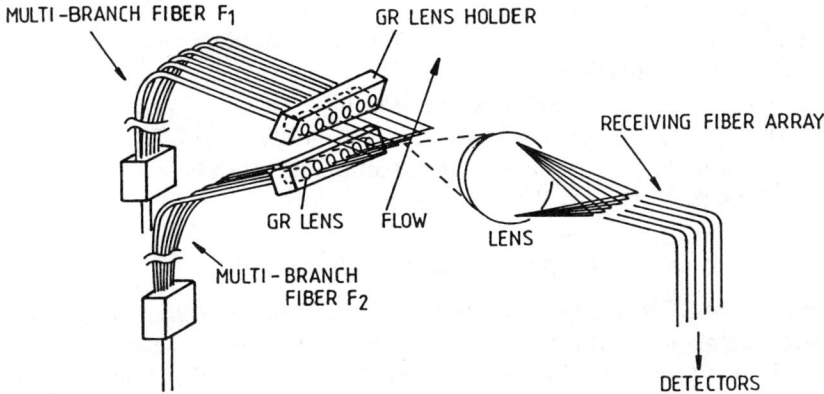

Fig. 8.23 Multi-point optical fiber LDV using fiber multi-branches (after [67]).

eigenaxes of a PM fiber. The shifted and unshifted light are split at the output of the fiber by a polarizing beam splitter to form the two beams for the Doppler difference channels. The scattered light is collected using a multi-mode optical fiber, which also acts as a spatial filter for the receiver. The use of a single fiber to deliver the two beams to the head provides a high degree of common mode rejection to environmental perturbations such as acoustic noise and temperature variations, and the incorporation of the Bragg cell before the fiber optic link ensures that no electrical power is required at the probe head.

To map an entire flow field using LDV instrumentation can be time consuming. Fiber optic systems allowing simultaneous measurements at several locations have been developed, such as that illustrated in Figure 8.23 [67]. The outputs from the two multi-mode fibers are each split into six beams at the fiber multi-branch. The beams are coupled into single-mode fibers, the outputs from which are collimated using GRIN lenses, and arranged to cross in the flow to form six discrete Doppler difference measurement channels. The scattered light is imaged onto an array of multi-mode fibers, arranged to each receive from a separate channel. Cross-talk between adjacent channels was found to be −10 dB.

8.5.3 Multi-component fiber optic laser Doppler velocimeters

Full characterization of the complex, 3D flows encountered in many engineering systems, for example turbomachinery, has long been the goal of research in flow measurement instrumentation. The potential of LDV for 3D flow measurement was realized soon after the first demonstration of the technique, with the first 3D LDV reported in the late 1960s. A large number of configurations have since been proposed and commercial instrumentation developed [68, 69]. Demands for improved fuel efficiency, reduced emissions and the requirement for high quality data to allow validation of complex computational fluid dynamic codes

286   FIBER OPTIC LASER ANEMOMETRY

is driving the development of new 3D LDV configurations. The use of optical fiber technology is key in the design and implementation of new multi-component LDV instrumentation.

The characterization of 3D flow requires the presence of three LDV channels in the flow, distinguished using wavelength [70–72], time [14, 73–75], frequency bias [71] or angularly [9]. The following section will discuss the implementation of optical fibers in 2D- and 3D-LDV configurations.

(a) Two-dimensional fiber optic laser Doppler velocimeters

In general 2D LDV configurations employ two Doppler difference channels to measure the orthogonal transverse velocity components. The beam configuration is shown in Figure 8.24. Beams 1 and 2 from a Doppler difference channel measure the velocity component in the x direction, while beams 3 and 4 measure the y component of the velocity. Light scattered from these two channels may be distinguished using beam pairs with different colors, orthogonal polarization, or differing frequency bias.

In wavelength division multiplexed (WDM) systems two distinct optical frequencies are employed, e.g. the 514.5 and 488 nm lines of an argon-ion laser. In general four lengths of PM fiber are used to deliver the two wavelengths to the probe head, the use of PM fiber ensuring optimum fringe visibility is maintained in the measurement volume.

Polarization discrimination requires that the beam pairs used to form the orthogonal measurement channels are themselves orthogonally polarized [76, 77]. Four PM fibers deliver light to the probe head. The fibers are oriented such that beams 1 and 2 have the same polarization, which is orthogonal to the polarization of beams 3 and 4. The collected scattered light is separated into the individual channels using a polarizing beam splitter in the probe head, and the resultant beams launched into two multi-mode fibers and delivered to separate detectors. Depolarization of the scattered light can result in cross-talk between channels.

A three-beam 2D Doppler difference LDV configuration using polarization discrimination is illustrated in Figure 8.25 [78]. The output from a laser is split and launched into three lengths of PM fiber. In two of them, the polarization is aligned with an eigenaxis. In the third the two orthogonal eigenaxes are equally

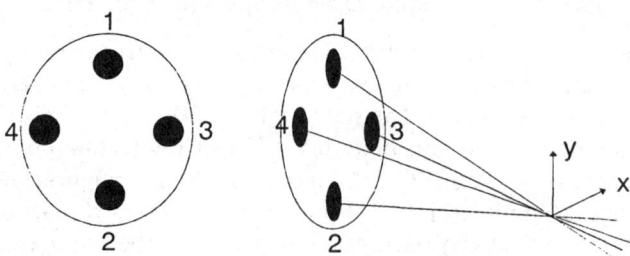

Fig. 8.24 Beam configuration for 2D laser Doppler anemometry.

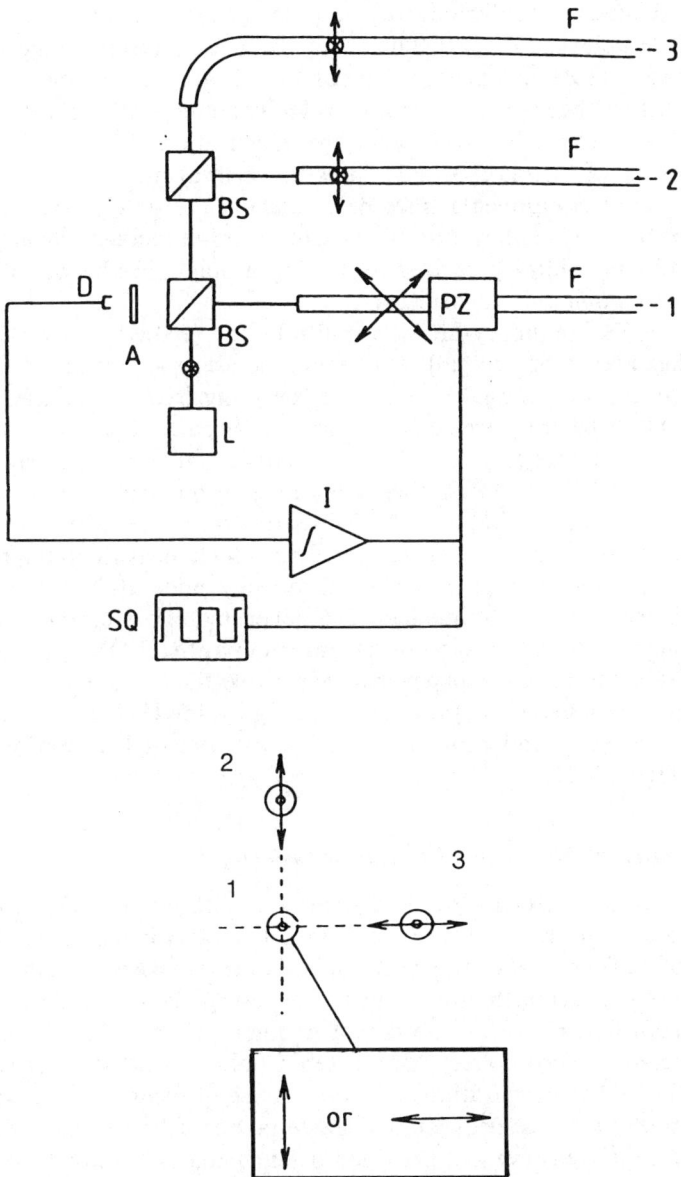

Fig. 8.25 Configuration for 2D laser Doppler anemometry using a polarization-based demultiplexing scheme (after [78]). A = analyzer, BS = 3 dB beamsplitter, D = detector, F = birefringent fiber, I = integrator, PZ = piezoelectric modulator, SQ = square wave generator. The lower figure shows the polarization states of the three beams as viewed from the measurement volume. Beams 2 and 3 are fixed.

populated, and an electrically controlled polarization state switch used to define the output polarization [79]. The beam and polarization configuration is shown in Figure 8.25. Beam 1 is used as a common beam forming measurement channels to measure the transverse x and y velocity components. The respective measurement channels are 'on' only when the polarization of beam 1 is aligned with one or other of beams 2 and 3. Thus, sequential measurements of the orthogonal velocity components may be made by alternately switching the polarization of beam 1, or they may be measured simultaneously by arranging for beam 1 to be circularly polarized and using a polarizing beam splitter to separate the two channels at the detectors.

A compact 2D frequency-division-multiplexed (FDM) laser Doppler anemometer has been reported [80]. The technique uses the pseudo-heterodyne method, discussed in section 8.4.3, to impose differing frequency biases on the two channels. Three beams were derived from the output of a frequency-modulated Nd:YAG miniature ring laser by a 1×3 fused coupler. The beams were guided to the probe head through fiber delay lines of differing length. A beam configuration similar to that in Figure 8.25 is used at the probe. The three beams cross in the flow to generate three fringe patterns, which measure different components of the flow. By virtue of the optical frequency modulation and the delay lines, three different carrier frequencies are generated for each channel. The carrier frequencies also allow directional discrimination. FDM may also be achieved utilizing the beat frequency between three fiber coupled distributed feedback laser diodes, using the same beam configuration [81], or by using frequency-modulated semiconductor laser diodes, as discussed in section 8.4.2, and shown in Figure 8.11.

(b) Three dimensional fiber optic laser Doppler velocimetry

Ideally, 3D flow characterization is performed with three LDV channels oriented to directly measure the three orthogonal velocity components. This is the situation often found in wind tunnel applications. However, in other applications, for example turbomachinery, optical access to the flow region may be severely limited, often requiring the use of a single probe head to deliver all illuminating beams and to collect the scattered light. To satisfy this requirement, many 3D LDV configurations rely on the use of measurement channels which are inclined with respect to the Cartesian coordinate system of the probe, such that the measurement provided by each channel contains information on all three orthogonal velocity components. A matrix transformation from the nonorthogonal measurement coordinate system to the Cartesian coordinate system then yields the orthogonal velocity components. In general the channels are distinguished using wavelength, via the use of three lines of the output from an argon-ion laser. In commercial instrumentation, PM fiber optic delivery systems are utilized. The two major commercial 3D LDV configurations are shown in Figure 8.26. In the two-headed system one of the heads contains a conventional 2D LDV beam configuration, while the other has a 1D

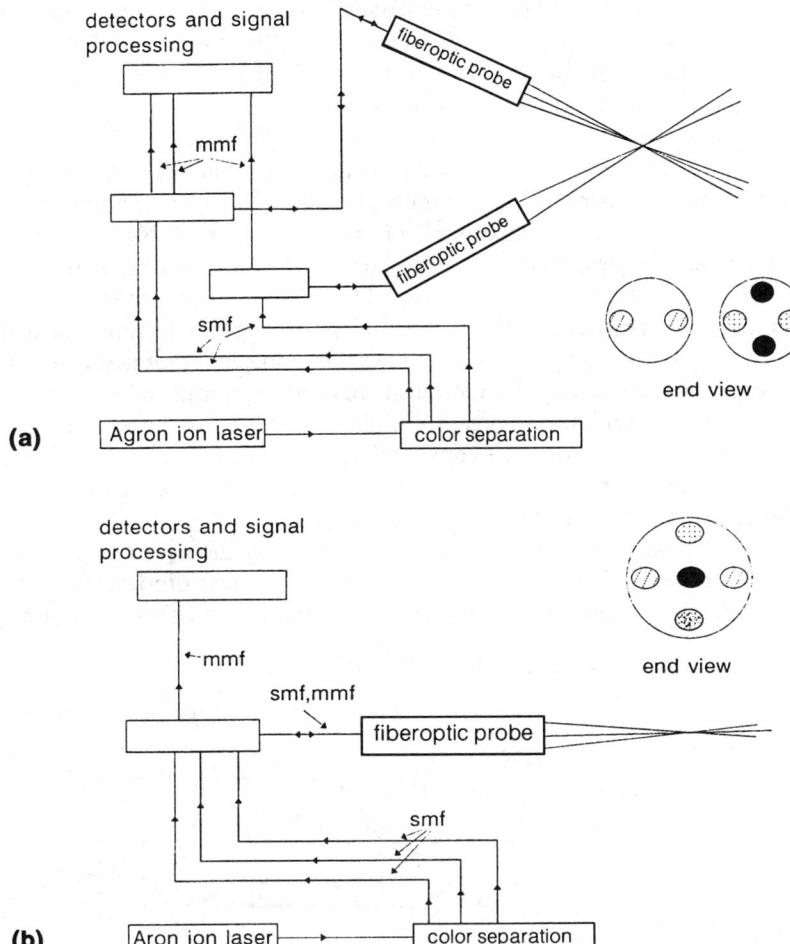

Fig. 8.26 (a) 3D laser Doppler anemometer; (b) single-headed 3D laser Doppler anemometer.

configuration [70]. The single-headed probe illustrated in Figure 8.26 uses either color, or a combination of color and frequency [71] separation to multiplex the three measurement channels. Two blue beams, derived from the output from an argon-ion laser, lie in the vertical plane, forming a single Doppler difference channel, measuring the transverse, y, velocity component. In the horizontal plane three green beams cross to form two Doppler difference channels, in which the central beam is common to the two channels. The measurements provided by the two green channels contain information on the transverse, x, velocity component and the on-axis, z, velocity component. The channels are distinguished by imposing different frequency shifts on the two outermost beams.

Time-division-multiplexed (TDM) 3D LDV is performed by sequentially pulsing the illumination for the three measurement channels. A particle passing through the measurement volume scatters light from each fringe pattern. The collected light is monitored on a single detector, yielding a train of pulses in which the pulses from each channel are interleaved. The channels are then separated by taking every third pulse from the pulse train.

A TDM scheme was reported by Carbonaro [82], using an argon-ion laser and a mechanically chopped configuration in which each channel was masked in turn. This instrument was restricted to measuring low Doppler frequencies by the limited speed of the mechanical chopper, and the Nyquist limit. Semiconductor laser diodes are ideally suited to TDM LDV, since they offer the capability for high frequency modulation. The practical implementation of TDM LDV may be achieved by the sequential pulsing of three separate laser diode sources, as reported by Dopheide [83]. This technique requires the use of a separate laser diode, power supply, temperature controller and pulse generator for each channel, and the channels have to be electronically synchronized.

The use of optical fibers allows the construction of a 3D TDM LDV configuration which uses a single optical source and a single detector to monitor all three channels [84]. A schematic diagram is shown in Figure 8.27.

The sequential pulsing of the channels is achieved by incorporating optical fiber delay lines into the fiber network. A single pulsed laser diode is launched into the fiber network. A 66/33 directional coupler sends one-third of the power

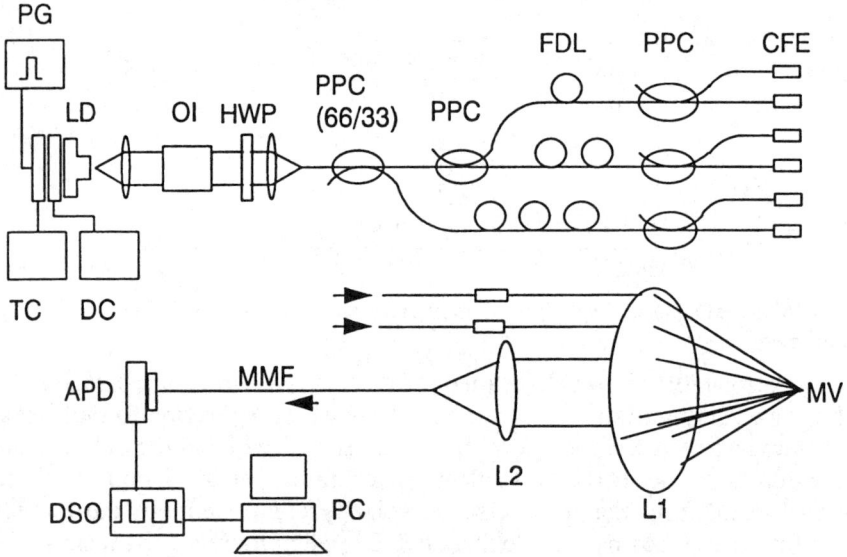

Fig. 8.27 Fiber optic based time division multiplexed 3D laser Doppler anemometer. PG = pulse generator, TC = temperature controller; DC = diode current supply, LD = laser diode, OI = optical isolator, HWP = half-waveplate, PPC = polarization-preserving coupler, FDL = fibre delay line, CFE = collimated fiber end, APD = avalanche photodiode, MMF = multi-mode fibre, DSO = digital storage oscilloscope, L1, L2 = lenses, PC = personal computer, MV = measurement volume (after [84]).

OPTICAL FIBER IMPLEMENTATIONS

into the first channel, while a 50/50 directional coupler splits the remaining power between the other two channels. The delay line lengths are selected such that the pulses emerge from each channel in turn, and that they are regularly spaced. The division of the output from a single source between the three channels clearly imposes a penalty in terms of power per channel. However, pulsed operation enables the laser diode to be driven at much higher peak powers [85]. A multi-mode optical fiber is used to collect the scattered light and deliver it to an APD.

The single-mode fiber network incorporated polarization state controllers to enable the polarization states of the output beams, to be matched and ensure optimum fringe visibility. A single-mode laser diode was biased at threshold and driven with square pulses at 33 MHz with a duty cycle of 20%. The light was launched into a 50/50 coupler to distribute the light between the channels. A 2.2 m delay line is incorporated in one of the channels, giving a 10 ns separation between the projected pulses. The fringes in the measurement volume had a spacing of 4.9 µm, which with the 33 MHz modulation frequency provides a measurable velocity limit of 80 m s^{-1}. Scattered light is collected using multi-mode fiber, and the pulses detected using an APD, and captured on a digital oscilloscope in single shot mode. The signal obtained from a typical particle transit through the measurement volume is shown in Figure 8.28.

The transformation from the nonorthogonal measurement coordinate system to the Cartesian coordinate system is very sensitive to the precise angular configuration of the probe. This may result in large uncertainties in the calculated orthogonal velocity components [86–88]. This limits the application in systems in which optical access is limited; for instance, in the two-headed

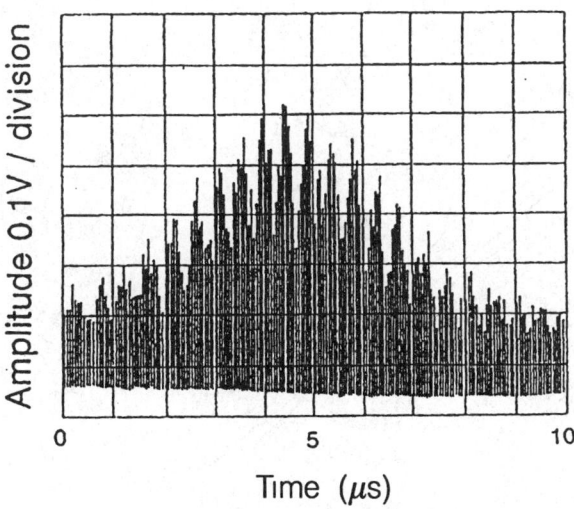

Fig. 8.28 Doppler signal obtained from the time division multiplexed laser Doppler anemometer shown in Fig. 8.27 (after [84]).

configuration shown in Figure 8.26, the angle subtended between the optical axes of the two heads has, in general, to be $>30°$ for accurate readings. For the single-headed configuration shown in Figure 8.26, accuracy considerations require careful calibration of the system using a spinning disk.

Fiber optic components facilitate the construction of a single-headed 3D LDV probe which measures the three orthogonal velocity components directly. The beam configuration is illustrated in Figure 8.29 [89, 90]. Five beams illuminate the flow, the two Doppler difference channels directly measuring the transverse components, while the slightly off-axis illumination direction of the reference beam channel results in the measurement predominantly containing the axial component with a small contribution from the transverse components. The use of off-axis illumination is aimed at reducing the effects of flare from solid surfaces, allowing the performance of near wall measurements, and reducing the length of the measurement volume.

The use of optical fibers is central to the operation of the probe, easing the limitations associated with bulk optic reference beam configurations, allowing the direct measurement of the on-axis velocity component.

Figure 8.30 shows a cross-section of the probe head. Light is delivered to the head by PM optical fiber, to ensure matched linear polarization states for the interfering beams, and thus optimum fringe visibility. The fiber ends are terminated with GRIN lens collimators, and the beams focused into the flow by the front lens. The outer portion of the lens ($f=200$ mm) collimates backscattered light, and the second lens launches this into the multi-mode fiber. The central portion of the bifocal lens ($f=40$ mm) launches scattered light into the PM fiber to act as the signal for the reference beam channel.

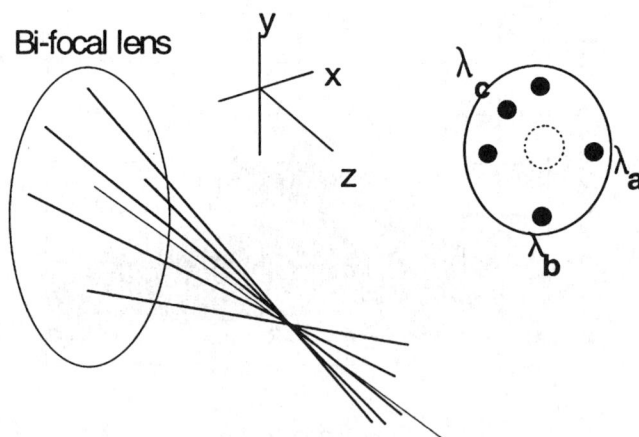

Fig. 8.29 Beam configuration of a single-headed fiber optic 3D laser Doppler anemometer employing a hybrid Doppler difference/reference beam geometry (after [89, 90]).

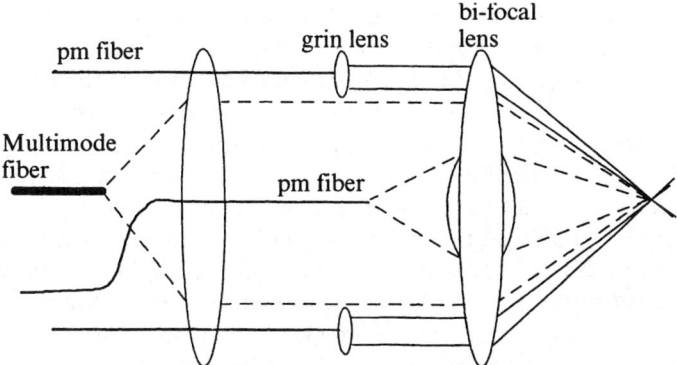

Fig. 8.30 Cross-section through the hybrid probe. pm = polarization-maintaining fiber. The focal lengths of the bifocal lens are 200 and 40 mm (after [89, 90]).

The three measurement channels are distinguished using wavelength-division-multiplexing, implemented using three 100 mW semiconductor laser diodes operating at slightly differing wavelengths around 800 nm.

Figure 8.31 shows a schematic of the layout of the launch/receive unit. The outputs from the laser diodes are coupled into PM fiber. The two Doppler difference channels incorporate fiber pigtailed 40 MHz Bragg cell frequency shifters

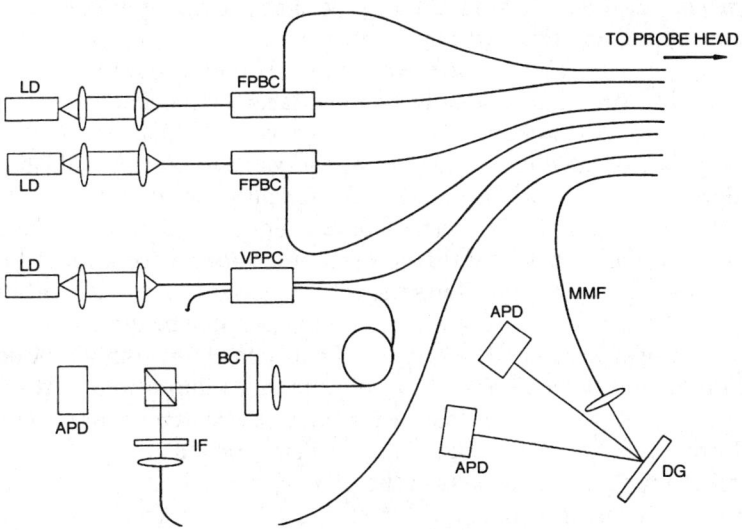

Fig. 8.31 Schematic of the launch/receive section of the anemometer. LD = laser diode, FPBC = fiber pigtailed Bragg cell, BC = Bragg cell, APD = avalanche photodiode, DG = diffraction grating, VPPC = variable split ratio polarization-preserving coupler, IF = interference filter, MMF = multi-mode filter. All unmarked fibers are polarization maintaining.

(discussed in section 8.4.1) which act as 50/50 beam splitters and allow direction discrimination. The diffraction efficiency of the Bragg cells is controlled via a voltage applied to an r.f. amplifier, allowing optimization of the intensity ratio of the beams. The reference beam channel incorporates a variable split ratio polarization-preserving directional coupler, set such that 95% of the power is guided to the probe head, and the remaining 5% used as the reference beam. This ratio can be adjusted to optimize the signal to noise ratio.

Scattered light is delivered to the detectors via a multi-mode fiber and a polarization-preserving fiber. The output from the multi-mode fiber is incident upon a diffraction grating to demultiplex the wavelengths of the Doppler difference channels. The two Doppler difference signals are then monitored on APDs. The signal for the on-axis measurement is derived from the PM return fiber, and mixed with a frequency shifted reference beam. A bandpass filter is used to remove the unwanted signals from the Doppler difference channels. The diffraction efficiency of the Bragg cell is again voltage controlled, providing a convenient method for controlling the signal to noise ratio.

8.5.4 Planar velocity measurement

So far LA systems that measure the flow velocity at a discrete point in space have been discussed. Over the past decade there has been increasing interest in techniques that allow simultaneous measurement over an extended area of the flow, as this reduces the time required to map out a large flow field, minimizing the expensive run time of test facilities, as well as providing information on the global flow characteristics. All planar measurement techniques illuminate the flow with a laser light sheet formed either by a combination of cylindrical and spherical optics, or using a rapidly scanned laser beam. Particle image velocimetry (PIV) is a technique that measures particle displacement between sequential pulses of the illuminating beam. An image of the flow field is captured on photographic film or a CCD camera for each pulse of the source, and sequential images are compared, via a variety of techniques, to determine the distance moved by each particle. Using this information, along with knowledge of the time elapsed between images, allows the velocity to be calculated. For low speed flow (<1 m s^{-1}), a c.w. laser, e.g. an argon-ion laser, can be used as a source and pulsed using a Bragg cell or Pockels cell. In this case single-mode optical fiber can be used to deliver the illumination, with the same limitations as discussed in section 8.3. For flows greater than a few meters per second a pulsed laser is required, for example a frequency-doubled YAG, producing 5–10 ns duration pulses at 532 nm with peak powers in the MW region.

At these high instantaneous power levels, fiber damage becomes a major problem. In principle, power transmission can be affected by the processes described in section 8.3. In addition, self-focusing of the guided beam can occur, increasing the power density at the fiber axis, and causing subsequent breakdown. Some of these problems may be overcome, using the technique shown in Figure 8.32, to allow fiber delivery of a pulsed light sheet of suitable quality for

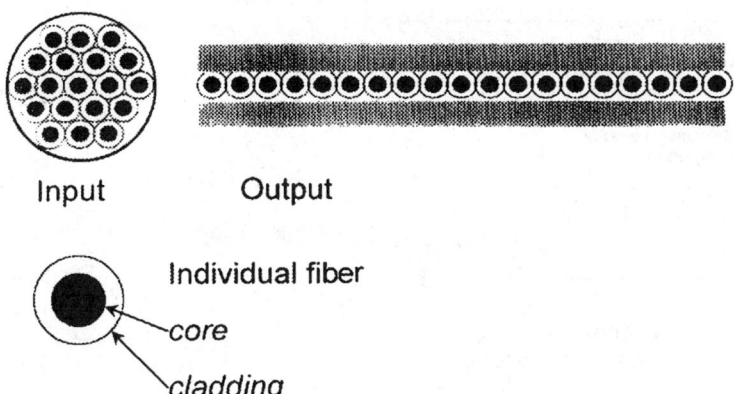

Fig. 8.32 Fiber delivery system for high power pulsed lasers (after [91]).

PIV [91]. Light is coupled into a fiber bundle consisting of 19 fibers, each with a 200 μm diameter pure fused silica core and a doped fused silica cladding of 220 μm outer diameter. At the input the fibers are configured as an array with a circular cross-section, and as a linear array at the output, to form, with additional optics, the light sheet. This system was found to be capable of delivering more than 20 mJ per pulse at 532 nm.

PIV may be considered as the extended field analog of LTA. The extended field analog of LDV is called Doppler global velocimetry (DGV) [92, 93]. However, interferometric measurement of the Doppler shift is replaced by an absorption cell, which acts as an edge filter, and transduces the Doppler frequency shift into intensity changes. A c.w. laser, usually argon-ion, is tuned to the midpoint of the slope of the transfer function of the absorption cell (iodine

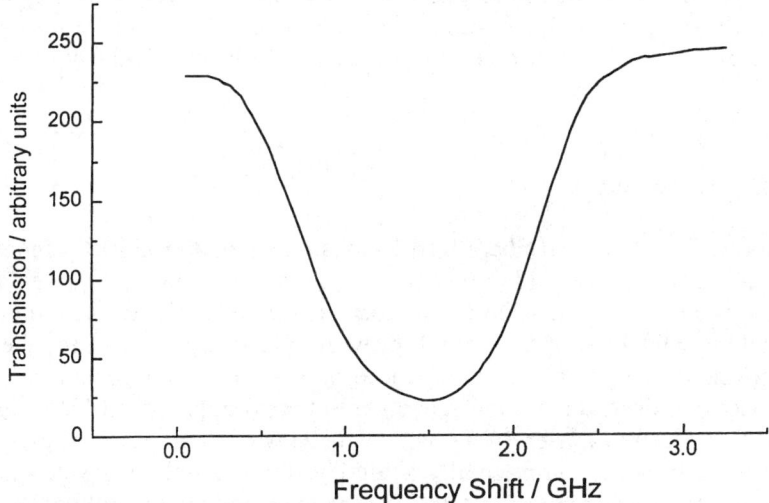

Fig. 8.33 Iodine absorption cell transfer function.

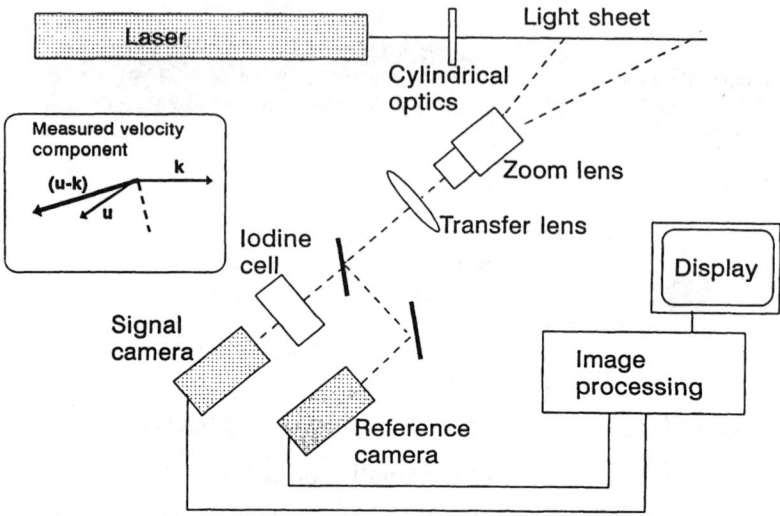

Fig. 8.34 Schematic of DGV arrangement (after [93]).

vapor for visible wavelengths), Figure 8.33. A positive shift in the optical frequency causes a reduction in transmission, while a negative frequency shift produces an increase in transmission. Directional discrimination is inherent in this technique. Intensity information is captured on two CCD cameras, as shown in Figure 8.34. The video output from the signal camera is divided by the output from the reference camera to normalize for nonuniform illumination and seeding densities. For this technique single-mode optical fiber can be used to deliver optical power to the beam-shaping optics. The linewidth of the laser must be narrower than the iodine absorption line (~2 GHz) and is typically 3 MHz with an intra-cavity etalon. The limitations on power transmission discussed in section 8.3 must be considered carefully, particularly the onset of SBS.

8.6 CONCLUSIONS

The introduction of optical fibers into LA systems has undoubtedly increased the versatility, ease of use and ruggedness of the instrumentation. These advantages have been incorporated into a number of commercially available systems [94]. Combining the fiber optic technology with solid state sources and detectors has led to the development of miniaturized multi-velocity component systems and has been demonstrated to offer the potential for simple optical TDM configurations. Further use of laser diodes will be made when high power short wavelength devices become commercially available, for example a device offering 295 mW at 690 nm has been reported [95]. The incorporation of robust, tunable in-fiber frequency shifters will enhance the performance of the instrumentation further. Further miniaturization of systems, required for some applications, may incorporate holographic elements combined with solid state sources and detectors and integrated optics [96].

REFERENCES

1. Tropea, C. (1995) Laser anemometry: recent developments and techniques. *Meas. Sci. Tech.* **16**, 605–19.
2. Cummins, H. Z. and Pike, E. R. (eds) (1977) *Photon Correlation Spectroscopy and Velocimetry*, NATO Advanced Studies Institute Series B23, Plenum.
3. Durst, F., Melling, A. and Whitelaw, J. H. (1976) *Principles and Practice of Laser Anemometry*, Academic Press.
4. Yeh, Y. and Cummings, H. Z. (1964) Localised fluid flow measurements with an He–Ne Spectrometer. *Appl. Phys. Lett.* **4**, 176–8.
5. Penny, C. M. (1969) Differential Doppler velocity measurements. *IEEE J. Quantum Electron.* **QE-5**, 318.
6. Vom Stein H. D. and Pfeifer, H. J. (1969) A Doppler difference method for velocity measurements. *Metrologia* **5**, 59–61.
7. Jackson, D. A. and Paul, D. M. (1970) Measurement of hypersonic velocities and turbulence by direct spectral analysis of Doppler shifted laser light. *Phys. Lett.* **32A**, 77–8.
8. Roehle, I. and Schodl, R. (1994) Evaluation of the accuracy of the Doppler global technique, *Optical Methods and Data Processing in Heat and Fluid Flow*, I Mech E, London.
9. Huffacker, R. M. (1970) Doppler detection system for gas velocity measurements. *Appl. Opt*, **9**, 1026–8.
10. Drain, L. E. (1980) *The Laser Doppler Technique*, Wiley, Chichester.
11. Mazumder, M. K. (1970) Laser Doppler velocity measurement without directional ambiguity by using frequency shifted incident beam. *Appl. Phys. Lett.* **16**, 462–4.
12. Tanner, L. H. (1972) A particle timing laser velocity meter, Aeronautical Engineering Department, The Queens University of Belfast (Report).
13. Schodl, R. (1976) Extension of the range of applicability of LDA by means of the laser-dual-focus (L2F)-technique. *The Accuracy of Flow Measurements by Laser Doppler Methods*, 480–9.
14. James, S. W., Lockey, R. A., Egan, D., Tatam, R. P. and Elder, R. L. (1996) 3D Fiber optic laser Doppler Velocimetry. *Proc. SPIE* **2839**, 323–34.
15. Van de Hulst, H. C. (1981) *Light Scattering by Small Particles*, Dover, New York.
16. Poyntz-Wright, L. J., Fermann, M. E. and Russell, P. St. J. (1988) Non-linear transmission and colour centre dynamics in gemanosilicate fibre at 420–540 nm. *Opt. Lett.* **13**, 1023–5.
17. Duffy, C. J. (1994) Stimulated Brillouin scattering in monomode optical fibers for sensing and signal processing applications. Ph.D. thesis, Cranfield University, Bedford, UK.
18. Smith, R. G. (1972) Optical power handling capacity of low loss optical fibres as determined by stimulated Raman and Brillouin scattering. *Appl. Opt.* **11**, 2489–94.
19. Tang, C. L. (1966) Saturation and spectral characteristics of the Stokes emission in the stimulated process. *J. Appl. Phys.* **37**, 2945–55.
20. Ippen, E. P. and Stolen, R. H. (1972) Stimulated Brillouin scattering in optical fibers. *Appl. Phys. Lett.* **21**, 539–40.

21. Duffy, C. J. and Tatam, R. P. (1993) An optical frequency shifter based on stimulated Brillouin scattering in birefringent optical fiber. *Appl. Optics.* **32**, 5966–72.
22. Labudde, P., Anliker, P. and Weber, H. P. (1980) Transmission of narrow band high power laser radiation through optical fibre. *Optics Comm.*, **32**, 385–90.
23. Yariv, A. and Yeh, P. (1984) *Optical Waves in Crystals*, John Wiley, Chichester.
24. Tatam, R. P. (1995) Optical fiber modulation techniques for single mode fiber sensors in *Optical Fiber Sensor Technology* (eds. K. T. V. Grattan and B. T. Meggitt), Chapman & Hall, London, 223–67.
25. Risk, W. P., Youngquist, R. C., Kino, G. S. and Shaw, H. J. (1984) Acousto-optic frequency shifter for single-mode fibres. *Opt. Lett.*, **9**, 309–11.
26. Pannell, C. N., Tatam, R. P., Jones, J. D. C. and Jackson, D. A. (1988) A fibre optic frequency shifter utilising travelling flexure waves in birefringent fibres. *J. Inst. Electron. Radio. Engrs.* **S8**, S92–8.
27. Kim, B. Y., Blake, J. N., Engan, H. E. and Shaw, H. J. (1986) All-fiber acousto-optic frequency shifter. *Optics Lett.* **11**, 389–91.
28. Birks, T. A., Farrell, S. G., Russell, P. St J. and Pannell, C. N. (1994) Fibre acousto-optic frequency shifter based on a null coupler. *Optics Lett.* **19**, 1964–66.
29. Sabert, H., Dong, L. and Russell, P. St J. (1992) Versatile acousto-optic flexural wave-modulator, filter and frequency shifter in dual-core fiber. *Int. J. Opto Electron* **7**, 189–94.
30. Fibre Pro, Donam Systems Inc., Korea.
31. Birks, T. A., Farrell, S. G., Russell, P. St J. and Pannell, C. N. (1994) Fibre acousto-optic frequency shifter based on a null coupler. *Opt. Lett.* **19**, 1964–66.
32. Cotter, D. (1983) Stimulated Brillouin scattering in monomode optical fibre. *J. Opt. Commun.* **4**, 10–19.
33. Cotter, D. (1982) Observation of stimulated Brillouin scattering in low loss silica fibre at 1.3 µm. *Electron. Letts.* **18**, 445–96.
34. Bayvel, P. and Giles, I. P. (1989) Linewidth narrowing in pumped all-fibre Brillouin ring laser. *Electron. Lett.* **25**, 260–2.
35. Khan, O. S. and Tatam, R. P. (1993) Fiber optic frequency shifter based on stimulated Brillouin scattering in a birefringent fiber ring resonator. *Optics Commun.* **103**, 161–8.
36. Culverhouse, D. O., Farahi, F., Pannell, C. N. and Jackson, D. A. (1989) Stimulated Brillouin scattering: a means to realise a tunable microwave generator or distributed temperature sensor. *Electron. Letts.* **25**, 915–16.
37. Kalli, K., Culverhouse, D. O. and Jackson, D. A. (1991) Fiber frequency shifter based on generation of stimulated Brillouin scattering in high-finesse ring resonators. *Optics Lett.* **16**, 1538–40.
38. Khan, O. S. and Tatam, R. P. (1993) Fiber optic frequency shifter based on stimulated Brillouin scattering in a birefringent fiber ring resonator. *Optics Commun.* **103**, 161–8.
39. Többen, H., Müller, H. and Dopheide, D. (1994) Brillouin frequency shift LDA. *Proc. of 7th International Symposium on Applications of Laser Techniques to Fluid Mechanics*, Lisbon, 14.2.1.
40. Freitags, I. and Welling, H. (1994) Investigation on amplitude and frequency noise of injection-locked diode-pumped Nd:YAG lasers. *Appl. Phys. B* **58**, 537–43.
41. For example: Millennia™ laser from Spectra Physics Lasers, Mountain View, California, USA, or the Verdi™ laser from Coherent Inc., Santa Carla, California, USA.
42. Martini, G. (1987) Analysis of a single-mode optical fibre piezoceramic phase modulator. *Opt. Quant. Electron.* **19**, 179–90.

43. Jones, J. D. C., Chan, R. K. Y. and Jackson, D. A. (1984) Designs of fibre optic systems for Doppler difference velocimetry. *Proc. of 2nd International Symposium on the Application of Laser Anemometry to Fluid Mechanics*, Lisbon, Portugal, 6.6.
44. Dandridge, A. and Goldberg, L. (1982) Current induced frequency modulation in diode lasers *Electron. Letts.* **18**, 302–3.
45. Ford, H. D., Atcha, H. and Tatam, R. P. (1993) Optical fibre technique for measurement of small frequency separations: applications to surface profile measurement using electronic speckle pattern interferometry. *Meas. Sci. Technol.* **14**, 601–7.
46. Jones, J. D. C., Jackson, D. A., Corke, M. and Kersey, A. D. (1984) *Optics and Lasers in Engineering* **5**, 127–39.
47. Jones, J. D. C., Corke, M., Kersey, A. D. and Jackson, D. A. (1982) Miniature solid-state laser Doppler velocimeter. *Electron. Letts.* **18**, 967–8.
48. Kato, S., Ichikaura, T. and Ito, H. (1996) Multipoint sensing laser Doppler velocimetry based on laser diode frequency modulation. *11th optical fibre sensors conference, OFS-11*, Sapporo, Japan, 606–9.
49. Többen, H., Müller, H. and Dopheide, D. (1996) Power and sensitivity of LDA systems by fibre amplifiers. *Proceedings of the 8th International Symposium on Applications of Laser Techniques to Fluid Mechanics*, Lisbon, 34.1.1–34.1.4.
50. Pannell, C. N., Midgley, J. H., Jones, J. D. C. and Jackson, D. A. (1988) Fibre-optic transit velocimetry using laser diode sources. *Electron. Letts.* **24**, 525–6.
51. Schodl, R. and Föster, W. (1988) A multi colour fiber optic laser two focus velocimeter for 3-dimensional flow analysis. *AIAA*-88-3034.
52. Tanaka, T. and Benedek, G. B. (1975) Measurements of the velocity of blood flow (*in-vivo*) using a fibre optic catheter and optical spectroscopy. *Appl. Opt.* **14**, 189–96.
53. Dyot, R. B. (1978) Fibre optic Doppler anemometer. *Microwaves, Optics and Acoustics* **2**, 13–18.
54. Meggitt, B. T., Boyle, W. J. O., Grattan, K. T. V., Palmer, A. W. and Ning, Y. N. (1990) Fibre optic anemometry using an optical delay cavity technique. *Proc. SPIE* **1314**, 321–9.
55. Kyuma, K., Tai, S., Hamanaka, K. and Nunoshita, M. (1981) Laser Doppler velocimeter with a novel optical fibre probe. *Appl. Opt.* **20**, 2424–7.
56. Nishihara, H., Matsumoto, K. and Koyama, J. (1984) Use of a laser diode and an optical fibre for a compact laser Doppler velocimeter. *Opt. Lett.* **9**, 62–4.
57. Nishihara, H., Koyama, J., Hoki, N., Kajiya, F., Muramoto, T. and Hironaga, K. (1984) Optical fibre laser Doppler velocimeter for pulsatile blood flows. *Laser Anemometry in Fluid Mechanics*, R. J. Adrain, D. F. G. Durao, F. Durst, H. Mishina and J. H. Whitelaw (eds), 335–45, Loadan, Lisbon.
58. James, S. W., Lockey, R. A., Egan, D. and Tatam, R. P. (1995) Fibre optic based reference beam laser Doppler velocimeter. *Opt. Comm.* **119**, 460–4.
59. Egan, D. A., James, S. W. and Tatam, R. P. (1997) A polarisation based optical fibre vibrometer. *Meas. Sci. Technol.* **8**, 343–7.
60. Toda, H., Haruna, M. and Nishihara, H. (1987) Optical integrated circuit for a fiber laser Doppler velocimeter. *J. Lightwave Technol.* **LT-5**, 901–5.
61. Pannell, C. N., Jones, J. D. C. and Jackson, D. A. (1994) The effect of environmental acoustic noise on optical-fibre based velocity and vibration sensors. *Meas. Sci. Technol.* **5**, 412–17.
62. Sasaki, O., Sato, T., Abe, T., Miuguchi, T. and Niwayama, M. (1980) Follow up — type laser Doppler velocimeter using single-mode optical fibres. *Appl. Opt.* **19**, 1306–8.

63. Knuhtsen, J., Olldag, E. and Buchave, P. (1982) Fibre optic laser Doppler anemometer with Bragg frequency shift utilising polarisation based single mode fibre. *J. Phys. E* **15**, 1888–91.
64. Jackson, D. A., Jones, J. D. C. and Chan, R. K. Y. (1984) A high power fibre optic laser Doppler velocimeter. *J. Phys. E: Sci. Instrum.* **17**, 977–80.
65. Chan, R. K. Y., Jones, J. D. C. and Jackson, D. A. (1985) A compact all optical fibre Doppler difference laser velocimeter. *Opt. Acta* **32**, 241–6.
66. Jones, J. D. C., Chan, R. K. Y. and Jackson, D. A. (1986) Design of fibre optic systems for Doppler difference laser velocimetry. *Laser Anemometry in Fluid Mechanics II*, R. J. Adrain, D. F. G. Durao, F. Durst, H. Mishina and J. H. Whitelaw (eds), 69–83, Loadan, Lisbon.
67. Nakatani, N., Maegwa, A., Izumi, T., Yamada, T. and Sakabe, T. (1986) Advancing multi-point optical fiber LDV's — vorticity measurement and some new optical systems. *Laser Anemometry in Fluid Mechanics II*, R. J. Adrain, D. F. G. Durao, F. Durst, H. Mishina and J. H. Whitelaw (eds), 3–18, Loadan, Lisbon.
68. Meyers, J. F. (1988) The elusive third component. *Finite Elements in Analysis and Design* **4**, 51.
69. Dancey, C. L. (1987) A review of three component laser Doppler anemometry. *Journal of Optical Sensors*, **2**, 437–69.
70. Yanta, W. J. (1979) A three dimensional laser Doppler velocimeter for use in wind tunnels. *ICASF Record* 294.
71. Stauter, R. C. (1993) Measurement of the three dimensional tip region flow in an axial compressor. *Journal of Turbomachinery* **115**, 496.
72. Dopheide, D., Strunk, V. and Krey, E. A. (1993) Three component laser Doppler anemometer for gas flow rate measurements up to 5,500 m^3/h. *Metrologia* **30**, 435–69.
73. Dopheide, D., Strunk, V. and Pfeifer, H. J. (1990) Miniaturized multicomponent laser Doppler anemometers using high-frequency pulsed diode lasers and new electronic signal acquisition systems. *Exp. Fluids* **9**, 309–16.
74. Carbonaro, M. (1984) Conception and design of a new type of three components laser Doppler velocimeter. Von Karaman Institute Technical Memorandum,
75. Dopheide, D., Rinker, M. and Strunk, V. (1993) High frequency pulsed laser diode application in multi-component laser Doppler anemometry, *Optics and Lasers in Engineering*, **18**, 135–43.
76. Nakatani, N., Tokita, M. and Yamada, T. (1984) LDV using polarisation preserving optical fibres for simultaneous measurement of two velocity components. *Appl. Opt.* **23**, 1686–7.
77. Nakatani, N., Tokita, M., Izumi, T. and Yamada, Y. (1985) Doppler velocimetry using polarisation-preserving optical fibres for simultaneous multidimensional velocity components. *Rev. Sci. Instrum.* **56**, 2025–29.
78. Pannell, C. N., Tatam, R. P., Jones, J. D. C. and Jackson, D. A. (1988) Two-dimensional fibre-optic laser velocimetry using polarisation state control. *J. Phys. E: Sci Instrum.* **21**, 103–7.
79. Tatam, R. P., Jones, J. D. C. and Jackson, D. A. (1986) Optical polarisation state control schemes using fibre optics or Bragg cells. *J. Phys. E: Sci. Instrum.* **19**, 711–17.
80. Czarske, J. W. and Müller, H. (1996) Two component directional laser Doppler anemometer based on a frequency modulated Nd:YAG ring laser and fibre delay lines. *Proceedings of the 8th International Symposium on Applications of Laser Techniques to Fluid Mechanics*, Lisbon, 34.2.1–34.2.6.

81. Müller, H., Wang, H. and Dopheide, D. (1996) Fibre optical multicomponent LDA system using the optical frequency difference of powerful DBR laser diodes. *Proceedings of the 8th International Symposium on Applications of Laser Techniques to Fluid Mechanics*, Lisbon, 34.3.1–34.3.4.
82. Carbonaro, M. (1984) Conception and design of a new type of three component laser Doppler velocimeter. Von Karman Institute Technical Memorandum.
83. Dopheide, D. *et al.* (1993) High frequency pulsed laser diode application in multicomponent laser Doppler anemometry. *Optics and Lasers in Engineering*, **18**, 135–43.
84. Lockey, R. A. and Tatam, R. P. (1997) Multi-component time-division multiplexed optical fibre Doppler anemometry. *IEE Proc. Optoelectronics* **144**, 168–75.
85. Yoo, J. S. *et al.* (1991) On the surface recombination velocity and output intensity limit of pulsed semiconductor lasers. *IEEE Photonics Technology Letters* **7**, 594.
86. Orloff, K. L. and Snyder, P. K. (1982) Laser Doppler anemometer measurements using non-orthogonal velocity components: error estimates. *Appl. Opt.* **21**, 39.
87. Morrison, G. L., Johnson, M. C., Swan, D. H. and Deotte, R. E. (1991) Advantages of orthogonal and non-orthogonal three dimensional anemometer systems. *Flow Meas. Instrum.* **2**, 89.
88. Snyder, P. K., Orloff, K. L. and Reinath, M. S. (1984) Reduction of flow measurement uncertainties in laser velocimeters with non-orthogonal channels. *AIAA Journal*, **8**, 1115.
89. James, S. W., Lockey, R. A., Egan, D., Tatam, R. P. and Elder, R. L. (1996) 3d fibre optic laser Doppler velocimetry. *Proceedings of the 8th International Symposium on Applications of Laser Techniques to Fluid Mechanics*, Lisbon, 28.2.1–28.2.6.
90. James, S. W., Tatam, R. P. and Elder, R. L. (1997) Design considerations for a 3d fibre optic laser Doppler velocimeter for turbomachinery applications. *Rev. Sci. Instrum.* **68**, 3241–6.
91. Jones, J. D. C., Anderson, D. J. and Greated, C. A. (1997) Fibre optic beam delivery systems for particle image velocimetry. *Opt. Laser Eng.* **27**, 657–74.
92. Meyers, J. F. and Komine, H. (1991) Doppler global velocimetry: a new way to look at velocity. *ASME 4th International Conference on Laser Anemometry, Advances and Applications*, 289–96.
93. Ford, H. D. and Tatam, R. P. (1997) Development of extended field Doppler velocimetry for turbomachinery applications. *Opt. Laser Eng.* **27**, 675–96.
94. For example, Dantec Measurement Technology, Denmark, TSI Inc., USA and Polytech GmbH, Germany.
95. Sharp Laboratories, Press Service June 1994.
96. Imam, H., Rose, B., Lindrold B. *et al.* (1996) Miniaturising and ruggedising laser anemometers. *Proceedings of the 8th International Symposium on Applications of Laser Techniques to Fluid Mechanics*, Lisbon, 40.2.1–40.2.7.

9
Fiber optic gyroscopes

J. Blake

Because of the richness of the technology involved, and the size of the market in which it can compete, the fiber optic gyroscope has acquired an uncommonly devoted following. It was first experimentally demonstrated in 1976 [1] and the first products became available in the early 1990s [2–7]. The first flight qualified fiber gyro products have been for attitude, heading, and reference systems (AHRS) applications which typically have performance requirements in the neighborhood of $1°$ h^{-1} drift rates, 500 ppm scale factor accuracy, and an angle random walk coefficient (RWC) of $0.05°$ $h^{1/2}$ (the RWC being defined as the r.m.s. angular error accumulated after integrating the gyro output for 1 h due to the presence of white noise). The fiber gyro has also been aggressively pursued for applications in both the low cost rate sensing and inertial grade navigation markets, but the introduction of products into these markets has been slower due to somewhat less favorable cost/performance tradeoffs.

The fiber gyro, like its older cousin, the ring laser gyroscope, is based on the Sagnac effect, namely, that counterpropagating light waves take slightly different times to traverse a loop rotating in inertial space. In the ring laser gyro, this time difference manifests itself as a frequency difference between counterpropagating lasing modes proportional to the rotation rate, whereas in the fiber gyro, it results in a phase shift between the counterpropagating waves proportional to the rotation rate.

This chapter begins with a derivation of the Sagnac effect from a special relativity point of view. Attention is then turned to a discussion of the most widely used optical circuit and signal processing techniques. The major error sources and methods for their suppression are then considered. Finally, two low cost alternative optical circuits, the depolarized approach and the 3×3 coupler approach, are described.

9.1 SAGNAC EFFECT IN AN OPTICAL FIBER LOOP [8]

Figure 9.1 shows a closed optical fiber loop of diameter, D, rotating in inertial space at Ω rad s^{-1} in the clockwise direction. A light wave is injected into the

Optical Fiber Sensor Technology, Vol. 2. Edited by K. T. V. Grattan and B. T. Meggitt.
Published in 1998 by Chapman & Hall, London. ISBN 0 412 782 901

loop at the common entry/exit point which is physically located at point A in inertial space at the time of entry. The light splits and propagates around the loop in both directions. Meanwhile, the common entry/exit point is also moving so that the light exits the loop at point B in inertial space. It is evident that the clockwise directed light wave must travel more distance in inertial space than the counterclockwise directed light wave to reach the exit point, even though they both travel through the identical number of glass molecules in the optical fiber. The times it takes for the clockwise and counterclockwise light waves to travel around the loop, τ_{cw} and τ_{ccw} respectively, are given by

$$\tau_{cw} = \frac{n_{cw}(L + \Delta L)}{c} \tag{9.1}$$

$$\tau_{ccw} = \frac{n_{ccw}(L - \Delta L)}{c} \tag{9.2}$$

where L is the circumference of the fiber loop (which comprises many turns of fiber), c is the vacuum speed of light, $\Delta L = nLD\Omega/2c$ is the distance the common entry/exit point has moved during the loop transit time, n being the rest refractive index of the glass, and n_{cw} and n_{ccw} are the indices of refraction seen by the clockwise and counterclockwise waves, respectively. The time of flight difference is found from equations 9.1 and 9.2 to be

$$\tau_{cw} - \tau_{ccw} = \frac{L}{c}(n_{cw} - n_{ccw}) + \frac{n^2 LD}{c^2}\Omega \tag{9.3}$$

to first order in Ω. The refractive indices, n_{cw} and n_{ccw}, are not the same because of the motion of the glass medium. They can be found by first finding the velocities of the counterpropagating waves, v_{cw} and v_{ccw}, and using the

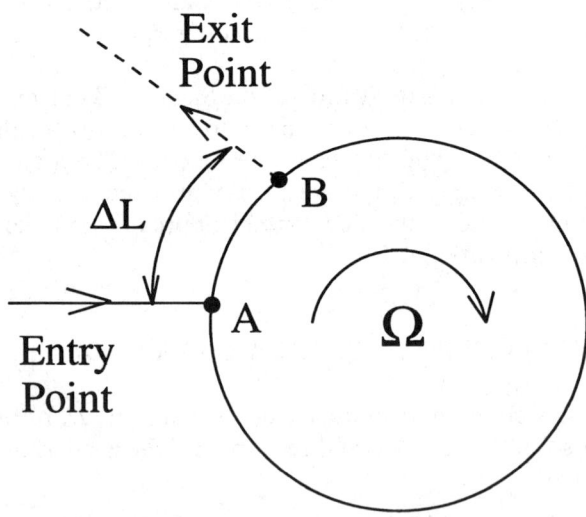

Fig. 9.1 Closed optical loop. Clockwise rotation rate $= \Omega$.

relations, $n_{cw,ccw} = c/v_{cw,ccw}$. The velocities of the counterpropagating waves in the moving glass are in turn found by relativistically adding or subtracting the velocity of the light in the medium at rest, c/n, to the velocity of the medium, $\Omega D/2$. The clockwise directed light wave travels with the comoving glass, and hence is speeded up by the dragging effect of the glass, whereas the counterclockwise directed light wave travels against the movement of the glass, and hence is slowed down by the dragging effect of the glass. The general expression for the relativistic addition of two velocities, v_1 and v_2, is

$$v_1 \oplus v_2 = \frac{v_1 + v_2}{1 + v_1 v_2/c^2} \tag{9.4}$$

Using $v_1 = c/n$, and $v_2 = \pm \Omega D/2$ we find that $v_{cw,ccw} = c/n \pm (1 - 1/n^2)\Omega D/2$ to first order in Ω. This partial dragging of the light by the moving medium is known as the Fresnel drag effect. The difference between the clockwise and counterclockwise refractive indices is then found to be

$$n_{cw} - n_{ccw} = (1 - n^2)\frac{D}{c}\Omega \tag{9.5}$$

Combining equations 9.3 and 9.5, we see that the time of flight difference for the counterpropagating light waves to travel around the loop is given by

$$\tau_{cw} - \tau_{ccw} = \left[(1 - n^2)\frac{LD}{c^2}\Omega + \frac{n^2 LD}{c^2}\Omega\right] \tag{9.6}$$

where the first term in equation 9.6 is the Fresnel drag term and the second is the coupler movement term. The index of refraction dependence exactly cancels, leaving

$$\tau_{cw} - \tau_{ccw} = \frac{LD}{c^2}\Omega \tag{9.7}$$

The increase in the relative motion of the loop due to n slowing down the light is exactly compensated by the dragging of the light by the glass medium [9].

This result can be cast in terms of a phase shift between the counterpropagating light waves, $\Delta\phi_r$,

$$\Delta\phi_r = \frac{2\pi c}{\lambda}(\tau_{cw} - \tau_{ccw}) = \frac{2\pi LD}{\lambda c}\Omega \tag{9.8}$$

Equation 9.8 has been derived using a circular loop; however, the result can be generalized for arbitrary shaped loops to

$$\Delta\phi_r = \frac{8\pi A}{\lambda c}\Omega \tag{9.9}$$

where A is the effective area enclosed by the loop.

As a numerical example, consider a fiber gyro using a 1.55 µm wavelength light source, and a 1 km coil wound on an average diameter of 7.5 cm. For these design choices, the scale factor, $2\pi LD/\lambda c$, which relates the optical phase shift, $\Delta\phi_r$, to the rotation rate, Ω, is about 1 s. A minimum detectable rotation rate,

$\Omega_{\min}$, of $0.01° \, h^{-1}$, which is a typical requirement for an inertial navigation grade gyroscope, thus corresponds to a minimum detectable phase shift, $\Delta\phi_{\min}$, of 48 nrad. This corresponds to a time of flight difference for the counterpropagating light waves to travel around the loop, $\tau_{cw} - \tau_{ccw}$, of 4×10^{-23} s. The distance in inertial space through which the common entry/exit point moves during the time the light travels around the loop, ΔL, is 9×10^{-15} m.

9.2 BASIC ELEMENTS OF THE FIBER GYROSCOPE

9.2.1 'Minimum reciprocal configuration'

The extremely small phase shifts the fiber gyro is required to resolve can only be measured because of the aid reciprocity provides. Reciprocity guarantees that the phase shift, loss, and polarization evolution a light wave suffers in traveling from point 1 to point 2 is exactly the same as would occur if the light wave were to travel from point 2 to point 1. Reciprocity remains valid when the propagation medium is linear, time invariant, nonrelativistic, and free of magnetic fields.

Figure 9.2 shows the 'minimum reciprocal configuration' of the fiber optic gyroscope [10]. Two beam splitters are used to ensure that the counterpropagating light waves see the loop splitter in the same way. Both waves go through and across the loop splitter, only at opposite ends of their respective journeys. The loop splitter may be implemented either using a fiber directional coupler (Figure 9.2(a)), or using an integrated optics Y-junction (Figure 9.2(b)) [11]. A

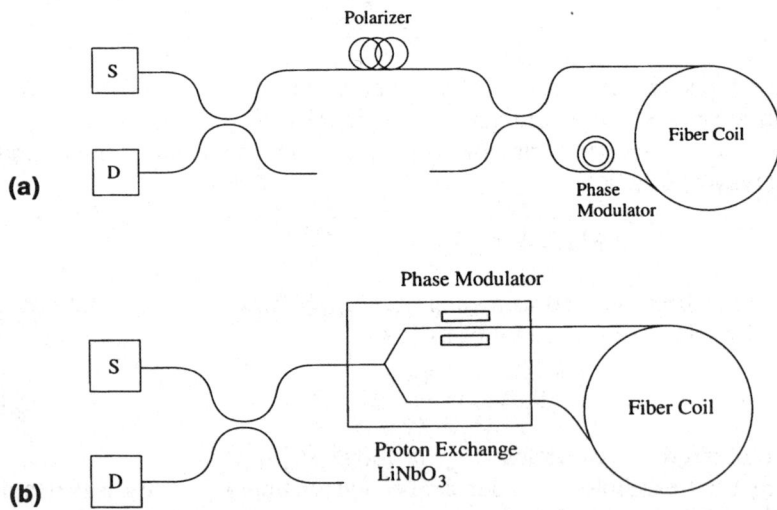

Fig. 9.2 Fiber gyro optical circuit using (a) a PZT phase modulator, and (b) an integrated-optics multi-function chip.

single-mode filter (single spatial mode and single polarization mode) in the common input/output path just to the left of the loop ensures that the counter-propagating light waves travel exactly the same path, only in reverse order, making the system reciprocal. This filter can be either a single-mode fiber polarizer, or an integrated optics polarizer. The integrated optics polarizer is commonly implemented by making the Y-junction loop splitter using proton-exchange $LiNbO_3$ technology, since waveguides made in this way only guide one polarization of light. The source splitter may, like the loop splitter, be a fiber directional coupler. Alternatively, it may be a simple macrobend tap [11] to divert some of the returning light to the photodetector. Ordinarily, the source and loop splitters are not implemented together with the polarizer onto a single integrated optics double-Y configuration because of the difficulty in obtaining sufficient spatial mode isolation [11].

9.2.2 Fiber coil

The clockwise and counterclockwise light waves which exit the loop and pass an ideal single-mode polarizer in the common input/output path of the fiber optic gyroscope are guaranteed to interfere in a perfectly reciprocal way. However, there is no guarantee that any light will pass the polarizer on the way out of the loop, unless something is done to ensure that its polarization state is at least partially aligned to the pass axis of the polarizer. Two effective solutions to this polarization-induced signal fading problem have been developed. The first is to use polarization-maintaining (PM) fiber throughout the loop [12]. This is the approach used in the first commercial products. The second solution uses a depolarizer within the loop to guarantee that one-half of the light is always aligned to the polarizer as it exits the loop [13]. The depolarized approach allows for the long fiber coil to be made using very low cost, telecommunications grade single-mode fiber.

9.2.3 Phase modulator

The next element of the fiber gyro optical circuit considered is the phase modulator. In its absence, the only phase shift between the interfering clockwise and counterclockwise directed light waves is that associated with the rotation of the instrument. The intensity of the light that falls on the detector, I_d, is, in this case, given by

$$I_d = \frac{I_0}{2}\{1 + \cos \Delta\phi_r\} \qquad (9.10)$$

Here I_0 is the intensity of the light falling on the detector in the absence of any nonreciprocal phase shift. Since, for the vast majority of applications, $\Delta\phi_r$ is a small number, a phase bias must be added to overcome the zero sensitivity and direction ambiguity associated with this raised cosine dependence. This phase bias is introduced by placing a phase modulator near one end of the loop as shown in Figure 9.2 [14]. The clockwise traveling light wave encounters the phase modulator at a time, T, later than the counterclockwise traveling wave, where $T = n_g L/c$, is the delay time of the light around the loop, $n_g = n - \lambda \, dn/d\lambda$ being the group index of the fiber. If a phase modulation signal $\Phi(t)$ is applied to the phase modulator, a phase difference modulation, $\Phi(t) - \Phi(t - T) \equiv \phi_m(t)$, appears between the two counterpropagating waves at the output of the loop. With this phase modulation activated in the optical circuit, the intensity of the light falling on the photodetector becomes

$$I_d = \frac{I_0}{2}\{1 + \cos[\Delta\phi_r + \phi_m(t)]\} \tag{9.11}$$

The task now is to choose $\Phi(t)$ to give the desired $\phi_m(t)$ so that $\Delta\phi_r$ can be easily and accurately retrieved.

The most straightforward solution, which forms the basis for the majority of the so-called 'open loop' designs, is to let $\Phi(t) = A \sin \omega_m t$. A fiber wound PZT modulator can easily provide this signal. In this case, $\phi_m(t)$ becomes $2A \sin[\omega_m T/2] \cos[\omega_m(t - T/2)]$. For mathematical convenience, we change the time base, $t - T/2 \equiv t'$, and define $2A \sin[\omega_m T/2] \equiv \phi_m$. Thus for the sinusoidally modulated fiber gyroscope, the detected intensity is

$$I_d = \frac{I_0}{2}\{1 + \cos[\Delta\phi_r + \phi_m \cos \omega_m t']\} \tag{9.12}$$

Figure 9.3 depicts the raised cosine response of the interference pattern, along with the intensity as a function of time for the sinusoidally modulated fiber gyro.

To show how the rotation information, $\Delta\phi_r$, can be recovered, we expand equation 9.12 into its Fourier components,

$$I_d = \frac{I_0}{2}\{1 + J_0(\phi_m)\cos\Delta\phi_r - 2J_1(\phi_m)\sin\Delta\phi_r \cos\omega_m t'$$
$$- 2J_2(\phi_m)\cos\Delta\phi_r \cos 2\omega_m t' + \cdots\} \tag{9.13}$$

where $J_n(\cdot)$ represents the nth order Bessel function of the first kind. In general, the intensity variations at the odd harmonics of the bias modulation frequency depend on $\sin \Delta\phi_r$, and the even harmonic variations depend on $\cos \Delta\phi_r$. Thus, for example, if the first harmonic is synchronously demodulated, the output obtained is $I_0 J_1(\phi_m) \sin \Delta\phi_r$, giving maximum sensitivity around zero rotation rate, and no direction ambiguity.

9.2.4 Light source

Thus far, the light reaching the detector has been considered to be a classical two-beam interference between the clockwise and counterclockwise light waves. In practice, however, there are many additional light waves reaching the detector, arising from (i) spurious paths through the loop associated with polarization cross-coupling points in combination with an imperfect input/output

SIGNAL PROCESSING

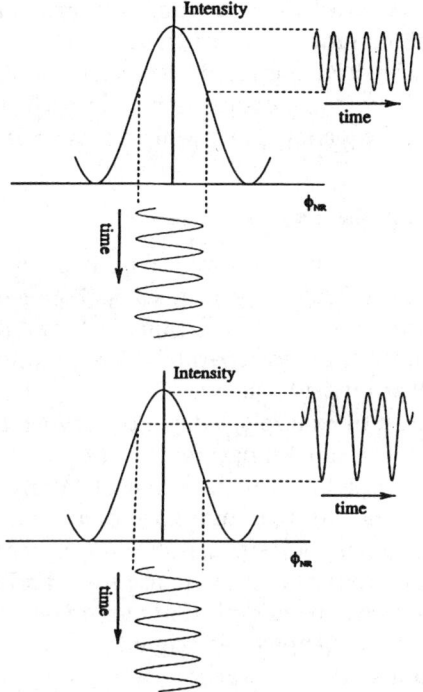

Fig. 9.3 Fiber gyro response to nonreciprocal phase shifts.

polarizer, and (ii) backscattered light. The coherence between these spurious light waves both with the main waves and each other, can be destroyed by using a broadband light source [12, 13, 15]. The two main waves retain their full coherence with each other, no matter how broadband the light source is, since they have traveled reciprocal paths. In this respect, the broader the linewidth of the source, the better.

Other important considerations for the light source are its noise properties and wavelength stability. It is important to maintain the stability of the mean wavelength of the light source since it directly enters into the relationship between the rotation rate and the nonreciprocal phase shift.

9.3 SIGNAL PROCESSING

9.3.1 Proper frequency

An important consideration in the design of the fiber gyroscope is the frequency of modulation, $f_m = \omega_m/2\pi$. The modulation depth applied to the system, $\phi_m = 2A \sin[\omega_m T/2]$, is maximized for a given drive amplitude, A, when $\omega_m T/2 = \pi/2$, or $f_m = 1/2T$. At this so-called 'proper frequency', the phase modulation applied to the clockwise traveling light wave is 180° out of phase with that applied to the counterclockwise traveling light wave. In addition to minimizing the modulator drive amplitude required to obtain a given modulation depth, a number of errors associated with an imperfect modulator are suppressed by the use of this frequency. These will be described in section 9.4.2.

9.3.2 Open loop signal processing considerations

The detected intensity, I_d, of the open loop output of the fiber gyro contains three parameters, I_0, ϕ_m, and $\Delta\phi_r$ (equation 9.12). To make an accurate determination of $\Delta\phi_r$, precautions against variations in I_0 and ϕ_m must be taken. A straightforward way to do this is to separately demodulate and compare three harmonics of the detected intensity such as DC, 1st, and 2nd, or 1st, 2nd, and 4th, or 2nd, 3rd, and 4th. For example, ϕ_m can be servoed to a fixed number by fixing the ratio of the 2nd to the 4th harmonic, $J_2(\phi_m)/J_4(\phi_m)$. Then the variations in the light intensity can be normalized out by taking the output to be the ratio of the 1st harmonic to the 2nd harmonic, yielding an output of $J_1(\phi_m)/J_2(\phi_m) \cdot \tan \Delta\phi_r$. Open loop signal processing schemes of this kind work well up to about 1 rad of rotation-induced phase shift, and the linearity is quite good for small rotation rates. Given typical minimum detectable phase shifts in the 0.1 µrad range, this type of open loop signal processing is useful to about seven orders of magnitude of rotation rates.

The use of open loop signal processing involves an inherent tradeoff between achieving scale factor accuracy and achieving low random noise performance. Consider the transimpedence amplifier detection circuit shown in Figure 9.4. To accurately control the modulation depth, and to properly normalize the output against changes in the light intensity, it is incumbent that each harmonic be demodulated accurately. This in turn requires that the gain of the amplifier be well stabilized for all three harmonics being used, which in practice means using a very wide bandwidth amplifier so that the gain is flat versus frequency for the

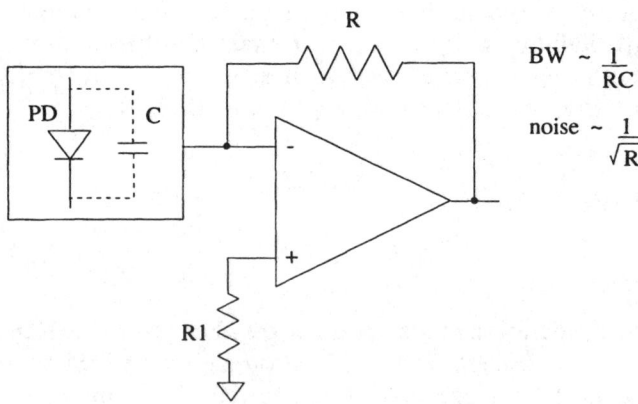

Fig. 9.4 Transimpedance photo-detection circuit.

harmonics being used. A wide bandwidth amplifier requires the use of a small feedback resistor ($BW \sim 1/RC$), which in turn yields a large Johnson noise current flowing in the resistor (r.m.s. current noise amplitude $= \sqrt{4kT/R}$).

This Johnson noise problem can be overcome by one of two methods. The first is to use a high power light source to boost the signal, rendering the Johnson noise relatively small. The second is to operate the gyro at a low modulation frequency to reduce the bandwidth requirement. However, this entails moving off the proper frequency in which case the instrument is extremely sensitive to modulator errors. It is interesting to note that the first commercial flight qualified fiber gyro (manufactured by Honeywell Inc.) uses an open loop signal processing scheme with a modulation frequency that is less than 10% of the proper frequency [2]. The use of this low frequency was directly motivated by the random noise requirement.

9.3.3 Closed loop signal processing

In closed loop signal processing, an extra nonreciprocal phase shift, equal and opposite to the rotation-induced phase shift, is induced between the counterpropagating light waves in addition to the usual bias modulation [14, 16, 17]. The odd harmonics of the usual bias modulation in the open loop output of the gyro are thereby nulled. The extra phase shift required to do this becomes the output, leading to an extraordinarily linear and stable scale factor compared to that achieved by using the raw open loop output. The techniques currently in use for doing this consist of complicating the modulation waveform. Since PZTs only operate well as narrow band devices, closed loop gyros use electro-optic modulators. At the time of this writing, scale factor linearities and stabilities of several ppm are being achieved in closed loop gyros.

Consider the basic equation for the detected intensity of the gyroscope:

$$I_d = \frac{I_0}{2}\{1 + \cos[\Delta\phi_r + \Phi(t) - \Phi(t-T)]\} \tag{9.14}$$

In closed loop gyros, $\Phi(t)$ contains two parts: (i) a phase modulation term, which could be a sine wave as was considered for the open loop gyro, but it can also be a square wave which is friendlier to digital signal processing, and (ii) a frequency shift term, which would normally be constructed using either a serrodyne waveform (Figure 9.5(a)) [18], a digital phase ramp (staircase type waveform — Figure 9.5(b)) [11], or a triangular 'dual ramp' waveform (Figure 9.5(c)) [19]. To mathematically develop this concept, consider the modulation signal to consist of a sine wave modulation on top of a frequency shift, Δf,

$$\Phi(t) = A\sin\omega_m t - 2\pi\Delta f t \tag{9.15}$$

In this case, the phase difference modulation becomes

$$\Phi(t) - \Phi(t-T) = \phi_m \cos\omega_m t' - 2\pi\Delta f T \tag{9.16}$$

Substituting equation 9.16 into 9.14, we obtain the detected intensity to be

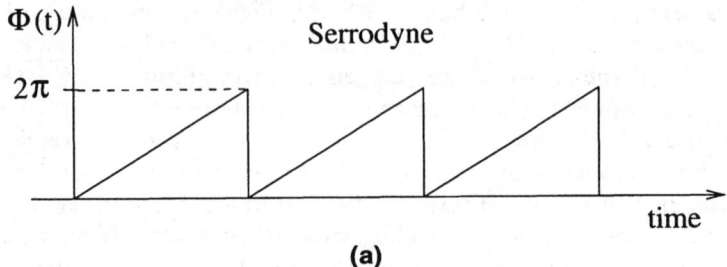

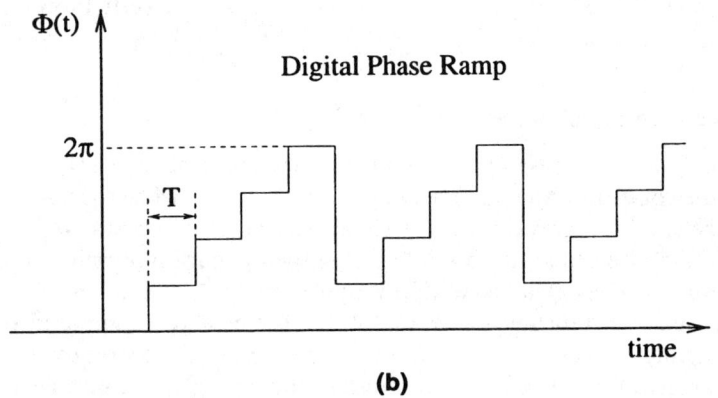

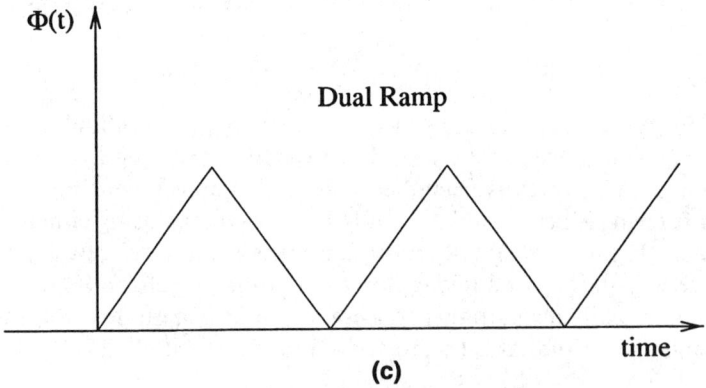

Fig. 9.5 Feedback signals waveforms for operating the fiber gyro closed loop: (a) serrodyne frequency shifter; (b) digital phase ramp; (c) dual ramp.

$$I_d = \frac{I_0}{2}\{1 + \cos[\Delta\phi_r - 2\pi\Delta f T + \phi_m \cos\omega_m t']\} \quad (9.17)$$

Now, the harmonic component of the detected intensity at ω_m, or any odd multiple is nulled if

$$\Delta\phi_r = 2\pi\Delta f T \quad (9.18)$$

independent of the light intensity, I_0, the modulation depth, ϕ_m, or the detector bandwidth. We see that the frequency shift necessary to null the odd harmonic components, Δf, is linear with respect to the rotation-induced phase shift, $\Delta\phi_r$. (Actually, a very slight nonlinearity in this relationship can exist for large phase shifts when a broadband light source is used owing to the attenuated raised cosine response of the interferometer stemming from the finite coherence of the light source [20].) Using the relations, $T = n_g L/C$ and $\Delta\phi_r = (2\pi LD/\lambda c)\Omega$ derived earlier for a circular gyro geometry, we find that the frequency shift necessary to null the odd harmonics is related to the rotation rate by

$$\Delta f = \frac{D}{\lambda n_g}\Omega \quad (9.19)$$

Even though this expression is independent of the loop length, L, the signal to noise ratio improves approximately linearly with loop length. Random noise in the system manifests itself as an uncertainty in the rotation-induced phase shift. Longer loop lengths relegate that uncertainty to correspondingly lower rotation rates. The improvement is only approximately linear, owing to the increased optical loss associated with longer loop lengths [21].

The angle through which the gyro rotates in a time, T_R, is simply ΩT_R, which from equation 9.19 is $(\lambda n_g/D)\Delta f T_R$. But the quantity $\Delta f T_R$ is just the number of resets a serrodyne frequency shifter undergoes in the time T_R. Thus

$$\text{rotation angle} = \frac{\lambda n_g}{D} (\text{No. of serrodyne resets}) \quad (9.20)$$

As a numerical example, the gyro having a diameter $D = 7.5$ cm, and a wavelength $\lambda = 1.55$ μm rotating at $0.01°$ h^{-1} is reconsidered. For these parameters, the frequency shift necessary to null the odd harmonic signals in the detected output is $\Delta f = 0.0016$ Hz. The rotation angle through which the gyro moves is found to be 30 μrad per serrodyne reset or 6.2 arcsec per reset. Smaller increments of angular rotations can still be measured, however, by noting the extent the phase of the serrodyne has ramped prior to it reaching 2π.

A major problem associated with using a serrodyne frequency shifter to close the loop of the fiber gyro is that the refractive index of the glass appears in the scale factor relating the frequency shift to the rotation rate. Since the refractive index of the fiber changes by about 10 ppm °C^{-1}, the scale factor of the gyro also exhibits this temperature dependence. One way to suppress this dependence is to use the digital phase ramp waveform (shown in Figure 9.5(b)) in place of the serrodyne frequency shifter. The time allotted to each step on the staircase is the loop delay time. Thus, one of the counterpropagating light waves always 'sees' a modulation signal exactly one 'step' behind the other. The phase difference between the two light waves is thus the step height and does not depend on the refractive index of the fiber. By setting the step height to be equal and opposite to the rotation-induced phase shift, the odd harmonics of the modulation signal are nulled and the loop is closed.

9.4 ERROR SOURCES

The measurement of the extremely small phase shifts ordinarily encountered in the fiber gyroscope require an extraordinary control over the various errors that can arise. A large percentage of the research effort on the fiber gyro has been devoted to understanding the various error sources and developing methods for their suppression. The major sources of error which affect the drift of the gyroscope are (i) polarization error, (ii) modulator-induced errors, (iii) thermal and mechanical errors, and (iv) magnetic field sensitivity. Two additional error sources, Kerr effect error, and backscatter error, gave large errors in the early days of the fiber gyro, but their importance diminished once it was recognized that broadband sources should be used [15, 22–25].

9.4.1 Polarization error

The analysis of the error associated with an imperfect polarizer in the common input/output port of the fiber gyroscope requires a calculation of the phase difference between the counterpropagating light waves starting and ending at a reference plane to the left of the polarizer, as shown in Figure 9.6. If the polarizer is perfect, the two paths considered are reciprocal, and hence, no polarization error exists. However, polarizers are not perfect, although the integrated optics and fiber types incorporated into fiber gyros typically exhibit better than 60 dB performance.

Detailed calculations show that polarization error nicely divides into two terms [26]; the so-called 'amplitude type error', $\Delta\phi_{amp}$ (sometimes known as 'Kintner error' [27]) is proportional to the amplitude extinction ratio of the polarizer, ε, and the so-called 'intensity type error', $\Delta\phi_{int}$, is proportional to the

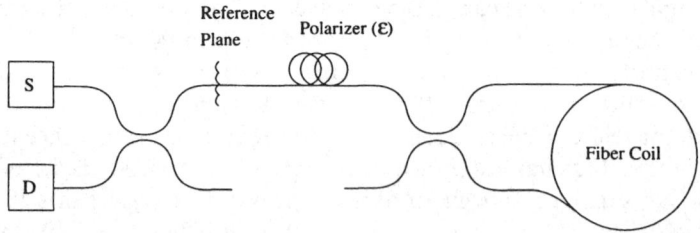

Fig. 9.6 Reference plane location for calculating polarization error in a fiber gyro.

intensity extinction ratio of the polarizer, ε^2. Errors proportional to the third and fourth powers of ε are usually considered negligible.

The physical mechanism of the amplitude type error is most easily illustrated for5 the PM gyro shown in Figure 9.7. It is caused by light entering through the reject axis of the polarizer and coupling to the pass axis within the loop via the coupling point, k_L. The magnitude of the error caused by the polarization coupling point within the loop is calculated by considering the four waves which exit the gyro in the pass axis of the polarizer. The two main waves (clockwise and counterclockwise) always traverse the gyro in the pass axis of the polarizer. Two spurious waves exist, originating from the left of the polarizer in its reject axis, passing through the polarizer in the reject axis and suffering an attenuation, ε, but coupling to the pass axis at the coupling point in the loop. In Figure 9.7, the spurious clockwise wave couples just after the light enters the loop, and the spurious counterclockwise wave couples just before exiting the loop. Since the counterclockwise traveling spurious wave traverses the whole loop in the reject axis, it becomes uncorrelated with the other waves and can be safely neglected. The clockwise wave, on the other hand, only traverses the polarizer and the beginning of the loop in the reject axis, and therefore may maintain some coherency with the main waves.

The error arises because of the interference between the spurious clockwise wave and the main counterclockwise wave; the interference between the spurious clockwise wave and the main clockwise wave is unmodulated and therefore falls outside the detection bandwidth of the demodulator. Keeping only the two main waves and the one spurious wave, the output intensity is written as

$$I_{out}(t) = |E_{pass}(t-T)e^{-j\Phi(t-T)} + E_{pass}(t-T)e^{-j\Phi(t)} + \varepsilon k_1 E_{reject}(t-T-\Delta\tau)e^{-j\Phi(t)}|^2 \quad (9.21)$$

where T is the loop delay time and $\Delta\tau$ is the excess time delay associated with the spurious wave path. The maximum phase error associated with the spurious wave is given by

$$\max \Delta\phi_{amp} = \varepsilon k_L \frac{|\langle E_{pass}(t)E^*_{reject}(t-\Delta\tau)\rangle|}{\langle |E_{pass}(t)|^2 \rangle} \quad (9.22)$$

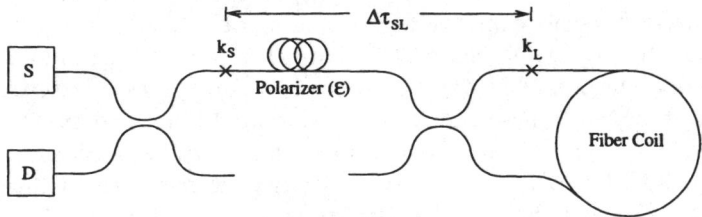

Fig. 9.7 Fiber gyro having polarization coupling points near the source, k_s, and within the loop, k_L.

Several points can now be made with regard to the design of the gyro to minimize this error.

1. A large birefringent element between the polarizer and the loop coupler makes for a large $\Delta\tau$, which suppresses this error by decorrelating E_{pass} and E_{reject} [28].
2. The error can be suppressed by minimizing the amount of light incident on the reject axis of the polarizer, i.e. setting E_{reject} as close to zero as possible. One method to do this is to use a highly polarized light source and then to align the axis of the source polarization to the polarizer.
3. The error can be minimized by arranging things so that E_{reject} is always uncorrelated with E_{pass}. For every optical source, there exists two orthogonal polarization states such that the fields contained in those states are uncorrelated [29]. For laser diodes, those two states are typically aligned to the geometrical axes of the diode. Thus aligning the geometrical axes of the laser diode to the axes of the polarizer helps to minimize this error [30]. Either (2) or (3) can be implemented by using PM elements between the source and the polarizer. When (2) is implemented, the maximum value of the correlation between E_{pass} and E_{reject} is given as

$$\frac{|\langle E_{\text{pass}}(t) E^*_{\text{reject}}(t-\Delta\tau)\rangle|}{\langle |E_{\text{pass}}(t)|^2\rangle} = k_S |\gamma(\Delta\tau_{\text{SL}})| \qquad (9.23)$$

where k_S is the amplitude of a polarization coupling point on the source side of the polarizer, and $|\gamma(\Delta\tau_{\text{SL}})|$ is the magnitude of the coherence function of the light, $\Delta\tau_{\text{SL}}$ being, specifically, the extra delay associated with the path between the two coupling points, k_S and k_L. When (3) is implemented, this correlation coefficient may be lower than this, depending on the coherence functions of the light emitted from the two orthogonal axes of the source. Substituting equation 9.22 into 9.23 yields

$$\max \Delta\phi_{\text{amp}} = \varepsilon k_S k_L |\gamma(\Delta\tau_{\text{SL}})| \qquad (9.24)$$

As a numerical example, assuming a 60 dB polarizer ($\varepsilon = 10^{-3}$), and 20 dB coupling points ($k_S = k_L = 10^{-1}$), then $|\gamma(\Delta\tau_{\text{SL}})|$ must be less than 10^{-1} to obtain a gyro with less than 1 µrad amplitude type polarization error performance. In fact, by carefully positioning the fiber splice points, it is fairly straightforward to guarantee that $|\gamma(\Delta\tau_{\text{SL}})|$ remains less than 10^{-3} for all significant pairs of coupling points. Thus, max $\Delta\phi_{\text{amp}}$ can quite easily be kept to less than 10 nrad.

Intensity type errors arise due to the interference between counterpropagating waves both of which (i) enter through the polarizer in its reject axis, couple to the pass axis within the loop, and exit through the pass axis of the polarizer, or (ii) enter through the polarizer in its pass axis, couple to the reject axis within the loop, and exit through the reject axis of the polarizer. In either case, the waves will be coherent with each other if there are pairs of polarization

ERROR SOURCES

cross-coupling points located symmetrically within the loop (Figure 9.8). Now, a pair of polarization cross-coupling points located at approximately symmetric positions within the loop creates both types of intensity error described above. It turns out that these two errors are 180° out of phase, so that no intensity type error exists when equal amounts of light impinge on the pass and reject axes of the polarizer [26]. It is found that

$$\Delta\phi_{\text{int}} = (1 - P_{\text{pass}}/P_{\text{reject}})k_1 k_2 \varepsilon |\gamma(\Delta\tau_{12})| \qquad (9.25)$$

where $1 - P_{\text{pass}}/P_{\text{reject}}$ is the fractional power imbalance of the light impinging on the polarizer and $\Delta\tau_{12}$ is the time delay asymmetry of the coupling points within the loop. For a numerical example, using the 60 dB polarizer and 20 dB polarization cross-coupling points assumed before, the intensity type error is easily kept well below 10 nrad even when only minimal attention is paid to the location of splice points.

There is some tension between the suppression of amplitude and intensity type errors, depending on the light source used. The amplitude type error is suppressed best when there is no light impinging on the reject axis of the polarizer, while the intensity type error is suppressed best when there is an equal amount of light impinging on both axes of the polarizer. In general, one should favor the suppression of the amplitude type error. Both errors can be suppressed simultaneously, however, by using a light source which has equal power in its two eigenaxes, but those two axes are aligned to the polarizer axes.

9.4.2 Modulator error

Loss modulation and waveform distortion associated with the phase modulator can greatly deteriorate the measurement capability of the fiber gyro. Although the intrinsic loss modulation associated with both PZT and integrated optic type phase modulators can be negligible, both types of modulators can exhibit polarization modulation, which, in combination with the polarizer in the optical circuit gives rise to a loss modulation in the overall circuit [2, 31].

The loss modulation problem is analyzed for the open loop configuration by considering the modulator to act on the wave as [32]

$$\text{modulator transfer function} = e^{jA\sin\omega t} \cdot [1 + \delta \sin \omega t] \qquad (9.26)$$

where δ is the parasitic amplitude modulation of the passing wave and is

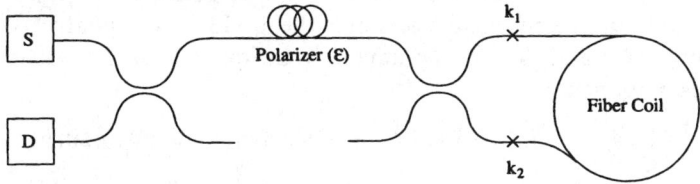

Fig. 9.8 Fiber gyro having polarization coupling points on opposite sides of the loop, k_1 and k_2.

assumed to be synchronous with the usual phase modulation, $A \sin \omega t$. The gyro response to phase and amplitude modulations is fundamentally different. The net phase modulation seen by the two counterpropagating light waves in the gyro loop is given by the difference between what each wave 'sees':

$$\text{net phase modulation} = A \sin \omega t - A \sin \omega(t - T) \tag{9.27}$$

T being the loop delay time, whereas the net amplitude modulation contained in the output after the two counterpropagating light waves are combined is given by the sum of what the two waves 'see':

$$\text{net amplitude modulation} = \delta \sin \omega t + \delta \sin \omega(t - T) \tag{9.28}$$

Writing $t = t' + T/2$ as before yields

$$\text{net phase modulation} = 2A \sin(\omega T/2) \cos \omega t' \tag{9.29}$$

and

$$\text{net amplitude modulation} = 2\delta \cos(\omega T/2) \sin \omega t' \tag{9.30}$$

From equations 9.29 and 9.30, it is evident that parasitic polarization and/or loss modulation yields a signal in quadrature to the main signal, and further, that it is nulled when the gyro is operated at its proper frequency, $\omega = \pi/T$. When the modulator is operated at the proper frequency, the amplitude modulation induced in the clockwise traveling light wave is 180° out of phase with that induced in the counterclockwise traveling light wave, so that when the two are combined, the two amplitude modulations cancel.

The second main error associated with the modulator is that a significant offset in the gyro output results from the presence of even harmonics of the bias modulation being present in the modulator waveform [33, 34]. Consider the gyro output where the sinusoidal bias modulation is corrupted by a second harmonic distortion:

$$I_{\text{out}} = \frac{I_0}{2}\{1 + \cos[\Delta\phi_r + \Phi_m \cos \omega t + \delta\Phi_2 \cos(2\omega t - \theta)]\} \tag{9.31}$$

Here, $\delta\Phi_2$ is the amplitude of the second harmonic distortion in the phase difference modulation. When the gyro is operated at its proper frequency, $\delta\Phi_2$ is identically zero, since the phase difference modulation, $\Phi(t) - \Phi(t - T)$, cannot contain even harmonics of the fundamental frequency, $f = 1/2T$. Thus equation 9.31 represents that case when the gyro is operated off its proper frequency, and there is second harmonic distortion in the modulating signal.

Assuming $\delta\Phi_2$ and $\Delta\phi_r$ to be small, equation 9.31 is found to have first harmonic components

$$I_{\omega\text{-signal}} = I_0\{J_1(\Phi_m)\Delta\phi_r + \delta\Phi_2[J_1(\Phi_m) - J_3(\Phi_m)]\sin\theta\} \tag{9.32}$$

and

$$I_{\omega\text{-quadrature}} = I_0\delta\Phi_2[J_1(\Phi_m) + J_3(\Phi_m)]\cos\theta \tag{9.33}$$

ERROR SOURCES

Usually, the second harmonic distortion follows a square law behavior with respect to the modulator drive, i.e. $\delta\Phi_2 \propto \Phi_m^2$ so that the bias offset in the gyro, Φ_{offset}, as a function of Φ_m follows the form (plotted in Figure 9.9)

$$\Phi_{\text{offset}} \propto \Phi_m^2 [J_1(\Phi_m) - J_3(\Phi_m)] \qquad (9.34)$$

Carrara's suggested options for suppressing this error [33] are to operate at $\Phi_m \approx 3.0$, where $J_1(\Phi_m) = J_3(\Phi_m)$, or to operate the gyro at a low modulation depth (e.g. $\Phi_m \approx 0.6$) to minimize the distortion.

An analysis of the θ dependence of this error reveals that it is modulator hysteresis that gives rise to the signal channel error, while a nonlinearity gives rise to a nonzero output in the quadrature channel.

For a gyro operated well off its proper frequency, the bias error is of the order of $\delta\Phi_2$. Thus, if the modulation waveform contains second harmonic distortion of the order -120 dB, an error of 1 μrad error in the nonreciprocal phase measurement results. The severity of the problem is appreciated when considering that conventional electronics instrumentation commonly contains second harmonic distortion levels of -60 dB to -80 dB, which would result in offsets in the gyro output of the order of $100-1000$ μrad.

On the other hand, for a gyro operated at 99.9% of its proper frequency, the level of $\delta\Phi_2$ seen for the same distortion in the modulator is reduced by a factor of 1000. Thus the modulation waveform requires a second harmonic distortion level only below -60 dB to achieve an offset less than 1 μrad. This is about how close to the proper frequency one can expect to remain over the operating environment of the gyro, since the proper frequency changes with temperature by 10 ppm °C^{-1}, commensurate with the change in the refractive index of the fiber glass.

In conclusion, the errors associated with an imperfect modulator are suppressed when the gyro is operated at its proper frequency. This is the case for all

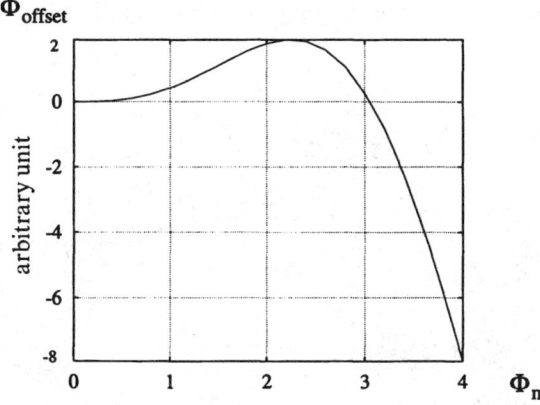

Fig. 9.9 Fiber gyro rate measurement offset as a function of the modulation depth, Φ_m, due to modulator hysteresis.

9.4.3 Mechanical and thermal gradient errors

Time-varying mechanical or thermal gradients in the fiber coil gives rise to a large nonreciprocal phase shift between the counterpropagating light waves [35]. Consider the case when a section of fiber on one side of the loop experiences a time-varying perturbation. The phase of the light passing through the perturbed section changes with time. The counterpropagating light waves encounter this section of fiber at different times, and thus 'see' different optical paths.

The solution to this problem is good packaging of the coil. Multiple metal housings are used to equalize the temperature around the coil. In addition, the fiber coil is wound starting from the middle of the coil, alternating the wind from each side. The idea is to ensure as much as possible that thermal and mechanical gradients across the coil affect the coil in a symmetric way, so as to minimize any differences between the clockwise and counterclockwise propagation paths. A popular coil wind is the so-called quadrupole wind [36], illustrated in Figure 9.10. Other more complicated winds that better ensure the symmetry of the coil are sometimes used for very high performance applications.

9.4.4 Random noise

The fiber gyro acts to encode the rotation rate it feels as intensity fluctuations in the light impinging on the photodetector and thus fluctuations in the electrical current flowing in the photodetector. The various modulation schemes that may be applied to the gyro do the encoding differently, but it is evident that fluctuations due to noise in either the light intensity or the photodetector current can yield a signal indistinguishable from that induced by rotation.

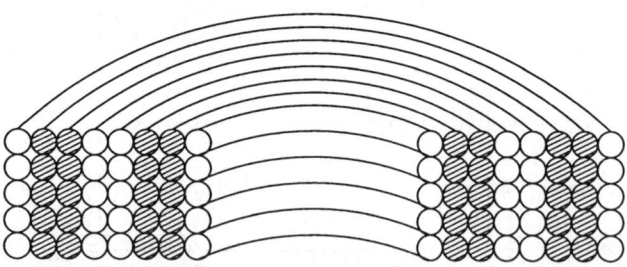

Fig. 9.10 Schematic of a quadrupole wound fiber gyro coil. Open layers represent fiber to the left of the loop center point, shaded layers represent fiber to the right of the loop center point.

The most important of the random noise sources are (i) light source intensity noise, (ii) shot noise, and (iii) thermal noise in the photodetector circuit.

The noise that dominates the random noise in the gyro output is largely a function of the light power falling on the detector. The thermal noise in the circuit is independent of the light power falling on the detector and typically dominates the light power dependent noise terms when the light power falling on the photodetector is less than 1 µW. The shot noise depends on the square root of the light power falling on the detector and typically dominates when the light power falling on the photodetector is in the 1–10 µW range. The intensity noise is proportional to intensity (as is the signal strength), meaning that the signal to noise ratio is independent of the light power falling on the detector when this noise source dominates. This is typically the case when more than 10 µW falls on the detector. However, it is important to note that the light source intensity noise can be measured at the source coupler dead end, and thus can be subtracted out from the main gyro signal [37].

When the light falling on the photodetector is polarized thermal light, as is desired for the suppression of Kerr effect error [23], it can be shown that the relative intensity noise, RIN, is given by [29]

$$\text{RIN} = \sqrt{2\tau_c} \text{ dB Hz}^{-1/2} \tag{9.35}$$

where τ_c is the coherence time of the light and is related to the coherence function, $\gamma(\tau)$, and the light power spectrum, $I(\nu)$, by the relations

$$\tau_c = \int_{-\infty}^{+\infty} |\gamma(\tau)|^2 \, d\tau = \frac{\int_0^\infty I^2(\nu) \, d\nu}{\left[\int_0^\infty I(\nu) \, d\nu\right]^2} \tag{9.36}$$

The intensity noise contribution is greater than the shot noise contribution when $\eta I_0 \sqrt{2\tau_c} > \sqrt{2e\eta I_0}$ where η is the photodetector responsivity (amps watt^{-1}), e is the electronic charge, and I_0 is the average light intensity falling on the detector as seen by the demodulation electronics. When the quantum efficiency of the detector approaches 1 electron/photon, the intensity noise exceeds the shot noise when more than one photon per coherence time impinges on the photodetector, i.e. when $I_0 > h\nu/\tau_c$, where $h\nu$ is the energy contained in each photon. Shot noise limited fiber gyros essentially operate a single photon at a time, i.e. on average, each photon arriving at the photodetector is uncorrelated with all the others.

Square wave modulated fiber gyros exhibit the important feature that the intensity noise can be greatly suppressed by properly choosing their modulation depth, ϕ_m [11]. The average intensity $\propto \cos^2(\phi_m)$ whereas the sensitivity (signal) $\propto \sin(2\phi_m)$. Thus the signal to intensity noise ratio $\propto 2\tan(\phi_m)$ which becomes infinite at $\phi_m = \pi$. Now, the signal itself is zero when $\phi_m = \pi$, so the presence of other noise sources prohibits the attainment of this infinite signal to noise ratio. However, large improvements in the overall signal to noise ratio can be achieved by operating near $\phi_m = \pi$.

9.5 DEPOLARIZED FIBER GYROS

Depolarized architectures for the fiber gyro are nearly as old as the 'minimum reciprocal configuration' [13, 38]. Its continuing attractiveness stems from its use of low cost telecommunications grade single-mode fiber for the coil instead of the more expensive PM fiber. However, because the depolarized fiber gyro was widely perceived to contain significantly more performance risk than the PM fiber gyro, most early product development efforts, and indeed the first products, used PM architectures.

The signal fading problem associated with the use of a nonpolarization preserving single-mode (SM) fiber in the loop can be overcome by inserting a depolarizer in the loop. The depolarizer distributes the polarization states of the individual wavelengths of the broadband source to guarantee that half the light falls on the polarizer pass axis on its return trip to the detector. Thus 3 dB excess loss is incurred as a trade for power stability and the use of an SM fiber loop. The depolarizer must be placed within the loop to avoid signal fading; the use of a depolarizer between the polarizer and the loop does not work. In this case the light waves pass the same depolarizer twice; the second pass can undo the depolarization accomplished during the first pass.

The oft-used Lyot depolarizer is shown in Figure 9.11. It comprises two sections of PM fiber, one twice the length of the other, joined with a 45° splice. Denoting the time delay differences, τ_d and $2\tau_d$, between the birefringent axes of each section, it can be shown that light having an arbitrary state of polarization is depolarized when the condition that

$$\gamma(\tau_d) = \gamma(2\tau_d) = \gamma(3\tau_d) = 0 \qquad (9.37)$$

is satisfied. Here $\gamma(\tau)$ is the coherence function of the light. This condition is easy to meet when the light source has a smooth broadband spectrum, since $\gamma(\tau)$ tends to zero for all large τ. However, forming a good depolarizer requires cutting the PM fiber lengths to precise values when the light source is a multimode laser diode which exhibits periodic peaks in the coherence function. Figure 9.12 shows example Lyot depolarizer length cut options for depolarizing multimode laser diode light [39]. Some safety margin must be kept between the delays of the depolarizer and the peaks in the coherence function since the

Fig. 9.11 Lyot depolarizer. τ_d and $2\tau_d$ represent the relative delays between the polarization eigenaxes for the respective fiber sections.

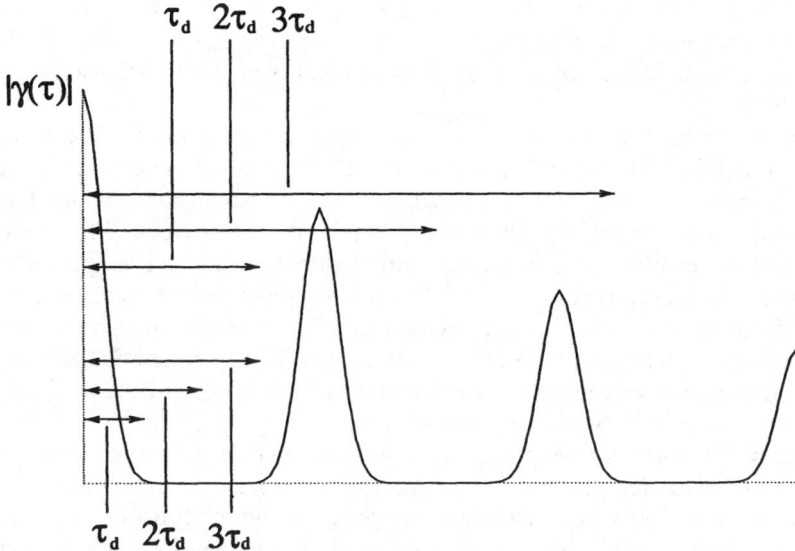

Fig. 9.12 Coherence function map representation of length cut options for a Lyot depolarizer to depolarize a multimode laser diode source.

depolarizer delays are temperature dependent and the SM coil contains some inadvertent delays associated with coil birefringence.

Although the presence of a depolarizer in the coil stabilizes the gyro against signal fading, in fact two depolarizers are required to stabilize the gyro against large magnetic field sensitivities [40]. Figure 9.13 shows the polarization evolution in a gyro having only one depolarizer. The clockwise traveling light wave evolves from a linear state at point A just after the polarizer to some definite (but unknown) polarization state, S_1 at point B, just before the depolarizer. After passing the depolarizer, at point C, the clockwise traveling light is depolarized. Depolarized light can be decomposed arbitrarily into orthogonal polarization states, S_2 and $S_2^\dagger$, which are chosen so that the light occupying the S_2 state evolves back to the linear state, x, upon propagating back to point A, and the

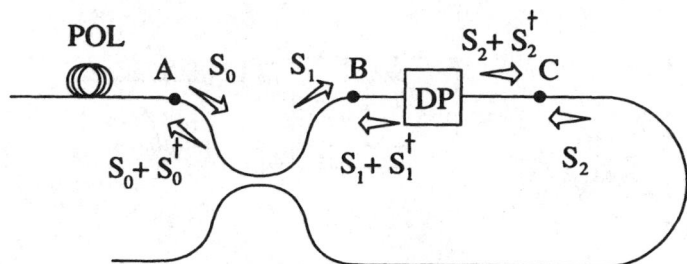

Fig. 9.13 Polarization evolution through the coil of a depolarized fiber gyro employing a single depolarizer. DP = depolarizer, POL = polarizer, S = state of polarization.

light occupying the state $S_2^\dagger$ evolves to the linear state, y, at point A. Thus the 'useful' clockwise traveling light, i.e. the light that reaches the detector, takes a definite, though unknown and environmentally sensitive, polarization path through the loop.

By reciprocity, the 'useful' counterclockwise traveling light wave takes the exact same polarization path through the fiber coil, only in reverse order. Thus, in a fiber gyro having a single depolarizer, the light which reaches the detector has taken a definite polarization path throughout the fiber coil. This may be a magnetically sensitive path since magnetic fields induce a nonreciprocal phase shift when the useful light is circularly polarized. Substantial cancellation of this effect exists since the polarization evolution of the useful light is random, and the direction of the light propagation with respect to any magnetic field reverses many times in a fiber coil. However, the residual error can still easily be tens or even hundreds of microradians in earth's field.

A second depolarizer added to the fiber circuit either between the polarizer and the loop coupler (Figure 9.14(a)) or on the other side of the loop (Figure 9.14(b)) acts to uniformly distribute the polarization evolution of the 'useful' light by wavelength, thus averaging out the magnetic field sensitivity. Depolarized gyros with two depolarizers have been observed to be orders of magnitude less sensitive to magnetic fields than similar gyros with only one depolarizer [40]. They are even less sensitive than PM fiber gyros.

When the optical power reaching the photodetector exceeds 1 photon per coherence time (several microwatts), depolarized fiber gyros exhibit a slightly higher random walk coefficient than comparable PM fiber gyros owing to the conversion of phase to amplitude noise in the depolarizer/polarizer combination that exists in the depolarized fiber gyro. The depolarizer/polarizer combination does not act to change the thermal nature of the light, but does act to

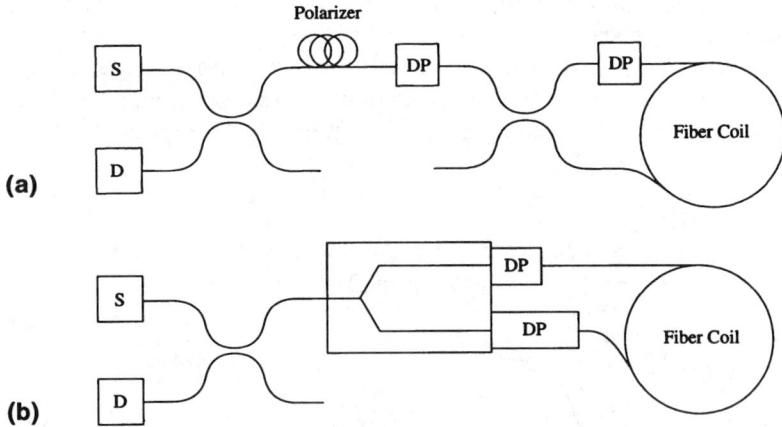

Fig. 9.14 Depolarized fiber gyro employing two depolarizers (DP): (a) one depolarizer within the loop and one between the polarizer and the loop coupler; (b) one depolarizer on either side of the loop.

channel the optical spectrum of the light that reaches the photodetector. The optical spectrum, $I(v)$, which determines the optical intensity noise, is the optical spectrum of the light falling on the detector, now different than that emitted by the light source. Assuming the simple case of a thermal source and a sinusoidally channeled spectrum falling on the detector, the coherence time of the light reaching the detector is 3/2 that of the coherence time of the light exiting the source ($\sqrt{1.5} = 1.21$), yielding a 21% increase in the random walk coefficient over a comparable PM fiber gyro.

9.6   3 × 3 FIBER GYRO

The 3 × 3 fiber gyro (shown in Figure 9.15) has been proposed as an ultra-low cost fiber gyro [41, 42]. It makes use of low cost telecommunications fiber as well as avoiding the use of a modulator and its attendant electronics. However, it is optically nonreciprocal, so its performance is severely limited.

The basic element of the 3 × 3 gyro is the 3 × 3 coupler shown in Figure 9.16. When three fibers are drawn together into a 3 × 3 coupler of perfect symmetry, there is a 120° phase shift associated with coupling from one fiber to another when the power split ratio is 1/3 : 1/3 : 1/3. Assuming (unrealistically) no polarization wander in the fiber coil, the intensity of the light falling on the two detectors is given by

$$I_1 = \tfrac{2}{9} I_s [1 + \cos(120° - \Delta\phi_r)] \tag{9.38}$$

$$I_2 = \tfrac{2}{9} I_s [1 + \cos(120° + \Delta\phi_r)] \tag{9.39}$$

where I_s is the source intensity. The normalized difference between these two detected intensities is

$$\frac{I_1 - I_2}{I_1 + I_2} = \frac{\sqrt{3}\sin\Delta\phi_r}{2 - \cos\Delta\phi_r} \approx \sqrt{3}\Phi_R \tag{9.40}$$

where the last approximation is for small $\Delta\phi_r$. Thus this ideal gyro yields a linear output for small rotation rates.

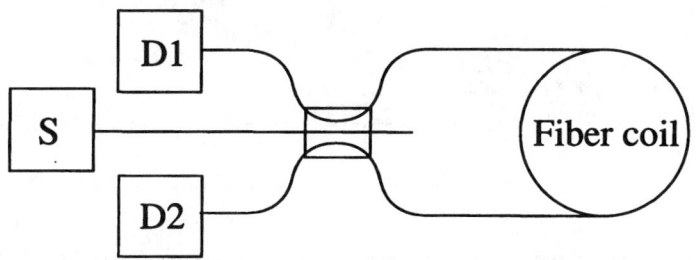

Fig. 9.15 3 × 3 fiber gyro optical circuit. S = light source, D = detector.

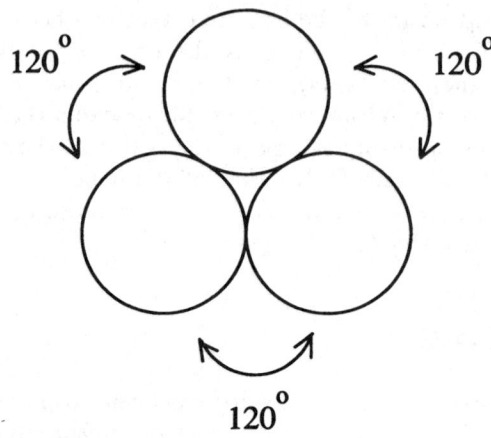

Fig. 9.16 3 × 3 coupler configuration.

To stabilize the 3 × 3 gyro output against polarization wander, one or more depolarizers should be introduced into the optical circuit [43], as shown in Figure 9.17. For this optical circuit, the intensity falling on the two detectors is given by

$$I_1 = \tfrac{1}{9} I_s [2 + \cos(120° + \Delta\phi_r)] \tag{9.41}$$

$$I_2 = \tfrac{1}{9} I_s [2 + \cos(120° - \Delta\phi_r)] \tag{9.42}$$

and the normalized difference intensity is given by

$$\frac{I_1 - I_2}{I_1 + I_2} = \frac{-\sqrt{3}\sin\Delta\phi_r}{4 - \cos\Delta\phi_r} \approx \frac{-\sqrt{3}}{3}\Delta\phi_r \tag{9.43}$$

where again the last approximation is for rotation rates corresponding to small nonreciprocal phase shifts. The accuracy of the 3 × 3 fiber gyro depends critically on the quality and stability of the 3 × 3 coupler. The drift of the gyro depends on the stability of the 120° phase shift between the adjacent coupler arms. It remains a promising approach for applications requiring drift stability not better than 100° h^{-1}.

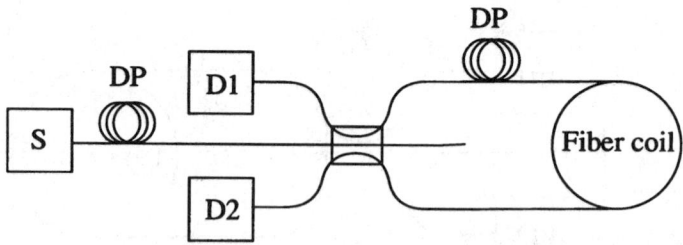

Fig. 9.17 3 × 3 depolarized fiber gyro configuration. DP = depolarizer, S = light source, D = detector.

REFERENCES

1. V. Vali and R. W. Shorthill, *Appl. Opt.* **15**, 1099 (1976).
2. J. Blake, J. Feth, J. Cox, and R. Goettsche, 'Design and test of a production open loop all fiber gyroscope', *Proc. SPIE* **1169**, 337 (1989).
3. H. Kajioka et al., 'Fiber optic gyro productization at Hitachi', *Proc. SPIE* **1585**, 17 (1991).
4. K. Sakuma, 'Fiber optic productization at JAE', *Proc. SPIE* **1585**, 8 (1991).
5. G. Pavlath, 'Productization of fiber gyros at Litton Guidance and Control Systems', *Proc. SPIE* **1585**, 2 (1991).
6. H. C. Lefevre, P. Martin, and J. Morisse, 'Fiber optic gyro productization at Photonetics', *Proc. SPIE* **1585**, 42 (1991).
7. W. Auch, M. Oswald, and R. Regener, 'Fiber optic gyro productization at Alcatel SEL', *Proc. SPIE* **1585**, 65 (1991).
8. This derivation follows that given by S. Ezekiel, S. P. Smith, and F. Zarinechti, in *Optical Fiber Rotation Sensing*, Chapter 1, Edited by W. K. Burns, Academic Press (1994).
9. H. J. Arditty and H. C. Lefevre, *Opt. Lett.* **6**, 401 (1981).
10. R. Ulrich, *Opt. Lett.* **5**, 173 (1980).
11. H. C. Lefevre, S. Vatoux, M. Papuchon, and C. Puech, 'Integrated optics: a practical solution for the fiber-optic gyroscope', *Proc. SPIE* **719**, 101 (1986).
12. W. K. Burns, C. L. Chen, and R. P. Moeller, *J. Lightwave Technol.* **LT-1**, 98 (1983).
13. K. Bohm, P. Marten, K. Petermann, E. Weidel, and R. Ulrich, *Electron. Lett.* **17**, 352 (1981).
14. R. F. Cahill and E. Udd, 'Phase nulling optical gyro', US Patent 4,299,490 (1981).
15. C. C. Cutler, S. A. Newton, and H. J. Shaw, *Opt. Lett.* **5**, 488 (1980).
16. J. L. Davis and S. Ezekiel, 'Techniques for shot-noise limited inertial rotation measurement using a multiturn fiber Sagnac interferometer', *Proc. SPIE* **157**, 131 (1978).
17. R. F. Cahill and E. Udd, *Opt. Lett.* **4**, 93 (1979)
18. A. Elberg and G. Schiffner, *Opt. Lett.* **10**, 300 (1985).
19. R. A. Bergh, 'Dual-ramp closed-loop fiber-optic gyroscope', *Proc. SPIE* **1169**, 429 (1989).
20. R. P. Moeller, W. K. Burns, and N. J. Frigo, *J. Lightwave Technol.* **7**, 262 (1989).
21. S. C. Lin and T. G. Giallorenzi, *Appl. Opt.* **18**, 915 (1979).
22. R. A. Bergh, H. C. Lefevre, and H. J. Shaw, *Opt. Lett.* **7**, 282 (1982).
23. R. A. Bergh, B. Culshaw, C. C. Cutler, H. C. Lefevre, and H. J. Shaw, *Opt. Lett.* **7**, 563 (1982).
24. N. J. Frigo, H. F. Taylor, L. Goldberg, J. F. Weller, and S. C. Rashleigh, *Opt. Lett.* **8**, 119 (1983).
25. K. Petermann, *Opt. Lett.* **7**, 623 (1982).
26. S. L. A. Carrrara, B. Y. Kim, and H. J. Shaw, *Opt. Lett.* **12**, 214–216 (1987).
27. E. C. Kintner, *Opt. Lett.* **6**, 154–156 (1981).
28. E. Jones and J. W. Parker, *Electron. Lett.* **22**, 54–56 (1986).
29. J. Goodman, in *Statistical Optics*, John Wiley & Sons, 1985.
30. S. L. A. Carrara, 'Drift reduction in optical fiber gyroscopes', Ph.D. dissertation, Stanford University (1988).
31. B. Szafraniec and J. Blake, *J. Lightwave Technol.* **12**, 1679 (1994).
32. E. Kiesel, 'Impact of modulation induced signal instabilities on fiber gyro performance', *Proc. SPIE* **838**, 129 (1987).

33. S. L. A. Carrara, 'Drift caused by phase modulator non-linearities in fiber gyroscopes', *Proc. SPIE* **1267**, 187–191 (1990).
34. B. Y. Kim, H. C. Lefèvre, R. A. Bergh, and H. J. Shaw, *Proc. SPIE*, **425**, paper 16, (1983).
35. D. M. Shupe, *Appl. Opt.* **19**, 654–655 (1980).
36. R. A. Bergh, 'All-fiber gyroscope with optical-Kerr effect compensation', Ph.D. dissertation, Stanford University (1983).
37. R. P. Moeller and W. K. Burns, *Opt. Lett.* **16**, 1902 (1991).
38. R. J. Fredricks and R. Ulrich, *Electron. Lett.* **20**, 330 (1984).
39. J. Blake, J. Feth, and B. Szafraniec, 'Configuration control of mode coupling errors', US Patent No. 5,377,283.
40. J. Blake, 'Magnetic field sensitivity of depolarized fiber optic gyros', *Proc. SPIE* **1367**, 81 (1990).
41. S. K. Sheem, *Appl. Phys. Lett.* **37**, 869 (1980).
42. G. Trommer, H. Poisel, W. Buhler, E. Hartl, and R. Muller, *Appl. Opt.* **29**, 5360 (1990).
43. G. Trommer, E. Hartl, and R. Muller, 'Progress in passive fiber optic gyroscope development', *Proc. SPIE* **2360**, 438 (1994).

10
Fiber gratings: principles, fabrication and properties

V. A. Handerek

10.1 INTRODUCTION: WHY FIBER GRATINGS?

Single mode fiber is often used for sensing when extreme sensitivity to the measurand is required. This is because this type of fiber permits the construction of guided wave interferometers directly from the fiber itself. Interferometers can be used to measure small phase changes in light transmitted through the sensing region. In simple, two-beam interferometers, this is achieved by comparing the phase of a light wave which has traversed a sensing path with the phase of another light wave originating from the same source but arriving via a protected, reference path. The phase difference can be measured with a sensitivity of $\sim 10^{-6}$ of a wavelength [1] and the path length for the measuring interaction can be millions of wavelengths long. This implies a possible measurement resolution for the optical path of one in 10^{12}! Simultaneously, the absence of free space optical paths between sources and detectors eliminates slow alignment drifts which could easily occur if bulk-optical interferometers had been used. However, in practice, these two-path interferometric sensors tend to need very stable, highly coherent sources with low phase noise in order to gain full advantage of their potential sensitivity. When such sources are used, absolute calibration of phase difference is normally not possible and a range limit arises because phase changes greater than π cannot be unambiguously distinguished from changes less than π. This is because the transfer function of these interferometers has a periodic response to phase difference, ϕ. In addition, the sensitivity of the output intensity to small phase changes around zero is itself zero. Optimum sensitivity occurs when $\phi = \pi/2$, so two-path interferometers need to be biased at this phase difference for best performance.

Multiple beam interferometers, on the other hand, divide the light into many separate paths. One way to make a component based on multiple beam interference is shown schematically in Figure 10.1. Here, each beam originates

---

Optical Fiber Sensor Technology, Vol. 2. Edited by K. T. V. Grattan and B. T. Meggitt.
Published in 1998 by Chapman & Hall, London. ISBN 0 412 782 901

330 FIBER GRATINGS

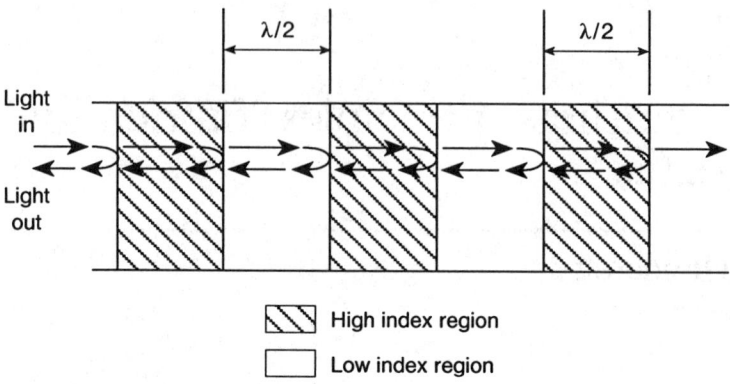

Fig. 10.1 Schmetic of a grating reflector.

from a different reflector in a set of equi-spaced, similar, weak reflectors. A set of reflectors like this is called a grating reflector and can be produced in an optical fiber by imposing a variation in the refractive index of the core periodically along the fiber axis. This can be achieved by making use of fiber photosensitivity [2]. At each interface between two regions of differing index, a small reflection occurs. The amplitude of the resultant reflected wave can be large and is determined by superposition of all of the individual reflected components. The reflection characteristics of a grating reflector show important differences from the two-path interferometer. First, the reflection response is not periodic. Light is reflected for only a very narrow wavelength range. As a result, the wavelength of peak reflection provides an absolute measure of the pitch of the index perturbation. Also, it is obviously unnecessary to provide phase bias in this case. These are both crucial advantages for the use of gratings in sensing.

This chapter begins by describing the phenomenon of photoinduced index change in fibers and briefly presents the microscopic mechanisms which have been proposed to account for it. In the next part of the chapter, the various grating types which have been demonstrated so far are introduced and their basic characteristics are discussed. The final part of the chapter gives the influences on fiber gratings of the key physical variables, temperature and strain.

10.2 FIBER PHOTOSENSITIVITY

A material is said to be photosensitive when its properties are changed by exposure to light. Common, everyday examples of photosensitive materials include photographic film and self-darkening sunglasses. The existence of photosensitivity in germanium-doped optical fibers has only been known since 1978, when the first in-fiber grating was accidentally made. The phenomenon itself is introduced in the following section.

10.2.1 Introduction to fiber photosensitivity

The first demonstration of photosensitivity in optical fibers was in 1978 by Hill *et al.*, who produced a Bragg reflection grating in an ordinary germanium-doped telecommunication fiber [2]. The grating was formed within minutes by visible laser light propagating in the fiber. This method for producing gratings is called the 'internal' method and is briefly described in section 10.3.1. The initial experiments showed that a self-organized modulation of the refractive index occurred in the core of the fiber with a period equal to half the wavelength of the writing beam in the glass medium. The topic remained principally an academic curiosity until 1989, when a major development in grating fabrication was made [3], using transverse exposure of the fiber with ultraviolet (UV) light. Until then, the reflected wavelength could not be freely chosen, and the experimental gratings had only been inefficiently formed by using visible light. With the advent of the new techniques, known as 'external' writing methods because the writing beam was no longer required to propagate along the fiber, this major problem of selection of the operating wavelength was eliminated, and simultaneously, the efficiency of the writing process was found to be several orders of magnitude higher for UV exposure than for visible exposure. Both of these features made the production of useful components much more practical than before, and this has led to many commercial applications, including new optical fiber sensors.

10.2.2 Physical mechanisms of fiber photosensitivity

Fiber photosensitivity was detected at first only in fibers containing germanium as a dopant. However, recent observations of photosensitivity in germanium-free fibers prove that the phenomenon in most optical fibers is similar under comparable irradiation conditions. Nevertheless, most models proposed in order to explain the photosensitivity of fibers are based on the well-known defects which exist in germanosilicate glass [4]. It seems that, whatever the fiber material, UV-absorbing defects catalyze the changes which lead to grating formation. In recent years, three principal microscopic mechanisms have been proposed for explaining the photosensitivity of optical fibers. These are the Kramers–Kronig model, the dipole model and the compaction models.

An optical absorption peak in a material is an example of resonant behavior at the optical excitation frequency. Any type of resonant behavior involves two separate kinds of response from an oscillator. These are firstly the amplitude response, which manifests itself as a large change in the magnitude of the oscillation developed in response to a forcing function, and secondly, the phase response, which describes the relative delay between the peak of each cycle of the oscillation in the oscillator and the peak of each cycle of the forcing function. These two features of resonance are plotted for a general oscillator in Figure 10.2. It is qualitatively seen from this figure that significant nonzero phase delays can be produced far from the resonant frequency, even when the amplitude response has ceased to show any signs of the existence of the resonance. In optical materials, the amplitude response is contained in the imaginary part of the complex refractive index and resonances are observed as absorption bands, while the phase response affects the real part of the complex refractive index, which can be affected very far from the absorption peak. These two responses are connected by the Kramers–Kronig relations [5, 6].

In germanium-doped silica optical fibers, the ideal molecular structure would have every Ge or Si atom purely covalently bonded directly only to oxygen atoms. However, in practice, some Ge atoms may be directly bonded to a Si or another Ge atom. This type of bond is called an oxygen-deficient bond and represents a structural defect in the glass matrix. These defects create an absorption band centered at 240 nm with a bandwidth of approximately 30 nm [7].

The first proposed model presented by Hand and Russell [8] is based on the Kramers–Kronig relations. In this model, the observed refractive index changes are linked to the changes in the absorption spectrum. In their model, Hand and Russell assumed that the dominant absorption bands were the oxygen-deficiency absorption bands centered at 240 nm [7], the Ge(1) defect band centered at 281 nm and the Ge(2) defect band centered at 213 nm [4]. According to their analysis, photoexcited electrons released from the oxygen-deficient germanium bonds (by a single UV photon or two visible photons) drift in the glass network until they are captured by Ge(1) and Ge(2) traps. The germanium oxygen-deficiency centers were estimated to be present in ~1% of all of the available germanosilicate bonds. At the time this model was proposed, it was able to explain the refractive index changes of up to 10^{-5}

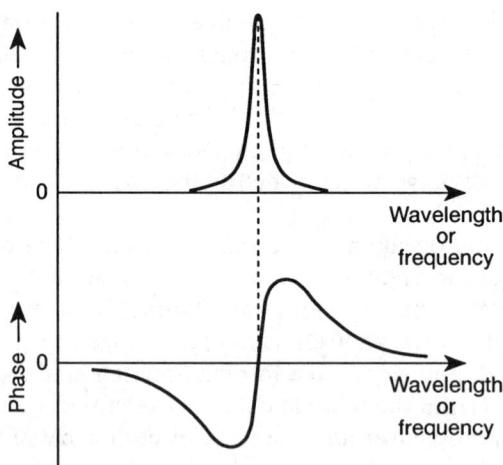

Fig. 10.2 Amplitude and phase response of a resonant system.

which had then been seen. However, more recently, refractive index changes of 10^{-3} and larger have been observed and this model alone cannot explain these much larger changes.

The second model which has been proposed to explain photosensitivity, the dipole model, is based on the fact that the photoinduced change of the refractive index is linked to the dipole moment of the generated defect centers, via the quadratic Pockels effect [8]. When the density of the dipoles is high enough, such that their spacing is less than the wavelength of propagating light, the system acts as a Rayleigh scattering medium with an effective index change. In this case, the internal electric field associated with the dipoles results in a change in the refractive index. However, the dipole density required to explain the large index changes noted above is once again implausibly large, so this model is also unable to account for all observations. However, one attractive feature of the dipole model is that it might explain the very strong photoinduced polarization couplers which can be written in highly-birefringent fibers.

The third model which has been proposed to explain photosensitivity is the compaction model. This model is based on the fact that compaction of pure silica samples has been observed under conditions of UV irradiation. The compaction of the samples is caused by the collapse of certain defect ring structures which have been observed in silica [9, 10]. The densification of the material causes a change in the refractive index. Experimental evidence of chemical changes and increased stress in the cores of irradiated fibers has been found, but at the time of writing, it is not clear what magnitude of refractive index change is directly attributable to this mechanism in typical fibers.

There is one type of photosensitivity, however, where structural change is definitely the major contributor to the overall index change. This occurs when extremely powerful, pulsed UV irradiation is used to write gratings, causing enormous changes in refractive index (of order 10^{-2}) and visible damage in the core of the fiber [11–13]. Single 20 ns pulses of intensity approximately 500 mJ cm^{-2} are used, creating local heating and shock in the core of the fiber. This photoinduced damage is so clearly different from the more usual changes caused by low intensity irradiation, of typically only a few mJ cm^{-2}, that the resulting gratings were given the description 'Type II' to distinguish them from the more normal 'Type I' gratings which had been observed previously.

Taken overall, the accumulated sum of experimental findings from Type I grating fabrication experiments do not fit exclusively with any of the three basic models, and it is clear that much remains to be done to elucidate the physics of fiber photosensitivity. Also, it often appears to be the case that more than one mechanism is operating at any time even in a single fiber. This is particularly obvious in the case of fibers having high Germania doping (~ 30 mole%), when Type I gratings are written. On continued exposure to the writing beam, a maximum reflectivity is reached and this is followed by a reduction in reflectivity to zero, followed by a renewed growth, normally reaching greater reflectivity than the original maximum. The second growth period is distinguished by a change in sign of the refractive index modification due to the exposure. Type I gratings always show an increase in the refractive index of the core after exposure, whereas the regrown grating is characterized by a reduction in the core index in the exposed region. This change from positive-going to negative-going refractive index modification caused by the UV exposure is believed to be due to a change in the nature of the dominant photosensitivity mechanism responsible for the appearance of the grating, and is called 'Type IIA' photosensitivity. Each of the identified photosensitivity types leads to gratings with different performance features after fabrication. These features will be compared later in this chapter.

A further complication in understanding photosensitivity appears because the photosensitive behavior of a fiber is not only dependent on its composition, but is also partly controlled by its processing history, suggesting that different mechanisms can be deliberately enhanced or suppressed. The most striking example of this is seen in a widely used technique for enhancing the photosensitivity of many types of fiber. This technique involves forcing hydrogen into the fiber core by diffusion. This is usually done by sealing the fiber inside a hydrogen-filled pressure chamber typically at ~ 100 bar for a period of a few days. If the fiber is irradiated soon after hydrogenation, the observed photosensitivity of some fibers is increased by more than an order of magnitude. However, the fiber must be irradiated before the hydrogen diffuses back out of the fiber, or the benefit of the treatment is lost. The largest photoinduced refractive index changes yet seen in optical fibers, up to 10^{-2}, have been produced in this way [14]. An explanation for the action of hydrogen in this context has been proposed by Abdulhalim [15], who suggests that hydrogen first increases the population of oxygen-deficiency centers by breaking $X—O—X$ bonds (where $X =$ Si or Ge) to form $X—OH$ before irradiation, and then stabilizes the defects created during irradiation so as to preserve the maximum change in index, which might otherwise be lost due to recombination back to the original defect center. Hydrogenation has several drawbacks for large-scale use. Obviously the process is slow and potentially dangerous. Also, additional attenuation is introduced into the fiber due to OH absorption bands. However, this last problem can be largely avoided by using deuterium instead of hydrogen as the diffusing species.

10.2.3 Photoinduced birefringence

Apart from the photoinduced change of the isotropic refractive index, it was discovered in 1985 by Parent *et al.* [16] that photoinduced birefringence could also be written into fibers by polarized radiation. Since the first observation of the effect, photoinduced birefringence has been reported in a variety of germanosilicate fibers. The resulting birefringence is always much smaller than the accompanying isotropic background index change. Typically, the induced birefringence is up to 8% of the overall index modification. This effect can lead to difficulties in some applications, but is likely to be useful in others, particularly for sensors. Photoinduced birefringence has been used, particularly in polarization-maintaining fibers, to form polarization couplers which operate in either the forward or backward direction. However, it has been found that different types of high-birefringence fibers exhibit different photosensitive behavior. At the time of writing, speculation continues as to the reasons for this unexpected and often contradictory behavior [17].

The origins of photoinduced birefringence are even less clear than those of isotropic index changes. Three main mechanisms have been proposed to account for the phenomenon. These are bleaching of original defects, stress modification and geometrical asymmetry.

An and Sipe first proposed a mechanism for the appearance of photoinduced birefringence [18] which suggests that the structural defects responsible for the 240 nm absorption band might be preferentially sensitive to the particular polarization alignment of a writing beam because of their own physical orientations. Defects aligned parallel to the polarization direction of the writing beam would be preferentially bleached, leading to a negative birefringence change. The predicted negative sign of the photoinduced birefringence agrees with observations [19].

The second mechanism for the production of photoinduced birefringence is particularly relevant to high-birefringence, polarization-maintaining fiber structures, where thermal stress levels can be expected to be much higher than for communications fiber. This is due to the strong asymmetry and also the often heavy dopant densities in the former types of fiber. It has been proposed that UV irradiation can reduce thermal stress in the core of the fiber by photolysis of strained defects [20]. Since the thermal stress will contribute to the intrinsic birefringence of the fiber, its reduction will inevitably modify that birefringence.

A third mechanism accounting for photoinduced birefringence applies only to the 'external' irradiation techniques. Ultraviolet light traversing the core of a fiber will suffer some absorption, depending on the fiber structure and the wavelength of the radiation. It can then be expected that the isotropic index change produced by the beam will be strongest on the side of the core facing the incoming beam. The resulting asymmetry in the isotropic refractive index modification will clearly lead to a type of form birefringence. This source of birefringence has been confirmed experimentally in circularly cored fiber [21].

It was pointed out at the beginning of this section that there is even less understanding of the mechanisms of photoinduced birefringence than of isotropic index change, however, the physical processes generating both effects are clearly strongly related. There is, for example, a strong similarity between the observation of type IIA photosensitivity and a recent observation of a sign change for the photoinduced birefringence in highly germanium-doped fiber [20]. While these fundamental investigations are continuing, there is also great interest in fabrication methods and applications of birefringent components and these will be examined later in this chapter.

10.3 FIBER GRATING TYPES AND PRINCIPLES

10.3.1 Classification of gratings

Fiber gratings can be classified in various ways according to their particular function. The action of each type of grating determines its suitability for different applications. We shall concern ourselves mainly with gratings which diffract guided light. Distributed structures can also be made to diffract light into or out of the fiber, as in the grating coupler which is used in integrated optics. Gratings can be divided into two main families: those which diffract light in the forward direction and those diffracting in the backward direction. Of these two types, reflection gratings were the first to be generated by Hill *et al.* [2] in 1978, while 11 years passed before photoinduced fiber gratings which operated in the forward mode were generated by Park and Kim [23]. Reflection gratings are normally designed to diffract light into the same spatial mode as the exciting mode. Gratings which diffract light in the forward direction generally act as intermodal couplers. The modes which are coupled can be either higher and lower order spatial modes or alternatively orthogonally polarized versions of a single spatial mode. Reflection gratings which diffract light into a higher order mode have also been demonstrated by Meltz *et al.* [3] while polarization-converting reflection gratings have also been demonstrated by Kamal *et al.* [24].

We shall briefly describe here the action of the various kinds of photogenerated fiber gratings. All of the gratings described are wavelength selective, i.e. they diffract light efficiently only within a narrow wavelength range. The first type of grating is the Bragg grating, which simply reflects light within the incident guided mode and is shown in Figure 10.3(a). The light reflected from this grating maintains its polarization state. The second kind of grating, called an intermodal coupler or a PRIME (photorefractive intermodal exchanger), couples light in the forward direction from one of the guided modes of the core into another [25]. The required pitch of the index perturbation is relatively long in this case. The polarization state of the light is not altered as it passes through the grating and this is shown in Figure 10.3(b). The third kind of grating, called a rocking filter or polarization coupler, couples light from one polarization mode of a highly birefringent fiber into the orthogonal polarization mode, again in the forward direction. Its action is represented by Figure 10.3(c). The fourth kind of grating, called a reflective polarization coupler or exchange Bragg grating, is similar to a rocking filter but operates in the backward direction. This type of grating reflects light from one polarization mode into a counter-propagating, orthogonally polarized mode and is shown in Figure 10.3(d). Finally, we mention here two types of grating which couple light into the radiation field. A blazed Bragg grating (Figure 10.3(e)) can couple light between a guided fiber mode and the radiation field so that light can be coupled into or out of the core at large angles in analogy to the grating couplers used in integrated optics. This type of grating is often called a 'fiber tap' and has short pitch perturbations. Coupling can also be achieved at small angles with respect to the core axis,

FIBER GRATING TYPES AND PRINCIPLES

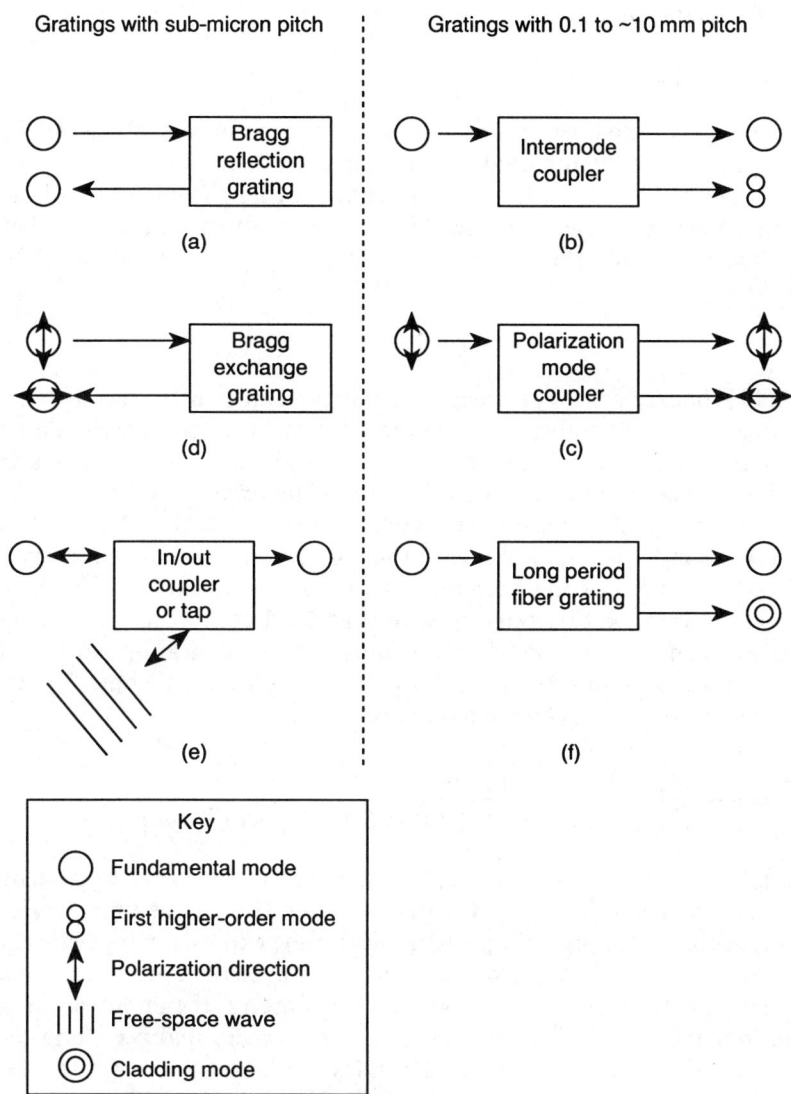

Fig. 10.3 Characteristics of various grating types.

when the effect is to excite cladding modes (Figure 10.3(f)). These types of gratings are called 'long period fiber gratings' or 'long period cladding mode couplers' because the pitch of the index perturbation is once again very long compared to a Bragg grating. Those gratings which involve long pitches (i.e. pitches which are very much greater than the operating wavelength) are relatively easy to make, compared to those which involve pitches of the same order as the operating wavelength.

10.3.2 Reflective gratings

(a) Bragg gratings

The reflectance of the fiber grating reflector shown in Figure 10.1 is determined by superposition of the reflected waves from each of the weak reflectors in the fiber. Clearly, the resultant wave has a maximum amplitude when all of these components are in phase. Thus, in this condition, the separation, Λ, between each adjacent pair of reflectors must be equal to half an optical wavelength. Hence

$$\Lambda = \lambda_0/2n_e \qquad (10.1)$$

where λ_0 is the free space wavelength of the most efficiently reflected light and n_e is the effective refractive index of the fiber at that wavelength. This is the Bragg condition for a grating reflector. Thus the pitch of the index perturbation determines the central reflection wavelength of the grating. In an optical fiber, the index difference between adjacent sections need not be very large in order to produce a strong reflection, provided that the grating contains enough reflectors. The magnitude of the reflection at each reflector can be described by a coupling coefficient, κ, between the incident and reflected waves. Often, the fiber attenuation can be neglected because most practical gratings are relatively short. Then it can be shown [6] that for a uniform grating of length L the reflection efficiency is described by the following equation:

$$\frac{I_R}{I_0} = \left| \frac{-\kappa \sinh(\kappa^2 - \Delta\beta^2)^{1/2}L}{i\Delta\beta \sinh(\kappa^2 - \Delta\beta^2)^{1/2}L + (\kappa^2 - \Delta\beta^2)^{1/2} \cosh(\kappa^2 - \Delta\beta^2)^{1/2}L} \right|^2 \qquad (10.2)$$

where $\Delta\beta = n_e k_0 - \pi/\Lambda$, $k_0 = 2\pi/\lambda_i$, I_R = reflected intensity, λ_i = wavelength of the incident wave and I_0 = incident intensity. $\Delta\beta$ is called a detuning parameter. Here, the second term on the right-hand side of the expression for $\Delta\beta$ is equal to the propagation constant of a wave at the wavelength of peak reflection, while the first term is the propagation constant of the input wave. If we assume that the refractive index varies in a sinusoidal manner along the axis of the fiber with the amplitude of the index perturbation being Δn, then the coupling coefficient is given by $\kappa = \Delta n k_0/2$.

Examples of the reflectance function of a grating of this type are shown as a function of wavelength in Figure 10.4. The exact shape of the curve depends on the magnitude of the peak reflectance, R_{MAX}, which can approach unity. At low reflectance, the main peak of the grating's reflection spectrum shows a gradual rise toward its maximum value. As the reflection efficiency approaches unity, the main reflection peak begins to approach a more rectangular shape, with very little change in reflectivity around the central wavelength. As the incident wavelength is reduced, a second reflection peak would appear when the wavelength becomes half of the value at the first peak, but in almost all practical cases, the fiber would have become multimode before this stage. As a rough guide, for

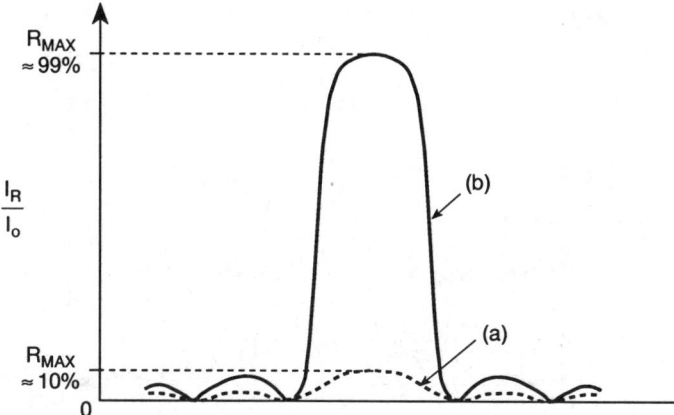

Fig. 10.4 Grating reflection spectra: (a) maximum reflectivity less than 10%; (b) maximum reflectivity approximately 99%.

gratings where the reflectance is small compared to unity, the width $\Delta\lambda$ of the reflection peak between the first zeroes is given approximately by

$$\Delta\lambda \approx \frac{\lambda_0^2}{n_e L} \tag{10.3}$$

Thus the wavelength selectivity of a grating is controlled principally by its length. The longer the grating, the more selective it will be. Numerous sidelobes are often visible on either side of the main reflection peak. Their magnitude is critically dependent on the variation of the index modulation depth along the grating. A uniform modulation depth leads to relatively strong sidelobes, which are often undesirable. However, the sidelobes can be greatly reduced by 'apodization' or 'shading' of the index modulation in order to reduce its magnitude smoothly to zero at each end of the grating. Figure 10.5 (from [26]) shows the reflection spectrum of a nominally uniform 0.84 mm long Bragg grating written in a standard telecommunication fiber.

(b) Other reflection gratings

Nonuniform gratings can also be designed and written into optical fibers. For example, gratings having a pitch which gradually changes along the length of the grating will reflect different wavelengths from different parts of the grating. These are known as 'chirped' gratings, and may offer advantages in some sensor situations, though their main use to date has been in dispersion compensation in high speed optical communication. Another type of nonuniform grating which may have uses in sensing is the 'sampled' or 'superstructure' grating, where the magnitude of the index modulation is changed periodically along the length of the grating, for example between zero and some fixed value. These types of grating can have complicated variations of reflection with wavelength.

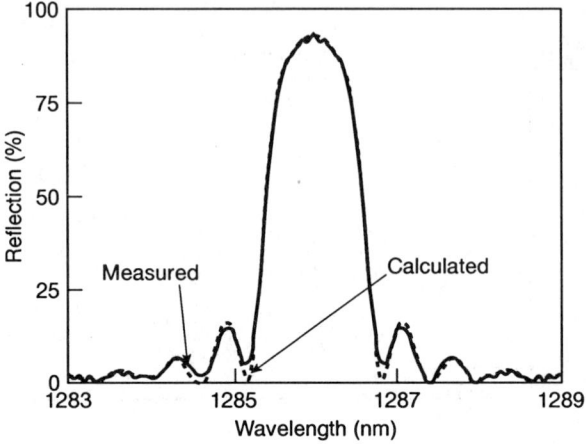

Fig. 10.5 Example of a practical grating reflection spectrum.

In the case where a grating is written in a birefringent fiber, the Bragg condition is satisfied at two different frequencies, one for each of the two orthogonal polarizations. This is a consequence of the fact that light polarized parallel to each of the two birefringence axes experiences slightly different respective refractive indices. The spacing δf of the two Bragg frequencies, for a fiber with a polarization beatlength L_b, is given by

$$\delta f = \frac{c}{nL_b} \tag{10.4}$$

where n is the mean effective refractive index of the two modes and c is the vacuum velocity of light. The polarization beatlength, L_b, is defined as the length of fiber in which the two orthogonally polarized linear eigenmodes accumulate a relative phase delay of 2π, or one full wave.

The exchange Bragg reflection grating acts as a polarization state converter that reflects light into the orthogonal polarization state of the incident guided spatial mode within a narrow wavelength range [24]. The principle on which the polarization conversion is based depends on a periodic rocking of the birefringence axes of the fiber with a pitch equal to that of a normal Bragg grating reflector of the same wavelength. The principle of operation is very similar to that of a Bragg grating, with the additional factor that as the forward traveling light propagates through each distinct birefringent section of fiber, with the birefringence axes tilted with respect to the incoming near-linear state of polarization, the phase retardation in that section of the fiber generates a small optical electric field component orthogonal to the polarization state of the incoming wave. Since each of these components is generated half a wavelength apart, the backward propagating waves interfere constructively. This type of grating can equally well be written in low or high birefringence fiber, since the fiber's own intrinsic birefringence plays no part in the principle of this grating.

FIBER GRATING TYPES AND PRINCIPLES 341

The expression for the reflectivity as a function of wavelength for a Bragg exchange grating is similar to that for the ordinary Bragg grating. The only difference is in the nature of the coupling coefficient, which is a function of the magnitude of the photoinduced birefringence modulation. To date, only low-efficiency coupling (a few percent) has been achieved.

10.3.3 Feedforward fiber gratings

The pitch of any grating which couples two propagating waves is always fixed by the spacing of the fringe pattern which the waves will produce when they interfere. Since all of the gratings considered here are passive in the sense that they scatter waves purely elastically (i.e. neither extracting energy from nor contributing energy to the waves), both waves will always have the same frequency, and so the only possible sources of phase variation between the waves can be differences in their propagation velocities or directions. Feedforward gratings by definition always couple waves which are traveling in very similar directions with similar velocities, so these gratings always employ a very long pitch compared to a reflective or a tap grating.

(a) Polarization couplers

Polarization couplers (or rocking filters) act as feedforward polarization state converters that couple light into the orthogonal polarization state of the incident guided spatial mode within a narrow wavelength range [27–29]. Polarization couplers are most useful in polarization-maintaining fiber, where the polarization modes are well controlled and stable, and the fibers have a relatively large and predictable value of birefringence. The fiber's intrinsic birefringence is used in this case to help to create the coupler. The polarization beatlength, L_b, is defined in section 10.3.2(b) as the length of fiber in which the two orthogonally polarized linear eigenmodes accumulate a relative delay of one full wave. A length $L_b/2$ of the fiber therefore produces a relative delay of half a wave between the eigenmodes. This half-beatlength section of fiber acts as a guided wave version of a bulk optic, half-wave plate. It is well known that when a linearly polarized wave passes through a half-wave plate, the wave's polarization direction is rotated by twice the angle between the original polarization vector and the preferred axis of the waveplate. Thus a train of half-wave plates can be set up where the fast axis of every second plate (e.g. all the even numbered plates in the train) are parallel and are rotated by a few degrees with respect to the first and all of the other odd numbered plates, whose axes are also aligned mutually parallel. A linearly polarized wave propagating through the series of plates will be rotated progressively through larger angles as the polarization rotation effect of each successive plate [22] is added to by the next. For example, if the rocking angle of each even numbered plate is θ, and the input polarization state is set parallel to one of the preferred axes of the first plate in the train, then the second plate will rotate the polarization in one sense by 2θ, the third will rotate the polarization state by 4θ in the opposite sense, and the fourth plate will again rotate the polarization in the first direction by 6θ and so on.

Polarization couplers are written by producing photoinduced birefringence at a spacing equal to the beatlength of the fiber at the desired coupling wavelength. The photoinduced birefringence is written with its axes at 45° with respect to the fiber's intrinsic birefringence axes and each exposure occupies approximately half a beatlength. The resultant birefringence of the exposed region will then have new axes slightly rocked with respect to those of the unexposed fiber. Usually the photoinduced birefringence (typically $\sim 10^{-6}$), is very small in magnitude compared to the fiber's own intrinsic birefringence (typically $\sim 10^{-4}$), so high efficiency polarization couplers often occupy tens of beatlengths, typically requiring physical lengths of 10 cm of fiber or more. Figure 10.6 shows the polarization coupling spectrum of an 11.5 cm long internally written polarization coupler [27]. In this case, the fiber's intrinsic birefringence was 1.29×10^{-4}, giving a polarization beatlength of 4 mm at 514.5 nm.

The coupling efficiency for a grating operating in the forward mode is given by

$$\eta = \frac{cc^*}{cc^* + (\Delta\beta_p/2)^2} \sin^2\left\{ L\sqrt{cc^* + \left(\frac{\Delta\beta_p}{2}\right)^2} \right\} \qquad (10.5)$$

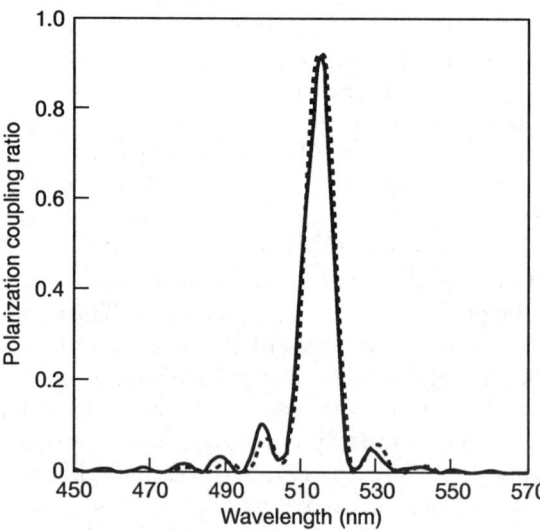

Fig. 10.6 Polarization coupling spectrum of an internally written rocking filter. The dotted line shows a theoretical fit to the measured result shown by the solid line.

where c = coupling coefficient and c^* represents its complex conjugate. The coupling coefficient is a function of the photoinduced birefringence modulation. Here L again represents the total length of the grating coupler and $\Delta\beta_p = bk_0 - \pi/\Lambda$ where b is the birefringence of the fiber, given by the difference between the effective refractive indices for the two principal, linearly polarized eigenmodes of the fiber, k_0 is the free space propagation constant of the incident radiation and Λ is the pitch of the perturbation, equal in this case to the polarization beatlength of the fiber at the wavelength of peak coupling.

In the case of phase matching ($\Delta\beta_p = 0$) equation 10.5 reduces to

$$\eta = \sin^2(cL) \tag{10.6}$$

In the absence of birefringence dispersion, the coupling bandwidth $\Delta\lambda_p$ of the grating can be approximated by the following relation:

$$\Delta\lambda_p = \lambda_0 \frac{L_b}{L} \tag{10.7}$$

(b) Intermodal couplers

Intermodal couplers, or photorefractive intermodal exchangers (PRIME) [25], are formed by inducing a periodic refractive index perturbation along the fiber length with a pitch that matches the intermodal beatlength determined by the propagation constant difference between the modes. This allows phase matched coupling between the selected fiber modes [23]. The location of the interference pattern of the two propagating modes within the cross-section of the fiber' core requires the refractive index grating not only to be written with the correct period for mode coupling but also to have the correct spatial distribution within the core.

So far, photoinduced couplers which couple light between the fundamental (LP_{01}) and the first higher order (LP_{11}) mode and also the fundamental (LP_{01}) and the LP_{02} mode [30] have been demonstrated. The properties of intermodal couplers can be analyzed using the coupled mode approach. It can be shown that the expression for the coupling efficiency between the two coupled modes is identical to the expression for the rocking filter. The only difference is that the coupling coefficient is a function of the overlap integral of the two modal fields.

(c) Long period fiber gratings

The final type of grating to be introduced here is the long period fiber grating, which is used to couple light from a guided mode, usually the fundamental LP_{01} mode of a fiber, to cladding modes [3]. These gratings are also known as 'long period cladding mode couplers'. The important factor controlling this coupling is a periodic modulation of the core to cladding refractive index difference of the fiber. The required pitch Λ of the index modulation is given by the difference between the propagation constant, β_g, of the relevant guided mode and that of

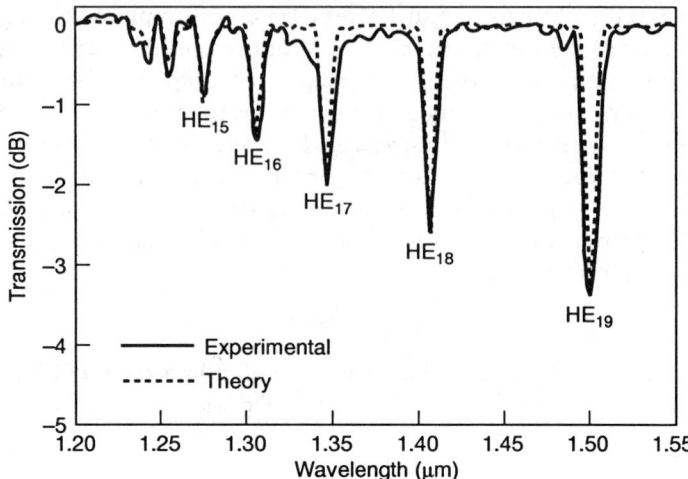

Fig. 10.7 Transmission spectrum of a long period grating (after [32]). The dotted line shows a theoretical fit to the measured result illustrated by the solid line.

the relevant cladding mode, β_c, at the desired coupling wavelength according to

$$\Delta\beta_1 = \beta_g - \beta_c = \frac{2\pi}{\Lambda} \qquad (10.8)$$

Typically, gratings are designed to couple from the fundamental mode of the core, and this most easily couples to the radially symmetric cladding modes HE_{1m}, where $m \geqslant 2$. In practice, $\Delta\beta_1$ is largely controlled by the magnitude of the core to cladding refractive index difference Δn. For a fiber with $\Delta n = 10^{-2}$ and a single mode operating wavelength of 1.5 μm, the approximate pitch Λ required for coupling will be of the order of 200 μm. The transmission spectrum of a fiber containing such a grating is shown in Figure 10.7 (after 32]).

10.4 FABRICATION OF FIBER GRATINGS

A brief description of the various writing techniques which have been used to generate fiber gratings will be presented in this section.

10.4.1 Internal method

Reflection gratings were originally fabricated by the internal method [2]. The grating is produced initially from a standing wave arising from interference between a forward traveling launched beam and the relatively weak reflected beam from the output end of the fiber. A gradual increase in the reflected power then occurs, typically over a period of a few minutes, due to a positive feedback process, until a maximum is achieved in the reflection efficiency. The basic disadvantage of this method is that these gratings can be operated efficiently only at the writing wavelength. It is also possible to use these gratings at other wavelengths by phase matching to higher order modes, but this remains restrictive with regard to operating wavelengths and loses the attraction of simple, single mode operation. This method has now been superseded by the external writing methods described below.

10.4.2 External methods

The most important development in grating fabrication which has led to the wide acceptance of fiber gratings as a key sensor component was the invention of the external writing approach [3].

(a) Holographic exposure

The first external writing method to be published was the holographic method. It was found that it is possible to write fiber gratings by exposing the fiber to an interference pattern of UV light close to the 240 nm oxygen-vacancy absorption of germanosilicate fibers [3]. A preselected period for the index modulation can be easily obtained by changing the angle of the two interfering beams. The holographic method is shown diagramatically in Figure 10.8.

The Bragg wavelength λ_0 of an externally written reflection grating is given by the expression

$$\lambda_0 = \lambda_w n_e / \sin \theta \tag{10.9}$$

where n_e is the effective modal refractive index of the fiber, λ_W is the writing wavelength of the fiber used to generate the grating and θ is the half-angle of the intersecting writing beams. It can be seen from the above expression that the Bragg wavelength can be controlled by adjusting the angle θ or the wavelength λ_W of the writing beams. Normally, the former is easier to control, because the laser sources used are often not tunable. Excimer lasers and frequency doubled argon lasers are the most popular tools for this task, since they offer high power at useful UV wavelengths. While most grating fabrication work has so far been done using 240 nm exposure, 193 nm radiation from excimer lasers has been shown to be very effective for writing useful index modulations in standard fibers and can be advantageous in preserving fiber strength compared to 240 nm excimer lasers [33].

Gratings have been generated by this method employing either free-space or solid path interferometers. External writing of reflection gratings in photosensitive fibers has many advantages over the intra-fiber writing technique as the Bragg wavelength and the peak reflection efficiency can be easily controlled by the periodicity and energy respectively of the fringes of the writing beam. The option of writing blazed gratings in fibers by tilting the angle of the

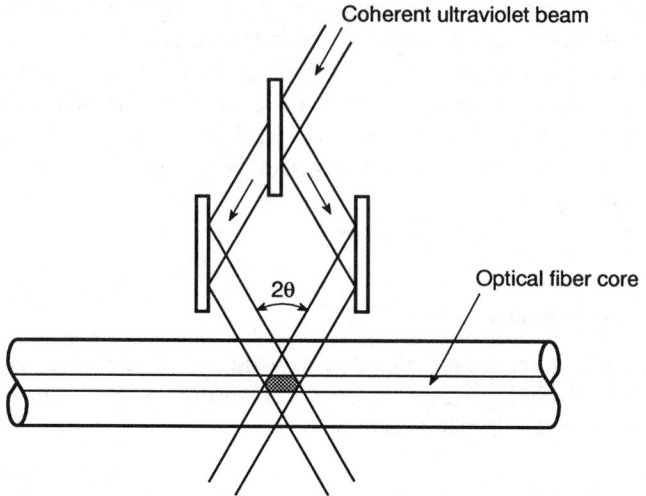

Fig. 10.8 Holographic method for writing reflection gratings.

interference pattern with respect to the fiber axis is also available with this method. Additionally, this method is noninvasive and the grating can be written at any position along the length of the fiber. However, there must be no outer polymer coating on the fiber during UV exposure. This is most easily accomplished by writing gratings during fiber manufacture, at the fiber drawing stage. This has been achieved using gratings written by single pulses from excimer lasers. Since the duration of a typical pulse is only ~20 ns, the motion of the fiber is negligible during the exposure. This approach could make grating production a very cheap process, provided that production volumes are large.

(b) Phase mask method

An alternative means for generating a fringe pattern in UV light is to use a phase mask. This is a transparent plate, normally of fused silica, with a ridge pattern etched into its surface. When illuminated at normal incidence by a collimated UV beam, the ridges diffract the beam as shown in Figure 10.9. The depth of the ridges is normally chosen for destructive interference in the zero-order beam which would otherwise pass straight through the plate. The light is then mainly diffracted into the $+1$ and -1 orders, forming an interference pattern beyond the mask. Fibers can be placed immediately behind the mask, or the interference pattern can be transferred to a more remote location by beam steering optics. The advantages of the phase mask method are first of all, that very highly reproducible gratings can be made in this way, since the periodicity of the interference pattern is guaranteed by the ridge structure of

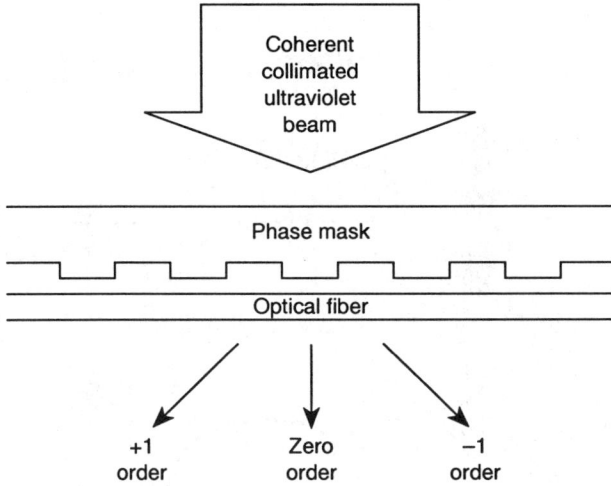

Fig. 10.9 Phase mask method for writing reflection gratings.

the mask itself, and does not depend on careful alignment of several separate components, unlike the holographic exposure method. Secondly, very long gratings can be made in this way, because long masks can be produced using standard lithographic techniques borrowed from semiconductor process engineering, and uniform writing beams can be translated over the masks to build up the desired length of grating. The phase mask method also lends itself well to chirped and apodized grating production. The major drawback of the phase mask method is that there is very little flexibility of the grating pitch which can be produced by any particular mask, so a variety of masks would be needed to write gratings operating at many different wavelengths.

(c) Point-by-point method

This method for writing gratings is based on the principle of exposing the fiber to UV light one point at a time [34]. The technique lends itself to gratings which are to be used in the feedforward mode since the intermodal beatlengths are a few orders of magnitude greater than the periods required for reflection gratings. The method is shown diagramatically in Figure 10.10. Each coupling point of the grating is exposed to a length which is equal to half the intermodal beatlength. This method is independent of the coherence length of the writing beam, and the advantage of this approach for long gratings is that it avoids the need for major expansion of the beam from the source laser. Stringent control of the translation of the fiber is required to obtain a constant grating period. The technique has also been applied to the fabrication of Bragg gratings, where this requirement is most exacting.

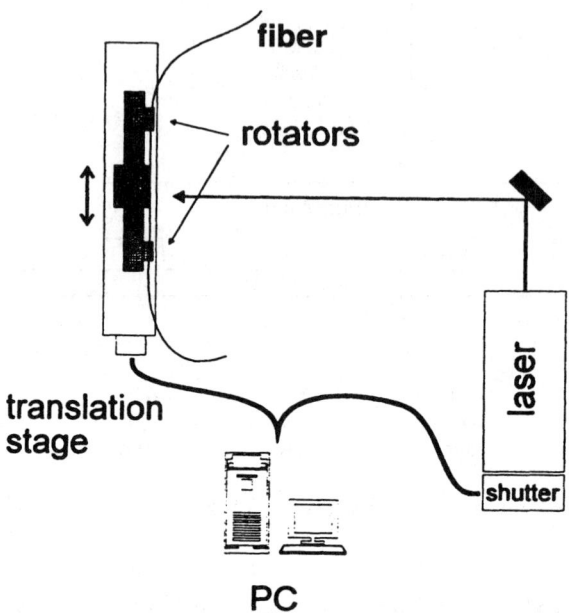

Fig. 10.10 Point-by-point method for writing intermodal couplers.

10.5 STABILITY OF PHOTOINDUCED FIBER GRATINGS

The thermal stability and mechanical reliability of Bragg gratings will be crucial in determining the applications in which they can be used. These issues are reviewed in this section.

10.5.1 Thermal stability

Type I gratings are reasonably stable at moderate temperatures [35, 36]. Often, no noticeable drop in grating reflectivity is observed up to a temperature of 200–300 °C when a grating is heated for a period of just a few minutes, however, if gratings are to be used commercially, then lifetimes of several years at temperatures up to 150 °C need to be guaranteed.

The lifetime of photogenerated gratings subjected to heat has been studied, and it is now possible to make predictions of grating performance which will satisfy commercial needs. The stability of gratings will be determined by the balance between the photosensitivity mechanisms which have been active during the writing process. The 'Type I' and 'Type IIA' mechanisms referred to earlier lead to gratings with distinctly different thermal behaviors. Gratings dominated by the 'Type I' process often follow a power-law decay according to [37]

$$R(t) = R_0\left(\frac{1}{1 + At^\alpha}\right) \tag{10.10}$$

where $R(t)$ = grating reflectance at time t, R_0 = grating reflectance immediately after fabrication and A and α are constants whose values depend on temperature and on the fiber used. The implication of this formula is that for a given temperature, most of the total reduction in reflectance occurs quickly, and thereafter the grating reflectance remains relatively stable. This behavior allows gratings to be made and then annealed by the manufacturer so as to provide stable performance in particular applications. When the maximum temperature to which a grating will be heated is no more than normal room temperature, even nonannealed gratings can often be expected to lose no more than 10% of their initial reflectance in a life span of 25 years.

While the thermal stability of Type I gratings is already very encouraging, the stability of Type IIA gratings has been observed to be even better, with 10% decay in the reflectance of these gratings being seen up to 300 °C, and correspondingly improved stability compared to Type I gratings up to 450 °C.

For higher temperatures, permanent changes start to occur until the grating reflectivity is quickly erased at about 600 °C, even for Type IIA gratings. For temperatures above 600 °C, Type II gratings offer useful life, but the optical quality of these gratings is poor, with serious scattering loss being their chief drawback. At the time of writing, research is proceeding on finding alternative ways to make gratings stable at high temperatures.

The coupling strength of photoinduced intermodal couplers written with UV light is reduced when the couplers are heated to 300–350 °C for approximately 1 minute. Complete erasure can be obtained after 1 minute at about 500 °C [34]. The thermal properties of photoinduced birefringence have been studied [27, 38] through experiments performed on the thermal stability of rocking filters. It was shown that couplers written externally with 266 nm light suffered only a small initial drop (10–20%) in their coupling efficiency measured after two hours at 240 °C. Upon repeated cycling of the temperature to the same maximum value no further drop in the coupling efficiency was seen.

10.5.2 Mechanical reliability

From the point of view of reliability, the important mechanical properties of fiber gratings include their intrinsic strength and their behavior at high temperatures. Silica optical fibers are intrinsically very strong, but prone to loss of much of their strength if they are not properly handled and protected by suitable coatings. Handling and protection procedures for use in the communications industry are well established, and fiber grating manufacturers can demonstrate good behavior of gratings for benign conditions. In contrast, for sensing applications, gratings may be placed under large, static strains and may also be simultaneously subject to high temperature. Careful design of transducers and the use of appropriate fiber protection, especially coating techniques, will therefore be vital for successful deployment of fiber grating sensors in demanding applications. Some loss of fiber strength can be expected to result from the operations required to write gratings, and some variations between different techniques can be expected. At the time of writing, one manufacturer has reported that the mean value of the breaking strength of its standard, commercially available gratings is well in excess of 1.38 GPa, based on the results of dynamic strength measurements.

The greatest test of grating reliability is to strain a grating at high temperature. Creep and hysteresis of grating responses have been reported at 650 °C. Hysteresis errors can be significantly reduced by using a strain cycling schedule, where a programmed series of strain impositions and releases are performed at the specified working temperature. However, creep characteristics are fundamental to the silica material and are not widely understood or quantified at 650 °C. At high temperatures, the germanium which defines the core of the fiber will begin to diffuse, leading to a change in the reflection properties of the grating over the long term. At 650 °C this problem is not great and for a typical fiber leads to a 10 pm shift in the Bragg wavelength after 2.8 years.

10.6 TEMPERATURE AND STRAIN SENSITIVITY OF FIBER GRATING SPECTRA

10.6.1 Bragg gratings

The potential of reflection gratings in optical fiber sensing was first studied by Meltz [39, 40]. The propagation constants of guided modes in an optical fiber are sensitive to both temperature and strain in the fiber. Both axial and transverse strains are important in principle, but in practice, most sensor applications place the gratings mainly under axial load. Here, the principal effects arise from the elongation (or compression) of the fiber and the change in the effective index of the guided modes through the photoelastic effect. Temperature changes also lead to changes in the refractive index of the material of the fiber via the thermo-optic effect, whose magnitude dominates over geometrical changes due to thermal expansion. All of these effects lead to changes in the Bragg wavelength of a grating written into a fiber. The coefficients which relate the temperature and strain changes to the change in the Bragg wavelength are given below. For a typical fiber, the fractional change of the Bragg wavelength for a temperature change δT and an axial strain ε is given by

$$\frac{\delta \lambda}{\lambda_0} = 8.9 \times 10^{-6} \delta T + 0.78 \varepsilon \qquad (10.11)$$

where δT is measured in °C and ε is in microstrain. The coefficient for temperature variations is more prone to vary between fibers than that for strain variations.

10.6.2 Polarization couplers

In strongly asymmetric fiber structures, such as polarization-maintaining fibers, significant birefringence changes can be caused by thermal stress. This thermally induced birefringence change causes the polarization coupling spectrum of a rocking filter to change with temperature. Likewise, axial strain leads to further changes in the peak coupling wavelength of a rocking filter. The coefficients linking the resonant wavelength of a rocking filter to changes in temperature and strain are less predictable than those for Bragg gratings, since they depend very much on the particular structure of the fiber being used. Values which have been measured in a particular case for an elliptically cored optical fiber [38] are given as follows:

$$\frac{\Delta \lambda}{\lambda_0} = -2.6 \times 10^{-4} \delta T - 2.2 \times 10^{-6} \varepsilon \tag{10.12}$$

where δT is measured in °C and ε is in microstrain.

REFERENCES

1. Jackson, D. A., Dandridge, A. and Sheem, S. K. 'Measurement of small phase shifts using a single mode optical fibre interferometer', *Optics Letters* **5**, 139 (1980).
2. Hill, K. O., Fujii, Y., Johnson, D. C. and Kawasaki, B. S. 'Photosensitivity in optical fiber waveguides: application to reflection filter fabrication', *Appl. Phys. Lett.* **32**, 647 (1978).
3. Meltz, G., Morey, W. W. and Glenn, W. H. 'Formation of Bragg gratings in optical fibres by a transverse holographic method', *Opt. Lett.* **14**(15), 823 (1989).
4. Friebele, E. J. and Griscom, D. L. 'Colour centres in glass optical fibre waveguides', *Mat. Res. Soc. Symp. Proc.* **61**, 319 (1986).
5. Saleh, B. E. A. and Teich, M. C. *Fundamentals of Photonics* Wiley, New York (1991).
6. Yariv, A. *Quantum Electronics* (2nd edn) Wiley, London (1975).
7. Yuen, M. J. 'Ultraviolet absorption studies of germanium silicate glasses', *Appl. Opt.* **21**, 136 (1982).
8. Hand, D. P. and Russell, P. St. J. 'Photoinduced refractive-index changes in germanosilicate fibres', *Opt. Lett.* **15**(2), 102 (1990).
9. Rothschild, M., Ehrlich, D. J. and Shaver, D. C. 'Effects of excimer laser irradiation on the transmission, index of refraction and density of ultraviolet grade fused silica', *Appl. Phys. Lett.* **55**(13), 1276 (1989).
10. Fiori, C. and Devine, R. A. B. 'Ultraviolet light radiation induced compaction and photoetching in amorphous thermal SiO_2', *Mat. Res. Soc. Symp.* **61**, 187 (1986).
11. Archambault, J. L., Reekie, L. and Russell, P. St. J. '100% reflectivity Bragg reflectors produced in optical fibres by single excimer laser pulses', *Elect. Lett.* **29**, 453 (1993).
12. Askins, C. G., Tsai, T. E., Williams, G. M., Putnam, M. A., Bashkansky, M. and Friebele, E. J. 'Fibre Bragg reflectors prepared by a single excimer pulse', *Opt. Lett.* **17**(11), 833 (1992).

13. Archambault, J. L., Reekie, L. and Russell, P. St. J. 'High reflectivity and narrow bandwidth fibre gratings written by a single excimer pulse', *Electron. Lett.* **29**(1), 28 (1993).
14. Lemaire, P. J., Atkins, R. M., Mizrahi, V. and Reed, W. A. 'High pressure H_2 loading as a technique for achieving ultra-high photosensitivity and thermal sensitivity in GeO_2 doped optical fibres', *Elect. Lett.* **29**, 1191 (1993).
15. Abdulhalim, I. 'Model for photoinduced defects and photorefractivity in optical fibres', *Appl. Phys. Lett.* **66**, 3248 (1995).
16. Parent, M., Bures, J., Lacroix, S. and Lapierre, J. 'Proprétés de polarisation des reflecteurs de Bragg induits par photosensibilité dans les fibres optique monomodes', *Appl. Opt.* **24**, 354 (1985).
17. Handerek, V. A. and Kanellopoulos, S. E. 'Photoinduced birefringence and applications in sensing', *Proc. Photosensitivity and Quadratic Nonlinearity in Glass Waveguides*, Portland, USA, Sept 9–11 (1995).
18. An, S. and Sipe, J. E. 'Polarisation aspects of two-photon photosensitivity in birefringent optical fibre', *Opt. Lett.* **17**, 490 (1992).
19. Bardal, S., Kamal, A. and Russell, P. St. J. 'Photoinduced birefringence in optical fibres: a comparative study of low birefringence and high birefringence fibres', *Opt. Lett.* **17**(6), 411 (1992).
20. Wong, D., Poole, S. B. and Sceats, M. G. 'Stress birefringence reduction in elliptical-core fibres under ultraviolet irradiation', *Opt. Lett.* **17**, 1773 (1992).
21. Vengsarkar, A. M., Zhong, Q., Innis, D., Reed, W. A., Lemaire, P. J. and Kosinski, S. G. 'Birefringence reduction in side-written photoinduced fibre devices by a dual exposure method', *Opt. Lett.* **19**, 1260 (1994).
22. Psaila, D. C., Ouellette, F. and DeSterke, M. C. 'Characterisation of photoinduced birefringence change in optical fibre rocking filters', *Appl. Phys. Lett.* **68**, 900 (1996).
23. Park, H. G. and Kim, B. Y. 'Intermodal coupler using permanently photoinduced grating in two-mode optical fibre', *Electron. Lett.* **25**(12), 798 (1989).
24. Kamal, A., Kanellopoulos, S. E., Archambault, J.-L., Russell, P. St. J. Handerek, V. A. and Rogers, A. J. 'Holographically written reflective polarisation filter in single-mode optical fibres', *Opt. Lett.* **17**(17), 1189 (1992).
25. Ouellette, F. 'Photorefractive intermodal exchangers in optical fibre', *IEEE J. Quant. Electron.* **27**(3), 796 (1991).
26. Limberger, H. G., Fonjallaz, P. Y. and Salathe, R. P. 'Spectral characterisation of photoinduced high efficient Bragg gratings in standard telecommunication fibres', *Electron. Lett.* **29**(1), 47 (1993).
27. Russell, P. St. J. and Hand, D. P. 'Rocking filter formation in photosensitive high birefringence optical fibres', *Electron. Lett.* **26**(22), 1846 (1990).
28. Kanellopoulos, S. E., Valente, L. C. G., Handerek, V. A. and Rogers, A. J. 'Photorefractive polarisation couplers in elliptical core fibres', *IEEE Photonics Technology Letters*, **3**(9), 806 (1991).
29. Hill, K. O., Bilodeau, F., Malo, B. and Johnson, D. C. 'Birefringent photosensitivity in monomode optical fibre: application to external writing of rocking filters', *Electron. Lett.* **27**(17), 1548 (1991).
30. Bilodeau, F., Hill, K. O., Malo, B., Johnson, D. C. and Skinner, I. M. 'Efficient, narrowband $LP_{01} - LP_{02}$ mode convertors fabricated in photosensitive fibre: spectral response', *Electron. Lett.* **27**(8), 682 (1991).
31. Bhatia, V. and Vengsarkar, A. M. 'Optical fiber long period grating sensors', *Opt. Lett.* **21**, 692–694 (1996).

32. Limberger, H. G., Varelas, D., Sayah, A., Salathe, R. P., Vasiliev, S. A. and Dianov, E. M. 'Post irradiation resonance wavelength adjustment of long period grating based loss filters', *Proc. IEE Colloquium on Optical Fibre Gratings* (1997/037), London, Feb. (1997).
33. Feced, R., Roe-Edwards, M. P., Kanellopoulos, S. E., Taylor, N. H. and Handerek, V. A. 'Mechanical strength degradation of UV exposed optical fibres', *Elect. Lett.* **33**(2), 157 (1997).
34. Hill, K. O., Malo, B., Vıneberg, K. A., Bilodeau, F., Johnson, D. C. and Skinner, I. 'Efficient mode conversion in telecommunication fibre using externally written gratings', *Electron. Lett.* **26**(16), 1270, (1990).
35. Meltz, G. and Morey, W. W. 'Bragg grating formation and germanosilicate fibre photosensitivity', SPIE vol. **1516**, International Workshop on Photoinduced Self-Organisation Effects in Optical Fibre, 185–199 (1991).
36. Morey, W. W., Dunphy, J. R. and Meltz, G. 'Multiplexing fibre Bragg grating sensors', *Fibre and Integrated Optics*, **10**, 351–360 (1992).
37. Erdogan, T., Mizrahi, V., Lemaire, P. J. and Monroe, D. 'Decay of ultraviolet-induced fibre Bragg gratings', *J. Appl. Phys/* **76**, 73 (1994).
38. Kanellopoulos, S. E., Handerek, V. A. and Rogers, A. J. 'Compact Mach–Zehnder fibre interferometer incorporating photoinduced gratings in elliptical core fibres', *Optics Letters* **18**(12), 15 June, 1013–15 (1993).
39. Morey, W. W., Meltz, G. and Glenn, W. H. 'Fiber optic Bragg grating sensors, *Fiber Optic and Laser Sensors VII, Proc. SPIE* **1169**, 98–108 (1989).
40. Morey, W. W., Meltz, G. and Glenn, W. H. 'Bragg-grating temperature and strain sensors', Springer Proceedings in Physics, *Optical Fibre Sensors*, **44**, 526 (1989)

11
Fiber Bragg grating sensors: principles and applications

Y.-J. Rao

11.1 INTRODUCTION

Following the early work on the formation of photogenerated gratings in germanosilicate optical fiber by sustained exposure of the core to the interference pattern produced by oppositely propagating modes of argon-ion laser radiation that was first reported in 1978 (Hill *et al.*, 1978), the pioneering sensor work at the United Technology Research Centre, which has been regarded as a milestone for in-fiber Bragg grating (FBG) sensors, was published 11 years later (Meltz *et al.*, 1989). Their side-writing technique makes a Bragg grating directly in the fiber core using a holographic interferometer illuminated with a coherent ultraviolet (UV) source. Versatility in the fabrication of FBGs has been gained from the fact that the Bragg wavelength is independent of the writing laser used. Subsequent to this initial work the interest in FBGs has increased considerably in recent years. There may be two main reasons for this:

1. The FBG has become a key passive device for applications in optical fiber telecommunications (Mizrahi, 1993), including wavelength-division multiplexing, fiber laser and amplifier pump reflectors (Ball and Morey, 1992; Farries *et al.*, 1992), gain flattening devices (Kashyap *et al.*, 1993), and dispersion compensation elements (Ouellette, 1991; Williams *et al.*, 1994);
2. It has been demonstrated that they have great potential for a wide range of sensing applications where quasi-distributed measurements for important physical quantities, such as strain, temperature, pressure, ultrasound, acceleration, high magnetic field, and force, are required (Morey *et al.*, 1989, 1992; Xu *et al.*, 1993; Foote *et al.*, 1996; Kersey and Marrone, 1994; Rao *et al.*, 1996d; Webb *et al.*, 1996; Theriault *et al.*, 1996; Bjerkan *et al.*, 1996).

Compared with other implementations of fiber-optic sensors, FBG sensors have a number of distinguishing advantages:

Optical Fiber Sensor Technology, Vol. 2. Edited by K. T. V. Grattan and B. T. Meggitt.
Published in 1998 by Chapman & Hall, London. ISBN 0 412 782 901

1. They can give an absolute measurement that is insensitive to fluctuations in the irradiance of the illuminating source, as the information is usually obtained by detecting the wavelength shift induced by the measurand;
2. They can be directly written into the fiber without changing the fiber diameter, making them compatible with a wide range of situations where small diameter probes are essential, such as in advanced composite materials for strain mapping, or the human body for temperature profiling;
3. They can be mass-produced at low cost, making them potentially competitive with conventional electrical sensors (Askins et al., 1994);
4. They can be multiplexed using similar techniques that have been applied for use with fiber-optic sensors, including wavelength-division-multiplexing (WDM), spatial-division-multiplexing (SDM), time-division-multiplexing (TDM), and their combinations (Kersey, 1993; Rao and Jackson, 1996a), making quasi-distributed sensing feasible in practice.

One of the most important applications of FBG sensors that has been demonstrated to date is for the so-called 'fiber-optic smart structures', where FBGs are embedded into the structure to monitor its strain distribution (Udd, 1995). Fiber-optic smart structure technology could in the future lead to structures that are self-monitoring and even self-scheduling of their maintenance and repair by the marriage of fiber-optic sensor technology and artificial intelligence with material science and structural engineering in the future. In order to see the significance of FBG sensors, a comparison between FBG sensors and other fiber-optic strain sensors is given in Table 11.1. It is seen that FBG sensors have more advantages than other fiber-optic strain sensors, making them ideal for

Table 11.1 Comparison of fiber-optic strain sensors

	FBG	FP	TM	PM
Linear response	Yes	Yes*	Yes*	Yes*
Absolute measurement	Yes	Yes[†]	Yes[†]	Yes[†]
Range to resolution	High	High	Low	Low
Sensor gauge length	Short	Short	Long	Long
Mechanical strength	High	Low[‡]	High	High
Multiplexing	Yes	Yes[§]	Yes[§]	Yes[§]
Mass production	Yes	Yes[¶]	Yes[¶]	Yes[¶]
Potential cost	Low	Low[∥]	Low[∥]	Low[∥]

Note: FP = Fabry–Pérot interferometric sensors (Kersey et al., 1983; Lee et al., 1992; Kist et al., 1984; Claus et al., 1993). TW = two-mode fiber optic sensors (Blake et al., 1987; Lu and Blaha, 1992). PM = polarimetric fiber-optic sensors (Varnham et al., 1983; Bock and Wolinski, 1991).
*Requires quadrature signal demodulation and it is difficult to achieve high stability in practice.
[†]Requires suitable signal processing, such as low-coherence interferometry (Rao and Jackson, 1996c).
[‡]Except for in-line fiber etalon strain sensors (Sirkis et al., 1993).
[§]Difficult if the number of sensors is large, unless spatial-division-multiplexing is used.
[¶]Requires some handling work either for cavity construction or for lead-in/lead-out fiber splicing, except for intrinsic Fabry–Pérot sensors made from two FBGs reported recently (Morey et al., 1991).
[∥]Good skills for fiber splicing or sensor assembly are needed.

fiber-optic smart structures. However, temperature compensation of the strain error caused by thermal fluctuation is essential for practical applications and this issue will be discussed later.

This chapter deals with an overview of the relevant FBG technology for sensor applications in several sections. Following the introduction, the essential sensing principles and properties of FBG sensors are provided in section 11.2. In section 11.3, various fabrication techniques for FBGs are outlined and section 11.4 introduces a range of interrogation techniques for FBG sensors, which can achieve high-resolution wavelength-shift detection. This chapter concludes with a summary and discussion of lively future developments of the FBG sensor technology.

11.2 PRINCIPLES AND PROPERTIES OF FBG SENSORS

11.2.1 Optical theory of in-fiber Bragg gratings

An FBG is written into a segment of Ge-doped single-mode fiber in which a periodic modulation of the core refractive index is formed by exposure to a spatial pattern of ultraviolet light in the region of 244–248 nm, as shown in Figure 11.1. This fabrication process is based on the photorefractive effect in the germania oxygen-vacancy defect band, which was observed in Ge-doped optical fibers by Hill *et al.* (1978). The lengths of FBGs are normally within the region of $1 \sim 20$ mm and grating reflectivities can approach $\sim 100\%$. When the FBG is illuminated by a broadband light source, a set of beams reflected from a set of partially reflecting planes formed by the periodic core index modulation interfere with each other. The interference is destructive unless each beam is in phase with all the others. According to Bragg's law which gives this condition, only one wavelength, i.e. the Bragg wavelength λ_B, is selected, which is given by the familiar equation

$$\lambda_B = 2n\Lambda \tag{11.1}$$

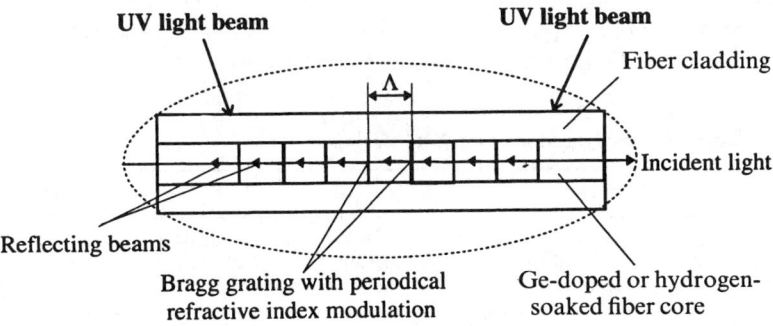

Fig. 11.1 Schematic diagram of in-fiber Bragg grating.

where n is effective core index of refraction and Λ the period of the index modulation.

The reflectivity at the Bragg wavelength can be estimated using the equation given by (Lam and Garside, 1981)

$$R = \tanh^2 \Omega \qquad (11.2)$$

where

$$\Omega = \pi n (L/\lambda_B)(\Delta n/n)\eta(V) \qquad (11.3)$$

The factor $\eta(V) \cong 1 - 1/V^2$, $V \geqslant 2.4$, is the fraction of the integrated fundamental mode intensity contained in the core (V is the normalized frequency of the fiber). It is seen that R is directly proportional to the grating length, L, and the index perturbation ($\Delta n/n$), which is normally determined by the exposure power and time of the UV radiation for a specified fiber.

The full width half maximum bandwidth, $\Delta\lambda$, of a grating is approximately given by (Russell et al., 1993)

$$\Delta\lambda = \lambda_B s \sqrt{\left(\frac{\Delta n}{2n}\right)^2 + \left(\frac{1}{N}\right)^2} \qquad (11.4)$$

where $s \sim 1$ for strong gratings (near 100% reflection) and $s \sim 0.5$ for weak gratings, and N is the number of grating planes. Because the change in the index is small, the main contribution to the linewidth change is attributed to the change in the modulation depth of the index perturbation.

Unlike conventional fiber-optic interferometric sensors illuminated by highly coherent lasers, FBG sensors generally require a broadband light source and a high resolution wavelength-shift detection system, which will be discussed later. From the light source point of view, a wide wavelength bandwidth and high optical power are normally required to achieve a large range to resolution for the wavelength-shift induced by the measurand. Most of the sources that have been used for FBG sensors are similar to those used for fiber-optic gyroscopes and low-coherence interferometric sensors (Lefevre, 1993; Jackson, 1994). They include edge-emitting light-emitting diodes (ELEDs), superluminescent diodes (SLDs), superfluorescent fiber sources (SFSs) and tunable fiber lasers (TFLs). Table 11.2 gives a comparison of these sources.

Table 11.2 Comparison of light sources for FBG sensors

	ELEDs	SLDs	SFSs	TFLs
Optical power*	1–10 µW	0.1–2 mW	1–10 mW	0.1–10 mW
FWHM bandwidth	40–100 nm	15–30 nm	20–40 nm	1–10 nm[†]
Device cost	Low	Medium	High	High

*Optical power coupled into single-mode fibre.
[†]Tunable wavelength range.

11.2.2 Sensing principles

FBG sensors have been reported for measurement of strain, temperature, pressure, dynamic magnetic field, etc. The FBG central wavelength will vary with the change of these parameters experienced by the fiber and the corresponding wavelength-shifts are given as follows.

(a) Strain

The wavelength shift, $\Delta\lambda_{BS}$, for an applied longitudinal strain $\Delta\varepsilon$ is given by

$$\Delta\lambda_{BS} = \lambda_B(1 - \rho_\alpha)\Delta\varepsilon \quad (11.5)$$

where ρ_α is the photoelastic coefficient of the fiber, given by

$$\rho_\alpha = \frac{n^2}{2}[p_{12} - \nu(p_{11} - p_{12})] \quad (11.6)$$

where p_{11} and p_{12} are the components of the fiber strain-optic tensor and ν is Poisson's ratio. For the silica fiber, the wavelength–strain sensitivities of 800 nm and 1.55 μm FBGs have been measured as ~ 0.64 pm $\mu\varepsilon^{-1}$ and ~ 1.15 pm^{-1} $\mu\varepsilon$ respectively (Morey et al., 1989; Rao et al., 1995). For the measurement of acceleration, ultrasound and force, equations 11.5 and 11.6 are still applicable as these measurands are converted from strain.

Linearly chirped FBGs have been demonstrated as dispersion compensation elements for high-speed optical fiber communication systems recently (Ouellette, 1991; Williams et al., 1994). These devices act as a wavelength bandpass filter in which the pitch of the grating is varied along the position of the grating length and the chirped FBG reflects a large number of wavelengths, i.e. optical frequencies, from different positions of the grating, as shown in Figure 11.2. The principle of using this chirped device for strain sensing is based on the effective change in reflection point, δb, which is given by (Kersey and Davis, 1994)

$$\delta b = -\frac{\lambda}{\Delta\lambda_c}B\zeta\Delta\varepsilon \quad (11.7)$$

where ζ is a constant determined by the photoelastic properties of the fiber and B is the grating length. λ is a fixed source wavelength and $\Delta\lambda_c = \lambda_{B_1} - \lambda_{B2}$ with $\lambda_{B_1} < \lambda_{B2}$ (Figure 11.2). For comparison, the optical length change for a length of fiber B is given by

$$\delta l = B\zeta\Delta\varepsilon \quad (11.8)$$

The ratio of equations 11.7 and 11.8 is thus expressed in the form

$$\frac{\delta b}{\delta l} = -\frac{\lambda}{\Delta\lambda_c} \quad (11.9)$$

It is seen that due to $\lambda \gg \Delta\lambda_c$, $\delta b \gg \delta l$, hence the chirped FBG gives a very large strain transduction amplification factor. From the high sensitivity point

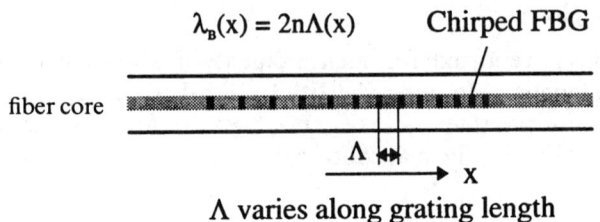
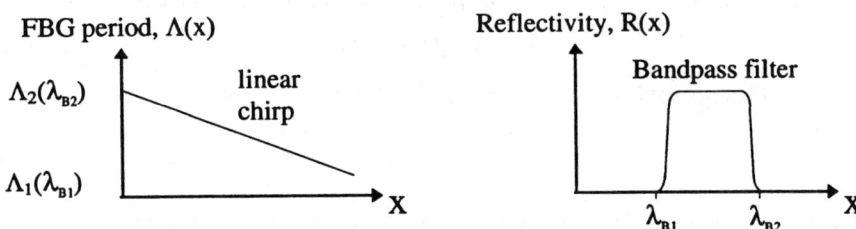

Fig. 11.2 Linearly chirped FBG sensor.

of view, linearly chirped FBGs have potential advantages for dynamic strain measurement. Another potential use of chirped FBGs is for distributed strain measurement by sequentially interrogating short sections along the grating with low coherence interferometry (Volanthen *et al.*, 1996).

(b) Temperature

For a temperature change of ΔT, the corresponding wavelength shift $\Delta\lambda_{BT}$ is given by

$$\Delta\lambda_{BT} = \lambda_B(1 + \xi)\Delta T \qquad (11.10)$$

where ξ is the fiber thermo-optic coefficient. For the silica fiber, the wavelength–temperature sensitivities of 800 nm and 1.55 μm FBGs have been measured with values of ~ 6.8 and ~ 13 pm °C^{-1}, respectively (Morey *et al.*, 1989; Rao *et al.*, 1995). Table 11.3 gives a summary of the wavelength–strain and wavelength–temperature sensitivities for FBGs with different wavelengths (Morey *et al.*, 1989; Xu *et al.*, 1994b; Rao *et al.*, 1995).

Table 11.3 Strain and temperature sensitivities of FBG sensors with different wavelengths

Wavelength (μm)	Strain sensitivity (pm με^{-1})	Temperature sensitivity (pm °C^{-1})
0.83	~ 0.64	~ 6.8
1.3	~ 1	~ 10
1.55	~ 1.2	~ 13

(c) Pressure

For a pressure change of ΔP, the corresponding wavelength shift $\Delta\lambda_{BP}$ is given by

$$\frac{\Delta\lambda_{BP}}{\lambda_B} = \frac{\Delta(n\Lambda)}{n\Lambda} = \left[\frac{1}{\Lambda}\frac{\partial\Lambda}{\partial P} + \frac{1}{n}\frac{\partial n}{\partial P}\right]\Delta P \qquad (11.11)$$

In the weakly guided single-mode region, the contribution to the fractional change in optical propagation delay arising from a small fractional change in fiber diameter due to the applied pressure is negligible when compared with the change in refractive index and physical length. The changes in physical length and refractive index are given by (Hocker, 1979; Morey *et al.*, 1992)

$$\frac{\Delta L}{L} = -\frac{(1-2v)P}{E} \qquad (11.12a)$$

$$\frac{\Delta n}{n} = \frac{n^2 P}{2E}(1-2v)(2\rho_{12} + \rho_{11}) \qquad (11.12b)$$

where E is the Young's modulus of the fiber. As a result of $\Delta L/L = \Delta\Lambda/\Lambda$ the normalized pitch-pressure and the index-pressure coefficients are given by

$$\frac{1}{\Lambda}\frac{\partial\Lambda}{\partial P} = -\frac{(1-2v)}{E} \qquad (11.13a)$$

$$\frac{1}{n}\frac{\partial n}{\partial P} = \frac{n^2}{2E}(1-2v)(2\rho_{12} + \rho_{11}) \qquad (11.13b)$$

By substituting equations 11.13a and 11.13b into 11.11, we obtain the wavelength–pressure sensitivity, given by

$$\Delta\lambda_{BP} = \lambda_B\left[-\frac{(1-2v)}{E} + \frac{n^2}{2E}(1-2v)(2\rho_{12} + \rho_{11})\right]\Delta P \qquad (11.14)$$

For a Ge-doped FBG at 1.55 µm $\Delta\lambda_{BP}/\Delta P$ was measured as -3×10^{-3} nm MPa^{-1} over a pressure range of 70 MPa (Xu *et al.*, 1993a). The pressure sensitivity has been enhanced by a factor of four by mounting the FBG in a hollow glass bubble to achieve mechanical amplification (Xu *et al.*, 1996).

(d) Dynamic magnetic field

An FBG can also be used for dynamic magnetic field detection with the Faraday effect to induce a slight change in the index of the fiber experienced by left and right circularly polarized light in an FBG (Kersey and Marrone, 1994). In the presence of a longitudinal magnetic field applied to the FBG, the index is changed for the two circular polarizations and in consequence two Bragg wavelengths are obtained, as shown in Figure 11.3, i.e.

$$\lambda_{B+} = 2n_+\Lambda \qquad (11.15a)$$

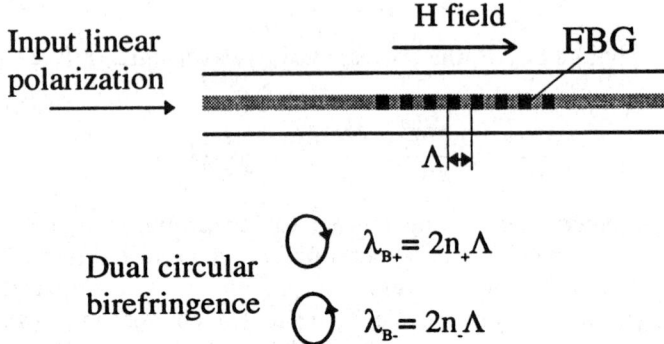

Fig. 11.3 Splitting of the Bragg resonance due to the circular birefringence induced by magnetic field.

$$\lambda_{B-} = 2n_- \Lambda \qquad (11.15b)$$

where the subscripts $+$ and $-$ represent the index and Bragg wavelength for right and left circularly polarized light at the FBG. For the silica fiber, the sensitivity of this effect is weak as it is determined by the inherent Verdet constant of the silica which is only $\sim 8 \times 10^{-1}\,\text{rad}\,\text{T}^{-1}\,\text{m}^{-1}$ at $\sim 1.3\,\mu\text{m}$ wavelength. The change in index is given by

$$n_+ - n_- = \frac{V_d H \lambda}{2\pi} \qquad (11.16)$$

where V_d and λ are the Verdet constant and the working wavelength. H is the applied magnetic field (tesla). Consequently, the resulting wavelength-shift is very small and it was detected with the interferometric interrogation scheme which can achieve a very high sensitivity (Kersey and Marrone, 1994). Based on this work, dynamic measurement of large current and voltage would be possible.

(e) Simultaneous strain and temperature measurement

For the quasi-static and static strain measurement, large temperature variations are common in practical applications. Hence the temperature compensation for the effect of ambient temperature fluctuation is required (Farahi et al., 1990). A range of techniques have been proposed to achieve this goal by measuring strain and temperature simultaneously.

(1) Reference FBG method. On a single measurement of the Bragg wavelength-shift it is impossible to separate the effects of changes in strain and temperature. Therefore a reference is required for temperature measurement. The most straightforward way is to use a separated, strain-free FBG as the temperature sensor to directly measure temperature of the strain sensor (Morey et al., 1992; Xu et al., 1994a). This reference FBG is located in the

same thermal environment as the strain sensor. The strain error caused by the temperature variation can be compensated to the first order by subtracting the wavelength-shift induced by temperature variation from the total wavelength-shift obtained with the strain sensor.

It has been found that a chirped FBG in a tapered optical fiber can be temperature-independent (Xu et al., 1995). The taper profile is designed such that the FBG is linearly chirped when tension is applied, creating a strain gradient along the FBG. By measuring the effective bandwidth variation rather than the Bragg wavelength change, the reflected intensity signal is insensitive to temperature. This special type of FBG sensor is attractive as no temperature compensation is required, but one problem is that at the tapered region of the fiber mechanical strength becomes much weaker and hence the fiber is easy to break. Furthermore, investigation might be needed to assess the measurement accuracy as any intensity fluctuation along the lead fiber would cause error.

(2) *Dual-wavelength superimposed FBGs method.* This method is based on the use of two sets of wavelength-shift data obtained from two superimposed FBGs written at the same location in the fiber (Xu et al., 1994b). Assuming that the wavelength-shifts in strain and temperature are linear, the Bragg wavelength-shift, $\Delta\lambda_B$, in response to a strain change, $\Delta\varepsilon$ and a temperature change, ΔT, is given by

$$\Delta\lambda_B = K_\varepsilon \Delta\varepsilon + K_T \Delta T \qquad (11.17)$$

where K_ε, K_T are the strain and temperature sensitivities of the FBG. Equation 11.17 assumes that the strain and temperature are essentially independent, which has been investigated to apply well for small perturbations (Morey et al., 1989). Hence, for the dual-wavelength superimposed FBGs, the strain and temperature can be obtained simultaneously using the following matrix:

$$\begin{pmatrix} \Delta\lambda_{B1} \\ \Delta\lambda_{B2} \end{pmatrix} = \begin{pmatrix} K_{\varepsilon 1} & K_{T1} \\ K_{\varepsilon 2} & K_{T2} \end{pmatrix} \begin{pmatrix} \Delta\varepsilon \\ \Delta T \end{pmatrix} \qquad (11.18)$$

where 1 and 2 represent the two wavelengths. The element of the **K** matrix can be determined experimentally by separately measuring the wavelength-shifts with strain and temperature. As long as **K** is known $\Delta\varepsilon$ and ΔT are easily determined. This concept has been demonstrated using two FBGs with central wavelengths of 0.85 and 1.3 µm.

(3) *Harmonics method.* Another way of obtaining two wavelengths data is to use the first harmonics of an FBG. For an FBG with high reflectivity, the index perturbation may be not a perfect sinusoidal profile due to possible overexposure, resulting in generation of the harmonics. The wavelength of the first harmonics is just twice the central wavelength of the FBG. An experiment has been carried out to verify this idea by using the first harmonics (0.78 µm) of a 1.56 µm FBG (Kalli et al., 1994).

For both schemes mentioned above, two sets of independent detection systems for the two wavelengths are required to obtain the corresponding wavelength-shifts. This makes the whole measurement system complicated and potentially expensive. On the other hand, the incompatibility of the two fibers operating at different wavelengths would be a problem for multiplexing of FBGs for quasi-distributed measurement.

(4) *Superimposed FBG and polarization-rocking filter method.* The all-fiber polarization-rocking filter was first demonstrated for use of polarization mode conversion in optical fiber telecommunication (Stolen *et al.*, 1984). An external point-to-point writing technique has been used to fabricate rocking filters in a photosensitive D-type polarization-maintaining fiber (Hill *et al.*, 1991). Based on this fabrication technique, a compact Mach–Zehnder fiber interferometer was formed by arranging two rocking filters on the same piece of fiber with a separation and has been used for temperature measurement (Kanellopoulos *et al.*, 1993). The temperature-induced resonant wavelength-shift is roughly two orders of magnitude greater than that of normal FBGs. As the wavelength-shift coefficients in temperature and strain for an FBG and a rocking filter are different, a matrix similar to K in equation 11.18 can be obtained to separate temperature and strain. By superimposing an FBG and a Mach–Zehnder interferometer formed by two rocking filters, this method has been demonstrated recently (Kanellopoulos *et al.*, 1995).

This scheme avoids the use of two wavelengths, but there are several constraints:

1. The physical length of the rocking filter is quite large compared with the FBG, hence limiting its application range;
2. The bandwidth of the rocking filter reported is much larger than that of the FBG, restricting the measurement accuracy and the number of sensors which can be multiplexed with WDM for quasi-distributed measurement;
3. The use of polarization-maintaining fiber and polarization-related devices would increase the cost and difficulty of multiplexing, especially when the number of sensors is large.

(5) *Combined FBG and long-period grating method.* Long-period fiber gratings were initially developed for optical fiber communications systems as band-rejection filters (Vengsarkar *et al.*, 1996). The basic difference between a long-period grating and a Bragg grating is that the periodicity of the long-period grating is typically several hundred times greater than that of the Bragg grating. This is because in order to couple from a forward propagating fundamental guided mode to forward propagation cladding modes the phase-matching vector is necessarily short, resulting in a long periodicity. During the strain and temperature property tests, it was discovered that long-period gratings have quite different strain– and temperature–wavelength response coefficients when compared with Bragg gratings (Bhatia and Vengsarkar,

1996). Therefore, simultaneous strain/temperature measurement could be achieved if the strain- and temperature-wavelength response coefficients of both the long-period grating and the Bragg grating could be known accurately. Preliminary results have been obtained to verify this idea (Patrick et al., 1996). Unlike the dual-wavelength superimposed FBGs method, this approach does not need two sets of independent detection systems for detecting changes of two widely separated wavelengths.

There are several concerns about this approach:

1. The bandwidth of the long-period grating is relatively large, which would limit measurement accuracy and the number of sensors which can be multiplexed with WDM;
2. The physical length (typically a few centimeters) of the long-period grating is much longer than that of the FBG, hence it may be inapplicable for some cases where the separation between two adjacent sensors, i.e. spatial resolution, is required to be small;
3. It may be difficult to use in an embedded version for smart structure applications, as the long-period grating may experience a significant nonuniform strain field along a grating length of a few centimeters;
4. As the long-period grating is extremely sensitive to bends in the fiber, separation of the wavelength changes caused by the bend and the longitudinal strain arises as a new problem.

(6) *Dual-diameter FBGs method.* It has been found that the strain responses to relative wavelength-shifts and the temperature responses to the weighted wavelength-shift differences of two FBGs with different cladding diameters are not the same (James et al., 1996). By fusion-splicing two FBGs with dual diameters, two sets of relative wavelength-shifts data are obtained for separating strain and temperature. The Bragg wavelengths of the two FBGs may differ by a few nanometers, allowing them to be measured independently with WDM. This scheme improves the WDM capacity over methods (4) and (5) mentioned above, but further investigation would be required to address the potential problem of low splice strength and high splice loss.

(7) *Combined FBG and in-line fiber etalon sensor method.* It has been demonstrated that an in-line fiber etalon sensor, formed by splicing a short segment of silica hollow-core fiber to two sections of single-mode fiber, has low temperature sensitivity and high mechanical strength, making such a type of sensor suitable for fiber-optic smart structures applications (Sirkis et al., 1993). With the combination of an FBG sensor and an in-line fiber etalon interferometric sensor, simultaneous strain and temperature measurement has been achieved due to their different strain and temperature response coefficients (Singh and Sirkis, 1996). As the return signals from cascaded FBG and etalon sensors mix together in spectral domain, it is impossible to completely separate them by using only WDM, resulting in relatively high cross-talk.

Table 11.4 Physical properties of normal in-fiber Bragg gratings

Ultimate strength (strain capability)	> 200 kpsi (2%)*
Failure time	> 1 million cycles[†]
Thermal stability	< 300 °C[‡]
Response time	< 1 µs[§]

*This strength reduction is caused by stripping, grating exposure and polyimide recoating when compared with a high-quality fiber with a strength of 800 kpsi.
[†]Obtained by laboratory simulation tests.
[‡]Permanent changes begin to occur in the grating reflectivity at this temperature.
[§]This is mainly limited by the acoustic transit time across the grating length.

Table 11.5 Optical properties of normal in-fiber Bragg gratings

Spectral linewidth	0.1 – 0.2 nm*
Reflectivity	0 – 100%
Excess loss	− 30 dB

*The spectral linewidth is, to a large degree, dependent on the grating length via an inverse relationship.

11.2.3 General properties of FBGs

The general properties of normal FBGs are summarized in Tables 11.4 and 11.5 respectively (Dunphy *et al.*, 1995).

11.3 FABRICATION OF FBGs

11.3.1 Summary of essential principles of FBG fabrications for sensor use

Advanced fabrication techniques are essential for obtaining high-quality, low-cost FBG sensor technology. In order to meet the required sensor specifications, an ideal FBG fabrication technique should have the following features:

1. *Flexibility*: reflectivities and central wavelengths of FBGs produced should be selectable for quasi-distributed measurement;
2. *Economical mass-production capability*: low-cost FBGs would be available if they could be mass-produced at a high speed;
3. *Good physical and optical qualities*: the mechanical strength of an FBG produced should not be degraded significantly after fabrication when compared with the strength of a good-quality fiber. A narrow spectral linewidth and a low excess loss are normally required to achieve high-resolution measurement;

4. *Good repeatability*: repeatabilities of both the central wavelength and the reflectivity of an FBG should be good enough in order to make FBGs standard devices under the condition of mass-production and interchangeable without calibration.

The fabrication techniques have been subjects of strong research interest owing to the driving force arising from communications and sensing applications. A number of schemes have been demonstrated to match the requirements mentioned above, which has led to successful commercialization of FBG fabrication very recently, as discussed in Chapter 10 by Handerek.

11.4 INTERROGATION TECHNIQUES

Precision measurement of the FBG wavelength-shift induced by the measurand is crucial for achieving good sensor performance. The general requirements for an ideal interrogation method are:

1. *High resolution with large measurement range* — typically a wavelength-shift detection resolution ranging from subpicometer to a few picometers is required for most applications. The range to resolution required is within $10^3:1 \sim 10^5:1$.
2. *Cost-effective* — the cost of an interrogation system should be competitive with conventional optical or electrical sensors potentially.
3. *Compatible with multiplexing* — an interrogation scheme should be able to cope with multiplexing topologies which can make the whole sensing system cost-effective.

Conventional spectrometers have a typical resolution of 0.1 nm, hence they are normally used for evaluation of the optical properties of FBGs during fabrication procedures rather than for precision wavelength-shift detection. Research on high-resolution interrogation methods has been a very active topic in recent years. A number of interrogation techniques have been reported for high-resolution wavelength-shift detection. According to the operating principles of the devices used for wavelength-shift detection, these techniques can be classified as edge filter, tunable filter, interferometric scanning and dual-cavity interferometric scanning methods (Kersey, 1993; Rao *et al.*, 1996a), which are introduced in the following section.

11.4.1 Principles of interrogation schemes

(a) Edge filter

This method is based on the use of an edge filter which has a linear relationship between wavelength-shifts and the output intensity changes of the filter, as

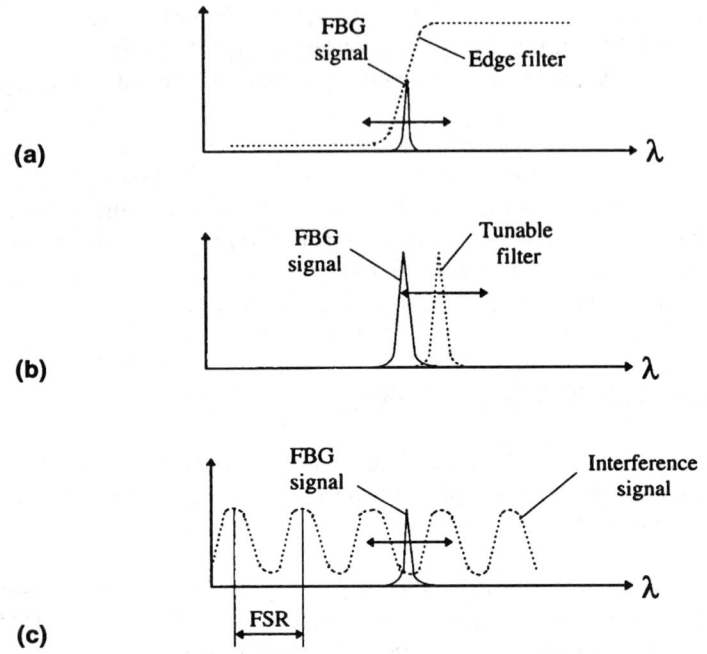

Fig. 11.4 (a) Principle of the edge filter method; (b) principle of the tunable filter method; (c) principle of the interferometric scanning method.

shown in Figure 11.4(a). By measuring the intensity change, the wavelength-shift induced by the measurand is obtained. The measurement range is inversely proportional to the detection resolution.

(b) Tunable filter

A tunable filter can be used to measure the wavelength-shift of the FBG and the output is a convolution of both the spectrum of the tunable filter and that of the FBG, as shown in Figure 11.4(b). When the spectrum of the tunable filter matches that of the FBG, the convolution equals 1, i.e. a maximum output occurs. By measuring this maximum point and the corresponding wavelength change of the tunable filter, the wavelength-shift of the FBG is obtained. The measurement resolution is mainly determined by the signal-to-noise ratio of the return FBG signal and both the linewidths of the tunable filter and the FBG. Normally, such an approach has a relatively high resolution plus a large working range.

(c) Interferometric scanning

The FBG wavelength-shift induced by strain and/or temperature can be detected with a scanned interferometer, which has been demonstrated for high-resolution dynamic and quasi-static strain measurement, termed the 'interferometric scanning method' (Kersey et al., 1992, 1993). The normalized interference signal from a scanned interferometer (SI), as shown in Figure 11.4(c), can be expressed as

$$I/I_0 = 1 + B\cos[\Delta\Phi_B + \phi(t)] \qquad (11.19)$$

where I_0 is the intensity of the incident light and B the visibility of the interference signal. $\phi(t)$ is the thermally induced phase drift in the SI. The SI acts as a wavelength scanner for FBGs when the optical path of the SI is modulated. The strain or temperature induced change in the reflected wavelength from an FBG produces a change in optical phase $\Delta\Phi_B$:

$$\Delta\Phi_B = -\frac{2\pi\Delta L_{SI}}{\lambda_B^2}\Delta\lambda_B = -\frac{2\pi\Delta L_{SI}}{\lambda_B}\xi_g\Delta Y \qquad (11.20)$$

where ΔY is the variation in strain or temperature applied to the FBG and ΔL_{SI} is the optical path difference (OPD) of the SI. ξ_g is the normalized FBG sensitivity for strain or temperature, given by

$$\xi_g = \frac{1}{\lambda_B}\frac{\delta\lambda_B}{\delta Y} \qquad (11.21)$$

Hence, the phase sensitivity in response to strain or temperature ($\Delta\Phi_B/\Delta Y$) is directly proportional to the OPD in the SI. By measuring $\Delta\Phi_B$ with the pseudo-heterodyne processing scheme (Jackson et al., 1982) the strain or temperature change can be determined.

The operational range of the FBG is set by the free spectral range (FSR) of the SI, which is given by

$$\text{FSR} = \frac{\lambda_B^2}{\Delta L_{SI}} \qquad (11.22)$$

It can be seen that the operational range is inversely proportional to the OPD in the SI. Therefore, there is a tradeoff between the sensitivity and the operational range; this is because in this method of signal recovery the unambiguous measurement range is equivalent to a 2π change in the SI. The system stability for quasi-static and static measurement is limited by $\phi(t)$. By incorporating a local reference FBG and using the phase difference between the sensing FBG and the reference FBG, the thermal drift can be compensated (Kersey et al., 1993).

(d) Dual-cavity interferometric scanning

A novel method using two sets of interferometric fringes obtained by stepping the SI from a long cavity to another short cavity has been proposed to enhance the unambiguous measurement range of the interferometric scanning scheme (Rao et al., 1996a). The optical phase output from the cavity with the larger OPD, i.e. range 1, gives the high-resolution measurement whilst the phase output from the cavity with the shorter OPD, i.e. range 2, is used for

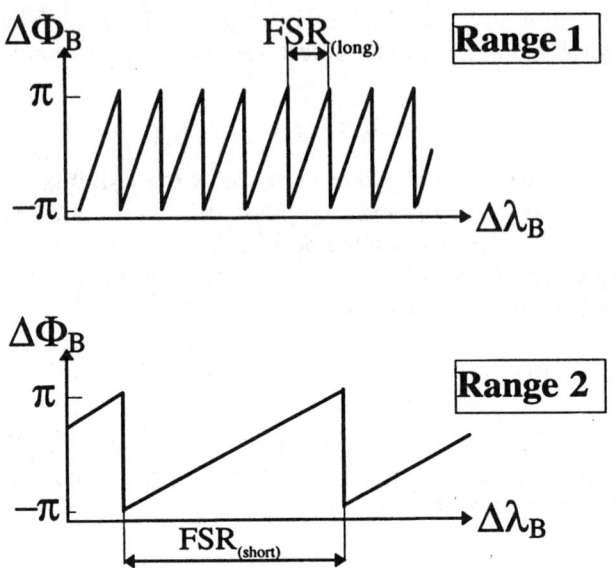

Fig. 11.5 Principle of the dual-cavity interferometric scanning scheme.

determining the number of the fringes obtained with the long cavity, as shown in Figure 11.5. The total absolute value of the phase change is thus given by $(\Delta\Phi_B + 2\pi N)$ (here N is the number of the interferometric fringe). The enhancement factor, M, in the unambiguous range is given by the ratio of the dual cavity lengths used in the stepped interferometer, i.e.

$$M = \frac{\text{FSR}_{\text{short}}}{\text{FSR}_{\text{Long}}} = \frac{L_{\text{SI}}}{L'_{\text{SI}}} \qquad (11.23)$$

where $L_{\text{SI}}, L'_{\text{SI}}$ are the cavity lengths of the stepped interferometer for the ranges 1 and 2, respectively, i.e. the long and the short cavities. In principle M could be very large as the cavity length of the stepped interferometer can be varied from a few hundred microns to a few hundred millimeters in a well-collimated interferometer, although in practice M is likely to be selected in the range of $10 \sim 100$.

Various signal processing aspects are discussed in the next chapter.

11.5 SUMMARY

The FBG sensor technology has proven to be a powerful tool for quasi-distributed measurement of strain and temperature and has found a number of applications in practice, such as cure monitoring of composite materials (Dunphy *et al.*, 1991), strain mapping of advanced composite materials (Simonsen *et al.*, 1992; Friebele *et al.*, 1994; Rao *et al.*, 1996b; Foote *et al.*, 1996), bridge strain monitoring (Measures *et al.*, 1994), measurement of high-voltage transformer

SUMMARY

winding temperatures (Hammon and Stokes, 1996), high-resolution temperature profiling for medical applications (Rao *et al.*, 1997), and high-*g* acceleration measurement (Theriault *et al.*, 1996). In this chapter, a systematic overview of the FBG sensor technology has been given, which indicates that rapid and significant progress has been made in fabrication techniques, signal processing and multiplexing techniques in recent years. FBG sensors have been explored for a wide variety of applications and in particular they have found important applications in strain monitoring of composite materials and structures, making the FBG sensor technology a very strong candidate for one rapidly developing field of optical fiber sensors—namely fiber-optic smart structures. Some instruments based on the edge filter method have been commercialized successfully. There is no doubt that more cost-effective instruments will be commercialized as the fabrication cost of FBG sensors will reduce considerably in the near future. More researchers and engineers have become involved in this area on FBG sensor technology, Tables 11.6–11.8 give a brief summary of the fabrication, interrogation and multiplexing techniques which have been discussed above.

Finally, there may be several key topics that need to be specifically addressed in the future:

1. Simultaneous measurement of strain and temperature for temperature compensation of embedded FBG strain sensors is still an active research project which may continue in the long term as there are few accurate and practical ways to achieve this goal, particularly in a multiplexed sensor system;
2. Distributed strain measurement with high spatial resolution exploiting low coherence interferometry could become an important technique for strain mapping of small structures and crack growing, and further extension of the measurement range would make this technique more useful;
3. Cost-effective fiber-optic multiplexing networks are required for use in sensing systems with a very large number of sensors (>100) for strain mapping of large structures, such as roads, aircraft, bridges, marine vehicles, and dams;

Table 11.6 Brief summary of FBG fabrication techniques

	TBI	PM	MF/PMS
Flexibility	Good	Medium*	Good
Repeatability	Medium†	Good	Good
UV laser coherence	High	Low	Low
Mass production	Yes	Yes	Yes
Potential fabrication cost	Low	Low	Low

Note: TBI = two-beam interferometer method; PM = phase-mask method; MF/PMS = moving fiber/phase-mask scanning method.
*Flexibility can be improved by using a lens to demagnificate the pitch of the phase-mask.
†It is difficult to maintain the mechanical stability of a two-beam interferometer in practice.

FIBER BRAGG GRATING SENSORS

Table 11.7 Brief summary of FBG interrogation techniques

	EF	TF	IS	DCIS
Range to resolution	$10^2 - 10^3$	$10^3 - 10^4$	$10^3 - 10^4$	$10^4 - 10^6$
Measurement speed	High	High	High	Medium*
Long-term stability	Good[†]	Good[†]	Good[†]	Good[‡]
WDM compatability	Low	High	High	High
Potential cost	Low	Medium	Low	Low

Note: EF = edge filter method; TF = tunable filter method; IS = interferometric scanning method; DCIS = dual-cavity interferometric scanning method.
*Measurement speed can be improved significantly by using two tandem interferometers with different cavity lengths (Rao et al., 1996c).
[†]Requires good temperature stabilization of the filter.
[‡]A reference FBG is required to compensate the thermal drift of the interferometric scanner.

Table 11.8 Brief summary of FBG multiplexing techniques

	TDM	WDM	SDM	SDM/WDM
Multiplexing capacity	Medium	Good*	Good	Very good
Spatial resolution	Low	High	High	High
Usage of optical power	Good[†]	Good	Medium[‡]	Good[‡]
Interchangeability	Low	Low	High	Medium
Potential cost	Low	Medium	Medium	Medium

Note: TDM = time-division-multiplexing; WDM = wavelength-division-multiplexing; SDM = spatial-division-multiplexing; SDM/WDM = spatial- and wavelength-division-multiplexing.
*In particular when combined with TDM scheme.
[†]Poor signal-to-noise ratio may occur when the number of FBG sensors in the network is large.
[‡]Usage of optical power can be more efficient if a fiber-optic switch is used to select the sensor channel sequentially, but the sampling rate reduces.

4. Extensive studies into the interrogation techniques for the extension of range to resolution are required in order to allow the FBG sensor technology to compete with traditional strain gauges or fiber-optic interferometric sensors;
5. Combinations of FBG sensors with other fiber-optic sensors would be an interesting area as one can complement their advantages for multi-parameter measurements (Rao and Jackson, 1996b; Singh and Sirkis, 1996);
6. A universal instrument may be required to be able to simultaneously interrogate static strain for load monitoring and dynamic strain for impact monitoring in a structure, as the measurement sensitivities for two types of strains may be quite different;
7. Chemical sensing may be another important application area of FBG sensors (Meltz et al., 1996; Asseh et al., 1996). Extensive theoretical and experimental studies may be needed to determine the relationships between the wavelength and intensity changes of FBG sensors and the refractive index change in the evanescent field of the fiber;

8. The concepts for FBG devices have been applied to photosensitive planar doped-silica waveguides (Kitagawa et al., 1994). The combination of FBG sensor technology with integrated optics technology offers the possibility of integration of the light source, passive optical devices, such as couplers, and multiple Bragg grating sensors, which may not only result in a significant reduction in the overall system cost, but also make the sensor more robust and compact for multi-parameter measurement.

REFERENCES

Askins, C. G., Putnam, M. A., Williams, G. M. and Friebele, E. J. (1994) Stepped-wavelength optical-fibre Bragg gratings arrays fabricated in line on a draw tower, *Opt. Lett.* **19**, 147–9.

Asseh, A., Sandgren, S., Ahlfeldt, H., Sahlgren, B., Stubbe, R. and Edwall, G. N. (1996) Evanescent refractive index sensor utilizing a narrow fibre Bragg grating and a tunable DFB laser, *Proc. SPIE* **2836**, paper 10.

Ball, G. A. and Morey, W. W. (1992) Continuously tunable single-mode fibre laser, *Opt. Lett.* **17**, 420–2.

Bhatia, V. and Vengsarkar, A. M. (1996) Optical fibre long-period grating sensors. *Opt. Lett.* **21**, 692–4.

Bjerkan, L., Hjelme, D. R. and Johannessen, K. (1996) Bragg grating sensor demodulation scheme using a semiconductor laser for measuring slamming forces of marine vehicle modes, *Proc. 11th Inter. Conf. on Optical Fibre Sensors* (Sapporo, Japan, 5. 1996, IEICE and IEEJ), pp. 236–9.

Blake, J. N., Huang, S. Y., Kim, B. Y. and Shaw, H. J. (1987) Strain effects on highly elliptical core two-mode fibres, *Opt. Lett.* **12**, 732–4.

Bock, W. J. and Wolinski, T. R. (1991) Temperature-compensated fibre-optic strain sensor based on polarization-rotated reflection, *Proc. SPIE*, **1370**, 189–96.

Claus, R. O., Gunther, M. F., Wang, A. B., Murphy, K. A. and Sun, D. (1993) Extrinsic Fabry–Pérot sensor for structural evaluations in *Applications of Fibre-Optic Sensors in Engineering Mechanics* ed. F. Ansari (Am. Soc. Civ. Eng., New York).

Dunphy, J. R., Meltz, G., Lamm, F. P. and Morey, W. W. (1991) Fibre-optic strain sensor multi-function, distributed optical fibre sensor for composite cure and response monitoring, *Proc. SPIE*, **1370**, 116–18.

Dunphy, J. R., Meltz, G. and Morey, W. W. (1995) in Udd, E. (ed) *Fibre-Optic Smart Structures* (New York: John Wiley & Sons) Ch. 10, pp. 271–85.

Farahi, F., Webb, D. J., Jones, J. D. C. and Jackson, D. A. (1990) Simultaneous measurement of strain and temperature: cross-sensitivity considerations, *J. Lightwave Technol.* **LT-8**, 138–42.

Farries, M. C., Ragdale, C. M. and Reid, D. C. (1992) Broadband chirped fibre Bragg filters for pump rejection and recycling in erbium doped fibre amplifiers, *Electron. Lett.* **28**, 487–9.

Foote, P. D., Read, I. J., Slack, T. J., Ball, A., Jackson, P., Rao, Y. J., Jackson, D. A., Bennion, I. and Zhang, L. (1996) Optical fibre Bragg sensors systems: a basis for aircraft health and usage monitoring submitted to *J. Smart Mater. & Struct.*

Friebele, E. J., Askins, C. G. and Putnam, M. A. (1994) Distributed strain sensing with fibre Bragg grating arrays embedded in CRTM composites. *Proc. SPIE* **2361**, 338–41.

Hammon, T. E. and Stokes, A. D. (1996) Optical fibre Bragg grating temperature sensor measurements in an electrical power transformer using a temperature compensated optical fibre Bragg grating as a reference, *Proc. 11th Inter. Conf. on Optical Fibre Sensors* (Sapporo, Japan, 5. 1996, IEICE and IEEJ), pp. 566–9.

Hill, K. O., Fujii, F., Johnson, D. C. and Kawasaki, B. S. (1978) Photosensitivity on optical fibre waveguides: application to reflection filter fabrication, *Appl. Phys. Lett.* **32**, 647–9.

Hill, K. O., Bilodeau, F., Malo, B. and Johnson, D. C. (1991) Birefringent photosensitivity in monomode optical fibre: application to external writing of rocking filters, *Electron. Lett.* **27**, 1548–50.

Hocker, G. B. (1979) Fibre-optic sensing of pressure and temperature, *Appl. Opt.* **18**, 1445.

Jackson, D. A., Kersey, A. D., Corke, M. and Jones, J. D. C. (1982) Pseudo-heterodyne detection scheme for optical interferometers, *Electron. Lett.* **18**, 1081–2.

Jackson, D. A. (1994) Recent progress in monomode fibre-optic sensors, *Meas. Sci. Technol.* **5**, 621–38.

James, S. W., Dockney, M. L. and Tatam, R. P. (1996) Independent measurement of temperature and strain using in fibre Bragg grating sensors. *Proc. 11th Inter. Conf. on Optical Fibre Sensors*, Sapporo, Japan, postdeadline paper: Fr3-3.

Kalli, K., Brady, G., Webb, D. J., Jackson, D. A., Zhang, L. and Bennion, I. (1994) Possible approach for simultaneous measurement of strain and temperature with second harmonics in a fibre Bragg grating sensor, *Colloquium on Progress in fibre optic sensors & their applications* (London, 5. 1994, IOP).

Kanellopoulos, S. E., Handerek, V. A. and Rogers, A. J. (1993) Compact Mach–Zehnder fibre interferometer incorporating photoinduced gratings in elliptical-core fibres, *Opt. Lett.* **18**, 1013–15.

Kanellopoulos, S. E., Handerek, V. A. and Rogers, A. J. (1995) Simultaneous strain and temperature sensing with photogenerated in-fibre gratings, *Opt. Lett.* **20**, 333–5.

Kashyap, R., Wyatt, R. and Campbell, R. J. (1993) Wideband gain flattened erbium fibre amplifier using a photosensitive fibre blazed grating, *Electron. Lett.* **29**, 154–6.

Kersey, A. D. (1993) Interrogation and multiplexing techniques for fibre Bragg grating strain-sensors, *Proc. SPIE* **2071**, 30–48.

Kersey, A. D., Jackson, D. A. and Corke, M. (1983) A simple fibre Fabry–Pérot sensor, *Opt. Commun.* **45**, 71–4.

Kersey, A. D., Berkoff, T. A. and Morey, W. W. (1992) High resolution fibre grating based strain sensor with interferometric wavelength shift detection, *Electron. Lett.* **28**, 236–8.

Kersey, A. D., Berkoff, T. A. and Morey, W. W. (1993) Fibre optic Bragg grating strain sensor with drift-compensated high-resolution interferometric wavelength shift detection, *Opt. Lett.* **18**, 72–4.

Kersey, A. D. and Davis, M. A. (1994) Interferometric fibre sensors with a chirped Bragg grating sensing element, *Proc. 10th Inter. Conf. on Optical Fibre Sensors* (Glasgow, UK, 10. 1994, SPIE) pp. 319–22.

Kersey, A. D. and Marrone, M. J. (1994) Fibre Bragg high-magnetic-field probe, *Proc. 10th Inter. Conf. on Optical Fibre Sensors* (Glasgow, UK, 10. 1994, SPIE) pp. 53–6.

Kist, R., Drope, S. and Wolfelschnelder, H. (1984) Fibre Fabry–Pérot (FFP) thermometer for medical applications, *Proc. 2nd Inter. Conf. on Optical Fibre Sensors*, Stuttgart, Germany, pp. 165–8.

Kitagawa, T., Bilodeau, F., Malo, B., Theriault, S., Albert, J., Johnson, D. C., Hill, K. O., Hattori, K. and Hibino, Y. (1994) Single-frequency Er^{3+}-doped silica-based planar waveguide laser with integrated photo-imprinted Bragg reflectors, *Electron. Lett.* **30**, 1311–12.

REFERENCES

Lam, D. K. W. and Garside, B. K. (1981) Characterisation of single mode fibre filters, *Appl. Opt.* **20**, 440–5.

Lee, C. E., Alcoz, J. J., Yeh, Y., Gibler, W. N., Atkins, R. A. and Taylor, H. F. (1992) Optical fibre Fabry–Pérot sensors for smart structures, *J. Smart Structures*, **1**, 123–7.

Lefevre, H. C. (1993) *The Fibre Optic Gyroscope* (Boston: Artech House).

Lu, Z. J. and Blaha, F. A. (1992) Application issues of fibre optics in aircraft structures, *Proc. SPIE* **1588**, 276–81.

Measures, R. M., Alavie, A. T., Maaskant, R., Ohn, M., Karr, S. and Huang, S. (1994) Bragg grating structural sensing system... for bridge monitoring, *Proc. SPIE* **2294**, 63–60.

Meltz, G., Hewlett, S. J. and Love, J. D. (1996) Fibre grating evanescent-wave sensors, *Proc. SPIE* **2836**, paper 9.

Meltz, G., Morey, W. W. and Glenn, W. H. (1989) Formation of Bragg gratings in optical fibre by a transverse holographic method, *Opt. Lett.* **14**, 823–5.

Mizrahi, V. (1993) Components and devices for optical communications based on UV-written-fibre phase gratings, *Tech. Digest Inter. Conf. on Optical Fibre Communications* (OFC'93, CA, USA) pp. 243–4.

Morey, W. W., Meltz, G. and Glenn, W. H. (1989) Fibre optic Bragg grating sensors, *Proc. SPIE* **1169**, 98–107.

Morey, W. W., Dunphy, J. R. and Meltz, G. (1991) Multiplexed fibre Bragg grating sensors, *Proc. SPIE* **1586**, 216–24.

Morey, W. W., Meltz, G. and Welss, J. M. (1992) Evaluation of a fibre Bragg grating hydrostatic pressure sensor, *Proc. 8th Inter. Conf. on Optical Fibre Sensors* (Monterey, CA, USA), postdeadline paper PD-4.

Ouellette, F. (1991) All-fibre filter for efficient dispersion compensation, *Opt. Lett.* **16**, 303–5.

Patrick, H., Williams, G. M., Kersey, A. D., Pedrazzani, J. R. and Vengsarkar, A. M. (1996) Strain/temperature discrimination using combined fibre Bragg grating and long-period grating sensors, *Proc. 11th Inter. Conf. on Optical Fibre Sensors* (Sapporo, Japan, 5. 1996, IEICE and IEEJ) pp. 96–9.

Rao, Y. J., Ribeiro, A. B. L., Jackson, D. A., Zhang, L. and Bennion, I. (1995) Combined spatial- and time-division-multiplexing scheme for fibre grating sensors with drift-compensated phase-sensitive detection, *Opt. Lett.* **20**, 2149–51.

Rao, Y. J. and Jackson, D. A. (1996a) Recent progress in multiplexing techniques for in-fibre Bragg grating sensors, *Proc. SPIE* **2895**, paper 514.

Rao, Y. J. and Jackson, D. A. (1996b) Universal fibre-optic point sensor system for quasi-static absolute measurements of multi-parameters exploiting low coherence interrogation, *J. Lightwave Technol.* **14**, 592–600.

Rao, Y. J. and Jackson, D. A. (1996c) Recent progress in fibre optic low-coherence interferometry, *Meas. Sci. & Technol.* **7**, 981–99.

Rao, Y. J., Jackson, D. A., Zhang, L. and Bennion, I. (1996a) Dual-cavity interferometric wavelength-shift detection for in-fibre Bragg grating sensors, *Opt. Lett.* **21**, 1556–8.

Rao, Y. J., Jackson, D. A., Zhang, L. and Bennion, I. (1996b) Strain sensing of modern composite materials with a spatial/wavelength multiplexed fibre grating network, *Opt. Lett.* **21**, 683–5.

Rao, Y. J., Jackson, D. A., Zhang, L. and Bennion, I. (1996c) Extended dynamic range detection system for in-fibre Bragg grating strain sensors based on two cascaded interferometric wavelength scanners accepted by *Meas. Sci. Technol.*

Rao, Y. J., Webb, D. J., Jackson, D. A., Zhang, L. and Bennion, I. (1996d) A novel high-resolution, wavelength-division-multiplexed in-fibre Bragg grating sensor system, *Electron. Lett.* **32**, 924–6.

Rao, Y. J., Webb, D. J., Jackson, D. A., Zhang, L. and Bennion, I. (1997) In-fibre Bragg grating temperature sensor system for medical applications accepted by *J. Lightwave Technol.* **15**, 779–85.

Russell, P. J., Archambault, J. and Reekie, L. (1993) Fibre gratings, *Phys. World*, **6**, 41–6.

Simonsen, H. D., Paetsch, R. and Dunphy, J. R. (1992) Fibre Bragg grating sensor demonstration in glass-fibre reinforced polyester composite, *Proc. SPIE* **1777**, 73–6.

Singh, H. and Sirkis, J. (1996) Simultaneous measurement of strain and temperature using optical fibre sensors: two novel configurations, *Proc. 11th Inter. Conf. on Optical Fibre Sensors* (Sapporo, Japan, 5. 1996, IEICE and IEEJ) pp. 108–11.

Sirkis, J. S., Brennan, D. D., Putnam, M. A., Berkoff, T. A., Kersey, A. D. and Friebele, E. J. (1993) In-line fibre etalon for strain measurement, *Opt. Lett.* **18**, 1973–5.

Stolen, R. H., Ashkin, A., Pleibel, W. and Dziedzic, J. M. (1984) In-line fibre-polarization-rocking rotator and filter, *Opt. Lett.* **9**, 300–3.

Theriault, S., Hill, K. O., Bilodeau, F., Johnson, D. C. and Albert, J. (1996) High-g accelerometer based on an in-fibre Bragg grating sensor, *Proc. 11th Inter. Conf. on Optical Fibre Sensors* (Sapporo, Japan, 5. 1996, IEICE and IEEJ) pp. 196–9.

Udd, E. (Ed) (1995) *Fibre Optic Smart Structures* (New York: John Wiley & Sons).

Varnham, M. P., Barlow, A. J., Payne, D. N. and Okamoto, K. (1983) Polarimetric strain gauge using high birefringent fibres, *Electron. Lett.* **19**, 699–700.

Vengsarkar, A. M., Lemaire, P. J., Judkins, J. B., Bhatia, V., Erdogan, T. and Sipe, J. E. (1996) Long-period fibre gratings as band-rejection filters. *J. Lightwave Technol.* **14**, 58–65.

Volanthen, M., Geiger, H., Cole, M. J. and Dakin, J. P. (1996) Measurement of arbitrary strain profiles within fibre gratings, *Electron. Lett.* **32**, 1028–9.

Webb, D. J., Surowiec, J., Sweeney, M., Jackson, D. A., Hand, J. M., Gavrilow, L., Zhang, L. and Bennion, I. (1996) Miniature fibre optic ultrasonic probe, *Proc. SPIE* **2839**, paper 8.

Williams, J. A. R., Bennion, I., Sugden, K. and Doran, N. J. (1994) Fibre dispersion compensation using a chirped in-fibre Bragg grating, *Electron. Lett.* **30**, 985–7.

Xu, M. G., Geiger, H. and Dakin, J. P. (1993) Optical in-fibre grating high pressure sensor, *Electron. Lett.* **29**, 398–9.

Xu, M. G., Archambault, J. L., Reekie, L. and Dakin, J. P. (1994a) Thermally-compensated bending gauge using surface-mounted fibre gratings, *Inter. J. Optoelectronics*, **9**, 281–3.

Xu, M. G., Archambault, J. L., Reekie, L. and Dakin, J. P. (1994b) Discrimination between strain and temperature effects using dual-wavelength fibre grating sensors, *Electron. Lett.* **30**, 1085–7.

Xu, M. G., Dong, L., Reekie, L., Tucknott, J. A. and Cruz, J. L. (1995) Temperature-independent strain sensor using a chirped Bragg grating in a tapered optical fibre, *Electron. Lett.* **31**, 823–5.

Xu, M. G., Geiger, H. and Dakin, J. P. (1996) Fibre grating pressure sensor with enhanced sensitivity using a glass-bubble housing, *Electron. Lett.* **32**, 128–9.

FURTHER READING

Anderson, D. Z., Mizrahi, V., Erdogan, T. and White, A. E. (1993) Production of in-fibre gratings using a diffractive optical element, *Electron. Lett.* **29**, 566–8.

Archambault, J. L., Reekie, L. and Russell, P. (1993) High reflectivity and narrow bandwidth fibre gratings written by single excimer pulse, *Electron. Lett.* **29**, 28–9.

Askins, C. G., Tsai, T. E., Williams, G. M., Putnam, M. A., Bashkansky, M. and Friebele, J. (1992) Fibre Bragg reflectors prepared by a single excimer pulse, *Opt. Lett.* **17**, 833–5.

Ball, G. A., Morey, W. W. and Glen, W. H. (1993) Standing wave monomode erbium fibre laser, *IEEE Photon. Technol. Lett.* **5**, 267–70.

Ball, G. A., Morey, W. W. and Cheo, P. K. (1994) Fibre laser source/analyzer for Bragg grating sensor array interrogation, *J. Lightwave Technol.* **12**, 700–3.

Brady, G. P., Hope, S., Ribeiro, A. B. L., Webb, D. J., Reekie, L., Archambault, J. L. and Jackson, D. A. (1994) Bragg grating temperature and strain sensors, *Proc. 10th Inter. Conf. on Optical Fibre Sensors*, Glasgow, UK, pp. 510–13.

Chang, I. C. (1981) Acousto-optic tunable filter, *Opt. Eng.* **20**, 824–9.

Cole, M. J., Loh, W. H., Laming, R. I., Zervas, M. N. and Barcelos, S. (1995) Moving fibre/phase mask-scanning beam technique for enhanced flexibility in producing fibre gratings with uniform phase mask, *Electron. Lett.* **31**, 1488–9.

Coroy, T., Ellerbrock, P. J., Measures, R. M. and Belk, J. H. (1995) Active wavelength demodulation of Bragg fibre-optic strain sensor using acousto-optic filter, *Electron. Lett.* **31**, 1602–3.

Davis, M. A., Bellemore, D. G. and Kersey A. D. (1994) Structure strain mapping using a wavelength/time division addressed fibre Bragg grating array, *Proc. 2nd Euro. Conf. on Smart Structures and Materials* (Glasgow, UK 10. 1994, SPIE) pp. 342–5.

Davis, M. A. and Kersey, A. D. (1994) All-fibre Bragg grating strain-sensor demonstration technique using a wavelength division coupler, *Electron. Lett.* **30**, 75–6.

Davis, M. A. and Kersey, A. D. (1995) Matched-filter interrogation technique for fibre Bragg grating arrays, *Electron. Lett.* **31**, 822–3.

Davis, M. A., Bellemore, D. G., Putnam, M. A. and Kersey, A. D. (1996) A 60 element fibre Bragg grating sensor system, *Proc. 11th Inter. Conf. on Optical Fibre Sensors* (Sapporo, Japan, 5. 1996, IEICE and IEEJ), pp. 100–3.

Dockney, M. L., James, S. W. and Tatam, R. P. (1996) Fibre Bragg gratings fabricated using a wavelength tunble laser source and a phase mask based interferometer *Meas. Sci. Technol.* **7**, 445–8.

Dunphy, J. R., Ball, G., D'amato, F., Ferraro, P., Inserra, S., Vannucci, A. and Varasi, M. (1993) Instrumentation development in support of fibre grating sensor array, *Proc. SPIE* **2071**, 2–11.

Geiger, H., Xu, M. G., Eaton, N. C. and Dakin, J. P. (1995) Electronic tracking system for multiplexed fibre grating sensors, *Electron. Lett.* **31**, 1006–7.

Hill, K. O., Malo, B., Bilodeau, F., Johnson, D. C. and Albert, J. (1993) Bragg gratings fabricated in monomode photosensitive optical fibre by UV exposure through a phase mask, *Appl. Phys. Lett.* **62**, 1035–7.

Jackson, D. A., Ribeiro, A. B. L., Reekie, L. and Archambault, J. L. (1993) Simple multiplexing scheme for a fibre-optic grating sensor network, *Opt. Lett.* **18**, 1192–4.

Kashyap, R., Armitage, J. R., Wyatt, R., Davey, S. T. and Williams, D. L. (1990) All-fibre narrow band reflection gratings at 1500nm, *Electron. Lett.* **26**, 730–2.

Kashyap, R., Wyatt, R. and Campbell, R. J. (1993) Wideband gain flattened erbium fibre amplifier using a photosensitive fibre blazed grating, *Electron. Lett.* **29**, 154–6.

Kashyap, R., Mckee, P. F., Campbell, R. J. and Williams, D. L. (1994) Novel method of producing all-fibre photoinduced chirped gratings, *Electron. Lett.* **30**, 996–7.

Kersey, A. D. (1995) in Udd, E. (ed.) *Fibre-Optic Smart Structures* (New York: John Wiley & Sons) Ch. 15, pp. 409–44.

Kersey, A. D. and Berkoff, T. A. (1992) Fibre-optic Bragg-grating differential-temperature sensor, *IEEE Photon. Technol. Lett.* **4**, 1183–5.

Kersey, A. D. and Dandridge, A. (1992) Low crosstalk code division multiplexing interferometric array, *Electron. Lett.* **28**, 351–2.

Kersey, A. D., Berkoff, T. A. and Morey, W. W. (1993) Multiplexed fibre Bragg grating strain-sensor system with a fibre Fabry–Pérot wavelength filter, *Opt. Lett.* **18**, 1370–2.

Koo, K. P. and Kersey, A. D. (1995) Fibre laser sensor with ultrahigh strain resolution using interferometric interrogation, *Electron. Lett.* **31**, 1180–2.

Lemaire, P. J., Atkins, R. M., Mizrahi, V. and Reed, W. A. (1993) High pressure H_2 loading as a technique for achieving ultrahigh UV photosensitivity and thermal sensitivity in GeO_2 doped optical fibres, *Electron. Lett.* **29**, 1191–3.

Martin, J. and Ouellette, F. (1994) Novel writing technique of long and highly reflective in-fibre gratings, *Electron. Lett.* **30**, 811–12.

Melle, S. M., Liu, K. and Measures, R. M. (1992) A passive wavelength demodulation system for guided-wave Bragg grating sensors, *IEEE Photon. Technol. Lett.* **4**, 516–18.

Melle, S. M., Alavic, T., Karr, S., Coroy, T., Liu, K. and Measures, R. M. (1993) A Bragg grating-tuned fibre laser strain sensor system, *IEEE Photon. Technol. Lett.* **5**, 263–6.

Prohaska, J. D., Snitzer, E., Rishton, S. and Boegli, V. (1993) Magnification of mask fabricated fibre Bragg gratings, *Electron. Lett.* **29**, 1614–15.

Rao, Y. J., Uttamchandani, D., Culshaw, B., Steer, P. and Briancon, J. (1993) Spread-spectrum technique for passive multiplexing of reflective frequency-out fibre optic sensors exhibiting identical characteristics, *Opt. Commun.* **96**, 214–17.

Rao, Y. J. and Jackson, D. A. (1995) A prototype multiplexing system for use with a large number of fibre-optic-based extrinsic Fabry–Pérot sensors exploiting low coherence interrogation, *Proc. SPIE* **2507**, 90–8.

Rao, Y. J., Kalli, K., Brady, G., Webb, D. J., Jackson, D. A., Zhang, L. and Bennion, I. (1995) Spatially-multiplexed fibre-optic Bragg grating strain and temperature sensor system based on interferometric wavelength-shift detection, *Electron. Lett.* **31**, 1009–10.

Rao, Y. J., Ribeiro, A. B. L., Jackson, D. A., Zhang, L. and Bennion, I. (1995) Combined spatial- and time-division-multiplexing scheme for fibre grating sensors with drift-compensated phase-sensitive detection, *Opt. Lett.* **20**, 2149–51.

Rao, Y. J., Hurle, B. A., Kalli, K., Brady, G., Webb, D. J., Jackson, D. A., Zhang, L. and Bennion, I. (1996) Spatially-multiplexed fibre-optic interferometric and grating sensor system for quasi-static absolute measurements, *Proc. 11th Inter. Conf. on Optical Fibre Sensors*, (Sapporo, Japan, 5. 1996, IEICE and IEEJ) pp. 666–9.

Rao, Y. J., Ribeiro, A. B. L., Jackson, D. A., Zhang, L. and Bennion, I. (1996) Simultaneous spatial, and time and wavelength division multiplexed in-fibre grating sensing network, *Opt. Commun.* **125**, 53–8.

Ribeiro, A. B. L., Rao, Y. J. and Jackson, D. A. (1994) Multiplexing interrogation of interferometric sensors using dual multimode laser diode sources and coherence reading, *Opt. Commun*, **109**, 400–4.

Ribeiro, A. B. L., Rao, Y. J., Zhang, L., Bennion, I. and Jackson, D. A. (1996) A combined spatial- and time-division-multiplexing tree topology for fibre grating sensor systems, *Appl. Opt.* **35**, 2267–73.

Rourke, H. N., Baker, S. R., Byron, K. C., Baulcomb, R. S., Ojha, S. M. and Clements, S. (1994) Fabrication and characterisation of long, narrowband fibre gratings by phase mask scanning, *Electron. Lett.* **30**, 1341–2.

Safin, S. A., Semenov, A. T., Shidlovski, V. R., Zhuchkov, N. A. and Kurnyavko, Y. V. (1992) High-power 0.82 µm superluminescent diodes with extremely low Fabry–Pérot modulation depth, *Electron. Lett.* **28**, 127–9.

Weis, R. S., Kersey, A. D. and Berkoff, T. A. (1994) A four-element fibre grating sensor array with phase-sensitive detection, *IEEE Photonics Technol. Lett.* **6**, 1469–72.

Xu, M. G., Geiger, H., Archambault, J. L., Reekie, L. and Dakin, J. P. (1993) Novel interrogating system for fibre Bragg grating sensors using an acousto-optic tunable filter, *Electron. Lett.* **29**, 1510–11.

12
Fiber Bragg grating sensors: signal processing aspects

Y. N. Ning and B. T. Meggitt

12.1 INTRODUCTION

12.1.1 Background

The development of advanced sensor systems is one of the most widely identified, key generic technology areas emphasized in recent strategic development reports. This interest has occurred due to a continual demand in engineering structures for improvements in integrity, reliability and safety along with the need for reduced maintenance costs. To achieve these aims, there is the desire to use advanced optical measurement techniques suitable for both laboratory developments and industrial condition monitoring applications. Conventional strain and temperature monitoring techniques, such as the foil and capacitance strain gauges and thermocouple temperature probes, are limited in their applicability since they are not easily multiplexed in large sensor arrays and are often susceptible to heavy electromagnetic interference (EMI) and measurement drift and offset especially when operated in hostile environments.

In contrast, optical in-fiber sensors such as Bragg grating sensors, can operate in noisy electrical environments and can be configured into multi-sensor arrays which can operate in harsh environments. Such sensors can be attached onto or embedded into metal or composite structures where constant monitoring or measurement is a critical requirement for structural integrity and health monitoring in modern engineering structures, since these sensors are small and compact in mass and form.

Optical fiber sensors using photo-induced Bragg gratings have recently emerged as important intrinsic fiber transducers to measure, for example, strain and temperature changes in a wide range of industrial applications, such as in integrity monitoring of advanced fiber-reinforced composite (AFRC) materials and health monitoring in engineering applications such as aerospace and civil engineering structures.

Optical Fiber Sensor Technology, Vol. 2. Edited by K. T. V. Grattan and B. T. Meggitt.
Published in 1998 by Chapman & Hall, London. ISBN 0 412 782 901

12.1.2 Bragg sensor properties

In the previous two chapters the fiber Bragg grating fabrication and operating characteristics were considered. Here we will review some of the more important properties of Bragg gratings in relation to their use in sensor networks. The holographic Bragg grating is formed by the introduction of a periodic refractive index modulation along a short optical fiber. This is achieved by use of an interference fringe pattern produced by a UV (~ 248 nm wavelength) interferometer system which writes the grating in the fiber core using the photo-refractive effect of the fiber. The periodic grating elements are then separated by a distance Λ and have an optical path length of $n\Lambda$, where n is the refractive index of the fiber core (~ 1.48). For a grating operating in the reflection mode, the characteristic Bragg grating equation is given by

$$\lambda_B = 2n\Lambda \tag{12.1}$$

where λ_B is the reflected Bragg wavelength. Any applied disturbance, such as temperature or strain, on the grating that affects either the refractive index n or the period Λ will cause a shift in the reflected Bragg wavelength. Temperature effects will cause both a change in grating pitch due to both thermal expansion and changes in the fiber refractive index. By a Taylor expansion on the characteristic Bragg relation above it can be shown that the fractional change in Bragg wavelength with temperature can be expressed as

$$\frac{\Delta\lambda_B(T)}{\lambda_B} = (\alpha + \xi)\Delta T \tag{12.2}$$

where α is the thermal expansion coefficient ($\alpha = \partial\Lambda/\Lambda\partial T$) and ξ is the thermo-optic coefficient ($\xi = \partial n/n\partial T$) of the fiber. Since the thermo-optic effect is about an order of magnitude greater than that of the thermal expansion effect, this effect is the dominant cause for changes in the Bragg wavelength with temperature changes and typical values for $\Delta\lambda_B$ in silica fibers operating at 1550 nm are ~ 13 pm °C^{-1}.

Similarly, the effect of longitudinal strain on the reflected Bragg wavelength can be considered. The analysis here is complicated by the fact that strain is a three-dimensional tensor field. Basically, increased longitudinal strain on a Bragg grating both increases the grating pitch due to fiber elongation and in addition it reduces the fiber refractive index due to the strain-optic (tensor) effect. From a more complex analysis it can be shown that for an applied strain,

$$\varepsilon = \frac{\Delta\Lambda}{\Lambda} \tag{12.3}$$

where $\Delta\Lambda$ is the grating pitch extension and Λ the original pitch length, the fractional change in Bragg wavelength with strain can be expressed as

$$\frac{\Delta\lambda_B(\varepsilon)}{\lambda_B} = (1 - p_e)\Delta\varepsilon \quad \text{where} \quad p_e = \frac{n^2}{2}(p_{11} - \nu(p_{11} + p_{12})) \tag{12.4}$$

where p_e is the effective photo-elastic coefficient, p_{11} and p_{12} are components of the strain-optic tensor and v is Poisson's ratio. For silica fiber operating at 1550 nm typical values for the change in Bragg wavelength with strain are $\Delta\lambda_B \sim 1.15\,\text{pm}\,\mu\varepsilon^{-1}$. As can be seen from equation 12.4, the fractional change in Bragg wavelength ($\Delta\lambda_B/\lambda_B$) is less than the corresponding change in fibre strain ($\Delta L/L$). For silica fiber, with $p_e \sim 0.24$, the fractional Bragg wavelength change is only $\sim 75\%$ of the corresponding strain change.

Since temperature and strain effects can be considered as mutually exclusive effects, when simultaneously acting on a fiber grating sensor their effects are additive, i.e.

$$\frac{\Delta\lambda_B(T,\varepsilon)}{\lambda_B} = (\alpha + \xi)\Delta T + (1 - p_e)\Delta\varepsilon \qquad (12.5)$$

It is seen from this relation that both temperature and strain effects will cause a shift in the reflected Bragg wavelength and therefore separation of these variables is not a simple matter. One method to achieve this separation (see Xu et al. [1]) is where two gratings having widely separated Bragg wavelengths (e.g. 850 and 1300 nm) are superimposed on the same region of fiber. It is seen that two linear equations can now be formed for a given temperature (ΔT) and strain ($\Delta\varepsilon$) change. These can be expressed in a matrix form as

$$\begin{pmatrix}\Delta\lambda_{B1}\\ \Delta\lambda_{B2}\end{pmatrix} = \begin{pmatrix}K_{\varepsilon 1} & K_{T1}\\ K_{\varepsilon 2} & K_{T2}\end{pmatrix}\begin{pmatrix}\Delta\varepsilon\\ \Delta T\end{pmatrix} \qquad (12.6)$$

The K elements of this matrix can be found by calibrating the system with known temperature and strain changes. For the measurement of unknown temperature and strain, transposing the matrix by inversion will then give the required temperature and strain values from measurements of the Bragg wavelength changes in the two grating elements. This approach is considered further in section 12.3.1.

In order to gain a feel for the relative sensitivities of the temperature and strain related Bragg wavelength changes, it is instructive to plot these two effects on a common axis as shown in Figure 12.1. Here, the change in Bragg wavelength with strain is plotted as a linear function over a strain change of 1% strain (10 000 µε). Also plotted is the Bragg wavelength shift resulting from a temperature change of $\pm 50\,°C$. It can be seen that temperature fluctuations greatly reduce the attainable resolution when considering small strain perturbations, but has a much smaller effect when considering much wider strain changes, with a $\pm 50\,°C$ change limiting the strain resolution to $\sim 5\%$ at a 1% full scale strain change.

12.1.3 Signal processing considerations

In an optical fiber grating based sensor system, the returned optical signal typically consists of a series of weak, narrow-band spectral components spread over an optical bandwidth of tens of nanometers. Therefore, it is necessary to employ

384 FIBER BRAGG GRATING SENSORS: SIGNAL PROCESSING

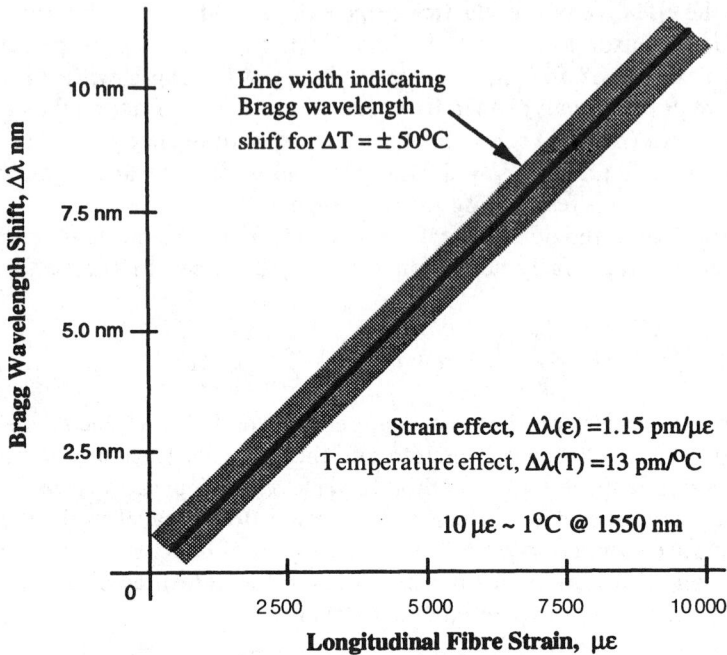

Fig. 12.1 Comparison of temperature and strain effects on Bragg grating wavelength.

a very efficient optical signal processing system to decode and monitor the variations in the Bragg wavelengths of interest. One of the key issues in designing an optical fiber sensing system based upon the Fiber Bragg grating (FBG) is the selection of a suitable wavelength detection scheme in terms of the measurement range, sensitivity, resolution, repeatability and response time. Other important considerations, especially in field applications, are the ability to multiplex large numbers of grating sensing elements in addition to the cost of ownership, installation and maintenance. A number of useful schemes for wavelength shift detection together with suitable multiplexing architectures have recently been proposed and investigated [1 – 5], as a result of the intensive research carried out in this field.

In this chapter, the main signal processing features of these schemes will be summarized and reviewed, together with some typical examples to illustrate their operating principles and performance and follows on from the previous chapters where the device development and operating principles were discussed in detail.

The Bragg fiber sensor processing architectures can be conveniently classified as being either (i) passive detection or (ii) active detection schemes depending on the type of elements employed in the processing configuration. The following processing schemes will be described using this classification. In addition, the signal processing can be carried out in both the spectral domain, where wavelength is measured directly, or in the phase domain where the phase of the optical carrier is monitored, with generally the later approach providing the higher sensitivity.

12.2 PASSIVE DETECTION SCHEMES

Generally speaking, passive wavelength detection schemes do not employ any type of wavelength modulation mechanism in the measurement system architecture. The wavelength variation of the reflected light from gratings is often detected with the use of passive optical components that exhibit some form of wavelength-dependent or wavelength-discrimination characteristics, such as optical filters and diffraction gratings. In this group of detection schemes, three examples are summarized to illustrate their basic working principles and operating performance.

12.2.1 Wavelength-dependent optical filter detection

Since the reflected Bragg wavelength varies with the change in measurand in the fiber grating sensor, any matched optical filter element operating about its linear, wavelength-dependent, absorption band-edge can be employed to convert the Bragg wavelength change into a corresponding change in transmitted intensity, from which the measurand change can be inferred. Based on this principle, Melle et al. [4] have reported a passive detection scheme, shown in Figure 12.2(a), where the output radiation reflected back from the Bragg grating was divided into two beams of equal intensity by using an optical beam-splitter, of which the split ratio had a negligible wavelength independence over the source wavelength range. One beam was then filtered by use of a wavelength-dependent filter, whilst the other was unfiltered to provide a reference to compensate or eliminate the effect of intensity variations due to uneven power distributions of the source spectrum, micro-bend attenuation in the lead and power fluctuations of the source.

If the two photodetectors used are identical, the transfer function of the filter can be represented with a linearized model where the back-reflected Bragg wavelength is a near Gaussian function of spectral width $\Delta\lambda$ and the central wavelength λ_B. The ratio of the two outputs, I_F and I_R, from the detectors can then be given by

$$\frac{I_F}{I_R} = A\left(\lambda_B - \lambda_0 + \frac{\Delta\lambda}{\sqrt{\pi}}\right) \quad (12.7)$$

where A and λ_0 are the slope and the central wavelength of the filter. Hence the value of this ratio varies linearly as the input wavelength λ_B shifts.

With the use of an infrared high-pass filter (RG830), which had a linear transmission function over a wavelength range from 815 to 838 nm, with a 50% transmittance at 828 nm, a strain measurement system was demonstrated as shown in Figure 12.2(a). The experimental results show that the ratio of the filtered to

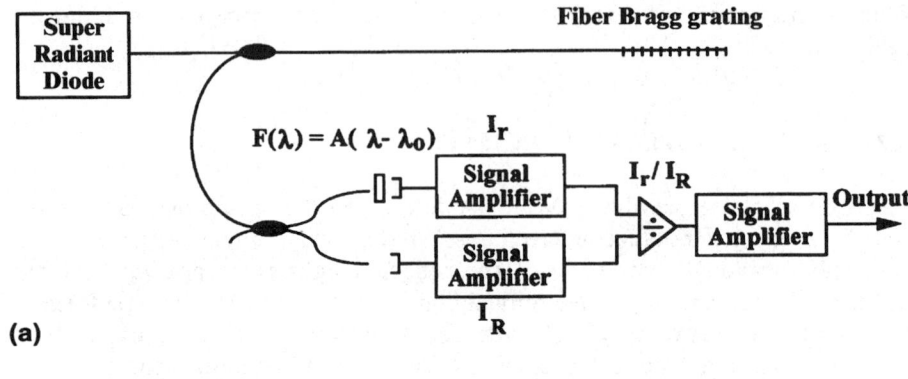

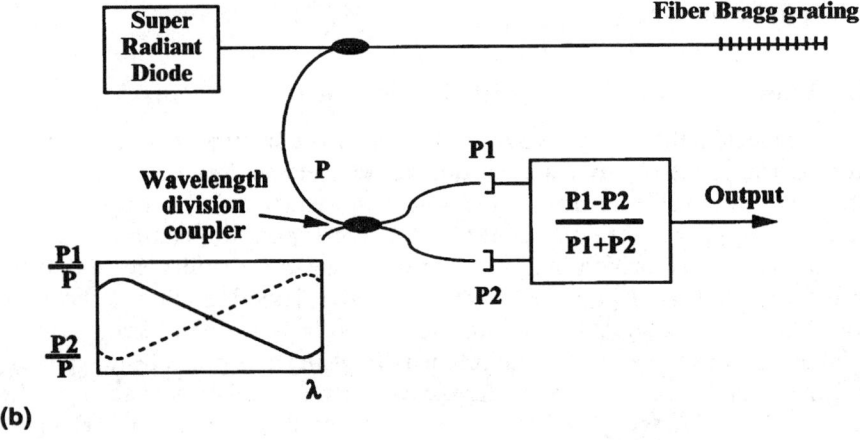

Fig. 12.2 (a) Scheme of fiber optic passive ratiometric wavelength detection system [4]; (b) diagram of strain sensor configuration using an FBG and wavelength division coupler [2].

reference light was a function of applied static strain over the range from -5500 to $+5500$ µε. The strain resolution was found to be 375 µε, corresponding to a resolution of about 3.4% over the full measurement range of the system employed.

It can be easily seen that the strain sensitivity of the system was dependent on the transmission slope of the filter band-edge. The higher the slope value, the larger the transmittance change per nanometer wavelength shift and hence the higher the sensitivity that can be obtained. However, the maximum measurement range, or dynamic range, of the system was dependent on the transmitting characteristics of the filter used. The main advantages of this measurement scheme are its requirements for relatively low cost components, along with a relatively simple system configuration and signal processing requirements.

12.2.2 Wavelength-dependent fiber coupler detection

Although the use of an optical filter to detect wavelength variations can give a modest strain resolution, for the application where large numbers of gratings are multiplexed the optical power level of the signal is relatively low and hence any optical power loss caused by the reflections or by the inherent attenuation of the filter needs to be considered.

An alternative scheme, proposed by Melle *et al.* [4] and demonstrated by Davis and Kersey [2], is that using an optical fiber coupler the coupling ratio of which varies with wavelength, allows variations in the reflected Bragg wavelength to be detected, as shown in Figure 12.2(b). Since the wavelength discrimination in this system is realized by using a wavelength-dependent 2 × 2 fiber optic coupler exhibiting a near linear variation in coupling ratio of 0.4 dB nm^{-1} across the wavelength range from 1520 to 1560 nm, also illustrated in Figure 12.2(b), the power loss is reduced and the strain resolution thus improved by some two orders of magnitude. In this scheme, the intensity variations were eliminated by taking the ratio of the difference to the sum of the outputs from the two detectors. With the use of this approach, a good linearity was observed over the measured range of +1050 µε with a static strain resolution of about 3 µε and a minimum detectable dynamic strain of 0.5 µε Hz$^{-1/2}$ at 10.5 Hz.

12.2.3 Plane grating and CCD array based analyzer detection

A further wavelength detection scheme, reported by Liu *et al.* [3], is based on the use of a prototype fiber optic CCD spectrometer, as shown in Figure 12.3, where the wavelength discrimination is achieved with the use of an optical diffraction grating. The spectrum of the back-reflected light from the Bragg sensors was diffracted by the optical grating and then detected by using a linear CCD array. The main problem in using such a device to discriminate the wavelength was that the accuracy of the wavelength measurement was sensitive to the relative position of the fiber-tip, as the incident angle of the beam at the

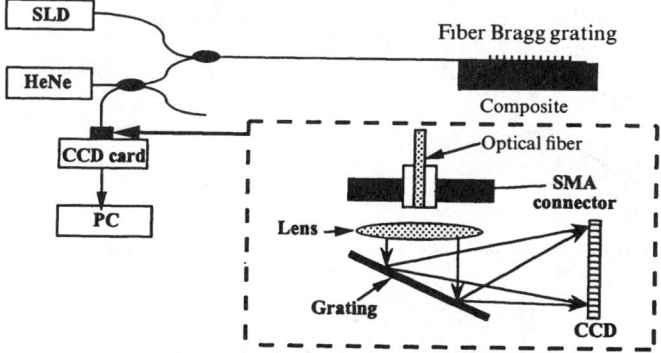

Fig. 12.3 Schematic diagram of the fiber grating strain measurement system [3].

optical grating changed as the position of the fiber-tip was altered. This difficulty was overcome by introducing an additional stabilized reference light source (HeNe, 632.8 nm) into the spectrometer system to provide a reference wavelength offset against which the variable Bragg wavelength (at 827 nm) was measured.

This wavelength detection system was used to measure the Bragg wavelength change from an FBG which was surface mounted onto a carbon fiber reinforced epoxy composite that was subjected to both quasi-static and dynamic loading conditions. In the quasi-static strain test, the measured strain sensitivity was about 0.65 pm $\mu\varepsilon^{-1}$ over a strain range of -2000 to $+3000\,\mu\varepsilon$, with a strain sensitivity of approximately 80 µε. As the response time of the system was less than 8 µs, this system was capable of measuring the relative reduction in stiffness of the composite as a function of fatigue cycles. From Figure 12.4, where the stress/strain data obtained after 266 000 cycles in the fatigue test are shown, it can be seen that an excellent correlation between the strain measured via an extensometer and that obtained with the use of the Bragg grating system. The FBG in this system survived the tensile and compressive fatigue loading tests for up to 700 000 cycles.

The main advantage of this wavelength detection scheme is that the signal-to-noise ratio of the detected spectral signal can be considerably improved through the use of a computer-based data acquisition and reduction system. Also the full broadband spectral region used can be monitored throughout the lifetime test period, and hence this scheme is able to provide additional information for the performance evaluation of the system and the materials under measurement.

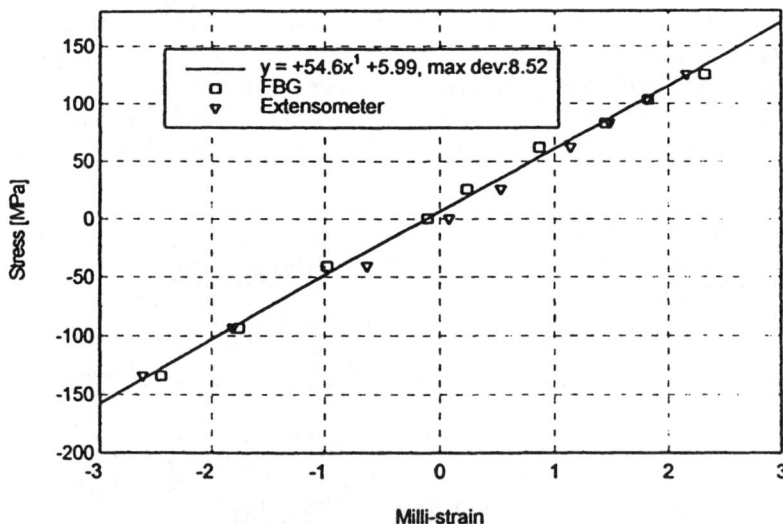

Fig. 12.4 Stress/strain curve after 266 000 cycles obtained with a surface-mounted extensometer and a surface mounted FBG sensor [3].

12.3 ACTIVE DETECTION SCHEMES

The active wavelength detection schemes used to detect Bragg wavelength changes in fiber grating based sensor systems often involve some type of wavelength-scanning mechanism. According to the method used for discriminating wavelength changes, the active detection schemes reported may be divided into three separate groups where the wavelength detection is achieved by the use of:

1. An optical spectral analyzer;
2. A tunable optical filter; and
3. An optical interferometer for the measurement of optical phase.

12.3.1 Optical spectral analyzer detection – temperature/strain separation

Since conventional spectral analyzers are common laboratory instruments having a good wavelength resolution and a wide spectral range, a considerable number of works have been carried out using such optical equipment to monitor the wavelength shifts induced by the sensor gratings, especially when the development was at the laboratory 'proof of principle' stage. Such instruments are particularly useful when wavelength division multiplexing schemes are used to interrogate a number of Bragg grating sensors of different central wavelengths [1, 5, 6].

One significant example here is in the separation of strain and temperature measurements from a single sensor element as reported by Xu et al. [1]. In this system, illustrated in Figure 12.5, the sensing element consisted of two superimposed Bragg gratings of differing grating pitch, written coincidentally on the same fiber section. Since these gratings had different optical parameters, i.e. with central wavelengths of 1298 and 848 nm, peak reflectivities of 70 and 55% and optical bandwidths (full width, half maximum (FWHM)) of 0.9 and 0.45 nm respectively, additional information can be obtained from the system. When the light from two broadband edge-emitting light-emitting diode (ELED) sources (of central wavelengths ~ 850 and ~ 1300 nm) were coupled to the fiber

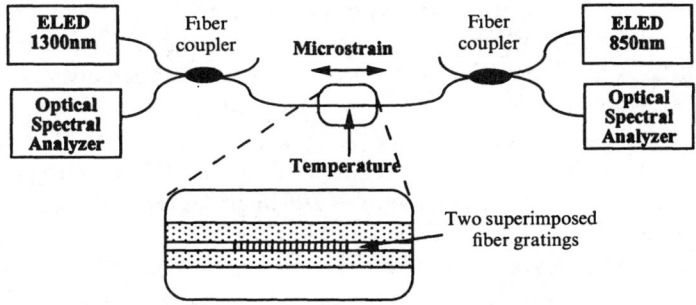

Fig. 12.5 Schematic diagram of the setup for simultaneous measurement of strain and temperature using two superimposed fiber gratings [1].

gratings, the reflected light was monitored with the use of a commercial optical spectrum analyzer, with the measurement of strain and temperature being carried out simultaneously by use of a matrix transform method.

Assuming that the strain and temperature changes of the two gratings are essentially independent of each other, the transfer function of the system can be described by the following matrix relation:

$$\begin{pmatrix} \Delta\lambda_{B1} \\ \Delta\lambda_{B2} \end{pmatrix} = \begin{pmatrix} K_{\varepsilon 1} & K_{T1} \\ K_{\varepsilon 2} & K_{T2} \end{pmatrix} \begin{pmatrix} \Delta\varepsilon \\ \Delta T \end{pmatrix} \qquad (12.8)$$

where $K_\varepsilon = \partial\lambda/\partial\varepsilon$, which is related to the Poisson's ratio of the fiber and $K_T = \partial\lambda/\partial T$, which is determined by the thermal expansion coefficient and the thermo-optic coefficient of the fiber.

For this system, if a known strain and a known temperature are applied to the grating pair, simultaneous measurement of the two Bragg wavelengths changes will allow the transfer function of the grating system to be determined. In this way, the elements of the K matrix can be experimentally obtained and the system can then be used to measure, simultaneously, unknown strain and temperature changes. The measured results showed that a typical error of $\sim 10\,\mu\varepsilon$ was evident over a strain range of $800\,\mu\varepsilon$ and an error of $\sim 5\,°C$ in temperature measurement over a temperature range of $50\,°C$.

In a multiplexed distributed strain sensing system reported by Friebele et al. [6], the spectra of the reflected light from a grating array was measured using a 1 m grating spectrometer and an optical multichannel analyzer (1 pixel corresponding to 0.04 nm). In this system, seven gratings, separated spectrally by 2 nm and spatially by 40 cm, were embedded in between the 9th and 10th cloth-layer of a 10-ply glass-fiber, vinyl-ester resin composite panel. The distance between the grating and the neutral axis of the section was about 5 cm. When the composite panel was subjected to three-point bending tests, the strain distribution was measured with a resultant error of $17\,\mu\varepsilon$.

12.3.2 Fiber Fourier transform spectrometer detection

Although the optical spectral analyzer can give useful spectral information about a grating sensor system, it is generally not suitable for use in industrial environments. A more practical solution is the use of a fiber Fourier transform spectrometer, shown in Figure 12.6, to detect the wavelength-encoded signal from a Bragg grating sensor system [7, 8]. Here, the optical fiber Michelson interferometer employed consisted of a 2×2 optical fiber coupler. The two fiber arms of the interferometer were ~ 200 m in length with one arm being wound around a piezoelectric driven stretcher capable of inducing about 10 cm fiber length change, giving a total optical path difference (OPD) variation of about 30 cm. When the light of interest was injected into the fiber spectrometer, an output interferogram was generated by linearly stretching the length of one fiber arm. The spectral distribution was then obtained by taking

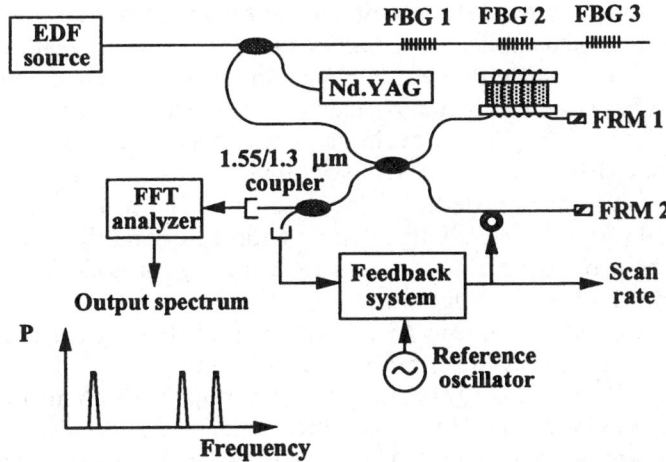

Fig. 12.6 Schematic of the FBG sensor system with fiber Fourier transform spectrometer [7].

the Fourier transform of the output interferogram. The interferogram spectral width was dependent on the coherence length of the radiation reflected from the Bragg grating. For a typical FBG with a central wavelength of 1.55 mm and a spectral bandwidth ~ 0.2 nm the coherence length of the reflected spectrum is ~ 0.5 cm. Hence the minimum interferometer OPD scan range (stretch) required to obtain a full output interferogram was > 2 cm.

The minimum detectable wavelength variation was affected by both the linearity of the scan and the polarization-fading noise introduced in the fiber interferometer. In order to achieve a highly linear scan in the interferometer OPD change, an additional stabilized source was used to provide an optical wavelength reference signal in order to correct the uniformity of the scan through a feedback network. The problem of polarization-fading was effectively overcome by the use of Faraday rotation mirrors. With these system refinements, a wavelength resolution of ~ 0.015 nm was demonstrated giving a strain resolution of $\sim 12\,\mu\varepsilon$ for a 1.55 μm grating.

12.3.3 Fabry–Pérot filter type detection

Among the various wavelength detection schemes recently reported for measuring the wavelength shifts associated with FBG sensing systems, the compact, electronically tunable, fiber pigtailed, Fabry–Pérot filter [9, 10] offers a number of advantages. Here, the free spectral range (FSR) and the bandwidth (FWHM) of such filters can be adjusted to suit different fiber grating parameters and required operating performances. In addition, the operational mode of the filter can provide either a linear wavelength scan mode or a wavelength lock-in mode in order to detect or track the movement in the central wavelength of the Bragg grating. Further, such bulk optic filter devices can be fiber-pigtailed in a ruggedized package to make them suitable for the interrogation of fiber-based sensing systems in field applications. One of the most attractive features of these devices is that the wavelength measurement and the wavelength demultiplexing operation can be performed within a single scan of the filter element.

A typical FBG sensing system using a compact electronically tunable fiber pigtailed Fabry–Pérot filter for the measurement of variations in the reflected Bragg wavelength was reported by Ning *et al.* [10]. The schematic diagram in Figure 12.7 shows the basic configuration of the FBG sensing system used in this work. The broadband source was a 1536 nm, signal-mode fiber, pigtailed ELED having a 15 μW output power spread over a 60 nm spectral width (FWHM). The optical receiver was a fiber pigtailed, high speed InGaAs PIN photo-detector with a 4 GΩ transimpedance amplifier.

The Fabry–Pérot filter (MF-100-40, Queensgate Instruments Ltd) is comprised of an electronically tunable, fiber pigtailed Fabry–Pérot interferometric cavity and a control unit which incorporates the capacitance bridge, piezo driver and digital interface. The FSR and the filter bandwidth (−3 dB) were measured to be 35.2 and 0.23 nm respectively. The insertion loss of the filter was about 3.6 dB. The transmission spectrum of the Fabry–Pérot filter can be tuned with the use of an external function generator connected to the scan input. The cavity spacing, which was related to the transmitted wavelength, can be measured by the internal capacitance micrometer and the results monitored at the wavelength monitor output.

The two fiber gratings used in the experiment had central wavelengths of $\lambda_{G1} = 1533.5$ nm and $\lambda_{G2} = 1549.5$ nm, reflectivities of $R_{G1} = 80\%$ and $R_{G2} = 81\%$ and bandwidths of $B_{G1} = 0.17$ nm and $B_{G2} = 0.15$ nm respectively. A typical transmission spectrum of the fiber grating is shown in Figure 12.8. Light from the ELED was divided by a 2 × 2 fiber coupler, and after reflection by the fiber gratings, the returned radiation was passed through the tunable filter

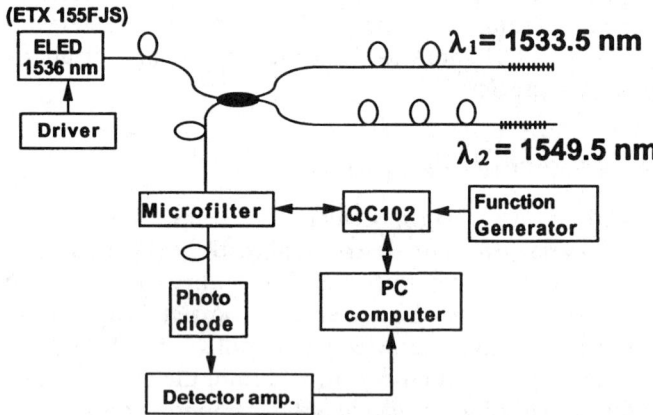

Fig. 12.7 Schematic of the FBG sensor system with Fabry–Pérot filter [10]. amp = amplifier.

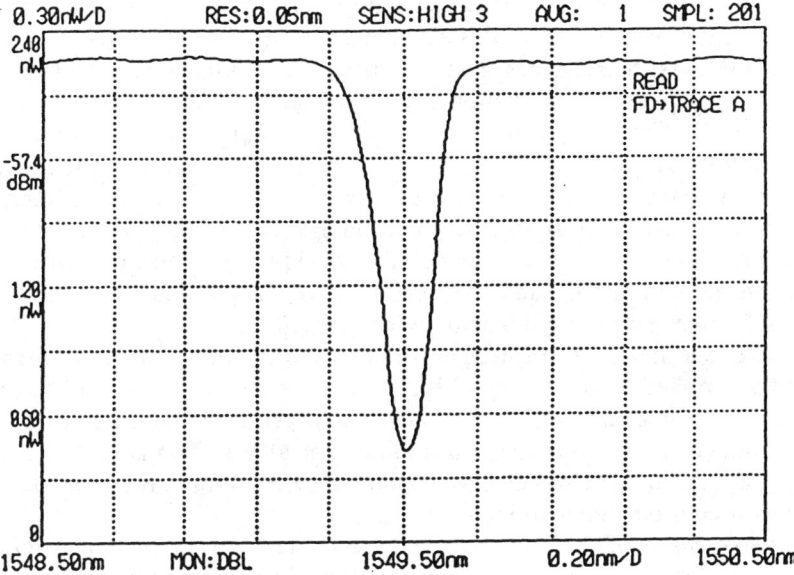

Fig. 12.8 A typical spectrum of the transmission from a grating, showing a peak reflectivity at 1549 nm. Horizontal scale 0.2 nm div^{-1} [10].

before being detected. The detected signal and the wavelength signal from the external wavelength monitor of the filter were simultaneously sampled with a high speed analog-to-digital (A/D) converter card and analyzed with the use of a microcomputer.

When the Fabry–Pérot filter was tuned to scan over the spectrum of the light reflected from a Bragg grating, the detected signal, S, at the output of the filter was a function of the cavity separation of the filter, and given by

$$S = \iint \frac{\exp[-(\lambda_0/\delta\lambda)^2]}{1 + F\sin^2(4\pi L/\lambda)} \, d\lambda \, dL \qquad (12.9)$$

where the cavity finesse $F = 4R/(1-R)^2$, R is the reflectivity of the filter, λ is the wavelength of the reflected spectrum, λ_0 is the central wavelength, $\delta\lambda$ the spectral bandwidth, and L is the separation of the cavity.

In computer simulations, the initial cavity separation of the filter was set to be 28.407 μm, giving a central peak transmission wavelength of 1545 nm, from which the separation was increased in a step size of 0.1 nm over of a displacement region of 40 nm, giving a scanned spectral range of 2.1 nm. The FSR of the cavity was 42 nm. The spectral width, $\delta\lambda$ and the central wavelength, λ_0, of the reflected light from the grating were set to be 0.2 and 1545 nm respectively. The wavelength, λ, was incremented from 1544 to 1550 nm ($\Delta\lambda = 6$ nm) within steps of 0.01 nm, covering the full range of the two fiber gratings.

394 FIBER BRAGG GRATING SENSORS: SIGNAL PROCESSING

It is easy to see that a larger value of filter bandwidth can result in a higher maximum intensity. However, this also increases the FWHM of the peak signal, and thus reduces the sharpness of the signal peak. Both the maximum value of the peak signal and the corresponding slope value of its first derivative were computed as a function of the filter bandwidth, and the results are shown in Figure 12.9. It can be seen that as the filter bandwidth increases, the maximum intensity increases but the slope value reduces gradually. Hence there is a 'tradeoff' between the peak intensity and the FWHM when selecting a suitable bandwidth value for the filter. Generally speaking, an optimized filter bandwidth can be obtained around the region where the two curves in Figure 12.9 cross each other, i.e. at about 0.3 nm in the case shown.

In the experiments performed, a triangular waveform with a frequency of 20 Hz was applied to the Fabry–Pérot filter and the amplitude of the driving waveform adjusted in such a way that the two peak signals from the detector were obtained in a half-scan period. The typical waveforms of the detected signal (trace A) and the wavelength signal (trace B) before and after the digital filtering process are shown in Figure 12.10.

The wavelength of the light reflected from the gratings was determined by undertaking the following steps. First, both the detected signal and wavelength signal were sampled and data read into the computer. Then the two sets of digital data were digitally filtered. Afterwards, the peak position of the detected

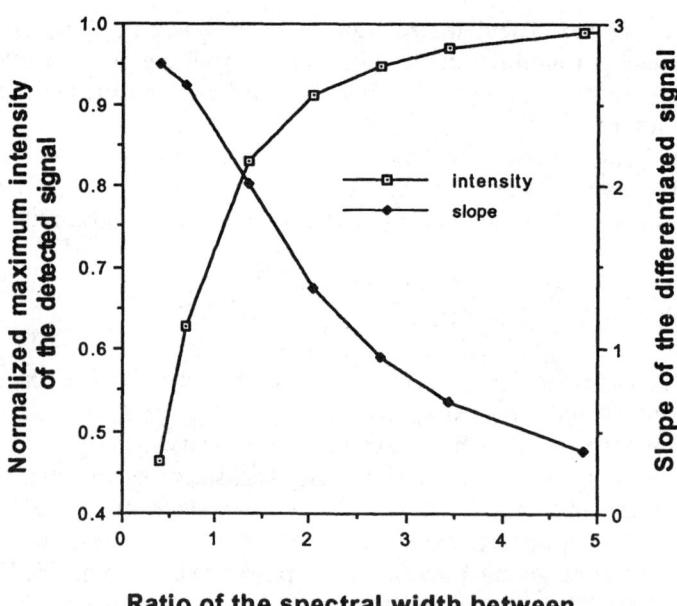

Fig. 12.9 Simulated results show the variations in both the maximum intensity and the slope of the differentiated signal as functions of the filter spectral bandwidth [10].

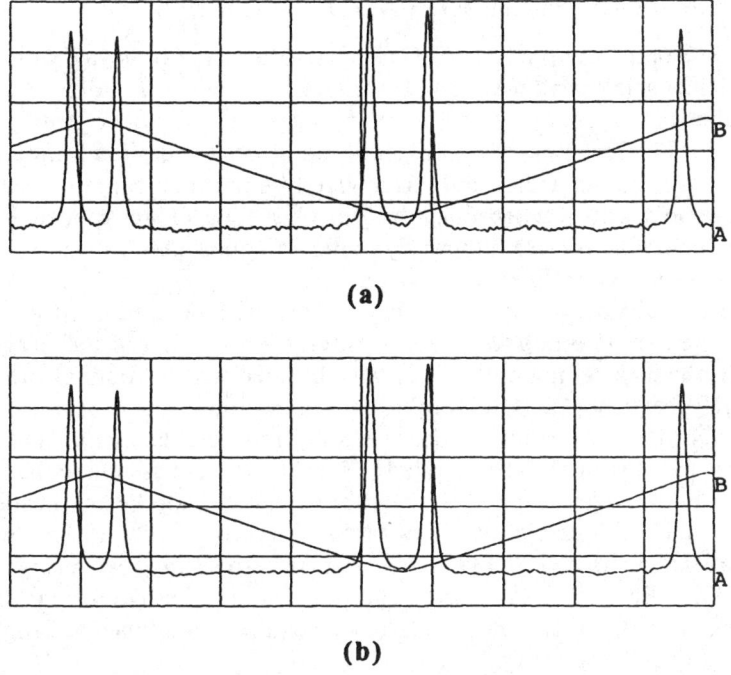

Vertical scale: 1 volt/div; Horizontal scale: 5 ms/div.

Fig. 12.10 Recorded waveforms obtained from the detector (trace A) and the wavelength monitor on the Fabry–Pérot filter (trace B) showing the two reflected wavelength peak signals being generated within both the upward downward scans: (a) before the digital filtering; (b) after the digital filtering. X axis: 1 volt div^{-1}; Y axis: 5 ms div^{-1} [10].

signal was found by identifying the maximum intensity of the signal peak, and finally, with the use of this peak position, the corresponding wavelength was determined. Thus, the value of the measured wavelength difference was calculated. For each measurement, about 10 readings were taken to give an averaged value of the wavelength difference, and the standard deviations for both the upward and downward scans respectively were obtained.

In order to calibrate the system, two gratings of known characteristic wavelengths were used to provide two stabilized wavelengths when kept unstrained in a temperature-controlled container. The repeatability of the system was measured and the results showed that the data points were scattered around an average wavelength difference of 16 nm over a range of ∼30 pm, where the typical standard deviation was then about ∼10 pm. If the data obtained from both the scanning modes were calibrated separately, the scattering effect was smaller, typically about 20 pm, with a standard deviation reduced to about 7 pm.

12.3.4 Acousto-optic tunable filter (AOTF) detection

The acousto-optic tunable filter (AOTF) is another type of device which can be used to detect small peak wavelength shifts associated with in-fiber Bragg gratings. Xu et al. [11] reported an interrogating system for fiber grating sensors using an AOTF. This type of tunable filter possesses the desired frequency-agile capability for random access and has a wide tuning range. Since its peak transmission wavelength is determined by the frequency of an applied RF drive signal, it is suitable for use in both dynamic and quasi-static strain monitoring and in addition for multiplexed sensing arrays.

One of the advantages of this device is that it can be used in either the scan mode or the lock-in mode to detect or track the variation of the Bragg wavelength. In the lock-in mode, the filter was dithered with a nominal value at the applied RF frequency, and an amplitude modulation of the optical carrier was thus produced and detected. Under the condition that the mean AOTF wavelength was proportional to the applied RF frequency and assuming the response curves for both the grating and the filter are symmetrical, the amplitude modulation of the optical carrier at the dithering frequency becomes zero when the mean wavelength of the AOTF coincided with the Bragg wavelength of the grating. With the use of this zero amplitude modulation condition, the mean optical wavelength of the filter was locked to the instantaneous Bragg wavelength of the in-fiber grating.

The experimental system used, as shown in Figure 12.11, was locked to the mean wavelength of the AOTF at the above condition, where the amplitude modulation was effectively zero. A single-mode fiber-pigtailed ELED (with a central wavelength of 1300 nm and bandwidth (FWHM) of 56 nm) was used as the broadband light source. A Bragg grating with a central wavelength of 1298 nm, reflectivity of 99% and a bandwidth of 1 nm was incorporated between

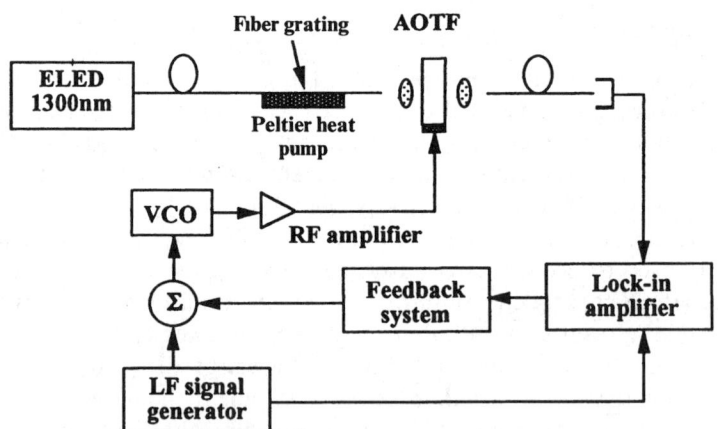

Fig. 12.11 Schematic diagram of the strain sensing system using an acousto-optic tunable filter [11].

the optical source and the AOTF which had a wavelength tuning range from 1.2 to 2.5 μm and a resolution of 4 nm. A mean RF drive frequency of 111.58 MHz and a dithering frequency of 360 kHz were applied to the AOTF. In the reported work, the wavelength variation of the Bragg grating was measured in terms of RF frequency shift, and a temperature sensitivity of -0.95 kHz °C^{-1} was demonstrated with this system.

12.3.5 Matched fiber Bragg grating pairs

A wavelength detection scheme proposed by Jackson *et al.* [12] was based on the so-called 'sensor–receiver grating pair'. The concept of this scheme is very simple; light reflected from each sensing grating is filtered with an identical receiver grating, either in the reflection mode [12] or in the transmission mode [13]. If a strain was applied to one sensor grating, the central wavelength of the reflected light varied in direct proportion to the strain, and generally did not match that of the receiver grating which was mounted at the network output end attached to a piezoelectric stretcher (PZT). However, by driving the PZT with a linear voltage ramp, the central wavelength of the receiver grating was scanned over the operational region of the sensor grating. Therefore, at the position where two wavelengths of the gratings coincide, a strong signal was detected and used to determine the variation of the sensing wavelength.

Since this wavelength detection scheme is operated in the spectral domain, it can also be used to demultiplex a sensor grating array using a wavelength division multiplexing (WDM) mode. With the use of matched receiving gratings in the reflection mode, a multiplexed Bragg grating system with a grating array was demonstrated [12], as shown in Figure 12.12. The measured results show that a resolution of 4.12 με was attainable for both quasi-static and periodic type measurands.

When the receiver gratings were used in the transmission mode [13], shown in Figure 12.13, they act as notch filters which strongly reject signals with matched wavelengths. In order to track the wavelength variation induced by the sensing gratings, the receiver gratings, which were mounted on separated

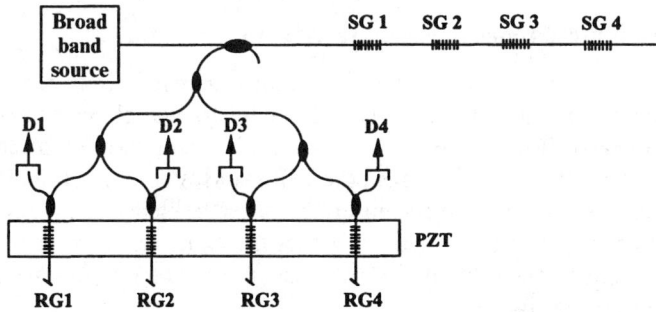

Fig. 12.12 Schematic diagram of the strain sensing system using matched grating pairs in reflection mode [12].

398 FIBER BRAGG GRATING SENSORS: SIGNAL PROCESSING

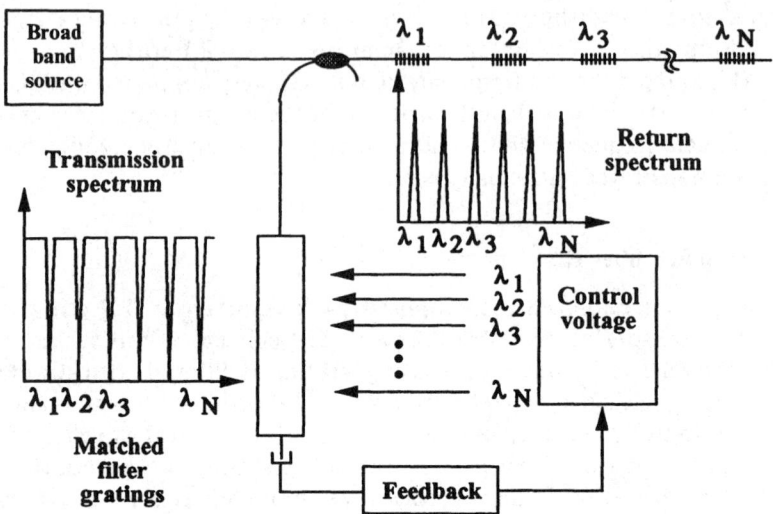

Fig. 12.13 Schematic diagram of the strain sensing system using matched grating pairs in transmission mode [13].

PZT-driven stretching devices, were modulated with a small dither signal respectively. The light transmitted through the (serial) filter grating array was detected by a single photodiode. The output signal was then fed to two lock-in amplifiers, each referenced to the dither signal applied to the filter gratings. The outputs from the two lock-in amplifiers were then summed with the dither signals and fed back onto the filter gratings. With the use of this feedback loop, the measurement system allowed the filter grating to accurately track strain-induced wavelength variations in the sensor gratings. The output signals from the lock-in amplifiers provided a voltage signal that was directly proportional to the applied strain. The experimental results showed that, with the use of this system, a strain measurement range of $\pm 100\,\mu\varepsilon$ and a dynamic strain resolution of $0.01\,\mu\varepsilon\,\text{Hz}^{-1/2}$ could be achieved.

12.3.6 Optical interferometer-based wavelength detection

One of the most sensitive wavelength detection schemes is that based on interferometric phase detection [14, 15]. Here, fiber Bragg wavelength variations are transposed into a shift in phase at the output of an unbalanced interferometer by passing the radiation reflected from the FBG sensor through the interferometer. In this scheme, the unbalanced interferometer behaves like a spectral filter with a raised cosine-transfer function. When the light reflected from the FBG sensor passes through the unbalanced interferometer, the output intensity of the interferometer can be given by

$$I = I_1 + I_2 + 2(I_1 I_2)^{1/2} |v(\tau)| \cos(\psi(\lambda) + \phi) \qquad (12.10)$$

where $\psi(\lambda) = 2\pi L/\lambda$ is the phase difference of the interferometer, L is its optical path difference (OPD), λ is the central wavelength of the reflected light from the Bragg grating sensor, I_1 and I_2 are the optical intensities reflected from the two arms of the interferometer respectively and $|v(\tau)|$ is the fringe visibility, which is proportional to the input intensity and interferometer arm imbalance.

For a quasi-static strain-induced wavelength variation $\Delta\lambda$, the corresponding phase variation $\Delta\phi$ of the output signal is given by

$$\Delta\phi = -\frac{2\pi L}{\lambda^2}\Delta\lambda \tag{12.11}$$

If the operational range of the sensor is defined in such a way that the total phase shift is limited to within $\Delta\phi = 2\pi$ (i.e. one interferometric fringe), the above equation can be rewritten as

$$\chi\frac{\Delta\lambda}{\lambda} = \frac{\Delta L}{L} = \varepsilon \tag{12.12}$$

where ε is the strain induced in the grating, $\chi = (1 - p_e)^{-1} \sim 1.32$ where p_e is the effective photoelastic coefficient giving $\Delta\lambda/\lambda = 0.76\varepsilon$. According to this equation, for the case of $\lambda = 1550$ nm, the relationship between the strain range (in μɛ) to be measured and the OPD (imbalance) of the interferometer can be determined, and the typical values are shown in Table 12.1.

If the value of the interferometer imbalance is set as 15.5 mm and for a central wavelength of 1550 nm, the relationship between the minimum detectable strain and the wavelength shift, hence phase shift, can be obtained and the typical values are shown in Table 12.2.

If the phase shift of the interferometric signal is measured with the use of a phase lock-in amplifier, different phase resolution of the lock-in amplifier will give different strain resolutions and some typical values are shown in Table 12.3.

For a dynamic strain-induced wavelength modulation $\Delta\lambda\sin\omega t$ on the sensor element, the corresponding change in phase shift is then given by

$$\Delta\phi = -\frac{2\pi\Delta L}{\lambda^2}\Delta\lambda\sin\omega t = -\frac{2\pi\Delta L}{\lambda^2}\gamma\Delta\varepsilon\sin\omega t \tag{12.13}$$

where $\Delta\varepsilon$ is the dynamic strain subjected to the grating, and γ is the strain-to-wavelength shift response (where $\gamma = \chi^{-1}$) of the grating. For example, for an interferometer with an unbalanced OPD of 14.6 mm, a strain responsivity γ of 1.15 pm με^{-1} and a wavelength of 1.55 μm, a sensitivity $\Delta\Psi/\Delta\varepsilon$ of

Table 12.1 Relationship between measured strain and range

Strain range ($\varepsilon = \Delta\lambda/\lambda$)	OPD required (ΔL) (mm)
1:100 (1×10^4 μstrain)	0.155
1:1000 (1×10^3 μstrain)	1.55
1:10 000 (1×10^2 μstrain)	15.5

Table 12.2 Relationship between minimal detectable strain and wavelength shift

$\Delta\varepsilon$ (μstrain)	Wavelength shift (nm)	Phase shift (rad)
1	1.55×10^{-3}	6.28×10^{-2} (3.6°)
1×10^{-3}	1.55×10^{-6}	6.28×10^{-5} (0.0036°)

Table 12.3 Different strain resolutions for typical values of phase resolution

Phase resolution (degrees)	Strain resolution
1.0	0.278 μstrain
0.1	0.0278 μstrain
0.01	2.78 nstrain

$0.045 \, \text{rad} \, \mu\varepsilon^{-1}$ is produced. If the dynamic resolution of a phase detection scheme is $\sim 10^{-6} \, \text{rad} \, \text{Hz}^{-1/2}$, which is typically achievable for interferometric sensors, a dynamic strain resolution of $\sim 20 \, \text{p}\varepsilon \, \text{Hz}^{-1/2}$ can be obtained. This high strain resolution is achieved due to the fact that the value of the phase shift is strongly amplified when a large OPD is used, as the phase variation of the output light from this interferometer is proportional to both the variation of the Bragg wavelength and the OPD of the interferometer. With the use of an optic fiber Mach–Zehnder interferometer of 10 mm (± 2 mm) fiber imbalance, a Bragg grating based strain sensor system with a dynamic strain resolution of $\sim 0.6 \, \text{p}\varepsilon \, \text{Hz}^{-1/2}$ at frequencies only >100 Hz has been demonstrated [14].

12.4 PHASE-DRIFT COMPENSATION IN INTERFEROMETRIC SYSTEMS

Although the value of the phase shift is amplified when a large OPD is used in the interferometric wavelength detection scheme discussed above, errors can be induced in measurements of the Bragg wavelength shift. These errors can result from fluctuations in the OPD due to environmental disturbances, such as temperature or mechanical vibration effects, since the phase change is governed by both the Bragg wavelength shift and OPD stability. This largely limits the scheme so far described to the measurement of the dynamic strains, e.g. vibration and acoustic emissions.

In order to achieve high wavelength accuracy for quasi-static strain measurement, the OPD imbalance in the interferometer must be stabilized. This can be achieved by stabilizing the average value of the OPD of the unbalanced interferometer by the use of a reference configuration or a feedback control system. A scheme designed to control the unbalanced interferometer was reported recently by Shi et al. [15], in which the stabilization of the OPD was achieved by using a stabilized reference light source along with a feedback control system to lock the average OPD at the selected value.

If there is a perturbation-induced OPD variation ΔL in the interferometer, according to equation 12.11, the phase shift $\Delta\phi_r$ in the output of the interferometer now becomes

$$\Delta\phi_s = -\frac{2\pi L}{\lambda_s^2}\Delta\lambda_s + \frac{2\pi}{\lambda_s}\Delta L \qquad (12.14)$$

Clearly, it is not possible to obtain the true value of $\Delta\lambda_s$ since only the value of $\Delta\phi_s$ can be measured. In order to eliminate the effect of ΔL fluctuations, an additional stabilized light source λ_r is employed to generate a reference beam in addition to the signal beam, as shown in Figure 12.14. The phase variation $\Delta\phi_r$ produced by the reference wavelength is then given by

$$\Delta\phi_r = \frac{2\pi}{\lambda_r}\Delta L \qquad (12.15)$$

In the measurement system, any phase changes detected by the reference wavelength due to disturbances other than from the applied carrier modulation (i.e. system noise), are used to generate an error signal which is then fed back in antiphase to the PZT mirror modulator to compensate for the noise fluctuations

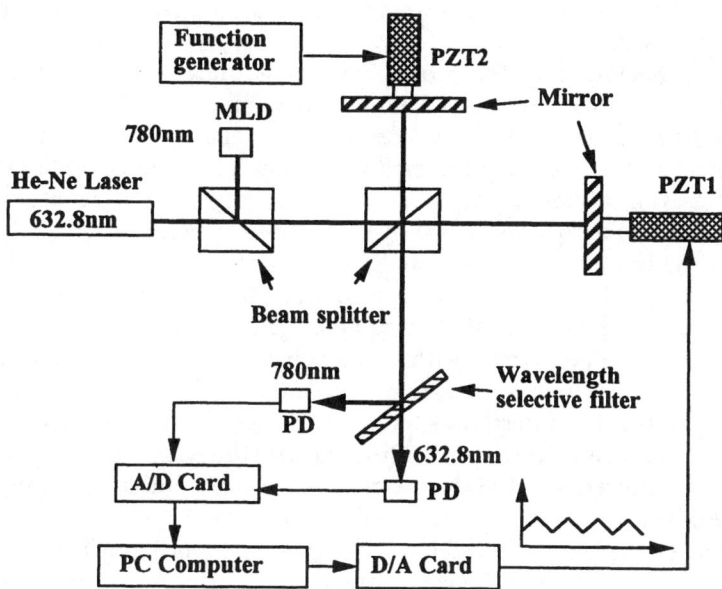

Fig. 12.14 Schematic diagram of the stabilized interferometric system using a reference light source. MLD = multimode laser diode, PZT1 and PZT2 = piezoelectric transducers, PD = photodiode, A/D and D/A = analog-to-digital and digital-to-analog conversion cards [15].

in the interferometer. A lock-in, stabilized measurement system is thus realized to dynamically suppress undesirable environmental fluctuations.

The operational principle of this scheme was demonstrated with the use of a Michelson interferometer as shown in Figure 12.14. The position of the mirror attached to PZT1 could be varied by a microcomputer via a D/A conversion board. In the experiment, the computer sent a triangular waveform signal to drive the PZT1 to cause the mirror to oscillate around the average OPD position. In addition, a feedback signal was also sent to the PZT1 to compensate the variation of the average fluctuations of the interferometer OPD. The frequency of the triangular signal was 300 Hz, limited by the speed of A/D and D/A cards. The environmental disturbance of the interferometer was simulated by adding a vibration signal to PZT2 using a function generator. A multi-mode laser diode with a central wavelength of 786 nm was employed as the signal source in this demonstration. By changing the driving current of the diode, a wavelength shift could be introduced to simulate the Bragg wavelength variation. An He–Ne laser of wavelength 632.8 nm was employed as the reference light source. Both the signal and reference beams were combined through the use of a beam-splitter and then launched into the Michelson interferometer which had an optical imbalance of about 4.4 mm. The outputs from the interferometer were then separated into the two original wavelengths with the use of a wavelength selective filter [16]. Two separate detectors were employed to convert the optical outputs into electronic signals, which were then sent to the computer via the A/D conversion board.

The basic operation of the processing scheme may be considered as being divided into the following steps. Initially, a computer program was used to generate a triangular waveform which was then sent to the PZT1 via the D/A card. The amplitude of this waveform was about 2.5 V, which makes the PZT1 vary over a range of about a quarter of the wavelength of the signal source (i.e. ~160 nm). The average DC value of the triangular waveform was about 5 V. Secondly, the outputs from the two detectors were sampled simultaneously by the A/D card and sent to the computer. Since the action to send the control code to drive the PZT1 and that to sample the corresponding intensities from the detectors were performed within one program loop, these signals may be considered as being in phase. The minimum intensity of the reference signal was then used to lock the interferometer at a fixed OPD by adjusting the control code of the average DC value. Further, the amplitude of the triangular waveform was monitored and the error signal was used to adjust the amplitude or AC value of the waveform.

The measured phase values from the signal beam as a function of time are shown in Figure 12.15 for the cases where the interferometer was locked and unlocked respectively, from which it can be seen that in the unlocked condition, the measured phase value varies randomly from 37 to 183° over the measurement period (~9 minutes), whilst when the average OPD of the interferometer is locked, only a 3° variation in phase over the same length of time can be seen.

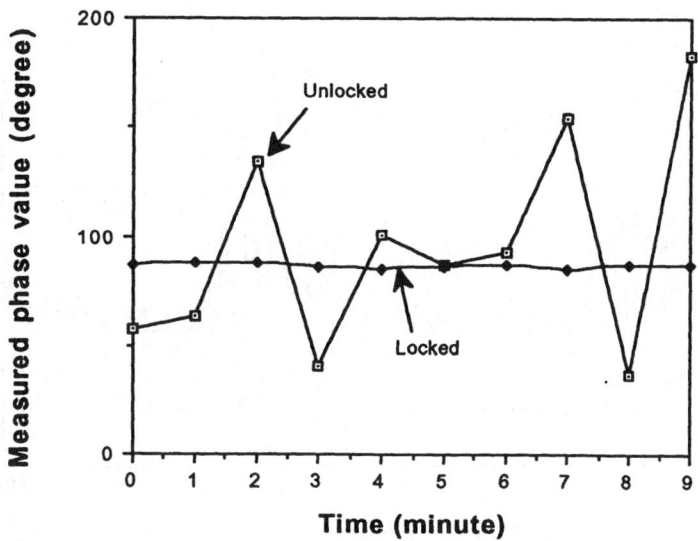

Fig. 12.15 Measured phase value of the interference fringe generated by the source wavelength as a function of time for the cases where the OPD is locked and unlocked respectively [15].

This result shows that the feedback control loop used reduced the fluctuation in the output phase by a factor of ~ 50.

Figure 12.16 shows the recorded spectrum of the outputs generated by the signal wavelength when noise due to a simulated 20 Hz vibration was induced on the mirror via PZT2. The upper trace is recorded from the unlocked interferometer, and the lower trace from the locked interferometer. From Figure 12.16, it can be seen that when the interferometer was unlocked it resulted in S/N ratio of $\sim 15\,\text{dB}$, while with the interferometer locked, the resulting S/N ratio was increased to $\sim 40\,\text{dB}$. This demonstrates the effect of noise suppression when using software-locking schemes to dynamically restrain fluctuations resulting from such environmental changes. In order to simulate the measurement of Bragg wavelength shifts, the drive current of the source laser diode (MLD) was varied to introduce wavelength changes in the signal beam. Under these conditions, the measured wavelength resolution was $\sim 1.2\,\text{pm}$.

In a further phase-drift compensation scheme reported recently [17], instead of using a reference light source, a local stable (shielded) reference FBG was employed in conjunction with the sensor grating. In this scheme, as shown in Figure 12.17, the sensing and reference fiber optic grating elements were placed in the output ports of an unbalanced fiber Mach–Zehnder interferometer. When the light from a broadband source was injected into the interferometer, each Bragg grating element reflected a portion of the interferometer output signal within its narrow-band spectrum, which was then monitored via tap-off couplers. Provided the OPD of the interferometer was shorter than the effective

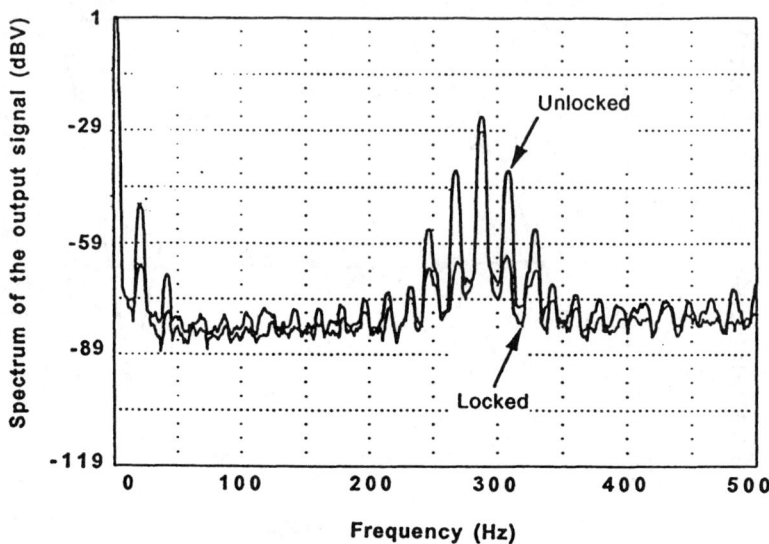

Fig. 12.16 Recorded spectrum of the output signal showing the induced noise at 20 Hz has been suppressed by the locking system [15].

coherence length of the Bragg wavelength reflected by the gratings, interference signals were generated at the two detector outputs. Therefore, any random variation in the average OPD of the interferometer due to environmental temperature drifts will affect both outputs in the same way if the wavelengths of the sensing and reference gratings are similar. The output of the reference detector can therefore be used to compensate the phase noise present in the sensor signal resulting from OPD fluctuations. This approach therefore provided a measure of the induced Bragg wavelength shift, free from any fluctuations in the unbalanced interferometer.

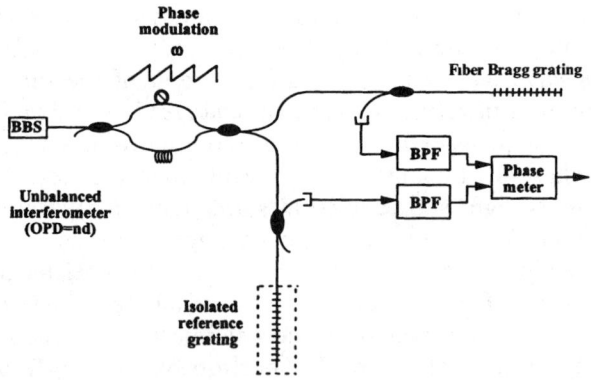

Fig. 12.17 Compensated Bragg-grating sensor with sensing and reference grating elements and interferometric wavelength-shift detection [17].

With the use of this system, a stability in the phase difference output of about $\pm 1°$ in the lock-in amplifier was observed, corresponding to an r.m.s. random drift of approximately $0.5\,\mu\varepsilon\,h^{-1}$. The sensitivity of the system to low-frequency strain perturbations was measured by applying $\sim 1\,\mu\varepsilon$ r.m.s. strain variation to the sensor grating at a frequency of 1 Hz. The signal-to-noise ratio of the component at 1 Hz was 44 dB, normalized to 1 Hz bandwidth, giving a minimum detectable strain perturbation of $\sim 6\,n\varepsilon\,Hz^{-1/2}$ at 1 Hz. The wavelength shift induced by the applied strain was approximately 1.2×10^{-3} nm r.m.s., which was less than 1% of the grating bandwidth.

12.5 WAVELENGTH ERRORS IN INTERFEROMETRIC DETECTION SCHEMES

As discussed above, in the interferometric wavelength detection scheme, the Bragg wavelength variation is monitored via the measurement of the phase variation of the interference fringe produced at the output of the interferometer. This is often achieved with the use of a phase lock-in amplifier. It is well known that when a lock-in amplifier is used in a measurement system, a stabilized frequency reference is often required. Typically, the lock-in amplifier generates its own internal reference signal by employing a phase-locked loop to lock at an external reference frequency, with respect to which phase variations of the carrier signal may be measured. When a lock-in amplifier is used in an FBG-based sensor system, the reference signal is often produced by filtering the interferometer OPD ramp signal. The carrier signal is usually generated by filtering the output signal from the interferometer with the use of a band pass filter (BPF) [18]. In this case, an accurate measurement can be achieved if the amplitude of the OPD ramp is set to a level equivalent to the Bragg wavelength of the FGS.

However, in a practical measurement system, the value of the OPD ramp amplitude cannot always be maintained to correspond directly to the Bragg wavelength, as the latter is constantly being altered by the measurand. As a result, a distortion in the waveform of the output signal will occur, and a measurement error may be introduced if the lock-in amplifier still operates at the fixed reference frequency. This so-called 'carrier frequency mismatch' problem considered above has been analyzed by Shi et al. [19] and its effect on the measurement accuracy has been simulated and discussed.

In general, in an interferometric wavelength detection scheme, if the value of the OPD ramp amplitude is equal to that of the Bragg wavelength, the output signal, S_0, of the lock-in amplifier is given by

$$S_0 = \frac{1}{2}\cos\left(\frac{2\pi L_0}{\lambda_0 + \delta\lambda}\right) \qquad (12.16)$$

where L_0 is the OPD of the unbalanced interferometer, λ_0 the central wavelength of the Bragg grating and $\delta\lambda$ the wavelength shift induced by the measurand. Hence any change in this wavelength shift will yield a corresponding change of the output signal.

However, if the value of the OPD ramp amplitude is not equal to that of the Bragg wavelength, a distortion will occur in the output waveform of the interferometer, as shown in Figure 12.18. If the lock-in amplifier is used to detect the phase variation of this carrier signal, the output signal, S, can then be expressed as [20, 21]

$$S = \frac{\lambda_0 - \Delta\lambda_0}{2\lambda_0 - \Delta\lambda_0 + \delta\lambda} \mathrm{sinc}\left[\pi\left(\frac{\Delta\lambda_0 + \delta\lambda}{\lambda_0 + \delta\lambda}\right)\right] \cos\left[\frac{2\pi L_0}{\lambda_0 + \delta\lambda} - \pi\left(\frac{\Delta\lambda_0 + \delta\lambda}{\lambda_0 + \delta\lambda}\right)\right] \quad (12.17)$$

where $\Delta\lambda_0$ is initial ramp deviation which is equal to the difference between the central Bragg wavelength λ_0 and the ramp OPD amplitude A (i.e. $\Delta\lambda_0 = \lambda_0 - A$). Hence a measurement error, E, will be introduced, and its value can be calculated by subtracting S_0 from S, and this is given by

$$E \approx \frac{1}{2}\mathrm{sinc}\left[\pi\left(\frac{\Delta\lambda_0 + \delta\lambda}{\lambda_0 + \delta\lambda}\right)\right]\cos\left[\frac{2\pi L_0}{\lambda_0 + \delta\lambda} - \pi\left(\frac{\Delta\lambda_0 + \delta\lambda}{\lambda_0 + \delta\lambda}\right)\right] - \frac{1}{2}\cos\left(\frac{2\pi L_0}{\lambda_0 + \delta\lambda}\right) \quad (12.18)$$

From equation 12.18, it can be seen that the measurement error is governed by two cosine functions with periods of $\delta\lambda_{P1} \approx 2\lambda_0^2/(2L_0 - \Delta\lambda_0 - \lambda_0)$ and $\delta\lambda_{P2} \approx \lambda_0^2/(L_0 - \lambda_0)$ respectively. When considering the conditions of $L_0 \gg \lambda_0$ and $L_0 \gg \Delta\lambda_0$, then $\delta\lambda_{P1} \approx \delta\lambda_{P2} \approx \lambda_0^2/L_0$, i.e. both the cosine functions have the same period. This indicates that the measurement error, E, is also a pseudo-periodical function with the same period, i.e. $\delta\lambda_P \approx \lambda_0^2/L_0$. Therefore, for different

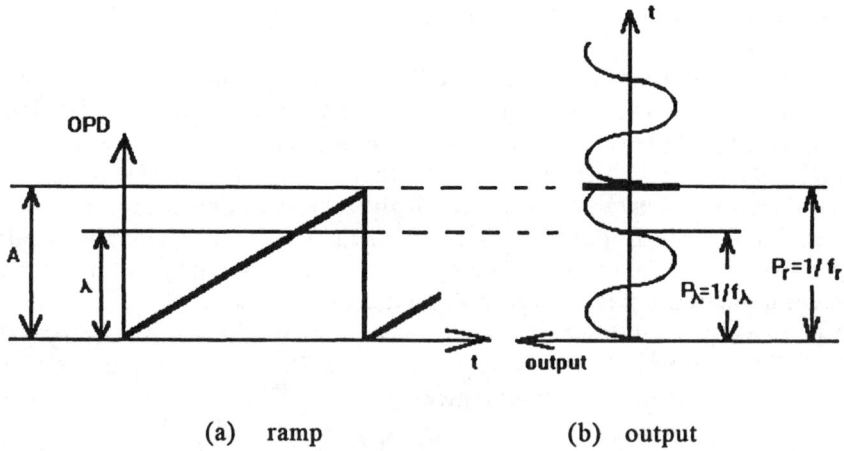

(a) ramp   (b) output

Fig. 12.18 Schematic of the relationship between the ramping amplitude and the distortion of the output signal of the interferometer. A is the ramping amplitude, λ is the Bragg wavelength, $P_r = 1/f_r$ is the ramp period, and $P_\lambda = 1/f_\lambda$ is the signal period corresponding to the Bragg wavelength in the distortion carrier signal [19].

values of wavelength shift, the measurement error induced will not be the same and hence cannot be completely eliminated by employing a relative measurement scheme. Furthermore, the period of the measurement error is also equal to the FSR of the sensor system.

If values for L_0 and λ_0 were chosen to be 1300 and 1.3 μm respectively, typical of optical fiber sensors, then the corresponding value for the FSR is ~ 1.3 nm. For systems configured using either a single fiber grating or a number of fiber gratings (of similar pitch and multiplexed in a time-division or spatial-division scheme) the simulated results show that when the Bragg wavelength is shifted from 1300 to 1301.3 nm, the measurement error varies as the value of the central wavelength increases, reaching a maximum value of ~ 3.25 pm, i.e. about 0.25% of the total measurement range. While for the case where a number of fiber gratings are multiplexed with the use of a WDM scheme, since different gratings have different central Bragg wavelengths, there will be a significant difference between the amplitude value of the ramp OPD and the values of each central Bragg wavelength, and as a result the value of the measurement error will also be relatively large. With the use of equation 12.18, the measurement error was calculated when the range of the total wavelength deviation, $\Delta\lambda = \Delta\lambda_0 + \delta\lambda$, was chosen to lie in a region from 0 to 20 nm. The calculated result is shown in Figure 12.19(a). It is clear that the measurement error is a pseudo-periodical function of the total wavelength deviation. If the peak-to-peak value of the measurement error is used to indicate the possible maximum measurement error, this continues to increase as the value of the total wavelength deviation increases. For example, when the initial ramp deviation is about 4 nm and the wavelength shift is 1.3 nm, the maximum measurement error is about 1% over a full FSR. If, however, the initial ramp deviation is increased to 20 nm, for instance, in the case where four gratings are multiplexed in a WDM scheme, the maximum measurement error will increase to about 5% of the full FSR, giving a maximum error of 65 pm. So if 1 με introduces a 1.15 pm wavelength variation, then a ~ 56.5 με measurement error can be induced with the use of a lock-in amplifier. It is clear that such a measurement error cannot be ignored if an accurate system is to be specified.

Further, for the case where a so-called double-phase lock-in amplifier is employed to detect the phase, the simulated results obtained are shown in Figure 12.19(b). It is valuable to note that the measurement error is not only a pseudo-periodical function with its peak-to-peak value being proportional to that of the total wavelength deviation, but it also has a slowly increasing linear DC component. However, since the DC component in the measurement error can be eliminated in AC strain measurements, a relatively large portion of the error can be removed in this case as the value of the DC component is larger than that of the peak-to-peak value. In comparison with the measurement error induced in the single-phase lock-in detection scheme, the double-phase lock-in scheme gives a smaller error (0.4% in a full FSR) for an AC strain measurement. On the other hand, it can be easily seen that the measurement

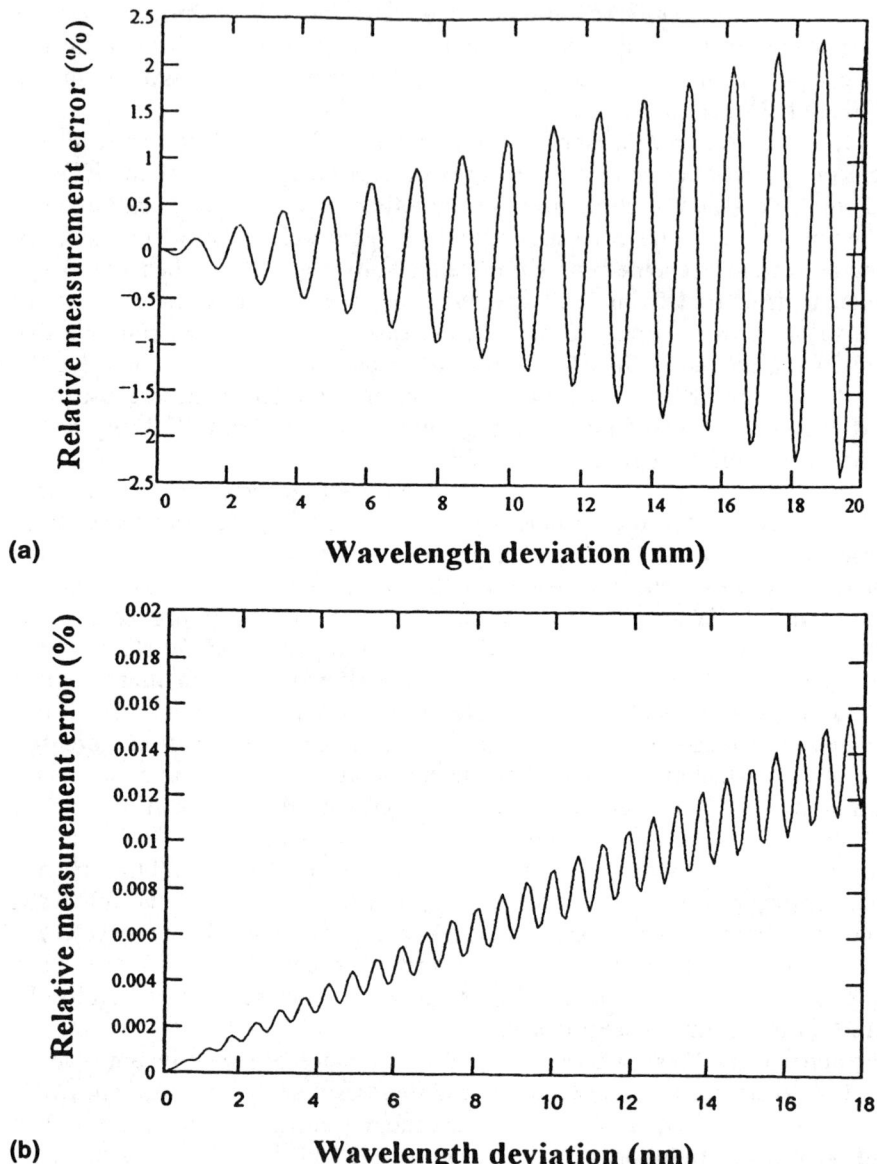

Fig. 12.19 Relative measurement error as a function of wavelength deviation with the use of: (a) a single-phase lock-in amplifier; (b) a double-phase lock-in amplifier [19].

error in the double-phase lock-in amplifier will change by two periods when the wavelength shifts by about one FSR, i.e. the period of the measurement error of the double-phase lock-in amplifier is twice that of the single-phase lock-in amplifier.

12.6 MULTIPLEXING SCHEMES FOR FIBER BRAGG SENSORS

As mentioned in the introduction, the ability to multiplex a large number of grating sensor elements is one of the key advantages in designing a fiber grating sensor system. As a result of intensive research over the past few years, a number of multiplexing schemes have been proposed and developed. The most commonly used multiplexing schemes are,

1. Wavelength division multiplexing (WDM);
2. Time division multiplexing (TDM);
3. Frequency division multiplexing (FDM); and
4. Spatial division multiplexing (SDM).

The combinations of two or even three such multiplexing schemes have also been reported for fiber grating sensing systems.

12.6.1 Wavelength division multiplexing (WDM)

One advantage in using a fiber grating network is that the grating elements can have different Bragg wavelengths, and hence, WDM schemes can be conveniently employed to address each individual grating element in the array. The most commonly used demultiplexing devices have been the optical spectrum analyzer [6], match grating pairs [12, 13] and wavelength tunable filters [22, 23]. Since the FSR of commercial Fabry–Pérot filters is generally larger than the spectral bandwidth of broadband light sources, such as the Er-doped fiber superfluorescent source, the number of grating elements that can be interrogated within the grating array is mainly determined by the bandwidth of the light source and the spectral region covered by the grating. For an Er-doped fiber superfluorescent source with a bandwidth of about 36 nm and a grating operating bandwidth of ± 3 nm, determined by the upper limit strain which can be safely applied to an FBG (approximately $\pm 2500\,\mu\varepsilon$), then six grating elements could be interrogated in one grating array. In order to increase the total number of sensor grating elements, other forms of multiplexing methods have to be used in conjunction with the WDM scheme.

A typical example of a wavelength multiplexed FBG strain-sensor system was reported by Kersey *et al.* [22]. In this system, shown in Figure 12.20, four grating elements were used to produce a four-sensor array. The demultiplexing device employed was a Fabry–Pérot tunable filter which was operated in its scanning mode with each scanning period covering the wavelength range occupied by the four grating elements. By taking the first derivative of the Bragg wavelength output signal and examining the shift in zero-crossing points of this function, corresponding to the peak in the Bragg spectrum, a strain resolution of $\pm 3\,\mu\varepsilon$ was obtained.

A tunable Fabry–Pérot filter has also been used to measure strain in a Bragg grating based fiber laser system [23], as shown in Figure 12.21. Here the

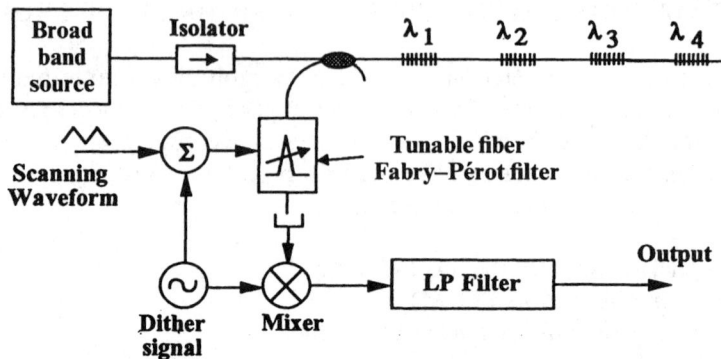

Fig. 12.20 Multiplexed FBG array with a scanning FFP demodulator [22].

fiber laser consisted of the traveling-wave form with the intra-loop, Fabry–Pérot, BPF acting as the tuning element. One cavity reflection point of the fiber laser was provided by the fiber-loop reflector, with the gain section formed with Er-doped fiber and a pump laser. The loop reflector operated in a unidirectional manner due to the inclusion of optical isolators. The other reflection point of the cavity was formed by one of the Bragg gratings in the grating array. Due to the inclusion of the wavelength filter element within the loop reflector, lasing operation of the system occurred only when the filter transmission band-pass was tuned to the wavelength of one of the grating elements. This selective lasing operation allowed each of the sensor gratings to be addressed separately and therefore the output signals from the grating array to be demultiplexed.

Although a Fabry–Pérot filter can be used to demultiplex the output signals reflected from a grating sensor array, the strain resolution is limited by the identification of peak position of each maximum signal intensity, or minimum detectable shift of the peak position [10]. In order to improve the strain resolution, interferometric detection methods can be employed together with other

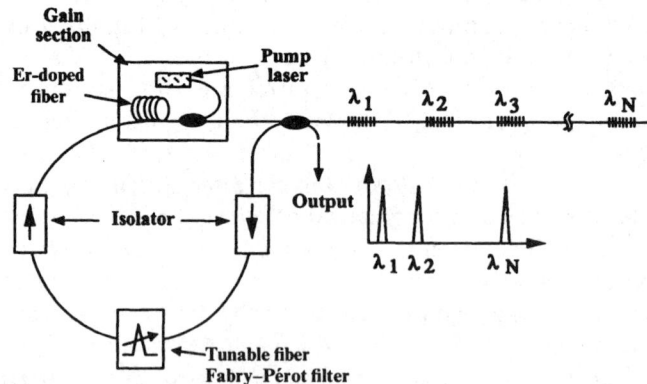

Fig. 12.21 Fiber loop-reflector laser with multiple FBG sensor feedback elements [23].

multiplexing schemes, such as TDM, band-pass WDM and FDM. For instance, with the use of a band-pass wavelength division multiplexer, Berkoff and Kersey [24] demonstrated an interferometric detection scheme, shown in Figure 12.22, where the returned wavelengths from an array of gratings were first phase carrier modulated by the interferometer and then separated by optical filtering before being detected and processed at the corresponding channel.

12.6.2 Time division multiplexing (TDM)

It is possible with the use of TDM to increase the number of measurable grating sensor elements. Further, by simultaneously employing an interferometric detection scheme, it is also possible to greatly increase the strain resolution of such measurements. Figure 12.23 shows an example of such a multiplexed sensor system [25] combining these two techniques, where eight grating elements were divided into two four-grating-element branches, which were coupled to either port of a 2 × 2 fiber coupler. The gratings in each branch were separated by 5 m of fiber, and the FBGs in one branch were temporally delayed with respect to the other by an additional 20 m of fiber. The pulsed broadband source was driven with a current pulse of 40 ns width with a 1:9 duty cycle. With one pulse injected into the grating network, a series of eight return pulses were produced at the output of the photodetector after passage through the unbalanced interferometer. Since a carrier modulation was applied to the interferometer imbalance, the output signal was suitable for demodulation using the lock-in amplifier (a technique discussed earlier). Each return pulse reflected from the individual grating elements can be uniquely discriminated by the use of either an electronic gate at the detector output and processed using a synthetic heterodyne demodulation scheme [25] or by use of a high speed electronic switching unit that diverts pulses from each sensor element into corresponding output channels [26].

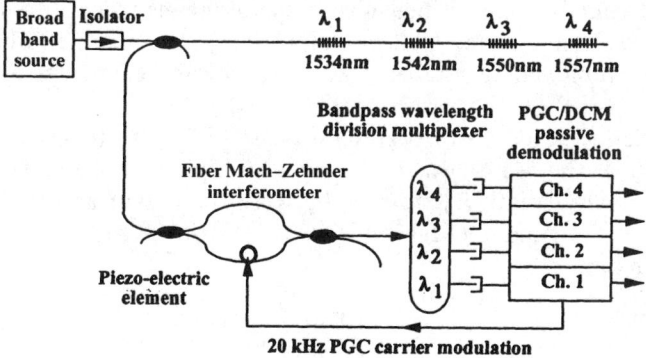

Fig. 12.22 Strain sensor system using a multiwavelength bandpass splitter and a two-beam interferometer [24].

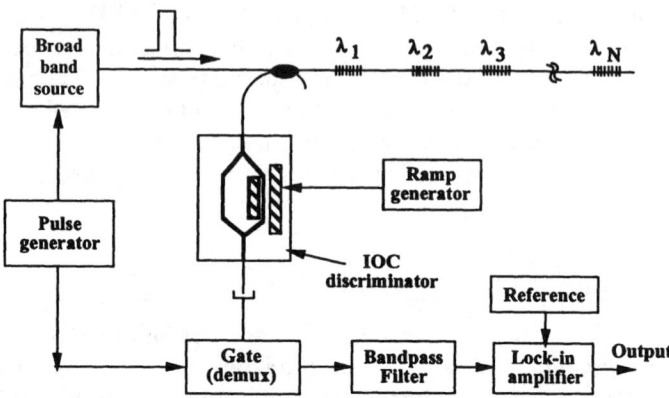

Fig. 12.23 Time division multiplexed fiber grating sensor system utilizing an integrated optic Mach–Zehnder interferometer [25].

12.6.3 Frequency division multiplexing (FDM)

In this fiber grating based laser sensor approach [27], the sensor gratings are used as feedback elements in a mode-locked fiber laser-cavity configuration. Figure 12.24(a) shows the operational principle of this detection scheme for a single grating element. The laser cavity was formed using an active fiber gain section (e.g. Er-doped fiber amplifier), a partial reflection mirror (e.g. the output port of the system) and an FBG as the sensing element. Since the sensor grating reflected with a relatively narrow spectral bandwidth (typically ~ 0.2 nm), the system was able to lase at the Bragg wavelength, and hence the lasing wavelength was used to determine changes induced by the strain or thermal perturbations in the FBG sensing element.

Since the output of the system was in the form of a wavelength variation, an additional wavelength detection scheme such as the interferometric detection approach [27] was needed to decode the output wavelength. One of the advantages of this scheme was that it provided a useful means to multiplexing FBG sensing elements when a mode-lock modulator [28] or a tunable Fabry–Pérot filter [23] were employed to address individual sensing elements in a grating array.

When a mode-lock modulator (MLM) was used in the system [28], the operating frequency of the MLM was set to be a multiple (m) of the cavity mode spacing. For a cavity length L_0, formed between the mirror and the FBG, the system locked at the fundamental mode ($m = 1$) with the mode-locked output frequency given by

$$f_1 = c/2nL_0 \tag{12.19}$$

The laser output comprises a pulse-train at frequency f_1, which can be used to address a selected FBG sensing element. The Bragg wavelength was determined

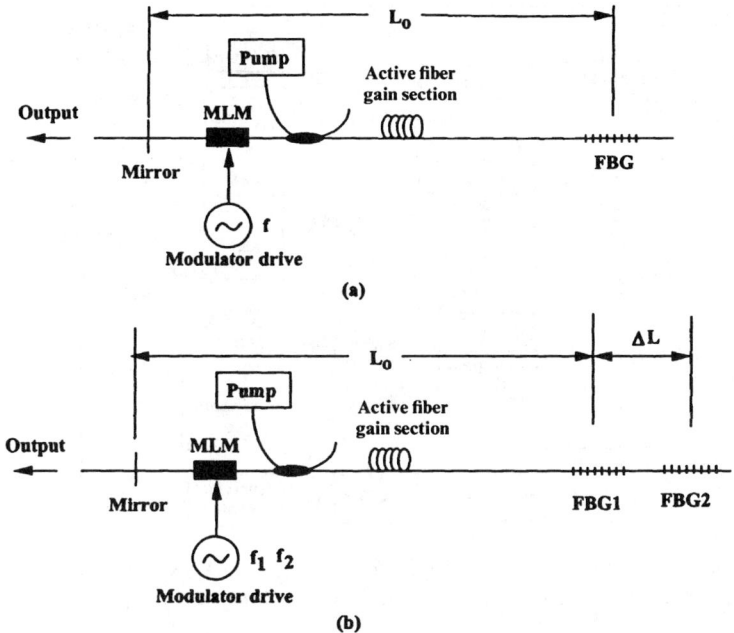

Fig. 12.24 (a) Single FBG-based fiber laser strain sensor; (b) selectively mode-locked fiber laser using two FBG elements [28].

by the pitch of the sensor grating, where strain or temperature induced wavelength shifts were detected using a suitable wavelength detection scheme.

If an additional sensor grating was added into the system, as shown in Figure 12.24(b), with the separation between the two gratings given set at ΔL, then the system can be forced to lase at the Bragg wavelength of the second grating provided the MLM is driven at the frequency f_2, given by

$$f_2 = \frac{c}{2n(L_0 + \Delta L)} \qquad (12.20)$$

Thus, by selecting different drive frequencies of the MLM, the sensor gratings can be addressed respectively and hence the demultiplexing of a fiber grating array can be achieved.

12.6.4 Spatial division multiplexing (SDM)

It is also possible to combine an interferometric wavelength detection scheme with a parallel-configured fiber network topology to form an SDM sensor system [29]. The schematic diagram of such a multiplexed sensor system is shown in Figure 12.25, where a pigtailed superluminescent diode having a bandwidth of 18.5 nm (centered at 827 nm) and an output optical power of > 1 mW was used as the light source. A Michelson interferometer with an OPD of about 0.9 ± 0.05 mm was used as the wavelength discriminator. With the use of four

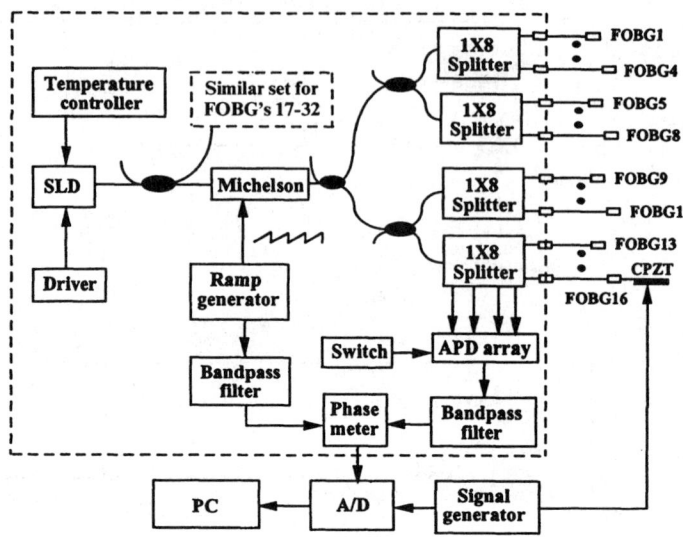

Fig. 12.25 Schematic diagram of multiplexing system. SLD = superluminescent diode, APD = avalanche photodetector, FOGB = fiber-optic Bragg grating [29].

2×2 optic fiber couplers and four 1×8 fiber-optic power-splitters, 32 identical Bragg gratings could be multiplexed to form an SDM sensor system, although only four channels were implemented in the demonstration. The 1×8 fiber-optic power-splitter consists of seven 2×2 optical power-splitters, forming eight output ports for sending and recovering light to and from sensing gratings and four ports for leading the returned light to a four-element avalanche photodiode (APD) array. Each sensing channel was addressed by using appropriate delay lines and a two-stage switch. The detected signal was then filtered by a BPF and its phase compared with that of the reference signal to yield the phase variation due to the Bragg wavelength change. The main advantages of this approach was that there was virtually no cross-talk between the channels and the operating wavelength range of each grating could be identical and unaffected by the number of grating elements multiplexed. With the use of this system, a strain resolution of $0.36\,\mu\varepsilon$ over a range of $1.5\,m\varepsilon\,Hz^{-1/2}$, and a temperature resolution of $0.036\,°C\,Hz^{-1/2}$ was reported.

12.7 SUMMARY

A considerable level of effort has been devoted to the research and development of FBG sensor systems over recent years and as a result, a number of wavelength detection schemes, together with multiplexing configurations, have been proposed and investigated. Examples of these schemes have been considered in a systematic way in order to reflect the activities and achievements in this field. From the technical point of view, the selection of any particular wavelength detection scheme for a specific application will be determined by the resolution, measurement range and frequency response required from the sensor system. In addition, the ability to multiplex a large number of elements into a grating network will also affect the selection of the measurement technique. Other important factors, from an applications point of view, are the installation and maintenance costs and the reliability and adaptability of the operating system. The transduction efficiency between the fiber Bragg sensor and the host matrix will also affect the system response and this aspect will need special attention before an FBG-based sensor system can be successfully applied to the engineering environment. Generally speaking, the advantage of wavelength selection makes FBGs naturally suitable to WDM techniques; however, the number of the grating elements is mainly limited by the bandwidth of the light source. By combining WDM and TDM techniques, it is possible to interrogate a larger number of grating elements in a sensor system, for example, a sensor system with 60 grating elements has recently been achieved with the use of a common source, optic fiber switches and a scanning narrow-band filter [30]. This high density multiplexing capability has important applications in the future monitoring of smart structure systems.

REFERENCES

1. M. G. Xu, J.-L. Archambault, L. Reekie and J. P. Dakin, 'Simultaneous measurement of strain and temperature using fibre grating sensors', *Proc. of the 10th Conf. on Optical Fibre Sensors*, 1994, 191–4.
2. M. A. Davis and A. D. Kersey, 'All-fibre Bragg grating strain-sensor demodulation technique using a wavelength division coupler', *Electron. Lett.*, **30**(1), 1994, 75.
3. T. Liu, G. F. Fernando, Y. J. Rao, D. A. Jackson, L. Zhang and I. Bennion, 'In-situ strain measurement in composite during fatigue testing using optical fibre Bragg gratings and portable CCD detection system', *PC'96*, 'Fibre Optic Sensors V', Beijing, China, 6–7, Nov. 1996, SPIE **2895**, 249–57.
4. S. M. Melle, K. Liu and R. M. Measures, 'A passive wavelength demodulation system for guided wave Bragg grating sensors', *IEEE Photonics Technology Letters*, **4**(5), 1992, 516.
5. M. G. Xu, J.-L. Archambault, L. Reekie and J. P. Dakin, *Electron. Letts*, **29**, 1993, 398–9.
6. E. J. Friebele, C. G. Askins, M. A. Putnam, A. A. Fosha, Jr, J. Florio, Jr, R. P. Donti and R. G. Blosser, 'Distributed strain sensing with fibre Bragg grating arrays embedded in CRTM® composites', *Electron. Lett.*, **30**(21), 1994, 1783–4.
7. M. A. Davis and A. D. Kersey, 'Fibre Fourier transform spectrometer for decoding Bragg grating sensors', *Proc. of the 10th Conf. on Optical Fibre Sensors*, 1994, 167–70.
8. M. A. Davis and A. D. Kersey, 'Application of a fibre Fourier transform spectrometer to the detection of wavelength encoded signal from Bragg grating sensors', *IEEE J. of Lightwave Technology*, **13**(7), 1995, 1289.
9. I. J. Read, 'High speed multiplexed interrogation of Bragg grating sensors using a Fabry–Pérot tunable filter', to be published in *Smart Materials and Structures*, 1996.

10. Y. N. Ning, A. Meldrum, W. J. Shi, B. T. Meggitt, A. W. Palmer, K. T. V. Grattan and L. Li, 'Design and performance of a Bragg grating sensing instrument using a tunable Fabry–Pérot filter to detect wavelength variations', Institute of Physics, Topic meeting on 'Tunable optical fiber filter and their applications', 6 Nov. London, UK.

11. M. G. Xu, H. Geiger, J.-L. Archambault, L. Reekie and J. P. Dakin, 'Novel interrogating system for fibre Bragg grating sensors using an acousto-optic tunable filter', *Electronics Letters*, **29**(17), 1993, 1510–11.

12. D. A. Jackson, A. B. Lobo Ribeiro, L. Reekie and J. L. Archambault, 'Simple multiplexing scheme for a fibre-optic grating sensor network', *Opt. Lett.*, **18**(7), 1993, 1192.

13. M. A. Davis and A. D. Kersey, 'Matched-filter interrogation technique for fibre Bragg grating arrays', *Electronics Letters*, **31**(10), 1995, 822–3.

14. A. D. Kersey, T. A. Berkoff and W. W. Morey, 'High-resolution fibre-grating based strain sensor with interferometric wavelength shift detection', *Electronics Letters*, **28**(3), 1992, 236–8.

15. W. J. Shi, Y. N. Ning, K. T. V. Grattan, A. W. Palmer and S. L. Huang, 'Novel wavelength measurement scheme using a stabilized interferometer system', *PC'96, 'Fibre Optic Sensors V'*, Beijing, China, 6–7, Nov. 1996, *Proc. SPIE* **2895**, 258–66.

16. Y. N. Ning, Q. Wang, A. W. Palmer and K. T. V. Grattan, Conference on Sensors and their Applications VII, Dublin, 10–13 Sept. 1995, *Sensors and their Applications VII* (Ed: A. T. Augousti) Institute of Physics Publishing, Bristol, UK, pp. 149–54.

17. A. D. Kersey, T. A. Berkoff and W. W. Morey, 'Fibre-optic Bragg grating strain sensor with drift-compensated high-resolution interferometric wavelength-shift detection', *Optics Letters*, **18**(1), 1993, 72.

18. A. D. Kersey, M. J. Marrone, K. P. Koo and A. Dandridge, 'Optically demodulated interferometric sensor system' *Proc. of the 10th Conf. on Optical Fibre Sensors*, 1994, 343–6.

19. W. J. Shi, Y. N. Ning, K. T. V. Grattan and A. W. Palmer, 'Analysis of the measurement error in wavelength shift measurement systems for optical sensor applications', *'Applied Optics and Optoelectronics'*, Reading Univ., UK, 1996, *Conf. Proc.*, 440–4.

20. Manufacturer's Catalogue, *Stanford Research Systems, Scientific and Engineering Instruments*, 1994–1995, pp. 145–55.

21. R. E. Best, *Phase-Locked Loop Theory, Design and Applications*, New York, McGraw-Hill, 1984.

22. A. D. Kersey, T. A. Berkoff and W. W. Morey, 'Multiplexed fibre Bragg grating strain-sensor system with a fibre Fabry–Pérot wavelength filter', *Optical Letters*, **18**(16), 1993, 1370–2.

23. A. D. Kersey and W. W. Morey, 'Multi-element Bragg-grating based fibre-laser strain sensor', *Electronics Letters*, **29**(11), 1993, 964–6.

24. T. A. Berkoff and A. D. Kersey, 'Fibre Bragg grating array sensor system using a band-pass wavelength division multiplexer and interferometric detection', *IEEE Photonics Technology Letters*, **8**(11), 1996, 1522–4.

25. T. A. Berkoff and A. D. Kersey, 'Eight element time-division multiplexed fibre grating sensor array with integrated-optic wavelength discriminator', *The Second European Conf. on Smart Structures and Materials*, Glasgow, 1994, 350–3.

REFERENCES

26. W. W. Morey, J. R. Dunphy and G. Meltz, 'Multiplexing fibre Bragg grating sensors', *SPIE* **1586** *Distributed and Multiplexed Fibre Optic Sensors*, 1991, 216–24.
27. K. P. Koo and A. D. Kersey, 'Fibre laser sensor system with interferometric readout and wavelength multiplexing', *Proc. of the 10th Conf. on Optical Fibre Sensors*, 1994, 331–4.
28. A. D. Kersey and W. W. Morey, 'Multiplexed Bragg grating fibre-laser strain sensor system with mode-locked interrogation', *Electronics Letters*, **29**(1), 1993, 123–4.
29. Y. J. Rao, K. Kalli, G. Brady, D. J. Webb, A. D. Jackson, L. Zhang and I. Bennion, 'Spatially-multiplexed fibre optic Bragg grating strain and temperature sensor system based on interferometric wavelength-shift detection', *Electronics Letters*, **31**(12), 1995, 1009–10.
30. M. A. David, D. G. Bellemore, M. A. Putnam and A. D. Kersey, 'A 60 fibre grating sensor system', *Proc. of the 11th Conf. on Optical Fibre Sensors*, 1996, 100–2.

Index

Page numbers appearing in **bold** refer to figures and page numbers appearing in *italics* refer to tables.

Absolute measurement
  Fabry–Pérot sensors *356*
  FBG sensors *356*
  polarimetric sensors *356*
  two-mode sensors *356*
Absorption
  bands
    edge **4**
    erbium 50
    oxygen-deficiency 333
    ytterbium ion 60
    peak 332
    saturable 69–70, *71*
Accelerometers 237
Accuracy, scale factor 310
Acoustic noise 248
Acousto-optic tunable filter (AOTF), detection 395–7, **396**
Active homodyne 243
Advanced fiber-reinforced composite (AFRC) 381
Aerospace 381
Airy function 174
Alignment 263, 264, 283
  drifts 329
Amplification, regenerative 77
Amplified spontaneous emission, *see* ASE
Amplifiers 99, 108
  lock-in 405
    double-phase 407
    in FBGs 405
  Raman 99
  rare-earth-doped 99
  telecommunications 99
Amplitudes
  input field 169
  modulation 178
  supermode fields 169
Anemometer
  beam **281**
  Doppler 282
    systems 270
  flow direction 269–70
  frequency-division-multiplexed (FDM) 288
  laser Doppler 2
  laser-dual focus (L2F) 265
  reference beam 267
Anemometry 238
  laser transit 277–9, **277**, **278**, **279**
  3D 279, **279**
Angular momentum 40
  see also LSJ levels
Applications
  engineering 263
  medical 371
  turbomachinery 288–94, **289**, **290**, **291**, **292**, **293**
Argon ion laser 47, 51, 58, 157
Argon lasers, frequency doubled 346
ASE 100, 181
  Er-doped 100
Asymmetry 335
Attenuation coefficient 139

Backscattering 277–8, 309
Bandwidth, limited 212
Beams
  blue 289
  diameter 266
  ratios 264
  splitters 306, 402
    polarization 246, 252

INDEX

Beat
  frequency 106, 141, 263
  length 273
    polarization 143, 342
  signal 104
    reducing 273
Bessel functions 227, **227**, 242–3
Bidirectional operation 107
Birefringence 47, 67
  control 252
  dispersion, absence of 343
  disturbance location **143**
  eigenstated 124
  high 333
    linear 124
  induced 247
  intrinsic 342
  magnitude 143
  minimizing 105
  and optical path difference 105
  and phase shift 251
  photoinduced 342
  time domain techniques 142–4, **143**
Blackbody sensor **5**
Bleaching 335
Blue lasers 56
Bragg
  cell 243, 248, 270–1, **270**, **271**, 294
    carrier frequency 253
    diffraction efficiency 294
    Doppler systems 270
    frequency shifts 270, 283–5, **284**, **285**
    modulation 21
    reference beam 283
    white light interferometry 254
  gratings 19, 21, 22, 101–3, **102**, 148, **148**, 160–1, 268, 336, 337–40, **338**, 351–2
    blazed 337
    equation 382
    holographic 382
    matched fiber pairs 397–8, **397**, **398**
    photoinduced 381
    TDM 139
  reflection grating 331
  reflectors 46–7, 51, 54, 79
  resonance, splitting **362**
  sensors 382–3

wavelength 345, 351, 357
  reflected 382
  reflectivity 358
  shift, phase change 400
Bragg's law 357
Broad band
  illumination 160
  interferogram 160
  light sources
    mechanical instability 13
    physical bulk 13
Broadening, homogeneous 40
Bulk diffraction grating 54
Bulk index, cladding 124

Cameras 215
  technology 215
Carrier frequency 132, 178, 243
  Bragg cell 253
  mismatch 405
Cavity
  coupled 75
  finesse 393
  lengths 103
  phase output 369
  separation of filter 393
  spacing 392
  standing waves 65, **65**
CDM 145, 152
  broad-band 154
Channel spectra processing 187–8
Charge injection device, *see* CID
Chemical sensing 372
Chirped pulse regenerative amplification 77
Chirping, gratings 161
Chirps 67–8, 141
  Bragg grating 68
  Bragg reflectors 77
  periodic frequency 142
CID, speckle applications 215
Cladding, bulk index 124
Classification
  Cartesian 6–8, **8**
  features 1
Closed loop
  homodyne 243
  signal processing 311–14, **312**
Codopants 60, 64

INDEX   421

Coefficients, index-pressure 361
Coherence 4, 184
  division multiplexing, *see* CDM
  lengths 185, 129, 146, 280
    and spectrum **130**
  modulation
    mono-mode fiber *24*
    multi-mode fiber *24*
    polarization-maintaining fiber *24*
    special fiber *24*
    two-mode fiber *24*
  spatial 10
  temporal 10
  time 181, 144
    source 171
  tracking 185, **186**
Coil, packaging 320
Collection apertures 264
Compact laser systems 39
Compaction, photosensitivity 331
Components, and modulators 131–2
Compound arrays 138
Continuous
  filters 52
  measurement 144–5, **146**
  waves 46
    frequency-modulated 238
    power 268
Contour
  intervals **217**, 219
  sensitivity 219
  two-wavelength 226
Contraction, lanthanide 40
Conversion
  efficiency 54
  polarization 341
Core
  damage 333
  diameter 16
  germanium in 268
  hydrogen in 334
  radius 14
  refractive index 124, 358
Coupled cavity 75
Couplers 241, 248–50, **248**, **249**
  3X3 325, **326**
  directional 211, 290–1, 168–70, **168**, 307
  fixed ratio 137

  highly birefringent 249–50
  intermodal 336, 343–5, 344, 336
  photoinduced 344
  polarization 333, 341–3, 351
  polarization-splitting 283
Coupling
  angle 169
  coefficient 133, 170, 344
    phase matched 133
  efficiency 271, 342
  low-efficiency 341
  ratios 133
  resonant 272
  strength 349–50
Creep 350
Cross-coupling, polarization 317
Cross-sensitivity, temperature 159
Crosstalk 142, 152, 153, 285
  polarization 158
Cure monitoring 370

D-shaped, specialized fibers 17, **18**
Damage
  core 333
  photoinduced 333
Decay, nonradiative 41
Decision evaluation 26–7
Decoding 117, 148
Decorrelation, speckle 219
Defects, UV-absorbing 331
Definitions
  intensity 120
  interferometry 122
  measurand 117
  multiplexed system 117–18
  optical fiber sensor 120
  photosensitivity 331
  soliton 67
Deformation 222–5, **224**, **225**
  dynamic measurements 225–32
Delay
  differential 170
  line 291
  phase 173
  propagation 184
  single 170
  time, loop 313–14, 318
Demodulation
  electronics 321

Demodulation (*contd*)
  interferometric 102, **102**
    FGB **102**
  pseudo-heterodyne **249**
Demultiplexing 409
  Fabry-Pérot tunable filter 409–10, **410**
Densification 333
Depolarization 286, 307
Depolarizer
  Lyot 322, **322**
  second 324, **324**
Detected
  light intensity 307, 308
  power 133, 134
  signals 256, **256**
Detection
  acousto-optic tunable filter (AOTF) 395–7, **396**
  active 389–400, **389, 391, 392, 393, 394, 395, 396, 397, 398,** *399*
  CCD array based 387–8, **387**
  Fabry-Pérot 391–5, **392, 393, 394, 395**
  Fourier transform spectrometer 390–1, **391**
  interferometric phase 398–400, *399*
  passive 385–8, **386, 387, 388**
  wavelength 391
    dependent 385–6, **386**, 387
Detectors, linear array 156
Detuning 339
Device cost
  ELEDs *358*
  SFSs *358*
  SLDs *358*
  TFLs *358*
DGV 295–6, **296**
Dielectric mirrors 46
Difference channel, Doppler 286
Difference intensity, normalized 326
Difference systems, Doppler 283–5, **284, 285**
Diffraction
  efficiency, Bragg cell 294
  grating 270
Digital phase ramp **312**, 313
Diodes
  continuous wave 214
  driving current 402

light-emitting 147–8
long wavelength 179
multimode laser 181–2
power emitted 182
superluminescent 148, 181, **190**
Dipole, photosensitivity 331, 333
Direct sensors 2
Directional coupler 168–70, **168**
  variable split ratio 211
Discrimination
  directional 269–76, **270, 271, 272,** *273,* **274, 275, 276**
  polarization 277, 279, 286
  wavelength 277, 279, 387
Dispersion 155, **155**
  atomic 67
  measuring 160
Dispersive optics 157
Displacement
  dynamic 225
  periodic 199
Distal face 124
Distortion, second harmonic 318–19
Distributed
  feedback (DFB) 64
  sensor **5**, 119, **119**
    mode coupling **5**
    quasi **5**
    Raman **5**
    Rayleigh **5**
Distribution, spectral 390–1
Dither 107–8, 150
  field 193
  low frequency 247
Dopants, germanium 268, 331, 334
Doped fiber, manufacture 41
Doping 66
  germanium 268, 331, 334
Doppler
  anemometer 282
  difference 157, **157**
    channel 286
    LDV 263–5
    measurement region 264
    signals 294
    systems 283–5, **284, 285**
    technique 262
    volume measurement **264**
  effect 126

INDEX 423

frequencies, measuring low 290
frequency shift 244, 262–2, **262**, 295–6
  velocity inferred 265
shift, modulation 21
spectrum 267
systems, Bragg cell 270
velocimeters
  multi-component fiber 285–9
  three-dimensional 288–94, **289, 290, 291, 292, 293**
  two-dimensional 286–8, **286, 287**
velocimetry 199, 279–85, **280, 281, 282, 284**
  fiber optic laser 279–85, **280, 281, 282, 284**
Double cladding 47, **48**
Down-lead, insensitivity 280
Drift
  gyro 326
  stability 326
Drive amplitude 309
Driving current, diodes 402
Dual cores 272
Dual ramp **312**, 313
Dynamic range 386

Edge filter, interrogation 367–70
Edge-emitting light-emitting diodes, *see* ELEDs
Effective pulse duration 144
Efficiency
  coupling 342
  reflection 337–9
Eigen polarization 111
Electro-optic holography (EOH) 226
Electromagnetic interference 381
Electronic pathways 43, 54
Electronic speckle pattern interferometry, *see* ESPI
Electronic tuning 53
Electronics, demodulation 321
Electrostriction 270
ELEDs
  device cost *358*
  fiber-pigtailed 396
  FWHM bandwidth *358*
  optical power *358*
Emission wavelength 37

Encoding 20–1
  plates **4**
  polarization 249–50, **250**
  rotation rate 320
Energy
  conservation 122
  gap 41
  levels 44
    neodymium **46**
    ytterbium ion 59, **59**
Engineering 381
Erbium
  absorption bands 50
  doped fiber 37, 39, 49–54, **50, 52**
  mode locking 75–7
  tuning 51
Error
  amplitude type 314–15, **315**
  bias 319
  intensity type 314, 316
  Kintner 314
  maximum phase 315
  mechanical 320
  minimizing 316
  modulation 317–20, **319**
  polarization 314–17, **314, 315, 317**
  sources 314–22, **314, 315, 317, 319, 320**
  strain 363
  thermal gradient 320
  wavelength 405–9, **406, 407**
ESA 48
  fluorozirconate 48–9
ESPI 207–14, 215
  endoscopic **214**
  in-plane 209
  out-of-plane 209
  static displacement **224**
Evanescence **4**
Excess loss, FBGs *366*
Excited state absorption, *see* ESA
Exposure, transverse 331
External
  irradiation 335
  writing 345
Extinction ratio 18
Extrinsic 2, **3**
  devices 272–3

Fabrication
 FBGs 355, *371*
 of gratings 345–8
  phase mask method 347–8, **347**
  point-by-point method 348, **348**
Fabry-Pérot 52, 80, 104, 105, 181
 cavity 181, 189
 detection 391–5, **392, 393, 394, 395**
 filters 43
 high finesse 199
 interferometers **123**, 124, 140, 174–5
  air spaced 174
  all fibre 174
  transfer function 174, 175
 sensors
  absolute measurement *356*
  interferometric **6**
  linear response *356*
  mass production *356*
  mechanical strength *356*
  multiplexing *356*
  potential cost *356*
  range of resolution *356*
  sensor gauge length *356*
 transmission spectrum 392
 tunable filter, demultiplexing 409–10, **410**
Failure time, FBGs *366*
Faraday
 effect 129, 194–6, **195**
 rotation 252
  mirrors 105–6, **106**, 172–4, **173**, 391
 sensors 2
Fast Fourier transform (FFT) 200, 291
FBGs 62, 99, 132
 see also Bragg gratings
 applications 356
 chirped 363
 dual wavelength 363
 dual-diameter method 365
 excess loss *366*
 fabrication 355, 366–7, *371*
 failure time *366*
 flexibility 366
 harmonics method 363–4
 and in-line fiber etalon 365
 interrogation techniques *372*
 lengths of 357
 linearly chirped 359, 360

and long-period grating 364–5
magnetic field 361–2
mass-production 356, 366
multiplexing 356, 409–14, **410, 411, 412, 413, 414**
 compatability 367
 techniques *372*
operational range 369
optical
 qualities 366
 theory 357–8, **357**
physical qualities 366
pressure 361
reference method 362–3
reflection wavelength 101
reflectivity *366*
repeatability 366
response time *366*
sensors 355
 absolute measurement *356*
 coherence length 391
 Fourier transform spectrometer **391**
 linear response *356*
 mass production *356*
 mechanical strength *356*
 multiplexing *356*
 potential cost *356*
 range of resolution *356*
 sensor gauge length *356*
spectral linewidth *366*
strain 359–60, *360*
 and temperature measurement 362
superimposed 364
temperature 360, *360*
thermal stability *366*
ultimate strength *366*
wavelength range 103
wavelength shift 368
 measurement 367
FDM 158
Feedforward, gratings 341–5
Fiber
 amplifers, Er-doped 412
 Bragg filters 51–2
 Bragg gratings, *see* FBGs
 coil 307
 core, intensity in 39
 delay 170
 gyroscopes 306–9, **306, 309**

3X3 325–6, **325**, **326**
　depolarized 322–5, **322**, **323**, **324**
　illumination 268
　loss 199
　optic interferometric sensor, *see* FOIS
　polarization-maintaining 277, 307, 341
　polishing ends 246
　processing history 334
　ring laser, Er-doped 108
　silica 359
　strength, loss of 350
　tap 337
Fibers
　addition of 245
　laser gyroscopes, *see* FLaGs
　multimode 9, 277
　numerical aperture 280
　path replacement 245–6
　polarization-maintaining 9, 15, **16**
　　circular 15, **16**
　　linear 15, **16**
　single-mode 9
　specialized 9
　two-mode 9
Filtering 246
　Fourier 219
　harmonics 249
Filters
　infrared high-pass 385
　notch 397
Finite element 210
Fizeau
　cavity 200
　interferometers 124
　reflective **153**
　sensors 152, **153**
　serially addressed 152
Fizeau/Fabry-Pérot 189, 191
FLaGs 106–12, **107**, **108**, **109**, **110**
　F-P cavity 111, **111**
　mode-locked 108, **108**
　see also ML-FLaG
Flow
　direction 265
　　anemometer 269–70
　low speed 294
　measurement, 3D 285
Fluorescence **4**
　time decay 22

Fluoride fibers, phonon energy 53
Fluorozirconate 37, 41
　erbuim doped 54
　ESA in 48–9
　thumium-doped 56
FOIS 178
　applications 188–202
　electrically passive 178
　high resolution 178
Fotonic sensor 2, 20
Fourier
　filtering 219
　transform spectrometer 148
　　detection 390–1, **391**
Fourier transform spectrometer, FBG
　　sensor **391**
Frames 26
Free spectral range (FSR) 104, 127, 199
Free-space, interferometers 346
Frequency
　division 142
　　multiplexing 150, 412–13, **413**
　　anemometer 288
　proper 309–10
　shifts 271, 272–3, **272**, 311
　　Bragg cell 270, 283–5, **284**, **285**
　　Doppler 244, 262–2, **262**
　　measuring 262
　　optical fiber 271–4, **272**
　　SBS 274
　　serrodyne **312**, 313
　spacing 60
　tracking 249
Frequency-modulated continuous wave
　　21, 132
Fresnel drag 305
Fringe 291
　central, identifying 187
　contour 216–17, **217**, **218**, 219
　gradient-of-displacement **224**
　interference 157, 241
　intervals 216
　pattern 256
　projection 215
　slope **221**
　speckle 208
　of superposition 23
　visibility 131
　Young's 157

Fringe (contd)
  zero-order 230
FWHM bandwidth
  ELEDs *358*
  SFSs *358*
  SLDs *358*
  TFLs *358*

Gain bandwidth 61
Gain competition 108
Gas turbines 279
Gaussian
  beams 264
  mode 47
  pump beam 38
  source profile 200
Germanium, oxygen-deficiency 333
Glass, germanosilicate 331
Gratings **4**
  Bragg exchange 341
  classification 336–7
  fabrication of 345–8
  feedforward 341–5
  long period 133, 337, 344–5, **344**
  nonuniform 340
  period 132–3
  photogenerated 336, 349, 355
  photoinduced 336
  pitch extension 382
  pitch of 341
  plane 387–8, **387**
  polarization coupler 336–7, **338**
    reflective 337
  producing, internal method 331
  reflectance 349
  reflection 336
    spectra **339**, 340, **340**
  reflective 337–41
  reflector 330, **330**
  rocking filter 336–7, **338**
  sampled 340
  stability 349–50
  superstructure 340
  types 336–45
  very long 347
  wavelength selective 336, 339
Green channels 289
GRIN lenses 285, 292
Group delay, measuring 160

Group velocity, dispersion 66
Guided beam 210
Gyroscopes 2
  closed loop 100
  coil, quadrupole wound 320, **320**
  detected intensity 311–13
  drift 326
  fiber 306–9, **306**, **309**
  fiber laser
    see FLaGs
    interferometric **112**
  fiber optic 99
  open loop 100
  reentrant fiber 99
  ring laser *see* RLG
Sagnac 176–7, 250

Harmonics
  filtering 249
  modulation 131
Heterodyne **228**, 229–31, 243–4, 265
  pseudo 244–5
  recovery 127–8
  signal processing 195, **195**
  synthetic 244
High resolution, OCDR 146
Hollow section, specialized fibers 17–18
Holmium-doped fibers 56–**7**
Holographic exposure 345–7, **346**
Homodyne
  active 243
  closed loop 243
  passive 242–3
  recovery 127–8
Hysteresis 350
  modulator 319

IFOG 100, **100**
  using ASE source **101**
Illumination 8–9
  broad band 9, 160
  direction 263
  N-mode 9
  narrow band 9
  pulsed laser 232
  single-mode 8
  stroboscopic **228**, 231–2, **232**
  two-mode 9
  two-wavelength 216

INDEX

In-fiber Bragg gratings, *see* FBGs
In-line fiber etalon, and FBGs 365
Index
  difference 337
  matching 70, 246
  pressure, coefficients 361
Infrared high-pass, filters 385
Insensitivity, down-lead 280
Insertion loss 74, 136
Integrated interference 64
Intensity 238, 239, 296, **296**
  definition 120
  modulation 120–1
    mono-mode fiber *24*
    multi-mode fiber *24*
    polarization-maintaining fiber *24*
    special fiber *24*
    two-mode fiber *24*
  multi-mode fiber **7**
  and phase difference 239
  and power 124
Interference
  destructive 357
  effects 184
  filters 52
  fringes 241
  speckle field 208
  two-beam 308
  unmodulated 315
Interferograms 217
  broad band 160
  dispersed, centroid 155
  parasitic 152–3
Interferometers
  broad band 129
  bulk optic 156
  definition 122
  delay in 177
  differential 176, **176**
  differentiating 198–9
  Fabry-Pérot **123**, 124, 140
  fiber 155
  Fizeau 124, 197, **197**
  Fizeau-type 160
  free-space 346
  guided wave 329
  history of 237
  hollow-core fiber 161
  holographic 215
  low-coherence 129–31, **130**, **131**, 191
  Mach-Zehnder 80, 122, **123**, 124–5, 137, 147, 170–1, **170**
    unbalanced 403
  measurement 254
  Michelson **123**, 124–5, 147, 171–2, **171**, 191, **401**, 402, 413–14
  multiple beam 329–30
  multiple wavelength 129–31, **130**, **131**
  multiplexing 142, **142**
  optical path difference 125
  optical phase information 104
  output 141, 242
  output spectrum 127, **127**
  passive optical 199
  path length imbalanced 275–6
  path-matched differential 23
  pigtailed 391–5
  polarimetric 176, **176**
  processing 188
  receiving 155–6, 280
  recovery 254–5
  reference phase 131
  Sagnac **123**, 125, 147, **147**, 176–8, 198, **198**, 250–3, **251**, **252**
  shearing, modified **225**
  signal phase 131
  single fiber 246–7, **247**
  solid path 346
  speckle 238
  spectral effects 126–7
  stabilizing 212
  tandem 131, 144, **145**, 183–4, 191, 151–4, **153**
  time division multiplexed 140, **140**
  transmitting 186
  two-beam 122–4, **123**, 176, 329, 181
  two-wavelength, low-coherence 185–7, **187**
  unbalanced 132, 398–9, 244
  white light 23, 180–1
Interferometric
  scanning 368–9
    dual-cavity 369–70, **370**
  sensor 3, **5**, **6**
    Fabry-Pérot **6**
    Mach-Zehnder **6**
    Michelson **6**
    mode coupling **6**

Interferometric (*contd*)
  sensor (*contd*)
    multi mode **6**
    polarization **6**
    ring resonator **6**
    Sagnac **6**
    single mode **6**
  vibrometers 245–57
Intermodal, couplers 343–5, 344
Interrogation 145, 367–70
  edge filter 367–70
  interferometric scanning 368–9
  Michelson interferometer 179–80, **179**
  techniques, FBGs *372*
  tunable filter 367–70
Intrinsic
  devices 2, **3**, 273–4
  sensors, applications **5**
Ionization, triple 49, **50**
Irradiance 120
  output 176
Irradiation
  external 335
  UV 333
Isotropic, refractive index 335

Jitter 274
Johnson, noise 311
Jones matrix 172, 176, 196

Kerr effect 19, 321
Kintner error 314
Knowledge based system 23–6, *24*, **25**
  operating principle 26
  test of 27
Kramers-Kronig 332–3
  photosensitivity 331

Ladder networks
  reflective **136**, **137**
  transmissive **136**, **137**
Ladders 136–7, **137**, 141
Lamps
  discharge 13
  low pressure mercury 13
Lanthanide, contraction 40
Lasers
  argon-ion 213, 268, 289, 294
  diodes

    high power 266
    multi-longitudinal 266
    multi-mode 402
    semiconductor, output 280
  Doppler velocimetry **4**, 192, **193**
    see also LDV
  distributed 192, **193**
  excimer 346
  frequency shift, measuring 104, **104**
  He-Ne 279
  linewidth 296
  transit anemometry, *see* LTA
  UV 133
  velocimetry 157–8, **157**, 212, **213**
    and TDM 158
Lathanide group, rare earths 39–40
LDV 261–5, **262**
  Doppler difference 263–5
  limitations 263
  receiving fibers 267
  reference beam 263–5, 282
  signal noise 263
Lead germanate 37
Lead silicate 37
LEDs
  extended 12
  pigtailed 12
Length
  interaction 269
  modulation
    mono-mode fiber *24*
    multi-mode fiber *24*
    polarization-maintaining fiber *24*
    special fiber *24*
    two-mode fiber *24*
Lenses, GRIN 285, 292
Lick-in mode 396
Light
  dragging 305
  efficiency 280
  pipe 245
  sources 308–9
    broad band 13
    coherence length 253
    intensity noise 321
    N-mode 11–12
    narrow-band 12–13
    noise properties 309
    single-mode 11

two-mode 11
  wavelength stability 309
  split 285
Linear response
  Fabry-Pérot sensors *356*
  FBG sensors *356*
  polarimetric sensors *356*
  two-mode sensors *356*
Linear retarder 132
Linearities, scale factor 311
Liquid core, specialized fibers 19
Lithium niobate 22
Loading 222
Location-multiplexed systems 118–19
Lock-in 107
Long sensors 156
Long-period grating
  see also LPG
  bandwidth of 365
  and FBGs 364–5
  length of 365
  periodicity 364
  strain field 365
Long-term drift 109, 111
Longitudinal mode hops 64
  spatial hole burning 61
Loops
  arbitrary shaped 305
  circular 305
  delay time 313–14, 318
  length 313
  phase-locked 243
  splitter 306
LPG 161–2
  designing 162
LSJ levels 40
  coupling 40
LTA 265–6, **267**
  receiving fibers 267
  variable beam separation 278
LVD
  3D 286
    time-division-multiplexed (TDM) 289–90
  instrumentation 285
Lyot depolarizer 322, **322**

Mach-Zehnder 122, **123**, 126

interferometers 80, **123**, 124–5, 137, 147, 170–1, **170**
interferometric sensor **6**
  modified 248, **248**
  phase sensitivity 124
  receiving 283
  sensors 152, **153**
  serially addressed 152
  tandem 183–4, **184**
  transmissive **153**
Macrobend tap 307
Magnetic
  field
    circular birefringence **362**
    detecting 80, 129
    FBGs 361–2
    longitudinal, FBGs 361
    measuring 106
    sensitivities 323
  probe sensors 194–6, **195**
Magnetometers 192–6
  axis 194
  gradient 194
  interferometric magnetostrictive 150
Magnetometry 158
Magnetostrictive transducer 193
Mass production
  Fabry-Pérot sensors *356*
  FBG sensors *356*
  polarimetric sensors *356*
  two-mode sensors *356*
Matched fiber pairs, Bragg grating 397–8, **397**, **398**
Measurands
  definition 117
  multiple 156–62
  multiplexed 118–19, **118**
  quasi-static 188–91, **189**, **190**
  reciprocal 125
  sensitivity 128–9, 329
Measurement
  blood flow 279
  conventional **7**
  dispersion 160
  distributed 4–5
  fiber intensity **7**
  flow, 3D 285
  frequency shift 262
  group delay 160

Measurement (*contd*)
  interferometer 254
  optically powered 7
  planar velocity 294–6, **295**, **296**
  single point 4
  slope 215–22, **217**, **218**, **219**, **220**, **221**, **222**
  static 222–5, **224**, **225**
  strain 125
    and temperature 158–62
  velocity 126, **126**
  vibration 226
Mechanical
  error 320
  reliability 350
  resonance 121–2
  strength
    Fabry-Pérot sensors *356*
    FBG sensors *356*
    polarimetric sensors *356*
    two-mode sensors *356*
Metal-glass, specialized fibers 18
Methane 53–4
Michelson 154–5
  interferometers **123**, 124–5, 147, 191, 390, **401**, 402
    interrogation 179–80, **179**
    unbalanced 179–80, **179**
  interferometric sensor **6**
  output recovery 255
  phase sensitivity 124
  two-beam 238–9, **238**, **239**, **240**
Microbend sensor **5**
Microresonators, silicon 122
Minimum reciprocal configuration 306–7, 322
ML-FLaG **110**
  polarization 109
  reciprocity 109
  short-term noise **110**
Mode
  control 111
  coupling
    distributed sensor **5**
    interferometric sensor **6**
  locking 66, 69, **73**
    additive pulse 75
    erbium 75–7
    erbium-doped 71, 74
  FBG sensors 103, **103**
  neodymium 70, 71
  neodymium-doped 74
  passive 70, 75
  praseodymium 75
  pulse energies 77
  telecommunications 74
  ytterbium 75
Modes, number of 16
Modulation
  amplitude 178
  Bragg cell 21
  categories 9–10
  coherence 17
    length 23
  cross-phase 67
  depth 309, 321, 340
  Doppler shift 21
  error 317–20, **319**
  frequency 121–2, 229
  harmonic 131
    frequency 141
  induced phase 177
  intensity 20, 120–1
  laser diode wavelength 216
  loss 317
  optical-path-difference 21
  parasitic amplitude 318
  phase 10, 311, 318
    difference 311
  polarization 20
  pure amplitude 109
  rate 22
  reference beam phase 229
  refractive index 22
  scheme 1
  self-phase 67
  serrodyne 131, 141
  signal 311
  sinusoidal 249, 308
  source 230
  techniques 19–23, 26
  time 22–3
  waveform 319
  wavelength 20, 121, 132
Modulators **42**, 43
  acoustic, *see* Bragg cell
  acoustic-optic 53
  amplitude 68, **69**, 70, 74

bulk phase 71
and components 131–2
cross-phase 70
drive frequencies 413
hysteresis 319
mode-lock (MLM) 412
optic phase 179–80, **179**
phase 68, **69**, 307–8, 61
piezoelectric 131, 274
self-phase 70
switch and phase 180
transfer function 317
Moiré 215
Mono-mode fiber 14–15
  coherence modulation *24*
  intensity modulation *24*
  length modulation *24*
  polarization modulation *24*
  refractive index modulation *24*
  time modulation *24*
  wavelength 15, *24*
Monochromator 148
Multi-mode fiber 3–4, 16–17
  choice of 267
  coherence modulation *24*
  intensity **7**
    modulation *24*
  length modulation *24*
  polarization modulation *24*
  refractive index modulation *24*
  time **7**
    modulation *24*
  wavelength **7**, *24*
Multiple wavelength
  illumination 129
  interferometry 129–31, **130**, **131**
Multiplexed
  fiber sensor 117, **117**
  sensors 119, **119**
  system, definition 117–18
Multiplexing
  cost-effective 371
  Fabry-Pérot sensors *356*
  FBG sensors *356*
  polarimetric sensors *356*
  sensors 151–4, **153**
  techniques, FBGs *372*
  topologies 188, 190
  two-mode sensors *356*

wavelength division 397
Mutual spatial coherence 124

Nd:YAG
  MISER 274
  ring laser 288
Neodymium
  absorption bands 45
  bidirectional 47
  doped fibers 45–9, **46**, 78, 79
  energy level **46**
  in fluorozirconate fibers 48
  laser oscillation 45
  optical path length 45
  output powers 48
  in silica fibers 48
  single frequency 62
  slope efficiency 47
  transitions, visible 49
  tuning 46, 47
Nested arrays 138, **138**
Noise
  acoustic 248
  environmental 193
  Johnson 311
  levels 257
  pseudo-heterodyne 245
  random 310, 320–2
  sensitivity 250
Notch filters 397
Numerical aperture 14
  fibers 280
Nyquist limit 290

OCDR 144–6, **145**, **146**
  high resolution 146
  spatial resolution 145
  strain-sensing 145–6
  tandem interferometers **145**
  transmissive **146**
OFDR 140–2, **141**, 143
OPD
  required *399*
  scan 155–6
  stability, phase change 400
Open loop 308
  signal processing 310–11
Optic phase modulator 179–80, **179**
Optical

432 INDEX

Optical (contd)
  access, limited 292
  attenuation 181
  circulator 136
  coherence domain reflectometry, see
    OCDR
  excitation 51
  fiber geometry 38–9
  fiber implementations 210–13
  fiber loop, Sagnac effect 303–6, **304**
  fiber ring, resonators 273
  fibers
    disadvantages 212
    silica 350
  frequency domain reflectometry, see
    OFDR
  imbalance 402
  Kerr effect 67
    nonlinear 108
  loops, closed **304**
  modulation 6–8, **8**, 9–10
  nonlinearities 37
  path difference 241, 398
    see also (OPD)
    and birefringence 105
    interferometers 125
    modulation 21
  path imbalance 178
  path lengths 66
    neodymium 45
  phase difference 123
  power
    ELEDs *358*
    SFSs *358*
    SLDs *358*
    TFLs *358*
  resolution, highest 199
  spectral analysis 389–90
  spectrum analyzer 199–200
  time domain reflectometry, see OTDR
Orbital angular momentum 40
  see also LSJ levels
Oscillation, steady-state 66
Oscillator
  response 332
  strength 40
OTDR 4–5, 138–40, **139**, 180
  delay resolution 143
  spatial resolution 144

  telecommunications 138
Ouput, powers 51, 56
Output
  maximum 127
  nonreciprocal 123
  power, low 63
  ratios of 385
  reciprocal 123
  signal 405
Oxygen-deficiency
  absorption bands 333
  germanium 333

Parallel
  networks 133–5, **134**, **135**
    reflective 134, **134**, **135**
    single detector 134, **134**, **135**
    transmissive 134, **134**, **135**
  topologies 188, 190
Parasitic, polarization 318
Particle image velocimetry (PIV) 294
Passive
  control 124
  homodyne 242–3
  signal processing 255, **255**
Path
  difference 240
  imbalance 130, 132, 141, 154–5, 244
  length 130, 251
    imbalance 142, 229, 231, 280
    scanning 131–2
Path-matched differential interferometry
  23
Peak power 78
Peak reflection, wavelength 330
Perturbations
  environmental 248, 283
  measuring 106
Phase
  bias 242, 307–8
  change
    Bragg wavelenth shift 400
    OPD stability 400
  delay 173
  difference 111, 329
    and intensity 239
    modulation 311
  drift 403
  generated carrier technique (PGC) 141

maps 226
  unwrapped 231
  mask method 347–8, **347**
    advantages 347
  modulation 311
  modulator 307–8
  perturbation 177
  recovery 127
  resolution *400*
  sensitivity 369
  shift 224, 251, *400*
    and birefringence 251
    Sagnac 111
  stepping 231
    recovery 127–8
  tracker 243
Phase-drift compensation 400–5, **401**, **403**, **404**
Phase-generated carrier (PGC) 178
Phonon energy
  fluoride fibers 53
  silica fibers 53
Photo-detection, transimpedance **310**
Photodetectors, identical 385
Photoelasticity **4**
  coefficient 359
Photoinduced
  birefringence 342
  Bragg gratings 381
  couplers 344
Photorefractive intermodal exchanger (PRIME) 336, 343–5
Photosensitivity 330, 331–6
  compaction 331
  definition 331
  dipole 331, 333
  enhancing 334
  Kramers-Kronig 331
Piezoelectric
  modulator (PZM) 274
  stretcher 397
  transducers 187, 196–7
  see also PZT
Pigtailed, interferometer 391–5
Pigtailing 214, 271, **271**, 293, **293**
Pitch-pressure 361
Planar velocity, measurement 294–6, **295**, **296**
Pockels cell 294

Pockels effect 333
Point-by-point 348, **348**
Polarimetric
  fiber 103–6, **105**
  interferometers 176, **176**
  sensors
    absolute measurement *356*
    linear response *356*
    mass production *356*
    mechanical strength *356*
    multiplexing *356*
    potential cost *356*
    range of resolution *356*
    sensor gauge length *356*
Polarization 14, 238, 246
  -maintaining fiber *24*, 307, 341
  aligning axes 80
  beam splitter 246
  beatlength 143, 342
  circular 106
  controllers 198
  conversion 341
  couplers 333, 341–3, 351
  cross-coupling 317
  discrimination 286
  eigen 111
  encoding 249–50, **250**
  extrinsic devices 272
  fading 391
  interferometric sensor **6**
  output 288
  parasitic 318
  photoinduced 335–6
  rotation 129
  single-mode fiber 291
  state of 210
  state switch 288
  states of (SOPs) 105
  unwanted 247
Polarization modulation
  mono-mode fiber *24*
  multi-mode fiber *24*
  polarization-maintaining fiber *24*
  special fiber *24*
  two-mode fiber *24*
Polarization-maintaining fiber 307, 341
  coherence modulation *24*
  D-type 364
  intensity modulation *24*

Polarization-maintaining fiber (*contd*)
  length modulation 24
  polarization modulation 24
  refractive index modulation 24
  time modulation 24
  wavelength 24
Polarization-splitting coupler (PSC) 283
Polarizers **42**, 43
  perfect 314
Polarizing, beam splitter 250, 252
Polished coupler 47
Population inversion 53
Population pathways **44**, **45**
Potential cost
  Fabry-Pérot sensors *356*
  FBG sensors *356*
  polarimetric sensors *356*
  two-mode sensors *356*
Power
  and intensity 124
  levels 294
  transmission 294
Power budget 134
  single detector 136
Power equalization 137, **137**
Power handling 268–9
Power transfer 143
Power transmission, limitation 268, 296
Praseodymium
  communications industry 59
  doped fibers 58–9, **58**
  fibers 71
  ion oscillation 59
  mode locking 75
  silica-based 59
Predicate calculus 26
Pressure, FBGs 361
Prism, dispersing 278
Probes 292
  head 292, **293**
Processed, signals 256, **256**
Processing
  channel spectra 187–8, 200–2, **201**
  interferometers 188
  pseudo-heterodyne 274–6, **275**, **276**
  signal 241–5, 309–14, **310**, **312**, 383–5
Production rules 26
Proper frequency 309–10
Proximal face 124

Pseudo-heterodyne 244–5, 249
  demodulation **249**
  limitations 245
  noise 245
Pulse energies, mode locking 77
Pulse train 68, 72
  dispersion 69
  wavelength 68
Pulses
  mode-locked 109, **109**
  ultrashort 74
Pulsing, sequential 290
Pump
  absorption 63
  beams 107
  light 47
  photons 55
  radiation 38
  wavelength 51
Pyrometers **4**
PZT 311
  fiber wrapped 211–12, 229

Q-switched fiber lasers 77–**78**
  fluorozirconate-doped 78
  silica-based 78
Q-switches **42**, 43
Quartz resonator 22
Quasi-distributed sensor 5, **5**

Raman
  amplifiers 99
  backscattering 180
  distributed sensor **5**
  scattering 268
    threshold 268
Ramp deviation 407
Ramp period 141
Ramping amplitude **406**
Random
  fluctuations 178
  noise 310, 320–2
Range of resolution
  Fabry-Pérot sensors *356*
  FBG sensors *356*
  polarimetric sensors *356*
  two-mode sensors *356*
Rare earths
  lathanide group 39–40

spectroscopy 39–41
trivalent ionization 40
Rare-earth-doped
 amplifiers 99
 specialized fibers 18–19
Rate modulation 22
Rayleigh
 distributed sensor **5**
 scattering 138–9, 333
Receiving
 fibers
  LDV 267
  LTA 267
 interferometers 155–6
Recent work 2
Reciprocity 306–7
Recovering signals 118
Recovery
 heterodyne 127–8
 homodyne 127–8
 interferometer 254–5
 phase-stepped 127–8
 signal 178
 strain and temperature 119
Reentrant fiber gyroscope 99
Reference beam, derivation 281
Reference method, FBGs 362–3
Reference phase, interferometers 131
Reflectance, gratings 349
Reflected, Bragg wavelength 382
Reflection
 efficiency 337–9
 gratings 336
 magnitude 337
 response 330
 and transmission **4**
Reflectivity
 biological tissues 192, **193**
 FBGs *366*
 maximum 334
Reflectometry
 low coherence **101**
 optical coherence domain 191–2, **192**
Reflectors
 grating 330, **330**
  reflection in 330
 loop 410
Refractive index 66
 changes in 333, 351

core 124
distribution 14
eigenmodes 143
of fiber 313
isotropic 335
modulation
 mono-mode fiber *24*
 multi-mode fiber *24*
 polarization-maintaining fiber *24*
 special fiber *24*
 two-mode fiber *24*
periodicity 270
raising 268
UV exposure 334
Relative intensity noise (RIN) 321
Reliability, mechanical 350
Resonance 332
Resonators 41–3, **42**
 finesse 199
 optical fiber ring 273
 ring, LVD applications 273
Response
 amplitude 332, **332**
 oscillator 332
 phase 332, **332**
 time, FBGs *366*
Ring laser gyroscopes, *see* RLG
Ring lasers, laser diode pumped 180
Ring resonator 64, 175, **175**, 199–200
 dynamic response **200**
 interferometric sensor **6**
 reflective 177, **177**
 resolution **200**
RLG 106–8, 111
 Brillouin fiber optic 108
 cavity, backscattering 107
Rocking angle 342
Rocking filter 341, **343**, 364
 bandwidth 364
 length of 364
Rod lens, graded-index 280
Rotation rate, encoding 320
Rotators, polarization state 194–6, **195**

Sagnac
 effect, optical fiber loop 303–6, **304**
 gyroscope 125, 250
 interferometers **123**, 125, 147, **147**,
  176–8, 198, **198**, 250–3, **251**, **252**

Sagnac (*contd*)
  interferometers (*contd*)
    modified **252**
  interferometric sensor **6**
  loops 252
    mirrors 47
  phase shift 111
  ring 198
Sampled, gratings 340
Sapphire, sensing element 191
Scanners, solid-state 156
Scanning time 192, **193**
Scattered light, collecting 291
Selection, temperature sensor 27–9
Semantic networks 26
Semiconductor diodes 47
Sensing
  chemical 372
  element 121
  principles 359–66, *360*, **360**, **362**
Sensitivity
  contour 219
  magnetic field 323
  measurand 128–9, 329
  strain 386
Sensors
  bend-loss 120
  crosstalk 142
  design 189–90, **189**
  detected power 133
  distributed 119, **119**
  dual parameter 159, 160
  Fizeau 152, **153**
  gauge length
    Fabry-Pérot sensors *356*
    FBG sensors *356*
    polarimetric sensors *356*
    two-mode *356*
  high pressure 189
  in-fiber Bragg grating (FBG) 355
  in-fiber grating 132–3
  interchangeable 189
  Mach-Zehnder 152, **153**
  microbending 120
  microresonator 121–2
  multiplexed 119, **119**
  multiplexing 151–4, **153**
  multipoint 79
  networks 188–91, **189**, **190**

optical fiber, definition 120
polarimetric 80, 128–9
reflective 133
strain 79
subcarrier-modulated 121–2
totally passive 4
and transducer 2
transduction 121
Sequential subtraction 225
Serial topologies 136, **136**
Serrodyne, modulation 131, 141
SFSs
  device cost *358*
  FWHM bandwidth *358*
  optical power *358*
Shearography 209, 211, **212**, 215, 222, **226**
  defect detection 222, **224**
  slope 226
  vibration analysis 227
Short-term noise 109, **110**, 111
  ML-FLaG **110**
Side-writing 355
Signal
  backscattered 139
  detected 256, **256**
  fading 322, 323
  modulation 311
  noise, LDV 263
  output 405
  phase, interferometers 131
  phase modulation 308
  processed 256, **256**
  processing 383–5, 109, 127–8, 154–5, 178–80, 182–4, 241–5, 309–14, **310**, **312**
    closed loop 311–14, **312**
    heterodyne 195, **195**
    open loop 310–11
    passive homodyne 255, **255**
    range of 248
  ramp modulation 275–6
  recovery 119, 178
  to noise ratio 253, 294
Silica
  based fibers 37
  fibers 359
    phonon energy 53

germanium-doped 332
optical fibers 350
Single
  detector
    parallel networks 134, **134**, 135
    power budget 136
  fibers 3–4
  frequency 60–5, **63**, **65**
    neodymium 62
    radiation 62
  wavelength 56
Single-mode fiber, polarization 291
SLDs 154, **190**
  device cost *358*
  FWHM bandwidth *358*
  optical power *358*
Slope efficiency 56
  neodymium 47
Slope measurement 215–22, **217**, **218**, **219**, **220**, **221**, **222**
Solid path, interferometers 346
Soliton 72
  definition 67
Sources
  arc lamp 221
  coherence time 171
  error 314–22, **314**, **315**, **317**, **319**, **320**
  fiber laser 180
  of illumination 6–8, **8**
  laser diodes 214
  low-coherence 180–1
  single frequency 180
  superfluorescent fiber 181
  white light 220–2, **222**
Spatial
  division multiplexing (SDM) 150–1, **151**, 413–14, **414**
  hole burning 61–2, 75
    eliminating 62
    limitations 64
    longitudinal mode hops 61
  resolution 191
    OCDR 145
    OTDR 144
Special fibers 17–19
  coherence modulation *24*
  D-shaped 17, **18**
  hollow section 17–18
  intensity modulation *24*

length modulation *24*
liquid core 19
metal-glass 18
polarization modulation *24*
rare-earth-doped 18–19
refractive index modulation *24*
time modulation *24*
wavelength *24*
Speckle
  decorrelation 219
  fields
    interference 208
    subtraction 208, 216
  fringes 208
  intensity 207
  white light interferometric 220–2, **222**, **223**
Spectral
  effects, interferometers 126–7
  linewidth, FBGs *366*
  side-bands 73
  width 66
Spectroscopy
  intensity fluctuation 199
  rare earths 39–41
Spectrum
  analyzer 247
  and coherence length **130**
Spin 40
  see also LSJ levels
Splitter, polarizing beam 281, **281**
Stability
  drift 326
  gratings 349–50
  thermal 349–50
State of polarization (SOP) 210
  controllers 210
Stimulated Brillouin scattering (SBS) 107, **107**, 175, **175**, 268
Stokes wave 268
Strain 359–60, *360*
  axial 125
  gauges 381
  mapping 370
  measurement, distributed 371
  optic effect 125
  ranges *399*
  resolution 386, 411, *400*
  sensitivity 386

Strain (contd)
  sensitivity (contd)
    and temperature 351–2
    sensors 79
    stress curve **388**
    and temperature 364, 383, **384**
      measurement 371, 158–62
      FBGs 362
Strain-induced, wavelength modulation 399
Strain-sensing, OCDR 145–6
Stress
  distribution 144
  modification 335
  strain curve **388**
Superfluorescent fiber sources, see SFSs
Superluminescent diodes, see SLDs
Supermode fields, amplitudes 169
Superradiant diodes 12–13
Superstructure, gratings 340
Surface
  profiles 215–22, **217**, 218, **218**, **219**, **220**, **221**, **222**
  roughness 192, **194**
Synthetic heterodyne 244

Tandem
  interferometers 131
    transfer function 183–4
  interferometry 151–4, **153**
Tap couplers 134
Target beam 245–6
TDM 411–12, **412**
  Bragg gratings 139
  and laser velocimetry 158
  LVD, implementing 290
  reflective ladder 140
  and WDM **149**
Telecommunications 48
  amplifiers 99
  fibers 71
  mode locking 74
  OTDR 138
Tellurite 37
Temperature 190–1
  applications 190–1
  cross-sensitivity 159
  dependence 313
  independence 363

  liquid nitrogen 49–50
  measurement 2
  probes 381
  sensitivity 125
  and strain 364, 383, **384**
    measurement 371
    sensitivity 351–2
    separation 389–90
    very high 191
Temporal spacing 66
Testing, nondestructive 196–9, **197**, **198**
TFLs
  device cost *358*
  FWHM bandwidth *358*
  optical power *358*
Thermal
  drift 156, 212–13
  gradient, error 320
  light, polarized 321
  stability 349–50
    FBGs *366*
Threshold 38
  power 269
Thulium
  absorption band 55
  doped fibers 54–6, **55**
  energy level **55**
  host fibers 54
Time
  coherence 181
  delay 322
  domain multiplexing, see TDM
  modulation 22–3
    mono-mode fiber *24*
    multi-mode fiber *24*
    polarization-maintaining fiber *24*
    special fiber *24*
    two-mode fiber *24*
  multi-mode fibers **7**
Time-of-flight 152
Timing, random 72
TMI 186
  OPD of 186
Topologies
  multiplexing 188, 190
  parallel 188, 190
Total internal reflection **4**
Tracking

INDEX

coherence 185, **186**
frequency 249
Transducers
  cylindrical shell 193
  design 350
  magnetostrictive 193
  piezoelectric 196–7
  and sensor 2
Transduction 5
  amplification 359–60
Transfer function 171
  Fabry-Pérot interferometer 174
Transit time 283
Transitions
  four-level 43, **44**
  lasing **58**
  multiphonon 41
  three-level 43, **44**, 49
  up-conversion 44, **45**
Transmission
  grating 255–6
  and reflection **4**
Transmissive, parallel networks 134, **134**, **135**
Transmitting interferometer, *see* TMI
Triangular, waveform 402
Trivalent ionization, rare earths 40
Tunable fiber lasers, *see* TFLs
Tunable filter, interrogation 367–70
Tuning
  electronic 53
  erbium 51
  filters **42**, 43
  neodymium 46, 47
  range 54
Turbomachinery 285
  applications 288–94, **289**, **290**, **291**, **292**, **293**
TV holography 207, **209**
Two-beam, interferometers 176
Two-mode fiber 16
  coherence modulation *24*
  intensity modulation *24*
  length modulation *24*
  polarization modulation *24*
  refractive index modulation *24*
  time modulation *24*
  wavelength *24*

Two-mode sensors
  absolute measurement *356*
  linear response *356*
  mass production *356*
  mechanical strength *356*
  multiplexing *356*
  potential cost *356*
  range of resolution *356*
  sensor gauge length *356*
Two-source technique 219–20, **220**
Two-wavelength, illumination 216
Types of fiber 9

Ultimate strength, FBGs *366*
Unbalanced interferometers 132
Unidirectional ring cavity **65**
Up-conversion 56, 57
UV exposure, refractive index 334

Variable ratio coupler 53
Vebert number 14
  number of modes 16
Vector quantities, detection of 119
Velocimetry **4**, 238, 251, **252**
  Doppler 199
  fiber use 245
Velocities
  flow 263
  low 275
  measurement 126, **126**
Verdet constant 195, 362
Vibrating target 240
Vibration
  fiber use 245
  measurement 2
  small amplitude 242
  time average 226–8, **227**, **228**
Vibrometers, interferometric 241, **241**, 245–57
Visibility
  fringes 152, 184
  interference patterns 239
  reducing 253

Walk-off 69
Waveforms
  modulation 319
  serrodyne 274, **275**
  triangular 394, **395**, 402

Wavelength
  Bragg 351, 357
  detection 391
  difference, reducing 129
  discrimination 387
  dispersion 66
  division multiplexing, see WDM
  dual oscillating 53
  mid-infrared, erbium 53
  modulation 121, 132
    strain-induced 399
  mono-mode fibers *24*
  multi-mode fibers **7**, *24*
  oscillating 56–7
  polarization-maintaining fiber *24*
  pulse 67, 68
  pump 273
  selection 52–3
  selective 136
  shift *400*
    temperature induced 364
  shift detection 384
  shortest 57
  single 56
  source, tuning 132
  special fiber *24*
  stability 100
    light source 309
  stabilizing 129
  two-mode fiber *24*
  variation 391

Waves
  counterpropagating
    interference 316
    velocities 304–5
  interference between 315
  WDM 20–1, **102**, 103, 147–50, **148**, **149**, 286, 293, 397, 409–11, **410**, **411**
    bandpass 411
    decoding 148
    systems 99
    and TDM **149**
  White light
    interferometry 23, 180–1, 253–7, **254**, **255**, **256**
      Bragg cell 254
      low-coherence 182–3
  Wollaston prism 211
  Working distance 257
  Writing methods 331

York technology 5
Young's modulus 361
Ytterbium
  ion
    absorption band 60
    energy levels 59, **59**
    fluorozirconate host fiber 60
    triply ionized 59
  mode locking 75
Ytterbium-doped fibers 59–60, **59**